FOOD SCIENCE AND TECHNOLOGY

SUPER AND NUTRACEUTICAL FOODS

COMPOSITION AND TECHNOLOGY

FOOD SCIENCE AND TECHNOLOGY

Additional books and e-books in this series can be found
on Nova's website under the Series tab.

FOOD SCIENCE AND TECHNOLOGY

SUPER AND NUTRACEUTICAL FOODS

COMPOSITION AND TECHNOLOGY

SANDRA TERESITA MARTÍN-DEL-CAMPO
JESSICA DEL PILAR RAMÍREZ-ANAYA
AND
ANABERTA CARDADOR-MARTÍNEZ
EDITORS

nova
science publishers
New York

Library of Congress Cataloging-in-Publication Data

ISBN: 978-1-53619-082-3

Published by Nova Science Publishers, Inc. † New York

CONTENTS

PREFACE

A balanced diet, providing essential nutrients is fundamental not only for human survival but to have good health. Good nutrition is a cornerstone of good health and wellbeing. Recently, the idea of good nutrition has become a fashion, and the terms superfood and functional food have emerged, mostly associated with marketing procedures.

Lifestyles have been changing over time; today, there are lower physical activity, less time to get equilibrium in meals, and the main productive activities require less movement and physical effort. These patterns predispose to the development of chronic degenerative diseases and demand constant monitoring of the factors that trigger them. One of the main factors to watch is diet, not only because of the control of caloric intake but also because of the need to guarantee a higher consumption of nutritious and non-nutritive substances to achieve metabolic regulation. This situation leads to the increased necessity of foods with low caloric density but with higher amounts of micro- and macronutrients and functional compounds.

Superfood has excellent caloric/nutrient rates, are rich in some micronutrients and natural substances with scientifically studied favorable activity over human health and lowering the risk for disease. It is well known that the inclusion of superfoods in a varied diet provides an extra supply of substances that maintain general wellbeing. Also, it helps in situations of increased nutrient demand such as medical, disease, lifestyles, smoking, consumption of drugs that interfere with nutrient absorption, or exposure to physical or chemical agents.

The foods considered now as superfoods have been part of the human diet for centuries. Most of these foods have been consumed locally, recognized for the nutritional value, and in most cases, are associated with cultural practices in different regions and countries. Some foods were exported from the endemic areas during colonization to various colonies, but others have been exported only recently.

Superfoods apport an important amount of essential nutrients such as proteins, fats, fibers, energy, as well as minerals and vitamins in few quantities of food. These superfoods are rich in essential nutrients that the body cannot produce, such as amino acids and fatty acids. Regular consumption of superfoods makes it possible to get the right nutritional balance helping to keep a healthy body weight and body systems function.

On the other hand, the term functional food is related to the term nutraceutical properties on superfoods. It has become relevant since, while nourishing, these foods provide additional health benefits associated with compounds presenting biological activities such as antioxidant, immunomodulant, prevention on metabolic syndrome, among others.

Traditionally, most of the superfoods are consumed in the fresh form locally in the production regions or the nearest markets since most of them present a short shelf life, and there is disponibility only some seasons, sometimes, in a really short period.

In some regions, traditional processing technologies have been used to preserve these foods and assure the supply during most of the year or to produce other traditional foods. To reach bigger markets, innovative processing technologies have been developed to preserve and process these superfoods safeguarding their biological activities.

This book shows information about superfoods, not traditionally described with this term, that also presents important nutraceutical properties. For each described food, the nutritional value, as well as the functional properties, are described. Additionally, information about bioactive compounds and the associated properties are described, as well as the effect of processing procedures on them.

The book is divided into three sections. The first section focuses on selected fruits, some eaten fresh, other only after a first transformation. The second section focuses on foods needing been transformed before human consumption, such as cereals, pseudocereals, and legumes. At the same time, the third section includes a set of products from animal and diverse origins.

SECTION 1. FRUITS

In: Super and Nutraceutical Foods
Editors: Sandra Teresita Martín-del-Campo et al.
ISBN: 978-1-53619-082-3

Chapter 1

AVOCADO: COMPOSITION AND PROCESSING TECHNOLOGIES

Jessica del Pilar Ramírez-Anaya[1,*], Adriana Cavazos-Garduño[2], Yanmei King-Loeza[3], Ma. Claudia Castañeda-Saucedo[4], Elia Herminia Valdez-Miramontes[5] and Jessica Elizabeth Pineda-Lozano[5,*]

[1]Departamento de Ciencias Computacionales e Innovación Tecnológica, CUSUR-Universidad de Guadalajara, Ciudad Guzmán, México
[2]Departamento de Farmacobiología, CUCEI-Universidad de Guadalajara, Guadalajara, México
[3]Facultad de Estudios Superiores, Universidad Nacional Autónoma de México, Cuautitlán Izcalli, México
[4]Departamento Ciencias del Naturaleza, CUSUR-Universidad de Guadalajara, Ciudad Guzmán, México
[5]Instituto de Investigaciones en Comportamiento Alimentario y Nutrición (IICAN), CUSUR-Universidad de Guadalajara, Ciudad Guzmán, México

ABSTRACT

Avocado (*Persea Americana* Mill) production has a significant impact on the economy and society being considered as one of the main crops worldwide. It has highly valuated sensorial characteristics and is recognized as a healthy fruit. Between the health benefits from its consumption, draws attention the dyslipidemia improvement; this effect has been tightly related with flesh chemical characteristics as the high content of unsaturated fatty acids, and the presence of functional compounds as sterols, carotenes and tocopherols. This fruit is produced mainly for fresh consumption. But, in the last years, the world production of avocado is raising continuously, with a concurrent increase in the rejected grade fruit. México is the leading producer and exporter, in this country, less than the fifth part of the production is destined to industrialization. This percentage

* Corresponding Author's E-mail: jessicar@cusur.udg.mx; jessica.pineda@cusur.udg.mx.

includes lower quality grade fruit, smaller sized avocados, and fruit that does not qualify for human consumption as a fresh product because of the level of surface damage. Successful and popular foods such as pure and more recently, extra virgin oil, are made from the edible fraction of avocado. However, with the increase of alternatives based on avocado flesh, problems as the short time of maturation and easy oxidation are more critical for producers. Those concerns lead to the need for improving the existing technology, and to develop new alternatives for avocado processing to preserve or enhance sensorial and functional characteristics.

Keywords: Avocado puree, Avocado oil, High pressure processing, Minimal processing, Oil extraction

INTRODUCTION: ORIGIN AND QUALITY OF AVOCADO

Avocado (*Persea americana* Mill) is a tropical species from the *Lauraceae* family. It originated in México, almost 70 species of the genus *Persea* are distributed in North, Central and South America, while 80 are located in East and Southeast Asia (Ding et al. 2007). Avocado's botanical classification is as a round, ovoid or pear shaped berry, which size depends on metabolic, ambient and cultivar factors, in the account of the different varieties, avocados can weight from 50g to 2 kg, all avocados have only one large, ovoid seed (Bost, Smith, and Crane 2013). According to its respiration pattern is classified as climacteric fruit, which is the reason behind its early harvest before ripeness, so that in the course of its commercial life it reaches maturity at the end of the supply chain (Barmore 1976).

The existence of different avocado varieties is related to commerce in Pre-Columbian America, with the arrival of the Spanish explorers to the continent avocado spread to Europe. At the start of the XX century, West Indian Garden collected material from Atlixco, Puebla, México, later named as "Fuerte" (Bost, Smith, and Crane 2013), during the 1930's Fuerte represented at least 75% of all fruit in the orchards. In 1920, Hass variety was developed, Rudolph Hass acquired and patented the variety in 1935, thus starting its propagation and soon replacing Fuerte as the preferred variety (Shepherd and Bender 2002). Today, Hass variety is widely popular for its high productivity and quality of fruits which bears an excellent taste and good transportation endurance, over 90% of the world´s production is Hass cultivar (Téliz 2007, SAGARPA 2016).

Avocado is one of the most important commercial commodities, having an essential impact on the economy and society. It is widely cultivated around the world, having great popularity as a healthy fruit, rich in antioxidants and fatty acids, being in the top five tropical fruit with more exports (Altendorf 2017). The avocado was expected in 2019 to have 2.3 million tons exported from a 918531 ha harvested area (FAO 2020). Mexico, as the leading producer exporter, supplies 58% of exports, while Peru, the second exporter accounts for 13% (FAO 2020). Mexico destines 78% of its exports to the USA, while Peru supplies mainly the European market. Other leading exporters of avocado are; Chile (192199 tons), Israel (95914 tons), Kenya (58730 tons), and South Africa (48003 tons) (FAO 2018).

Avocados are primarily grown for fresh fruit consumption. As can be seen in Table 1, the harvested fruits must cover the minimal requirements stated at international and local levels.

Table 1. Minimal requirements of fresh fruit fit for human consumption, according to the regulation of leading avocado exporters

	International CODEX STAN 197-1995 (FAO 1995)	**Mexico NMX-FF-016-SCFI-2016 (SCFI 2016)**	**Peru PNTP 011-018-2014 (INDECOPI 2014)**
Maturation	Fruits must have reached physiological maturity to ensure completion of commercial ripening.	Fruits must have reached physiological maturity	Fruits must have reached physiological maturity to ensure completion of commercial ripening.
Dry matter content (%)	≥ 21	≥ 21.5	≥ 21
Oil content (% d.w.)	-------	------	≥ 9.25
Fruit integrity	Fruit free of rotting and deterioration, clean. Free of pest and its damages. Free of damage by high or low temperatures. With no presence of abnormal moisture.	Fruit free of rot, pest, and disease. No high or low temperatures damage. Manually harvested, using scissors, net, and clean transporting plastic boxes. Never touch the ground.	Wholesome, with no deterioration, clean. Free of pest and damages. No high or low temperatures damage. No abnormal moisture. Stalk must be less than 10mm, clean cut off.

d.w.: dry weight basis.

Dry matter determination is used by the avocado industry to determine maturity and oil content, as strong correlations between these factors have been found. Fruit harvested with a dry matter content of <20% are likely to have poor sensory attributes on eating, recommended minimum dry matter percentages to harvest and when the avocado is considered mature have been decided in every exporting country. Another parameter for avocado exploitation is the oil content after postharvest maturation, the avocado pulp can reach 18 to 33% of lipid content and the seed or pit approximately 2%. The oil content in the flesh differs between varieties being higher in Hass avocados (Wong, Eyres, and Ravetti 2014).

Between 1999 to 2014, worldwide avocado production has increased by 52%. With this rise in production, there is a concurrent supply of reject grade fruit. In Mexico, only 16% of the production is destined to industrialization (Economía 2012), this percentage includes lower quality grade fruit, smaller sized avocados, and fruit that does not qualify for human consumption as a fresh product because of the level of surface damage (Table 2).

There is growing competition for rejected fruit for the production of foods based on avocado. From pulp are made the most successful and popular avocado foodstuffs such as puree, salads, oil, among others. However, there is growing attention on the recuperation of avocado processing wastes. Seed and peel have shown to be a potential source of antioxidants, antimicrobials, colorants and thickener agents, which can be used as food additives; also are considered as a source of bioactive molecules which can be used as food, pharmaceutical and cosmetic ingredients (Colombo and Papetti 2019, Ayala-Zavala et al. 2011).

Table 2. Nomenclature and commercial quality standards by class for fresh avocados fit for human consumption, according to the regulation of leading avocado exporters

International **CODEX STAN 197-1995** **(FAO 1995)**	**Mexico** **NMX-FF-016-SCFI-2016** **(SCFI 2016)**	**Peru** **PNTP 011-018-2014** **(INDECOPI 2014)**
EXTRA Fruits of superior quality. Characteristics of the variety (size, shape and smell). Free of defects. Accept slight defects in the surface while do not affect the appearance. The fruit that has a stalk must have it intact.	EXTRA (SUPREMA) Characteristics of the variety (color, smell, shape and size). Uniform appearance and size. Without defects higher than $2cm^2$.	EXTRA Characteristics of the variety (size, shape and smell). Free of defects. Accept slight defects in the surface that do not affect the appearance of the produce. The stalk, if present, must be intact.
CLASS 1 Fruit of good quality. Characteristics of the variety. Slight defects are accepted ($\leq 4cm^2$ of the surface). Defects must not affect the flesh. The stalk, if present, maybe slightly damaged.	CLASS I Characteristics of the variety (color, smell, shape and size). Good aspect in general. Some defects are permitted (< 6 cm^2 of the surface).	CLASS 1 Fruit of good quality. With characteristics of the variety, minor defects in color and shape, sunburns and slight skin defects ($\leq 4cm^2$ of the surface). Defects must not affect the flesh. Optional stalk with slight damage.
CLAS 2 Fruits that do not qualify for higher classes. Accepts defects in shape, coloring, sunburns and skin (≤ 6 cm^2 of the surface). The stalk, if present, may be damaged.	CLASS II Characteristics of the variety (color, smell, shape and size). Damage is accepted ($\leq 50\%$ of the surface).	CLASS 2 Fruits that do not qualify for higher classes. Accepts defects in shape, coloring, and skin (≤ 6 cm^2 of the surface). The stalk, if present, may be damaged.

Avocado Pulp Composition

Fat is the main constituent in the avocado pulp, representing 14.66% of the total weight. From the lipids present, 9.8% are composed by monounsaturated fatty acids (MUFAs), 2.12% are saturated fatty acids (SFAs), 1.81% are polyunsaturated fatty acids (PUFAs), and contains zero cholesterol. Also includes 2% of protein and 6.7% of fiber. The 8.53% are carbohydrates, being present mainly glucose, fructose, sucrose, galactose and starch. It has minerals like calcium, iron, magnesium, phosphorous, potassium, sodium, zinc, copper, manganese, selenium and fluoride. Are present vitamins as ascorbic acid, those from the B complex, A, K, and E. It has been reported that regular avocado consumption, helps to guarantee the supply of dietary fiber, potassium, magnesium and vitamins K, E and those from B complex (USDA 2011).

The avocado pulp contains high levels of bioactive phytochemicals with antioxidant activity and lipophilic character. Some of them are α, β, γ, and δ-tocopherol (Table 3).

Table 3. Tocopherol content in the avocado pulp

Compound	Content (µg/100g w.w.)	References
α-tocopherol	2548-3278	(Lu et al. 2005, Lu et al. 2009, Bressani, Rodas, and Ruiz 2009, Monge-Rojas and Campos 2011, Stanizzi et al. 2015)
β-tocopherol	50	(USDA 2011)
γ-tocopherol	199-746	(Lu et al. 2005, Monge-Rojas and Campos 2011)
δ-tocopherol	50	(USDA 2011)

w.w.: wet weight basis.

Other antioxidants also present are carotenoids, being lutein and β-carotene the most abundant (Table 4). The carotenoids and tocopherols have clinical properties against cell proliferation; tocopherols additionally, has antiinflamatory activity. The α-tocopherol shows recognized activity in cancer prevention, angiogenesis modulation, and the improvement of the vascular epithelium.

Table 4. Content of carotenoids in the avocado pulp

Compound	Content (µg/100g w.w.)	References
Lutein	213-842	(Lu et al. 2005, Lu et al. 2009, Bressani, Rodas, and Ruiz 2009, USDA 2011)
α-carotene	19-89	(Monge-Rojas and Campos 2011, Lu et al. 2005, Lu et al. 2009, USDA 2011)
β-carotene	10-139	(Lu et al. 2005, Lu et al. 2009, Bressani, Rodas, and Ruiz 2009, Monge-Rojas and Campos 2011, USDA 2011)
β-cryptoxanthin	3-32	(Monge-Rojas and Campos 2011, Lu et al. 2005, Lu et al. 2009, Yamaguchi and Uchiyama 2004, USDA 2011)
Neoxanthin	5-0.53	(Lu et al. 2009)
Neochrome	0.04-0.16	(Lu et al. 2009)
Chrysanthemaxanthin	0.31-2.72	(Lu et al. 2009)

Polyphenol are compounds with recognized antioxidant activity. In avocado, its total and individual contents (Table 5) vary among cultivation environments, varieties, and maturation stages (Di Stefano et al. 2017). But, between fruit structures, the pulp shows the lowest total content than seed, peel and leaf (Jimenez et al. 2020). Avocado polyphenols have a wide range of properties and applications in the alimentary, clinical and cosmetic industries.

Sterols are abundant in pulp predominating β-sitosterol over campesterol and stigmasterol (Table 6). The sterols exert a clinical effect in the management of serum cholesterol (USDA 2011, Ortiz-Viedma et al. 2018). Has been proved that avocado pulp consumption is associated with health benefits. Diets which include avocado modulates the postprandial vasoconstriction effect in healthy people and reduce cardiovascular biochemical risk parameters in overweight or obese patients. In the previous diagnosis of dyslipidemia, it reduces the levels of total cholesterol (TC) and low-density lipoproteins (LDL); and raise the high-density lipoproteins (HDL) levels. The mentioned modifications also have been seen in patients with type 2 diabetes; with an additional improvement of triglyceride status, and an increase of the hypoglycemic and hypoinsulinemic effect (Wang et al. 2015, Li et al. 2013, Silva Caldas et al. 2017, Park, Edirisinghe, and Burton-Freeman 2018).

Table 5. Content, properties and applications of polyphenolic compounds of avocado pulp

Compound	Content (mg/100g d.w.)	Properties and Applications	References
Apigenin	1.56	Clinical: antioxidant, cell antiproliferative, cancer prevention. Alimentary: supplement.	(Poovarodom et al. 2010, Madunić et al. 2018)
Catechin	3.3	Clinical: antioxidant.	(Figueroa et al. 2018, Kosińska et al. 2012, Tremocoldi et al. 2018, Rodríguez-Carpena, Morcuende, and Estévez 2011, Saavedra et al. 2017,)
Caffeic acid	0.25	Clinical: antioxidant, cell antiproliferative. Alimentary: antimicrobial. Cosmetic: hair growth.	(Hurtado-Fernández, Carrasco-Pancorbo, and Fernandez-Gutierrez 2011, López-Cobo et al. 2016, Poovarodom et al. 2010, Saavedra et al. 2017, Zhang et al. 2020)
Gallic acid	198.57	Clinical: antioxidant, reduction of plasma lipids, cancer prevention. Alimentary: antimicrobial, antioxidant, improvement of rheological properties.	(Hurtado-Fernández, Carrasco-Pancorbo, and Fernandez-Gutierrez 2011, Monika and Geetha 2015, Poovarodom et al. 2010, Limpisophon and Schleining 2017, Widsten et al. 2014)
Luteolin	165	Clinical: prevention of cancer and Alzheimer´s progression, antioxidant.	(Figueroa et al. 2018, Kwon 2017, Monika and Geetha 2015)
Protocatechuic acid	0.037	Clinical: antioxidant. Alimentary: antimicrobial. Cosmetic: hair growth.	(Figueroa et al. 2018, López-Cobo et al. 2016, Pahua-Ramos et al. 2012, Poovarodom et al. 2010, Terasawa, Sakakibara, and Murata 2006, Widsten et al. 2014)
Quercetin	0.04-5.25	Clinical: antioxidant, reduction of plasma lipids, cancer prevention. Alimentary: antimicrobial. Cosmetic: hair growth.	(Figueroa et al. 2018, Hurtado-Fernández, Carrasco-Pancorbo, and Fernandez-Gutierrez 2011, Kosińska et al. 2012, Monika and Geetha 2015, Poovarodom et al. 2010, Widsten et al. 2014)
Rutin	141.79	Clinical: antioxidant, antidiabetic, antimicrobial, antiinflammatory, antiallergenic, cell antiproliferative, cancer prevention.	(Figueroa et al. 2018, López-Cobo et al. 2016, Pahua-Ramos et al. 2012, Monika and Geetha 2015, Gullón et al. 2017)
Vanillic acid	4.79-5.61	Clinical: antioxidant, reduction of plasma lipids, cell antiproliferative. Cosmetic: hair growth.	(Monika and Geetha 2015, Pahua-Ramos et al. 2012, Hurtado-Fernández, Carrasco-Pancorbo, and Fernandez-Gutierrez 2011, Poovarodom et al. 2010)

d.w.: dry weight basis.

Table 6. Content of phytosterols in avocado pulp and oil

Compound	Source	Content (mg/100g w.w.)	References
β-sitosterol	Pulp, oil	76-262.9	(USDA 2011, Mo et al. 2013)
Campesterol	Pulp, oil	5-40.43	(USDA 2011, Mo et al. 2013)
Stigmasterol	Pulp, oil	2-17.26	(USDA 2011, Mo et al. 2013)
Cycloartenol	Oil	34.71	(Mo et al. 2013)
Lupeol	Oil	1.81	(Mo et al. 2013)

w.w.: wet weight basis.

The Technology of Avocado Pulp Products

Puree

Avocado puree formulation may include stabilizers as gums, alginates, polyphosphates; preservatives like sodium or potassium sorbate and benzoate, salt and spices. The general sequence of operations to obtain paste includes fruit washing and sanitation, destoning and peeling, chunk, mix, and vacuum packaging in pouches or trays under modified atmosphere conditions (Jacobo-Velázquez and Hernández-Brenes 2012, Jacobo-Velázquez et al. 2013, Jacobo-Velázquez and Hernández-Brenes 2010). Another pulp product is the "reduced calorie avocado paste." It is a subproduct of the avocado oil extraction by microwave squeezing or cold pressing of microwave dried avocado paste. That could be a good source of protein and carbohydrates (Wang et al. 2019) and attenuates the effects of hypercholesterolemic high fructose diet in rats (Pahua-Ramos et al. 2014). The first production method has been patented but not yet commercialized, and the resulting pulp has 30% of the original oil content (Ortiz et al. 2004).

Two central problems are associated with the formulation and production of avocado pulp products, namely: enzymatic browning and lipid oxidation, both of them reduce the shelf life significantly. Most studies focus on the inactivation of polyphenoloxidase (PPO) and lipoxygenase (LOX) enzymes. Whose activity affect the color, taste and nutritional value of avocado products. Nevertheless, polygalacturonase (PG) can affect the texture and the rheological properties during storage. This effect has been documented in avocado halves and chunks, guacamole, avocado puree, and sauces (Bermejo-Prada et al. 2014).

PPO enzymatic browning spoilage is conventionally controlled using chemical compounds such as sodium bisulfite, ascorbic, isoascorbic and citric acids (Olaeta 2003). It has been proved that, during processing, the addition of extracts as rosemary, thyme and grapefruit-seed effectively prevent color alterations from PPO activity. Avocado refrigerated paste (4°C) could be preserved for more than 30 days efficiently by addition of scallion and garlic extracts showing more significant antibrowning effect than Brussel sprout, white cabbage and cauliflower extracts (Bustos, Mazzobre, and Buera 2015).

Some antioxidants are used to preserve lipid oxidative stability. Most of them are synthetic, like buthilhidroxitoluene, butilhidroxianisol and tertiary butilhidroxiquinone (Jacobo-Velázquez et al. 2013). α-tocopherol, ascorbic acid and ethylenediaminetetraacetic acid (EDTA) also have been assayed in minimally processed avocado puree; where was found that α-tocopherol was effective in preserving lipid fraction in combined methods (Elez-Martinez et al. 2005).

Conventional and Emerging Technologies to Improve Puree Quality

Technological processing of puree have been including emerging technologies to decrease the microbial and enzymatic spoilage during elaboration or once the product was packed (Vargas-Ortiz, Rodríguez-Jimenes, et al. 2017). Although thermal processes are effective for microbial inactivation, they introduce changes in product quality and cause off-flavor generation, texture softening, and the destruction of colors and vitamins (Wang et al. 2016).

Hurdle technology is used as a gentle but effective preservation alternative for the production of safe, stable, nutritious, tasty and economical foods. Its application is more prevalent now, because the principles of major preservative factors for foods (e.g., temperature, pH, a_w, competitive flora), and their interactions, became better known (Singh and Shalini 2016). In avocado puree processing, the addition of 300 mg/kg of sorbic acid can extend the stability during storage periods up to four months. Vacuum packaging as well as storage at 4°C also have a significant influence over the control of yeast and mold populations and could preserve avocado purees during 112 days without the addition of antimicrobial compounds. The addition of maltose to reduce a_w to 0.96 results in slightly more stable purees but would impart important changes to the product palatability (Soliva-Fortuny et al. 2004).

The use of microwaves in paste and guacamole processing has been assayed as a bleaching alternative since the application of high temperatures for short times results in minimum damage. The aim has focused on the destruction of PPO and the improvement of color and lightness. For example, after recovering low-calorie paste from oil extraction of dehydrated avocado pulp, microwave treatment (11.0 W/g during 80 s) raised the temperature to 90°C and reduced PPO activity to 20% during the four weeks of storage. When microwaving at 633 W for 30 s avocado pulp previously diluted with water and adjusted to pH 5.5, was shown that negligible losses of chlorophyll occur and lightness increases. The prevention of extensive changes in flavor during microwaving has been studied by the addition of avocado leaves (Zhou et al. 2016, Guzmán-Gerónimo, López, and Dorantes-Alvarez 2008).

In most cases, the application of ultrasound as nonthermal technology has resulted in positive effects on food quality, such as texture, while ensuring the microbial safety with minimal damage to nutritional and organoleptic properties of heat sensitive materials (Charoux, O'Donnell, and Tiwari 2017). Conversely, in avocado puree ultrasound treatment may affect the quality. Have been reported changes in color and increases in the viscosity of the puree mainly due to the solubilization of the cell wall material (Bi et al. 2015). Changes in nutrient contents have been reported. Loss of E vitamin can reach 79% with ultrasonic power densities ranging from 55 to 5000 W/L at 23 or 40°C; it has been postulated the addition of tocopherol acetate to control vitamin content during processing (Fernandes et al. 2016).

For avocado puree production, the flash vacuum expansion shows good results. Allows the extraction of bioactive compounds and the reduction of enzyme activity, leading to a rise in the stability of the finished product, improvement of the nutritional quality, and reduction of the oxidation processes (Vargas-Ortiz et al. 2017). During flash vacuum, avocados are steam heated (20 min). Immediately, fruits are disintegrated when introduced into a vacuum chamber (3 kPa). This last effect takes place from the instantaneous vaporization of the constituent water. Then, the peel and seed are removed, and the puree is packaged in hermetic plastic containers. During refrigerated storage at 5°C, polyphenolic content increases, PPO activity is inhibited, and the color and luminosity of the puree remain stable for at least 15 days without adding preservatives (Vargas-Ortiz, Rodríguez-Jimenes, et al. 2017).

High hydrostatic pressure (HHP) is an emerging physical disinfection technology with effective application to fruits and vegetables (Deng et al. 2019); involves treating foodstuff at high pressure (100-800 MPa) conveyed by liquids (typically water) for an appropriate period. This non-thermal technology can achieve the same standards of food safety of heat pasteurization and meet consumer requirements for fresher tasting, minimally processed foods. Applying HHP can inactivate pathogenic and spoilage microorganisms and enzymes, as well as modify structures with little or no effects on the nutritional and sensory quality of foods (Wang et al. 2016).

HHP is the most effective non-thermal technology used in processing the avocado pulp. Its industrial use has contributed to the rise in consumption and development of ready to eat avocado products (Wang et al. 2016). Commercial processing of avocado puree is usually carried at 600 MPa for 3 min at 23°C. The shelf life of puree products under refrigeration is at least 20 days.

After puree HHP treatment, the activity of PPO, LOX and PG enzymes lowers to a residual value of 46%, 58% and 0%. However, the first two reactivates at 10 to 15 days of refrigerated storage to return to initial values at day 45. Other changes during refrigerated storage of HHP puree includes pH reduction along the first 20 days; steady-state of the lactic acid bacteria counts (Jacobo-Velázquez and Hernández-Brenes 2010). Low increase in color parameter a* correlated with PPO activity; increase of the sensorial perception of rancid flavors parallel with a rise of hexanal, and the increment of free fatty acids although lipidic profile remains constant (Jacobo-Velázquez et al. 2013). HHP also changes the bioavailability of functional compounds; as the increase in total extractable carotenoids (56%) followed by a decrease during storage (40 days at 4°C). There is a reduction in lipophilic and hydrophilic radical absorbance capacities, with a rise during storage (Jacobo-Velázquez and Hernández-Brenes 2012).

Minimal or Fresh Cut Processed Avocados

Minimal or fresh-cut processing of fruits and vegetables has gained importance over recent decades due to the availability of convenience products to consumers (Yousuf et al. 2020). Peeled, sliced, shredded, trimmed, and washed fruits and vegetables may be regarded as fresh-cut produce (Francis et al. 2012). Main concerns on avocado minimal and fresh-cut processed are microbial spoilage, changes in color, and oxidative processes. While the treatment of avocado pulp/guacamole has been a commercial success, production of acceptable not pulped products like slices or halves has been a challenge (Woolf et al. 2013). A sequence to produce avocado minimally processed slices includes halve ripe avocados longitudinally, destone, peel, slicing longitudinally, ends remotion, and vacuum package into polyethene with nylon barrier bag pouches (Woolf et al. 2013).

One of the strategies to improve minimally processed foods is the edible coating using biological or chemical materials as a coating layer on the surface of the product by dipping or spraying the coating solution. The edible coat function as a barrier on the fruit surface, modifying the internal gas atmosphere, and decreasing water loss. Also, gives antibrowning effect improving the appearance of fruits by maintaining the color and glossy surface. Can preserve the phytonutrients of fresh/fresh-cut fruits, and reduce microbial spoilage (Maringgal et al. 2019). It has been used to extend the preservation of avocado slices minimally processed and frozen. In avocado slices, the PPO activity has been inhibited by antibrowning agents (4-hexylresorcinol, sodium D-isoascorbate, L-glutathione and sodium hexametaphosphate) applied into a gellan gum coating (Mendoza-Gómez et al. 2017). In other fresh fruit and vegetable products has been assayed the addition of plant extracts and essential oils in coating formulations to delay deteriorative changes from respiration and microbiological spoilage (Saxena, Sharma, and Maity 2020).

Immersion treatments are frequently used. For example, for the conservation of Pollock avocado, halves have been immersed in solutions of 1% ascorbic/1% citric acids at pH 5 together with refrigerated storage at 7°C. The procedure was enough to get an inhibitory effect of enzymatic browning and preserving color after 15 days (Gómez-López 2002).

Pulse light treatments (3.6, 6.0 and 14 J/cm^2) reduce aerobic mesophilic microorganisms and inhibited the proliferation of yeasts and molds prolonging shelf life up to 15 days at 4°C. Hue is better maintained in association with the high chlorophyll retention and did not increase the rancidity processes (Aguiló-Aguayo et al. 2014).

HHP processing at 600 MPa of avocado slices vacuum packaged reduced up to 50% peroxidase activity but increased that of PPO. Induced changes in the cell wall structure like the disruption of the cellular network and coalescence of oil vesicles. The respiratory activity was reduced to 50% and ethylene production to 80% without visible color changes, but the application was considered immediately irrelevant to HPP industries (Woolf et al. 2013).

Dried Products

Drying has become necessary because most fruits are highly perishable owing to their high moisture content. In addition to preservation, the reduced weight and bulk of dehydrated products decrease packaging, handling and transportation costs. Quality changes associated with the drying of fruit products include physical, sensory, nutritional, and microbiological. Drying process of avocado has been scarcely studied; but the general operations included during avocado dehydration are: washing and sanitation, peeling and destoning, pureeing, mixing of optional ingredients, drying and packing. (Omolola, Jideani, and Kapila 2017).

Drying under circulating air has been studied as pretreatment for oil extraction. In the case of Fortuna avocado were used temperatures of 40 or 70°C until moisture values of 5-6.5%. It was seen that temperature and drying method barely altered the oil composition. As drying temperature increased, the concentration of functional compound lowered as the case of squalene, α-tocopherol and β-sitosterol; but other as cycloartenol and campesterol raised (dos Santos et al. 2014).

Drying in spouted bed has been applied on avocado modified pulps. The modification was carried out by mixing with glucose and fructose, soluble starch, citric pectin, citric acid, olive oil, mango fibres and distilled water to adjust the contents of reducer sugar, starch pectin acidity, lipid, fiber and moisture, respectively. Conditions assayed included 2.5 kg of polyethylene inert particles and 50 g of pulp, for 40 min, at 70°C. It was seen that inert particles could contaminate the final product by attrition effect, and the sugar concentration cause increase in retention values affecting the efficiency of avocado pulp powder production (Medeiros et al. 2002).

Although freeze drying technology better preserves sensorial and nutritional characteristics, high operation costs lower its industrial application. After freeze dry at -40°C and 0.133 Pa, linoleic acid content diminishes (Castañeda-Saucedo et al. 2014), and the dried product has higher browning index than fresh avocado pulp (Souza et al. 2015). Freeze-dried avocado has been used as an amorphous foodstuff in powder mix. The formulation included carrier agents to avoid agglomeration and caking (maltodextrin and inulin). From the tests was seen that the additives were not necessary to improve stability (Stępień, Witczak, and Witczak 2020).

The development of powdered blends employing spray drying of avocado drinks has been done whit formulations that include ingredients as maltodextrin (0 to 23%), refined sugar, and cow milk. Spray drying of liquid formulations at temperatures from 80 to 120°C and small droplet sizes, results in lower final moisture contents, yields over 45%, and a greener and less dense product. Use of maltodextrin also preserve contents of protein, ascorbic acid, and phenolic compounds (Dantas et al. 2018). In spray dried avocado paste, has been proved that the inclusion of monoacylgliceride as an emulsifier, and a mix of tertiary butylhydroquinone (TBHQ) with citric acid as antioxidants, lower rancidity during storage at 6 and 12°C (Grajales-Lagunes et al. 1999).

Emulsifiers

Avocado mesocarp and seed contains molecules with tensoactive properties. The tensoactive compounds identified are phospholipids (Table 7), from which, phosphatidylinositol also shows biological activities as cell antiproliferative and cancer prevention (Takenaga et al. 2008).

Table 7. Content of phospholipids in the avocado pulp

Compound	Content (% w.w.)	References
Phosphatidic acid	0-10	(Takenaga et al. 2008, Züge et al. 2017)
Phosphatidylethanolamine	40-55	(Takenaga et al. 2008, Züge et al. 2017)
Phosphatidylglycerol	7	(Takenaga et al. 2008, Züge et al. 2017)
Phosphatidylcholine	20-36	(Takenaga et al. 2008, Züge et al. 2017)
Lisophosphatidylcholine	0-10	(Takenaga et al. 2008, Züge et al. 2017)
Phosphatidylinositol	1-15	(Takenaga et al. 2008, Züge et al. 2017, van Dieren et al. 2011)

w.w.: wet weight basis.

Phospholipids can be recovered to be used as food emulsifiers to form stable oil/water (O/W) emulsions. When phospholipids are added in a 2 wt% proportion, the resulting emulsions are pseudo-plastic fluids and present typical gel-like behavior. Those with 50 and 60% oil phase show the highest stability during creaming (Züge et al. 2017).

Cold Pressed Avocado Oil

As almond, olive, hazelnut and macadamia oils, avocado oil is considered as a specialty oil with health properties. Its high content of MUFAs has a prophylactic effect of cardiovascular diseases, atherogenesis, and dyslipidemia, as has been described before for avocado pulp (Tan 2019).

The composition of avocado oil varies according to cultivar, geographical condition, maturity level and extraction method (Manaf et al. 2018). Oil from various avocado cultivars such as Hass, Fuerte, and some Indonesian varieties is composed mainly by MUFAs. Between them, oleic acid (C18:1) is the most abundant, which, together with palmitoleic (C16:1) sum over 55 to 65% of the total content. In the SFAs fraction, palmitic acid (C16:0) is the most abundant (22 to 34%), while the presence and amount of stearic (C18:0) is variable. The former ranges from 0.65 to 1.04% in Hass, Fuerte, Shepard, Merah and Ijo avocado oils, but has also been reported as no detected in Hass avocado oil. Other saturated fatty acids, such as margaric and myristic acids has been occasionally identified (Manaf et al. 2018). A MUFAs to SFAs ratio of 1.51 to 2.59 was measured. High ratio values are generally viewed as beneficial to humans (Woolf et al. 1999). PUFAs as linoleic (C18:2) reaches 14 to 21%. In comparison, linolenic (C18:3) ranges between 1.1 to 12.24%, Hass avocado oil presents higher content of oleic (64%) and total unsaturated fatty acids (~85%) than oils from other cultivars (Manaf et al. 2018).

Have been identified important functional compounds like antioxidants such as carotenes, tocopherols and polyphenols whit inhibitory growth effect of cancer and prostate cellular lines. Additionally, it contains sterols, being β-sitosterol the most abundant, which reduce intestinal cholesterol absorption, increase immune response and have an antiinflammatory effect (Ding et al. 2007). Contents of alpha, beta, gamma and delta tocopherols range from 40 to 45, 0.8 to 2.0, 3.8 to 4.2 and 0.04 to 0.08 mg/kg, respectively (Manaf et al. 2018). While total phenolic, tartaric esters and flavonol content round, respectively, 11.77, 0.31 and 0.2 mg (+) catechin, caffeic acid, and quercetin equivalents/g d.w. (Espinosa-Alonso et al. 2017). Keep the avocado's peel increase phenolic content in pressed oil and reduce extra waste disposal (Santana et al. 2019). Like other vegetable oils, avocado oil is sensitive to oxidation resulting in rancidity and subsequent presence of undesirable flavors and quality losses throughout storage (Gunstone and Norris 1983).

Avocado oil is produced from the pulp instead of the seed (unlike conventional seed oils). The flesh of Hass avocado can contain as much as 32% oil (fresh weight) depending on the time of harvest; this is a maximum, and the average at the peak of the processing season is approximately 18% (Wong, Eyres, and Ravetti 2014). This oil can be consumed in its crude form without the necessity of refining. The oil can be graded into extra virgin, virgin or pure, depending on the extraction methods and conditions (Tan and Ghazali 2019). Avocado oil used to be obtained by mashing the pulp in water, then heating and skimming off the supernatant oil. Later, the oil extraction was done from dried fruits using solvents (Reddy, Moodley, and Jonnalagadda 2012). The industrial avocado oil extraction is done by solvents, by pressing, or most recently, by centrifugation. The third option avoids the use of organic solvents (Costagli and Betti 2015).

Cold-pressed avocado oil is unrefined, and so retains the flavor and color characteristics of the fruit flesh. It is produced by mechanical methods at temperatures under 50°C, with no solvents (Woolf et al. 2009). The first attempt to develop a method to produce cold-pressed oil to attain high-quality edible oil was made back in the 1990s (Eyres, Sherpa, and Hendriks 2001). Steps for the cold extraction of avocado oil are almost the same as those for olive oil extraction. Includes fruit manual or mechanical peeling and destoning, followed by grinding, mixing and malaxing (Costagli and Betti 2015). In some cases, the avocado paste is partially dehydrated by convective, microwave or freeze dry methods before oil extraction (Santana et al. 2019). Extraction is done by pressing or centrifuging. For large operators, the use of horizontal decanting centrifuges at 3000 to 4000 rpm is a standard to separate the oil phase from the solids and wastewater. The oil phase is passed through polishing disc centrifuge to remove residual water. The process requires minimal addition of water, and no other additives are necessary (Costagli and Betti 2015, Yang et al. 2018). The goal of the aqueous extraction methods is to produce high-quality oils ready for consumption requiring no further processing. Any inferior quality oils produced are generally sent for further refining, bleaching and deodorizing and sold as lower grade oil (Wong, Eyres, and Ravetti 2014).

During centrifuging and pressing, avocado oil yield used to be less than 38%, depending on fruit variety (Schwartz et al. 2007). Despite the poor performance, oil quality is better, because mechanical methods eliminate the need for organic solvents and prolonged temperature increases. In this way, characteristic color is preserved, the formation of free and trans-fatty acids is avoided, and the oxidation of unsaturated fatty acids is lowered. Keep those standards is a priority during the production of edible cold-pressed avocado oil as a result of the worldwide increase of its culinary use (Woolf et al. 2009).

Malaxing is a determinant step in ease of oil recovery and yield improvement. Is an operation employed to rupture subcellular oil compartments (idioblasts) of the parenchyma cells, and facilitates the process of droplet oil coalescence. Malaxing consists of slow mixing, from 60 to 180 min at temperatures from 45 to 50°C (Yang et al. 2018). Those conditions destabilize the highly stable fine dispersions inside cells. These emulsions are rounded by lipoproteic membranes or lipophilic solids of the avocado paste that absorbs part of the oil (Santana et al. 2015). When softer ripe avocado fruit is processed, occur more parenchyma cell rupture in mesocarp tissue, increasing the yield of oil when cold press method is used. However, when overripen fruits are used, the percentage of free fatty acids and peroxide value rises, lowering the oil quality (Yang et al. 2019).

The rupture of the idioblasts can be improved using processing aids like the addition of water, calcium salts, talc, and acidifiers (chlorhydric or phosphoric acids), discrete temperature increases, or adding industrial enzymes like cellulase, hemicellulase or pectinase (Santana et al. 2015, Schwartz et al. 2007, Buelvas-Salgado, Patiño-Gómez, and Cano-Salazar 2013). Water addition is commonly used to disperse the pulp into a mash and to reduce de viscosity before pumping or mixing operations (Martínez-Padilla, Franke, and Juliano 2017). During oil extraction, the storage and loss moduli of avocado paste have a solid-like behavior no matter the treatment or dilution. Extended shearing or water dilution until 72-94% moisture, diminish both moduli, while sonication with high power ultrasound or megasonics increases it (Martínez-Padilla et al. 2018). Pulp dilution followed by sonication increase PPO activity from 25 to 187% with dilution ratios of 1:2 to 1:9; this increase might be related to the decrease in particle size (Bi et al. 2015).

Novel Technologies to Improve Oil Recovery

Some novel techniques as microwaves and high-frequency ultrasound have been mainly studied for preliminary operations improvement (dehydration and malaxation) to assist the extraction process of avocado oil. While production employing compressed fluid-based techniques are alternatives to rising yields of oil by the direct improvement of the extraction operation.

Microwave squeezing method was studied at laboratory level by (Ortiz et al. 2004). In this method, the fruits are peeled and destoned, and the pulp is microwave heated (900 W, 2450 MHz, 95°C), immediately afterwards the oil is extracted by squeezing manually in a cloth mesh. The process yielded 67% of the oil and gave the best quality compared with hexane and acetone extraction. A similar yield was obtained when microwaves were used (1000W, 2450 MHz, maximum power, 70°C, and 11 min) for dehydrate a thin layer of pulp that was grounded to a fine powder. The oil obtained from the microwaved pulp powder, had higher quality using soxhlet extraction with hexane, than that obtained from the sonicated pulp (Reddy, Moodley, and Jonnalagadda 2012). In other studies, the microwave dried pulp has also been extracted by expeller pressing (Santana et al. 2015).

Application of ultrasound technology has been assayed in pulp conditioning and malaxation at laboratory level and pilot scale. High-frequency ultrasound conditioning of avocado puree can enhance oil separation and decrease malaxation time without impacting oil quality. At the laboratory level, low (18 + 40 kHz) and high (2 MHz) frequency ultrasound treatments of no malaxed avocado puree improve extractability from 45% by 15 to 24%. An additional oil recovery, with higher extractability after 2 MHz treatments and longer times, had been obtained depending on the fruit maturity and oil content. Oils from sonicated purees at high frequencies show peroxide and free fatty acid values below the industrial specification level and increased content of total phenolics (Martínez-Padilla et al. 2018).

At the pilot plant scale, ultrasound also showed a positive effect when used between puree malaxation and centrifugation steps. After one minute application of high power ultrasound (1.73 MHz) to fresh pulp paste diluted with water (1:1), it was observed that virgin avocado oil extractability increased in 40% over the process without ultrasound. The fatty acid composition and the content of tocopherols and polyphenols were not affected by the ultrasound application (Segura et al. 2018).

The use of compressed fluid-based techniques as supercritical fluid for oil extraction has been mainly explored to produce oil with higher biological activity, augmenting the recovery of functional compounds (Herrero et al. 2013). Supercritical extraction with CO_2 is an alternative to the use of organic solvents. The CO_2 is inert, cheap, abundant in nature, no inflammable, no explosive, suitable for thermolabile compounds, and easy to eliminate at low pressure (King 2014). Oil extraction from avocado pulp dried in an oven at 45°C and powdered, has been done whit supercritical CO_2. Yields ranged from 40 to 64% when were assayed pressures from 250 to 450 bar, during 120 to 150 min, at 40-45°C. The obtained oil had a fatty acids profile close to that of the oil extracted by centrifugation (Abaide et al. 2017). When done by subcritical CO_2 at 68 bar, during 7.5 min, at 27°C; the extraction also leads to a virgin avocado oil with good oxidative quality, and retention of nutty and grassy flavors (Tan et al. 2018). Oil yield with both methods is superior to pressing, to ultrasound-assisted extraction, or soxhlet extraction using hexane, ether and ethanol. The sub or supercritical CO_2 extraction of avocado oil guarantee strong antioxidant activity and high concentration of functional molecules like sterols, lycopersene, and α and γ-tocopherols (Reddy, Moodley, and Jonnalagadda 2012). Extraction with compressed liquefied petroleum gas (0.5 MPa) was assayed with good results, providing higher extraction rates (60%), lower solvent consumption, and better concentrations of functional molecules than supercritical CO_2. However, while CO_2 is considered as generally recognized as safe (GRAS), liquefied petroleum gas (LPG) is highly flammable, is toxic, being a colorless odorless liquid when it is pressurized. When it evaporates into a gas, high concentrations can act as suffocating agent. Thus, the process extraction with LPG requires more safety, attention and control (Abaide et al. 2017).

Protected-Delivery Technologies of Avocado Oil

The functional compounds present in avocado oil are too sensitive to environmental factors like temperature, light and oxygen. An alternative to protect those compounds is encapsulation. This technology prevents the deterioration and loss of biological activity employing a cover. Also provides physicochemical stability and even modifies its bioavailability. There are numerous encapsulation methods and systems; some of them include emulsions, the use of biopolymers for inclusion in gel, or the formation of solid structures. These systems have also been developed in different size scales, using micro- and nano-carriers. In either case, the kind of system developed will depend on the characteristics of the bioactive compound of interest, the final formulation, and the area of application, mainly in food and pharmaceutical (McClements 2018, Sarabandi, Gharehbeglou, and Jafari 2019).

Emulsions have been developed to encapsulate avocado oil on a nanometric scale. These O/W nanoemulsions protect against environmental conditions, oxidation, also modify the physical properties and enhance solubility. In O/W nanoemulsions, the lipophilic compounds are incorporated into the dispersed phase as small particles or droplets (20-200 nm of diameter) surrounded by a layer of emulsifier. The selected emulsifier influences the stability during storage and the final size of the particle. In this way, avocado oil nanoemulsions formed in ultrasound equipment were stabilized by soy lecithin (natural emulsifier) and tween 80 (synthetic emulsifier) in concentrations of 10%. The particle sizes obtained ranged between 103 and 249 nm; and presented stability during storage of 15 and 10 days using tween 80 and soy lecithin, respectively. The nature of the emulsifier also influences the oxidative stability of the oil in the nanoemulsion, the use of tween 80 reduces lipid oxidation; however, soy lecithin provides less oxidation stability, probably because phospholipids themselves are susceptible to lipid oxidation (Arancibia et al. 2017).

In addition to the selection of suitable emulsifiers, should be used low oil concentrations (5-10%) for the formation of stable nanoemulsions. When it is necessary to incorporate a more considerable amount of oil, just can be formed emulsions. For the encapsulation of avocado oil at high concentrations (50%), have been assayed stabilizers as cellulose nanofibrils and sorbitan monostearate (Span 60) in levels of 1.0 and 3.5%. Additionally, it is used specific technology as rotor-homogenizers for the formation process to obtain more stable emulsions (30 days) (Franco et al. 2020).

The inclusion in a gel is another alternative for the protection of avocado oil. However, when dehydration steps are carried out after gel preparation, it is possible to obtain other systems. The resulting porous structures are known as cryogel, xerogel, and aerogel, when drying was done, respectively, by freeze drying, hot air, or supercritical CO_2. An assayer procedure begins with the oleogel formation using 11% of oil and 54% of cassava starch as a stabilizer; then, it was retrograded at different temperatures and, finally, freeze or vacuum dried (3.3 kPa at 60°C). The resulting cryogel and xerogel stability were evaluated throw oil encapsulation capacity, being higher in cryogels retrograded at -18°C (Pérez-Monterroza, Chaux-Gutiérrez, and Nicoletti 2018).

The microencapsulation using spray drying is a widely used method; it is an alternative to protecting the bioactive compounds in avocado oil powdered products. These products have the potential to be incorporated into different formulations. In this method are used stable emulsion with the oil in the core, the core material is mixed with a polymer which will serve as wall material and finally the mixture is contacted with a hot gas for seconds allowing solvent removal and microcapsule formation (Plazola-Jacinto et al. 2019). The polymers used as wall materials for encapsulation have been Arabic gum, maltodextrin, modified octenylsuccinic anhydride maltodextrin and whey protein isolate. The operational conditions in the spray drying equipment have been for the inlet and outlet temperatures of the drying air of 110-180 and 62 to 102°C respectively, airflow to the drier chamber at a rate of 20-800 mL/min and a feed flow rate of 5-20 mL/min. The range of particle sizes obtained has been from 2.2-4 μm. They have achieved encapsulation efficiency of 18.3 to 66%, the encapsulation using this methodology has been carried out with the purpose of protection from deterioration due to external factors and in the case of oils mainly to lipid oxidation (Bae and Lee 2008, Plazola-Jacinto et al. 2019, Dantas et al. 2018, Sotelo-Bautista et al. 2020, Permal et al. 2020).

ACKNOWLEDGMENTS

Jessica del Pilar Ramírez-Anaya thanks the FODECYJAL-COECYTJAL (project no. 8063-2019).

REFERENCES

Abaide, Ederson R., Giovani L. Zabot, Marcus V. Tres, Rafael F. Martins, Jean L. Fagundez, Lucielle F. Nunes, Susanne Druzian, Juliana F. Soares, Valéria Dal Prá, and Juliana R. F. Silva. 2017. "Yield, Composition, and Antioxidant Activity of Avocado Pulp Oil Extracted by Pressurized Fluids." *Food and Bioproducts Processing* no. 102:289-298.

Aguiló-Aguayo, Ingrid, Gemma Oms-Oliu, Olga Martín-Belloso, and Robert Soliva-Fortuny. 2014. "Impact of Pulsed Light Treatments on Quality Characteristics and Oxidative Stability of Fresh-cut Avocado." *LWT-Food Science and Technology* no. 59 (1):320-326.

Altendorf, Sabine. 2017. "Perspectivas Mundiales de las Principales Frutas Tropicales." ["World perspectives of the main tropical fruits"]. *http://www.fao.org/fileadmin/templates/est/COMM_MARKETS_MONITORING/Tropical_Fruits/Documents/Tropical_Fruits_Spanish2017.pdf.*

Arancibia, Carla, Natalia Riquelme, Rommy Zúñiga, and Silvia Matiacevich. 2017. "Comparing the Effectiveness of Natural and Synthetic Emulsifiers on Oxidative and Physical Stability of Avocado Oil-based Nanoemulsions." *Innovative Food Science & Emerging Technologies* no. 44:159-166.

Ayala-Zavala, J. Fernando, Violeta Vega-Vega, Cindy Rosas-Domínguez, Hugo Palafox-Carlos, Jose A. Villa-Rodriguez, M. Wasim Siddiqui, Jorge E. Dávila-Aviña, and Gustavo A. González-Aguilar. 2011. "Agro-industrial Potential of Exotic Fruit byProducts as a Source of Food Additives." *Food Research International* no. 44 (7):1866-1874.

Bae, E. K., and S. J. Lee. 2008. "Microencapsulation of Avocado Oil by Spray Drying Using Whey Protein and Maltodextrin." *Journal of Microencapsulation* no. 25 (8):549-560.

Barmore, Charles R. 1976. *Proceedings of the I. International Tropical Fruit Short Course: The Avocado.* Edited by J. W. Sauls, R. L. Phillips and L. K. Jackson. Miami, USA: Fruit Crops Department, Florida Cooperative Extension Service, Institute of Food and Agricultural Sciences, University of Florida.

Bermejo-Prada, Ana, Sandy Van Buggenhout, Laura Otero, Ken Houben, Ann Van Loey, and Marc E Hendrickx. 2014. "Kinetics of Thermal and High-pressure Inactivation of Avocado Polygalacturonase." *Innovative Food Science & Emerging Technologies* no. 26:51-58.

Bi, Xiufang, Yacine Hemar, Murat O. Balaban, and Xiaojun Liao. 2015. "The Effect of Ultrasound on Particle Size, Color, Viscosity and Polyphenol Oxidase Activity of Diluted Avocado Puree." *Ultrasonics Sonochemistry* no. 27:567-575.

Bressani, Ricardo, Brenda Rodas, and Ana, S. Ruiz. 2009. *La Composición Química, Capacidad Antioxidativa y Valor Nutritivo de la Semilla de Variedades de Aguacate. Tegucigalpa (Guatemala): Fondo Nacional de Ciencia y Tecnología-FONACYT*, Universidad del Valle de Guatemala-UVG. [*Chemical composition, antioxidant capacity and nutritional value of the seed from avocado varieties (Guatemala): National Foundation of Science and Technology FONACYT*, University of Guatemala Valley-UVG].

Buelvas-Salgado, Gustavo A, J. H. Patiño-Gómez, and J. A. Cano-Salazar. 2013. "Evaluación del Proceso de Extracción de Aceite de Aguacate Hass (*Persea americana* Mill) Utilizando Tratamiento Enzimático." *Revista Lasallista de Investigación* no. 9 (2). ["Evaluation of the Extraction Process of Hass Avocado Oil (Persea americana Mill) Using Enzyme Treatment." *Lasallian Research Journal*]

Bustos, M. Cecilia, María F. Mazzobre, and M. Pilar Buera. 2015 *LWT-Food Science and Technology* no. 61 (1):89-97.

Castañeda-Saucedo, M. Claudia, Elia H. Valdés-Miramontes, Ernesto Tapia-Campos, Adriana Delgado-Alvarado, Ana C. Bernardino-García, María R. Rodríguez-Ramírez, and Jessica P. Ramirez-Anaya. 2014. "Efecto del Liofilizado y del Proceso de Producción en la Composición Química y el Perfil de Ácidos Grasos de la Pulpa de Aguacate." *Revista chilena de nutrición* no. 41 (4):404-411. [Effect of lyophilization and production process on the chemical composition and the fatty acid profile of the avocado pulp. *Chilean Journal of Nutrition.* 41(4), 404-411].

Codex 1995. Codex Standard for Avocado, CODEX STAN-197-1995 C.F.R.

Colombo, Raffaella, and Adele Papetti. 2019. "Avocado (*Persea americana* Mill.) By-products and Their Impact: From Bioactive Compounds to Biomass Energy and Sorbent Material for Removing Contaminants. A review." *International Journal of Food Science & Technology* no. 54 (4):943-951.

Costagli, Giacomo, and Matteo Betti. 2015. "Avocado Oil Extraction Processes: Method for Cold-pressed High-quality Edible Oil Production Versus Traditional Production." *Journal of Agricultural Engineering* no. 46 (3):115-122.

Charoux, Clémentine M. G. Colm P. O'Donnell, and Brijesh K. Tiwari. 2017. "Ultrasound processing and food quality." In *Ultrasound: Advances for Food Processing and Preservation*, edited by D. Bermudez-Aguirre, 215-235. London, United Kingdom: Elsevier.

Dantas, Denise, Matheus Augusto Pasquali, Mário Cavalcanti-Mata, Maria Elita Duarte, and Hugo M. Lisboa. 2018. "Influence of Spray Drying Conditions on the Properties of Avocado Powder Drink." *Food chemistry* no. 266:284-291.

Deng, Li-Zhen, Arun S. Mujumdar, Zhongli Pan, Sriram K. Vidyarthi, Jinwen Xu, Magdalena Zielinska, and Hong-Wei Xiao. 2019. "Emerging Chemical and Physical Disinfection Technologies of Fruits and Vegetables: A Comprehensive Review." *Critical reviews in food science and nutrition*: 1-28.

Di Stefano, Vita, Giuseppe Avellone, David Bongiorno, Serena Indelicato, Roberto Massenti, and Riccardo Lo Bianco. 2017. "Quantitative Evaluation of the Phenolic Profile in Fruits of Six Avocado (*Persea Americana*) Cultivars by Ultra-high-performance Liquid Chromatography-heated Electrospray-mass spectrometry." *International journal of food properties* no. 20 (6):1302-1312.

Ding, Haiming, Young-Won Chin, A. Douglas Kinghorn, and Steven M. D'Ambrosio. 2007. "Chemopreventive Characteristics of Avocado Fruit." *Seminars in cancer biology* no. 17 (5):386-394.

dos Santos, Marco A. Z. Tatiana V. R. Alicieo, Claudio M. P. Pereira, Guillermo Ramis-Ramos, and Carla R. B. Mendonça. 2014. "Profile of Bioactive Compounds in Avocado Pulp Oil: Influence of the Drying Processes and Extraction Methods." *Journal of the American Oil Chemists' Society* no. 91 (1):19-27.

Economía, Secretaría de. 2012. *Monografía del Sector Aguacate en México: Situación Actual y Oportunidades de Mercado.* Edited by Dirección General de Industrias Básicas. México: Dirección General de Industrias Básicas. [*Monography of the Avocado Sector in Mexico: Actual situation and market opportunities: General Direction of Basic Industries*].

Elez-Martinez, Pedro, Robert C. Soliva-Fortuny, Shela Gorinstein, and Olga Martin-Belloso. 2005. "Natural Antioxidants Preserve the Lipid Oxidative Stability of Minimally Processed Avocado Purée." *Journal of food science* no. 70 (5):S325-S329.

Espinosa-Alonso, L. Gabriela, Octavio Paredes-López, Maribel Valdez-Morales, and B. Dave Oomah. 2017. "Avocado Oil Characteristics of Mexican Creole Genotypes." *European Journal of Lipid Science and Technology* no. 119 (10):1600406.

Eyres, Laurence, Nimma Sherpa, and Glenn Hendriks. 2001. "Avocado Oil: A New Edible Oil from Australasia." *Lipid Technol* no. 13 (4):84-88.

FAO. 2018. FAOSTAT. FAO http://www.fao.org/faostat/en/#data/QC (Retrieved April 2020).

FAO. 2020. *Major Tropical Fruits - Preliminary Market results 2019*. Rome.

Fernandes, Fabiano A. Valéria S. Oliveira, Wesley F. Gomes, and Sueli Rodrigues. 2016. "Degradation Kinetics of Vitamin E during Ultrasound Application and the Adjustment in Avocado Puree by Tocopherol Acetate Addition." *LWT-Food Science and Technology* no. 69:342-347.

Figueroa, Jorge G. Isabel Borrás-Linares, Jesús Lozano-Sánchez, Rosa Quirantes-Piné, and Antonio Segura-Carretero. 2018. "Optimization of Drying Process and Pressurized Liquid Extraction for Recovery of Bioactive Compounds from Avocado Peel By-product." *Electrophoresis* no. 39 (15):1908-1916.

Francis, G. A., A. Gallone, George John Nychas, J. N. Sofos, G. Colelli, M. L. Amodio, and G. Spano. 2012. "Factors Affecting Quality and Safety of Fresh-cut Produce." *Critical reviews in food science and nutrition* no. 52 (7):595-610.

Franco, Talita-Szlapak, Dana C. Martínez-Rodríguez, María F. Jiménez-Soto, Rosa M. Jiménez-Amezcua, Maite Renteria-Urquíza, Eduardo Mendizábal-Mijares, and Graciela I. Bolzon de Muniz. 2020. "Production and Technological Characteristics of Avocado Oil Emulsions Stabilized with Cellulose Nanofibrils Isolated from Agroindustrial RÑLesidues." *Colloids and Surfaces A: Physicochemical and Engineering Aspects* no. 586:124263.

Gómez-López, Vicente M. 2002. "Inhibition of Surface Browning, Cut Avocado." *Journal of food quality* no. 25 (4):369-379.

Grajales-Lagunes, A, Hugo S. Garcia-Galindo, O. Angulo-Guerrero, and J. A. Monroy-Rivera. 1999. "Stability and Sensory Quality of Spray Dried Avocado Paste." *Drying technology* no. 17 (1-2):318-326.

Gullón, Beatriz, Thelmo A. Lú-Chau, María T. Moreira, Juan M. Lema, and Gemma Eibes. 2017. "Rutin: A Review on Extraction, Identification and Purification Methods, Biological Activities and Approaches to Enhance its Bioavailability." *Trends in food science & technology* no. 67:220-235.

Gunstone, Frank D. and Frank A. Norris. 1983. *Lipids in foods: chemistry, biochemistry and technology*. Elmsford, USA: Pergamon Press.

Guzmán-Gerónimo, Rosa I. Mercedes G. López, and Lidia Dorantes-Alvarez. 2008. "Microwave Processing of Avocado: Volatile Flavor Profiling and Olfactometry." *Innovative food science & emerging technologies* no. 9 (4):501-506.

Herrero, Miguel, María Castro-Puyana, José A. Mendiola, and Elena Ibañez. 2013. "Compressed Fluids for the Extraction of Bioactive Compounds." *TrAC Trends in Analytical Chemistry* no. 43:67-83.

Hurtado-Fernández, Elena, Alegría Carrasco-Pancorbo, and Alberto Fernandez-Gutierrez. 2011. "Profiling LC-DAD-ESI-TOF MS Method for the Determination of Phenolic Metabolites from Avocado (*Persea americana*)." *Journal of Agricultural and Food Chemistry* no. 59 (6):2255-2267.

INDECOPI. 2014. *Proyecto de Norma Técnica Peruana, Palta Recomendaciones. PNTP 011.018 C.F.R.* Edited by Instituto Nacional de Defensa de la Competencia y de la Protección de la Propiedad Intelectual. Peru. [*Project of Peruvean Technical Regulations. Avocado Recommendations. PNTP 011.018 C.F.R.* Edited by National Institute for Defense of Competition and the Protection of Intellectual Property. Peru].

Jacobo-Velázquez, Daniel A. Gerardo Castellanos-Dohnal, Porfirio Caballero-Mata, and Carmen Hernández-Brenes. 2013. "Cambios Bioquímicos Durante el Almacenamiento de Puré de Aguacate Adicionado con Antioxidantes Naturales y Procesado con Alta Presión Hidrostática." *CyTA-Journal of Food* no. 11 (4):379-391.

Jacobo-Velázquez, Daniel A. and Carmen Hernández-Brenes. 2012. "Stability of Avocado Paste Carotenoids as Affected by High Hydrostatic Pressure Processing and Storage." *Innovative Food Science & Emerging Technologies* no. 16:121-128.

Jacobo-Velázquez, Daniel A. and Carmen Hernández-Brenes. 2010. "Biochemical Changes during the Storage of High Hydrostatic Pressure Processed Avocado Paste." *Journal of Food Science* no. 75 (6):S264-S270.

Jimenez, Paula, Paula Garcia, Vilma Quitral, Karla Vasquez, Claudia Parra-Ruiz, Marjorie Reyes-Farias, Diego F. Garcia-Diaz, Paz Robert, Cristian Encina, and Jessica Soto-Covasich. 2020. "Pulp, Leaf, Peel and Seed of Avocado Fruit: A Review of Bioactive Compounds and Healthy Benefits." *Food Reviews International*: 1-37.

King, Jerry W. 2014. "Modern Supercritical Fluid Technology for Food Applications." *Annual Review of Food Science and Technology* no. 5:215-238.

Kosińska, Agnieszka, Magdalena Karamać, Isabel Estrella, Teresa Hernández, Begoña Bartolomé, and Gary A Dykes. 2012. "Phenolic Compound Profiles and Antioxidant Cpacity of *Persea americana* Mill. Peels and Seeds of Two Varieties." *Journal of Agricultural and Food Chemistry* no. 60 (18):4613-4619.

Kwon, Youngjoo. 2017. "Luteolin as a Potential Preventive and Therapeutic Candidate for Alzheimer's Disease." *Experimental Gerontology* no. 95:39-43.

Li, Zhaoping, Angela Wong, Susanne M. Henning, Yanjun Zhang, Alexis Jones, Alona Zerlin, Gail Thames, Susan Bowerman, Chi-Hong Tseng, and David Heber. 2013. "Hass Avocado Modulates Postprandial Vascular Reactivity and Postprandial Inflammatory Responses to a Hamburger Meal in Healthy Volunteers." *Food & Function* no. 4 (3):384-391.

Limpisophon, Kanokrat, and Gerhard Schleining. 2017. "Use of Gallic Acid to Enhance the Antioxidant and Mechanical Properties of Active Fish Gelatin Film." *Journal of food science* no. 82 (1):80-89.

López-Cobo, Ana, Ana M. Gómez-Caravaca, Federica Pasini, María F. Caboni, Antonio Segura-Carretero, and Alberto Fernández-Gutiérrez. 2016. "HPLC-DAD-ESI-QTOF-MS and HPLC-FLD-MS as Valuable Tools for the Determination of Phenolic and Other Polar Compounds in the Edible Part and By-Products of Avocado." *LWT* no. 73:505-513.

Lu, Qing-Yi, James R. Arteaga, Qifeng Zhang, Sergio Huerta, Vay Liang W. Go, and David Heber. 2005. "Inhibition of Prostate Cancer Cell Growth by an Avocado Extract: Role of Lipid-soluble Bioactive Substances." *The Journal of nutritional biochemistry* no. 16 (1):23-30.

Lu, Qing-Yi, Yanjun Zhang, Yue Wang, David Wang, Ru-po Lee, Kun Gao, Russell Byrns, and David Heber. 2009. "California Hass Avocado: Profiling of Carotenoids, Tocopherol, Fatty Acid, and Fat Content during Maturation and from Different Growing Areas." *Journal of agricultural and food chemistry* no. 57 (21):10408-10413.

Madunić, Josip, Ivana Vrhovac-Madunić, Goran Gajski, Jelena Popić, and Vera Garaj-Vrhovac. 2018. "Apigenin: A Dietary Flavonoid with Diverse Anticancer Properties." *Cancer Letters* no. 413:11-22.

Manaf, Yanty Noorzianna, Agustinus P. Rahardjo, Yus A. Yusof, Mohd N. M. Desa, and Bangun P Nusantoro. 2018. "Lipid Characteristics and Tocopherol Content of the Oils of Native Avocado Cultivars Grown in Indonesia." *International Journal of Food Properties* no. 21 (1):2758-2771.

Maringgal, Bernard, Norhashila Hashim, Intan S. M. Amin-Tawakkal, and Mahmud T. Muda-Mohamed. 2019. "Recent Advance in Edible Coating and its Effect on Fresh/fresh-cut Fruits quality." *Trends in Food Science & Technology*.

Martínez-Padilla, Laura Patricia, Lisa Franke, and Pablo Juliano. 2017. "Characterisation of the Viscoelastic Properties of Avocado Puree for Process Design Applications." *Biosystems Engineering* no. 161:62-69.

Martínez-Padilla, Laura Patricia, Lisa Franke, Xin-Qing Xu, and Pablo Juliano. 2018. "Improved Extraction of Avocado Oil by Application of Sono-physical Processes." *Ultrasonics Sonochemistry* no. 40:720-726.

McClements, David Julian. 2018. "Encapsulation, Rotection, and Delivery of Bioactive Proteins and Peptides Using Nanoparticle and Microparticle Systems: A review." *Advances in Colloid and Interface Science* no. 253:1-22.

Medeiros, M. F. D., Sandra C. S. Rocha, O. L. S. Alsina, C. E. M. Jerônimo, U. K. L. Medeiros, and A. L. M. L. Da Mata. 2002. "Drying of Pulps of Tropical Fruits in Spouted Bed: Effect of Composition on Dryer Performance." *Drying Technology* no. 20 (4-5):855-881.

Mendoza-Gómez, Violeta, Montserrat Calderón-Santoyo, Pedro U. Bautista-Rosales, Rosa I. Ortiz-Basurto, Darvin E. Jimenez-Sánchez, and Juan A. Ragazzo-Sánchez. 2017. "Antibrowning Agents Added in Polymeric Coating Applied to Avocado Fruit Slices." *Interciencia* no. 42 (12):812-817.

Mo, Shunyan, Linlin Dong, W. Jeffrey Hurst, and Richard B. van Breemen. 2013. "Quantitative Analysis of Phytosterols in Edible Oils Using APCI Liquid Chromatography–tandem Mass Spectrometry." *Lipids* no. 48 (9):949-956.

Monge-Rojas, Rafael, and Hannia Campos. 2011. "Tocopherol and Carotenoid Content of Foods Commonly Consumed in Costa Rica." *Journal of Food Composition and Analysis* no. 24 (2):202-216.

Monika, Padmanabhan, and Arumugam Geetha. 2015. "The modulating Effect of *Persea americana* Fruit Extract on the Level of Expression of Fatty Acid Synthase Complex, Lipoprotein Lipase, Fibroblast Growth Factor-21 and Leptin–A Biochemical Study in Rats Subjected to Experimental Hyperlipidemia and Obesity." *Phytomedicine* no. 22 (10):939-945.

Olaeta, J. A. 2003. Industrialización del Aguacate: Estado Actual y Perspectivas Futuras. Paper Read at Proceedings V World Avocado Congress, at Granada-Málaga, Spain.

Omolola, Adewale O, Afam I. O. Jideani, and Patrick F. Kapila. 2017. "Quality Properties of Fruits as Affected by Drying Operation." *Critical reviews in food science and nutrition* no. 57 (1):95-108.

Ortiz-Viedma, Jaime, Alicia Rodriguez, Camila Vega, Fernando Osorio, Bruno Defillipi, Rául Ferreira, and Jorge Saavedra. 2018. "Textural, Flow and Viscoelastic Properties of Hass Avocado (*Persea americana* Mill.) During Ripening Under Refrigeration Conditions." *Journal of Food Engineering* no. 219:62-70.

Ortiz, M. Alicia, A. Lidia Dorantes, M. Juvencio Gallndez, and S. Elizabeth Cárdenas. 2004. "Effect of a Novel Oil Extraction Method on Avocado (*Persea americana* Mill) Pulp Microstructure." *Plant Foods for Human Nutrition* no. 59 (1):11-14.

Pahua-Ramos, M. Elena, Leticia Garduño-Siciliano, Lidia Dorantes-Alvarez, German Chamorro-Cevallos, Julieta Herrera-Martínez, Obed Osorio-Esquivel, and Alicia Ortiz-Moreno. 2014. "Reduced-Calorie Avocado Paste Attenuates Metabolic Factors Associated with a Hypercholesterolemic-high Fructose Diet in Rats." *Plant foods for human nutrition* no. 69 (1):18-24.

Pahua-Ramos, M. Elena, Alicia Ortiz-Moreno, Germán Chamorro-Cevallos, María D. Hernández-Navarro, Leticia Garduño-Siciliano, Hugo Necoechea-Mondragón, and Marcela Hernández-Ortega. 2012. "Hypolipidemic Effect of Avocado (*Persea americana* Mill) Seed in a Hypercholesterolemic Mouse Model." *Plant Foods for Human Nutrition* no. 67 (1):10-16.

Park, Eunyoung, Indika Edirisinghe, and Britt Burton-Freeman. 2018. "Avocado Fruit on Postprandial Markers of Cardio-metabolic Risk: A Randomized Controlled Dose Response Trial in Overweight and Obese Mmen and Women." *Nutrients* no. 10 (9):1287.

Pérez-Monterroza, E. José, Ana M. Chaux-Gutiérrez, and Vânia R. Nicoletti. 2018. "Encapsulation of Avocado Oil in Amylose Solution from Cassava Starch." *Journal of Food Processing and Preservation* no. 42 (5):e13594.

Permal, Rahul, Wee L. Chang, Brent Seale, Nazimah Hamid, and Rothman Kam. 2020. "Converting Industrial Organic Waste From the Cold-pressed Avocado Oil Production Line into a Potential Food Preservative." *Food chemistry* no. 306:125635.

Plazola-Jacinto, C. P. V. Pérez-Pérez, S. C. Pereyra-Castro, L. Alamilla-Beltrán, and A. Ortiz-Moreno. 2019. "Microencapsulation of Biocompounds From Avocado Leaves Oily Extracts." *Revista Mexicana de Ingeniería Química* no. 18 (3):1261-1276.

Poovarodom, Sumitra, Ratiporn Haruenkit, Suchada Vearasilp, Jacek Namiesnik, Milena Cvikrová, Olga Martincová, Aviva Ezra, Milan Suhaj, Pramoj Ruamsuke, and Shela Gorinstein. 2010. "Comparative Characterisation of Durian, Mango and Avocado." *International journal of food science & technology* no. 45 (5):921-929.

Reddy, Mageshni, Roshila Moodley, and Sreekanth B. Jonnalagadda. 2012. "Fatty Acid Profile and Elemental Content of Avocado (*Persea americana* Mill.) Oil–effect of Extraction Methods." *Journal of Environmental Science and Health, Part B* no. 47 (6):529-537.

Rodríguez-Carpena, J. German. David Morcuende, and Mario Estévez. 2011. "Avocado By-products as Inhibitors of Color Deterioration and Lipid and Protein Oxidation in Raw Porcine Patties Subjected to Chilled Storage." *Meat science* no. 89 (2):166-173.

Saavedra, Jorge, Andrés Córdova, Rosa Navarro, Paulo Díaz-Calderón, Claudia Fuentealba, Carolina Astudillo-Castro, Lea Toledo, Javier Enrione, and Lena Galvez. 2017. "Industrial Avocado Waste: Functional Compounds Preservation by Convective Drying Process." *Journal of Food Engineering* no. 198:81-90.

SAGARPA. 2016. Avance de Siembras y Cosechas. *Resumen nacional por estado*. Mexico: SAGARPA. [Advance of Plantings and Harvests. National Summary by State. *Statistic Yearbook.* México: SAGARPA].

Santana, Isabelle, Luciana M. F. dos Reis, Alexandre G. Torres, Lourdes M. C. Cabral, and Suely P. Freitas. 2015. "Avocado (*Persea americana* Mill.) Oil Produced by Microwave

Drying and Expeller Pressing Exhibits Low Acidity and High Oxidative Stability." *European journal of lipid science and technology* no. 117 (7):999-1007.

Sarabandi, Khashayar, Pouria Gharehbeglou, and Seid Mahdi Jafari. 2019. "Spray-drying Encapsulation of Protein Hydrolysates and Bioactive Peptides: Opportunities and Challenges." *Drying Technology*: 1-19.

Saxena, Alok, Loveleen Sharma, and Tanushree Maity. 2020. "Enrichment of Edible Coatings and Films with Plant Extracts or Essential Oils for the Preservation of Fruits and Vegetables." In *Biopolymer-Based Formulations*, 859-880. Elsevier.

SCFI. 2016. *Productos Alimenticios no Industrializados para Uso Humano - Fruta Fresca-Aguacate Hass (Persea americana) Mill.* Edited by SEGOB. México: Diario Oficial de la Federación. [*Non-Industrialized Food Products for Human Use. -Fresh fruits- Hass Avocado (Persea americana) Mill.* Edited by SEGOB. Mexico: Official Journal of the Federation.

Schwartz, Marco, J. A. Olaeta, Pedro Undurraga, and V. Costa. 2007. *Mejoramiento del Rendimiento de Extracción del Aceite de Palta (aguacate).* Paper read at VI World Avocado Congress (Actas VI Congreso Mundial del Aguacate) at Vina Del Mar, Chile. [*Improvement of extraction of avocado oil.* Paper presented at the Proceedings of the VI World Avocado Congress (Acts of the VI World Avocado Congress), Vina Del Mar, Chile.

Segura, Nadia, Miguel Amarillo, Natalia Martinez, and Mará Grompone. 2018. "Improvement in the Extraction of Hass Avocado Virgin Oil by Ultrasound Application." *J. Food Res* no. 7:106-113.

Shepherd, John, and Gary Bender. 2002. "A history of the avocado industry in California." *Calif Avocado Soc Yearb* no. 85:29-50.

Silva Caldas, A. Paula, Larissa Oliveira Chaves, Letícia Linhares Da Silva, Dayane De Castro Morais, and Rita C. Gonçalves Alfenas. 2017. "Mechanisms Involved in the Cardioprotective Effect of Avocado Consumption: A Systematic Review." *International journal of food properties* no. 20 (sup2):1675-1685.

Singh, Shiv, and Rachana Shalini. 2016. "Effect of Hurdle Technology in Food Preservation: A Review." *Critical reviews in food science and nutrition* no. 56 (4):641-649.

Soliva-Fortuny, Robert C. Pedro Elez-Martínez, Mercè Sebastián-Calderó, and Olga Martín-Belloso. 2004. "Effect of Combined Methods of Preservation on the Naturally Occurring Microflora of Avocado Purée." *Food Control* no. 15 (1):11-17.

Sotelo-Bautista, Martin, Luis A. Bello-Perez, Rosalía A. Gonzalez-Soto, Jorge Yañez-Fernandez, and Jose Alvarez-Ramirez. 2020. "OSA-maltodextrin as Wall Material for Encapsulation of Essential Avocado Oil by Spray Drying." *Journal of Dispersion Science and Technology* no. 41 (2):235-242.

Souza, Danilo Santos, Luanda Gimeno Marques, Edelvio de Barros Gomes, and Narendra Narain. 2015. "Lyophilization of Avocado (*Persea americana* Mill.): Effect of Freezing and Lyophilization Pressure on Antioxidant Activity, Texture, and Browning of Pulp." *Drying technology* no. 33 (2):194-204.

Stanizzi, Antonio, Manuela Bottoni, Matteo Torresetti, Anna Campanati, and Giovanni Di Benedetto. 2015. "Topical Use of α-tocopherol Acetate in Delayed Wound Healing." *International Wound Journal* no. 12 (6):746-747.

Stępień, Anna, Mariusz Witczak, and Teresa Witczak. 2020. "Moisture Sorption Characteristics of Food Powders Containing Freeze Dried Avocado, Maltodextrin and Inulin." *International Journal of Biological Macromolecules* no. 149:256-261.

Takenaga, Fumio, Kaori Matsuyama, Shin Abe, Yasuyoshi Torii, and Shingo Itoh. 2008. "Lipid and Fatty Acid Composition of Mesocarp and Seed of Avocado Fruits Harvested at Northern Range in Japan." *Journal of Oleo Science* no. 57 (11):591-597.

Tan, C. Xuan. 2019. "Virgin Avocado Oil: An Emerging Source of Functional Fruit Oil." *Journal of functional foods* no. 54:381-392.

Tan, C. Xuan, Gun Hean Chong, Hazilawati Hamzah, and Hasanah Mohd Ghazali. 2018. "Characterization of Virgin Avocado Oil Obtained Via Advanced Green Techniques." *European journal of lipid science and technology* no. 120 (10):1800170.

Tan, C.Xuan, and Hasanah Mohd Ghazali. 2019. "Avocado (*Persea americana* Mill.) Oil." In *Fruit Oils: Chemistry and Functionality*, 353-375. Springer.

Téliz, Daniel & Keeve, C. 2007. "Comercio Internacional del aguacate." In *El Aguacate y Su Manejo Integrado*, edited by D. & Mora Téliz, A. Mexico: Ediciones Mundi Prensa.

Terasawa, Naoko, Miki Sakakibara, and Masatsune Murata. 2006. "Antioxidative Activity of Avocado Epicarp Hot Hater Extract." *Food science and technology research* no. 12 (1):55-58.

Tremocoldi, M. Augusta, Pedro L. Rosalen, Marcelo Franchin, Adna Prado Massarioli, Carina Denny, Erica R. Daiuto, Jonas A. Rizzato Paschoal, Priscilla Siqueira Melo, and Severino Matias de Alencar. 2018. "Exploration of Avocado By-products as Natural Sources of Bioactive Compounds." *PLoS One* no. 13 (2).

USDA. 2011. Avocados, Raw, All Commercial Varieties. U.S. Department of Agriculture.

van Dieren, Jolanda M. Ytje Simons-Oosterhuis, H. C. Raatgeep, Dicky J. Lindenbergh-Kortleve, Margaretha E. H. Lambers, C. Janneke van der Woude, Ernst J. Kuipers, Gerry T. Snoek, Ron Potman, and Hamida Hammad. 2011. "Anti-inflammatory Actions of Posphatidylinositol." *European Journal of Immunology* no. 41 (4):1047-1057.

Vargas-Ortiz, Manuel, Guadalupe Rodríguez-Jimenes, Marco Salgado-Cervantes, and Dominique Pallet. 2017. "Minimally Processed Avocado through Flash Vacuum-expansion: Its Effect in Major Physicochemical Aspects of the Puree and Stability on Storage." *Journal of Food Processing and Preservation* no. 41 (3):e12988.

Vargas-Ortiz, Manuel, Adrien Servent, Guadalupe Rodríguez-Jimenes, Dominique Pallet, and Marco Salgado-Cervantes. 2017. "Effect of Thermal Stage in the Processing Avocado by Flash Vacuum Expansion: Effect on the Antioxidant Capacity and the Qualitaty of the Mash." *Journal of Food Processing and Preservation* no. 41 (5):e13118.

Wang, Chung-Yi, Hsiao-Wen Huang, Chiao-Ping Hsu, and Binghuei Barry Yang. 2016. "Recent Advances in Food Processing Using High Hydrostatic Pressure Technology." *Critical Reviews in Food Science and Nutrition* no. 56 (4):527-540.

Wang, Jia-Shui, An-Bang Wang, Xiao-Ping Zang, Lin Tan, Bi-Yu Xu, Hai-Hong Chen, Zhi-Qiang Jin, and Wei-Hong Ma. 2019. "Physicochemical, Functional and Emulsion `Properties of Edible Protein from Avocado (*Persea americana* Mill.) Oil Processing By-products." *Food chemistry* no. 288:146-153.

Wang, Li, Peter L. Bordi, Jennifer A. Fleming, Alison M. Hill, and Penny M. Kris-Etherton. 2015. "Effect of a Moderate Fat Diet with and Without Avocados on Lipoprotein Particle Number, Size and Subclasses in Overweight and Obese Adults: a Randomized, Controlled Trial." *Journal of the American Heart Association* no. 4 (1):e001355.

Widsten, Petri, Cristina D. Cruz, Graham C. Fletcher, Marta A. Pajak, and Tony K. McGhie. 2014. "Tannins and Extracts of Fruit Byproducts: Antibacterial Activity against Foodborne Bacteria and Antioxidant Capacity." *Journal of agricultural and food chemistry* no. 62 (46):11146-11156.

Wong, Marie, Laurence Eyres, and Leandro Ravetti. 2014. "Modern Aqueous Oil Extraction—Centrifugation Systems for Olive and Avocado Oils." In *Green vegetable oil processing*, edited by Walter E. Farr and Andrew Proctor, 19-51. Urban, USA: AOCS Press.

Woolf, Allan B. I. B. Ferguson, L. C. Requejo-Tapia, L. Boyd, W. A. Laing, and A. White. 1999. "Impact of Sun Exposure on Harvest Quality of 'Hass' Avocado Fruit." *Revista Chapingo Serie Horticultural* no. 5:352-358.

Woolf, Allan B. Reginald Wibisono, Judie Farr, Ian Hallett, Lena Richter, Indrawati Oey, Mark Wohlers, Jing Zhou, Graham C. Fletcher, and Cecilia Requejo-Jackman. 2013. "Effect of High Pressure Processing on Avocado Slices." *Innovative Food Science & Emerging Technologies* no. 18:65-73.

Woolf, Allan, Marie Wong, Laurence Eyres, Tony McGhie, Cynthia Lund, Shane Olsson, Yan Wang, Cherie Bulley, Mindy Wang, and Ellen Friel. 2009. "Avocado Oil." In *Gourmet and Health-promoting specialty oils*, edited by RA Moreau and K Kamal-Eldin, 73-125. Urbana, USA: AOCS.

Yamaguchi, Masayoshi, and Satoshi Uchiyama. 2004. "β-Criptoxanthin Stimulates Bone Formation and Inhibits Bone Resorption in Tissue Culture in Vitro." *Molecular and Cellular Biochemistry* no. 258 (1-2):137-144.

Yang, Shuo, Ian Hallett, Hyunah E. Oh, Allan B. Woolf, and Marie Wong. 2019. "The Impact of Fruit Softening on Avocado Cell Microstructure Changes Monitored by Electrical Impedance and Conductivity for Cold-pressed Oil Extraction." *Journal of Food Process Engineering* no. 42 (4):e13068.

Yang, Shuo, Ian Hallett, Ria Rebstock, Hyunah E. Oh, Rothman Kam, Allan B. Woolf, and Marie Wong. 2018. "Cellular Changes in "Hass" Avocado Mesocarp during Cold-pressed Oil Extraction." *Journal of the American Oil Chemists' Society* no. 95 (2):229-238.

Yousuf, Basharat, Vinayak Deshi, Burhan Ozturk, and Mohammed Wasim Siddiqui. 2020. "Fresh-cut Fruits and Vegetables: Quality Issues and Safety Concerns." In *Fresh-Cut Fruits and Vegetables. Tehnologies and mechanisms for safety control*, edited by Md Wasim Siddiqui, 1-15. San Diego, USA: Academic Press.

Zhang, Nana, Yanfang Chen, Yueliang Zhao, Daming Fan, Lijie Li, Bowen Yan, Guanjun Tao, Jianxin Zhao, Hao Zhang, and Mingfu Wang. 2020. "Caffeic Acid Assists Microwave Heating to Inhibit the Formation of Mutagenic and Carcinogenic PhIP." *Food Chemistry* no. 317:126447.

Zhou, Linyan, Chia Ying Tey, Gokhan Bingol, and Jinfeng Bi. 2016. "Effect of Microwave Treatment on Enzyme Enactivation and Quality Change of Defatted Avocado Puree during Storage." *Innovative Food Science & Emerging Technologies* no. 37:61-67. doi: https://doi.org/10.1016/j.ifset.2016.08.002.

Züge, Luana C. Bosmuler, Helayne Aparecida Maieves, Joana L. Meira Silveira, Vitor Renan da Silva, and Agnes de Paula Scheer. 2017. "Use of Avocado Phospholipids as Emulsifier." *LWT-Food Science and Technology* no. 79:42-51.

In: Super and Nutraceutical Foods
Editors: Sandra Teresita Martín-del-Campo et al.
ISBN: 978-1-53619-082-3

Chapter 2

CACAO: AN ANCESTRAL SOURCE OF HEALTHY COMPOUNDS

Blanca I. Maldonado-Guevara*
Escuela de Ingeniería y Ciencias,Tecnologico de Monterrey,
Querétaro, México

ABSTRACT

The constant need to find products that contribute to people's well-being has led to cacao revaluation and its byproducts in recent years. An important amount of investigations are developed around the stages of industrialization of cacao. Most of those works focus on understanding the chemical reactions that occur and their effect on the content of nutrients and the quality and acceptance of the final products.

Other investigations are related to the effects of consuming cacao and its byproducts and its impact on health. This increasing attention has also led to efforts to recover Theobroma cacao´s native species and develop better sowing and production conditions. .

Cacao is not only an important source of energy, but a source of bioactive compounds, like polyphenoles, methylxanthines, vitamins, and minerals, among others. All of those compounds have proved beneficial effects on health.

The concentration of those compounds, their bioavailability and, therefore, their impact on health, depends on different variables such as origin variables; for instance, cocoa variety, cultivation area, fertilization, cultivation health (presence of pests), etc.; processing variables: stages, and processing conditions (for instance; temperature, time, and moisture content); consumption variables: form and frequency of consumption of the byproduct; and finally variables related to consumer health: normal, borderline, or with the disease. Generally, each stage of cacao processing is, to a greater or lesser extent, related to the loss of biomolecules, especially antioxidants. The consumption of less processed cocoa forms, such as nibs, minimally processed drinks, and dark chocolate (90%), will yield more significant health benefits.

Keywords: cacao, antioxidants, processing variables, health benefits

---

* Corresponding Author's E-mail: bimaldon@tec.mx.

INTRODUCTION

The cacao tree belongs to the Malvaceae family, subfamily Sterculioideae, which includes 22 different species, cacao having the highest commercial importance. Theobroma is found wild in the rain forest of the western hemisphere from 18°N to 15°S, that is, from the southern part of Mexico to the Amazon jungle at altitudes from 100 to 1,000 feet above sea level. Theobroma grows in areas where temperatures are relatively uniform through the year (between 68°F and 82°F), and having constant high humidity. The tree´s rain requirements depend on the soil characteristics, such as its porosity and humus´s content. The minimum necessary rainfall for Theobroma is about 39 inches equally distributed throughout the year, being the optimal 59–79 inches. In these conditions, the cacao tree requires porous soil that allows the drain of the water.

The cacao tree has shallow roots, hence they are often planted with corps like banana, rubber, or palm. Those crops protect the young Theoborma trees from strong winds and also provide them with shade.

a)

b)

Figure 1. a) Cacao tree. b) Cacao pods.

Cacao trees are branched trees with simple leaves, reaching heights of 20-40 feet (Figure 1A); it has oblong leathery leaves of 12 inches long. Its flowers are ill-scented or odorless; These flowers are present throughout the year but have a significant flowering only twice a year during June and July. Flower colors vary from white, pink, yellow to bright red, depending on the tree´s variety.

The cacao tree takes four years to produce fruit in the form of elongated pods (Figure 2). A single tree produces up to 70 fruits per year. The pods ripen in less than six months to a length up to 35 cm (14 inches) and a width of 12 cm (4.7 inches). The pods range in color from bright yellow to deep purple. The pods contain 20 to 60 seeds or cacao beans arranged around pod´s long axis. The oval seeds are about 1-inch-long and are covered with a sweet sticky white pulp (mucilage) (Figure 2).

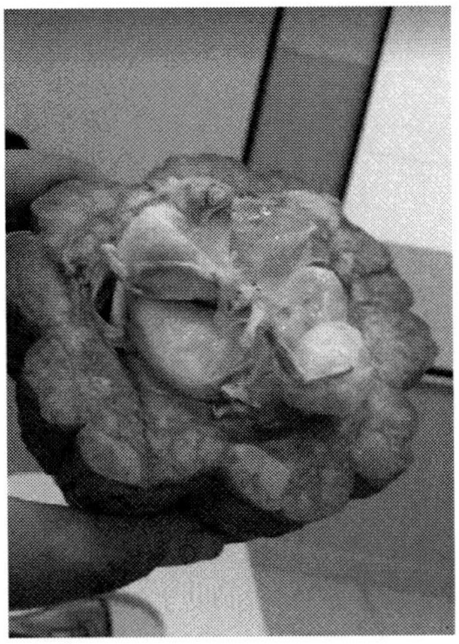

Figure 2. Cacao pod cross-section.

The cacao tree has a wide genetic diversity, both wild and cultivated. The types of cacao are traditionally classified into three genetic groups: Creole (fine grade), Forastero (bulk grade), and Trinitario (fine grade) (Redovniković et al. 2009).

Creole variety is the finest cacao, the beans from this variety have a unique bouquet of flavors and aromas (Oracz, Zyzelewicz, and Nebesny 2015) (Redovniković et al. 2009). This cacao requires special conditions for growing and is the most susceptible to diseases. This cacao grows in the south of Mexico, Central America, and the northern part of South America and has the least genetic diversity.

Forastero is the variety with the highest genetic diversity and has high commercial importance due to 95% of the commercialized cacao is from this variety. This cacao tree is strong and disease-resistant, but its flavor and taste are less aromatic than the fine varieties. National cacao is genetically considered a forastero variety, but unlike these, National cacao can be considered a third fine variety of cacao bean besides Creole and Trinitario. Ecuador's economic history has been linked to National cacao since the pre-Columbian age (Solorzano et al. 2012).

Trinitario species is a hybrid of beans of Creole from Trinidad and Forastero from the upper region of the Amazon basin (Oracz, Zyzelewicz, and Nebesny 2015). Trinitario was grown initially in Trinidad but currently grows in Ecuador and Venezuela. This cacao tree is resistant to environmental conditions and is a fine variety.

Creole, Trinitario, and Nacional varieties provide the most prized, high-quality cacao beans but represents only 5% of the world´s commercialized cacao.

Most cacao trees are hybrids of mixed origins, and difficulty can be grouped in the original varieties.

Chemical Components of Cacao and Its Effect on Health

Historically, human´s have searched for foods to consume that contain nutrients and contribute to well-being. Cacao, coffee, and tea are the most common examples of those products. Cacao has gained interest in recent years as a superfood due to its content of biologically active components that have potentially beneficial effects on human health, preventing and combating many diseases. Those components are mainly antioxidants, including polyphenols, tocopherols, and methylxanthines (Oracz, Zyzelewicz, and Nebesny 2015). Besides antioxidants, cacao contains peptides and minerals, which also have significant effects on health.

Polyphenols

Polyphenols are among the most numerous and widely distributed groups of substances in the plant kingdom; polyphenols are among the main classes of secondary metabolites. They include a range of simple phenolic molecules like phenolic acids to highly polymerized compounds with high molecular weight as tannins (Bravo 1998). Polyphenols are responsible for the sensory properties of the food-derived products and beverages, especially color and taste. Polyphenols are the most abundant antioxidants on a diet (Scalbert, Johnson, and Saltmarsh 2005), their daily intake is highest among the antioxidants, even over vitamin C, and E and can be as high as 1 g/day. Polyphenols are widely found in fruits, vegetables, cereals, dry legumes, chocolate, and beverages, such as tea, coffee, or wine (Scalbert et al. 2005). Polyphenols are the main antioxidants in unfermented cocoa beans (2%w/w) (Di Mattia et al. 2017).

Flavonoids are one of the most studied polyphenols because they are the major active nutraceutical ingredients in plants. Those are the most ubiquitous compounds of the group. Over 8,000 varieties of flavonoids have been identified (Tapas, Sakarkar, and Kakde 2008).

Major cocoa flavonols are (-)-epicatechin, (-)-catechin, and procyanidins (Afoakwa, 2008) (Figure 3). These flavonols are water-soluble compounds that are susceptible to enzymatic and non-enzymatic oxidation.

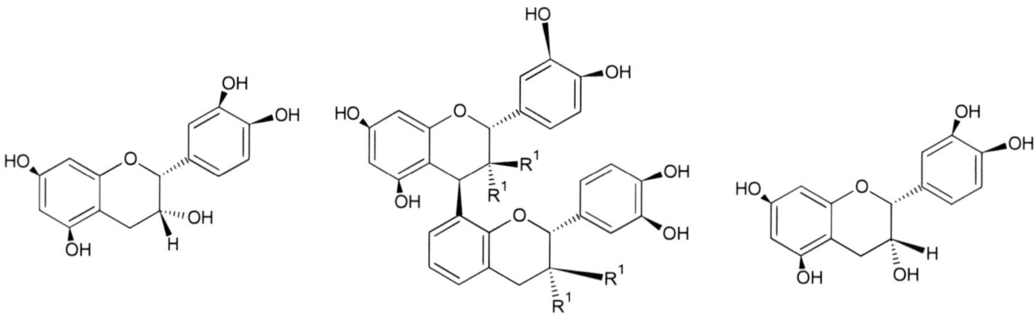

a) (-)-catechin b) procyanidin (4β->6) dimers c) (-)-epicatechin

Figure 3. Chemical structure of the major cacao flavonols.

Cacao Polyphenols and Its Effect on Health

Flavonoids, the most common polyphenols on cacao, are potent antioxidants and metal chelators. Are associated with health for having antiulcer activity. In the cardiovascular system, flavonoids act as a vasorelaxant agent. Flavonoids have antibacterial, antifungal, antiviral, hepatoprotective and anti-inflammatory activities; and have antidiabetic, antiatherosclerotic; anti-thrombogenic and cardioprotective effects, among other properties.

Flavonols inhibit lipid peroxidation and improve the obesity-related inflammation in mice (Franco, Oñatibia-Astibia, and Martínez-Pinilla 2013). Flavonols also have an important role in the grown of beneficial gut microbes (e.g., Lactobacilli) and the decrease of less beneficial ones like Clostridia (Franco, Oñatibia-Astibia, and Martínez-Pinilla 2013).

Tocopherols

Tocopherols are important antioxidants of fat-soluble nature. Are found in peanuts, asparagus, tomatoes, and carrots. Animal fats also contain tocopherols, but in a much lower amount than vegetable sources. Tocopherols are a group of 8 different homologs having a hydroxylated ring system (chromanol ring) with a phytol chain (Decker et al. 2010). Tocopherols are synthesized only in photosynthetic organisms and act as protective components (Duncan and Chang 2012).

Cacao Tocopherols and Its Effect on Health

Tocopherols have antioxidant effect, maintain the cellular structure, and contribute to the functions of the muscles, blood vessels, and the nervous system. Tocopherols acts as a messenger of the cell′s oxidative state. Are associated with the stabilization of cell membranes of the immune system. Also are related to increasing the production of immunoglobulins. Tocopherols have effects on preventing aging and cancer, among other effects on health (Fernández et al. 2002).

Methylxanthines

Methylxanthines are a group of substances produced by plants and animals that are no widely studied like polyphenols and tocopherols. These compounds belong to the family of purines. The most popular methylxanthines in cacao are caffeine, theobromine, and theophylline (Franco, Oñatibia-Astibia, and Martínez-Pinilla 2013) (Figure 4).

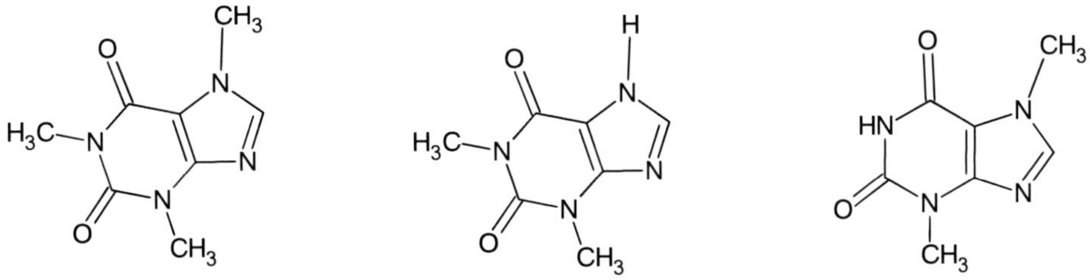

a) Caffeine b) Theophylline c) Theobromine

Figure 4. Major methylxanthines present in cacao (Jalil and Ismail 2008).

The most abundant methylxanthine in cacao is theobromine, and theophylline is found in less concentration. The methylxanthines content on cacao beans is influenced by the species of cacao, varietal, and hybrid. Table 1 list the content of the main methylxanthines on some theobroma varieties and some byproducts of cacao.

Table 1. Content of methylxanthines on some cacao varieties and different common byproducts

Material	Methylxantines (mg/kg)		
	Theobromine	Caffeine	Theophylline
Creole cacao defatted	12400	11300	-
Trinitario defatted	33200	6300	-
Forastero defatted	36200	1300	-
Nacional defatted	35400	2400	-
Raw ground paste	33000	5600	200
Roasted ground paste	36000	330	BLD
Cacao butter	140	400	BLD
Dark chocolate	5000-7500	625-875	-

BLD: Below limit of detection (Franco, Oñatibia-Astibia, and Martínez-Pinilla 2013, Apgar, Tarka, and Stanley 1998, Peláez, Bardón, and Camasca 2016).

Cacao Methylxanthines and Its Effect on Health

Besides the polyphenols, methylxanthines are the most active components of cacao. They are recognized for their antioxidant capacity and their potential benefits for human health, like increasing beneficial gut microbes (e.g., Lactobacilli) and decreasing the less beneficial ones (e.g., Clostridia) (Franco, Oñatibia-Astibia, and Martínez-Pinilla 2013).

Methylxanthines are cough suppressants (without side effects), lung function enhancers, and bronchodilators in asthma patients. Methylxanthine therapy is used to treat central apnea by stimulating the central nervous system and respiratory muscle function; and with a psycho-stimulant effect (Franco, Oñatibia-Astibia, and Martínez-Pinilla 2013). Methylxanthines are associated with a low incidence of neurodegenerative diseases such as Parkinson's and Alzheimer's (Franco, Oñatibia-Astibia, and Martínez-Pinilla 2013).

Peptides

Cacao is also rich in proteins (Table 1). The content of proteins is 10–15% of the dry weight of unfermented cocoa beans (Barišić et al. 2019). Peptides are responsible for the flavor precursor formation (Jalil and Ismail 2008). Cacao contains four types of proteins: albumins, globulins, prolamin, and glutelin (Barišić et al. 2019). Albumin constitutes the major protein fraction, and albumin and globulin fractions accounted for 52% and 43% of total cacao bean proteins, respectively (Jalil and Ismail 2008).

Bioactive peptides were reported to possess antihypertensives, antithrombotic, hypocholesterolemic and hypotriglyceridemic, and antiobesity effects (Jalil and Ismail 2008).

Minerals

Minerals contribute to health benefits attributed to cacao and cacao products. The content of some minerals is shown in Table 2. Cocoa and cocoa byproducts are rich in magnesium. The content of magnesium in dark chocolate (99% cacao) has been reported as 252.2 mg/100 g, which represents 67.0% of Nutrient Reference Values (NRV)(Cinquanta et al. 2016) (Jalil and Ismail 2008). Magnesium is an essential mineral for optimal metabolic function and may reduce diabetes risk. Studies have shown magnesium´s effectiveness in treating eclampsia and preeclampsia, arrhythmia, severe asthma, and migraine (Guerrera, Volpe, and Mao 2009). Magnesium is a cofactor of more than 300 metabolic reactions in the body. It plays a critical role in nerve transmission, cardiac excitability, neuromuscular conduction, and muscular contraction (Guerrera, Volpe, and Mao 2009).

Cocoa and cocoa products are also rich in copper, an essential element required as a cofactor and structural component of numerous metalloenzymes (Stern 2010).

Iron is a mineral present in cacao and cacao byproducts. Dark chocolate (90% cacao) contains 10.9 mg/100 g, which represents 80.3 of Nutrient Reference Values (NRV) (Cinquanta et al. 2016).

Selenium, a mineral present in cacao, is an essential micronutrient with the antioxidant ability to scave free radicals and inhibit lipid peroxidation (Jalil and Ismail 2008). Dark chocolate (99% cacao) was reported to have 0.1 mg per100 g (Cinquanta et al. 2016).

Table 2. General composition of major components of cacao and its byproducts (100g of product)

Product	Energy (Kj)	Moisture (%)	Cude Protein (%)	Ether extract (%)	Total CHS* (%)	Sugar (%)	Fiber (TDF) (%)
Cacao beans (raw, unfermented)	1519.0	5.3	72.8	2.2	14.0		1.1
Cacao Creole with hull	2308.0	3.6	15.0	47.0	26.1		
Cacao pulp		85.5				12.5	
Cacao juice (raw)		85.7	7.2	3.5		22.3	
Cacao nibs (dried, fermented)		7.5					
Cacao liquor	1552.0	10.7	1.8	4.2	81.0		
Powdered cocoa	1301.0	7.4	19.8	5.2	46.3		33.2
Cacao powdered chocolate	1541.0	1.0	6.5	3.4	77.7		3.6
Cacao fat	1350.0	6.7	22.9	4.7	52		ND
Dark chocolate (without sugar)	2171.0	3.2	11.8	38.7	36.2		15.0
Tascalate (50% cocoa, 50% corn)	1769.7	2.4	5.5	4.4	85.2	6.5	9.3
Cacao beans (raw, unfermented)	63.0	287.0	397.0	100.0		50.0	10.0
Cacao Creole with the hull	134.0	538.0					
Cacao pulp			256.5	33.0			
Cacao juice (raw)	188.5	63.5	958.0	82.5			
Cacao liquor	249.0	74.0					
Powdered cocoa	128.0	734.0	1524.0	499.0		0.12	0.0
Cacao powdered	949.0	971.0	712.0	83.0	3.8	0.11	0.0
Cacao fat	372.0	471.0					
Dark chocolate (without sugar)	134.0	15.0	397.0	100.0	1.4	50	10
Tascalate (50% cocoa, 50% corn)	87.5	153.4	272.8	74.6			

* CHS: Carbohidratos, TDF: Total dietary fiber.

Instituto Nacional de Nutrición (INCMNSZ), (KYB Anvoh, Bi, & Gnakri, 2009).

Zinc is also present in chocolate; dark chocolate contains 3.5 mg/100 g (Cinquanta et al. 2016). Zinc deficiency is related to damage to the fetus (Walsh et al. 1994). It is known to stimulate at least 100 different enzymes. Its deficiency in children is associated with problems in growth and increasing the risk of infections.

Besides the bioactive components mentioned above, cocoa and its derivatives contain other macronutrients, vitamins, and minerals essential for human health. The content of some of those compounds is shown in Table 2.

Factors Affecting the Chemical Composition of Cacao and Its Byproducts

The content of bioactive compounds (polyphenols, methylxanthines, and tocopherols), major components (proteins, fat, sugars, etc.), vitamins, minerals, and other important chemical compounds (such as vitamins and organic acids), on cacao and it's byproducts; and therefore, the real benefits on health and well-being depend on a wide variety of factors. One of the most important is the tree´s genetic diversity; for instance, the creole cacao variety has a lower (2/3) content of phenolic compounds than the forastero variety. The climatic or agronomic conditions of growing the cacao tree, the crop health (diseases), and post-harvest management like storage time and storage conditions are also important in the final chemical composition of cacao (Redovniković et al. 2009).

Once the cocoa seeds are harvested, they are subjected to fermentation and drying operations to stabilize and extend the shelf life of beans, besides generating compounds that precursors their sensory properties. Although being relatively simple, these first stages are carried out using different procedures that involve variations in time, temperature, and airflow, among other factors that directly affect the chemical composition of the fermented, dehydrated, and husked cocoa seeds.

From this point, all of the processing steps will influence the chemical composition of the cacao byproducts. Each of the processing steps will decrease or increase the content and availability of its bioactive components. Besides, the addition of ingredients (i.e., probiotics, sugars, species, etc.) affect the cocoa byproduct´s final composition depending on the ingredient´s kind, and the amount added.

Nutrient content, bioavailability, and, therefore, its effect on health for each cacao byproduct must be analyzed individually even if it belongs to the same category of products due to the differences in raw material composition and variability of conditions on the production process.

Described below are the main steps in the production process and its effect on the final product.

Raw Cacao Beans

Cacao fruit contains about 30–40 seeds covered by a mucilaginous pulp. Those nibs are rich in antioxidants. Polyphenols are the main antioxidants in unfermented cacao beans, and they account for approximately 2% (w/w). Cacao contains several classes of phenolic compounds; the most important are: flavanols (37%), proanthocyanidins (58%), and anthocyanins (4%) (Di Mattia, Sacchetti, Mastrocola, & Serafini, 2017). Cacao beans are rawly consumed in countries like Mexico, frequently added with chili powder (Figure 5).

Figure. 5. Raw cacao beans added with chili.

Cacao Juice

Cacao seeds are surrounded by mucilaginous pulp, and when the fruit is open, the juice is obtained. In general, one liter of cacao juice can be produced from 10 kg of fresh cacao beans (Koutoua Anvoh, Guéhi, Beugré, Kinimo, & Gnakri, 2010). This juice of sweet and acidic taste is mostly composed of water. The primary nutrients are described in Table 2. (KYB Anvoh, Bi, & Gnakri, 2009). This juice also contains some organic acids like citric acid (9.14 ± 0.64 mg/L) and malic acid (3.6 ± 0.050mg/L), among other nutrients. This juice, which traditionally constitutes a waste with an important impact on the environment, has been recently valued (Figure 6). It has gained popularity as a beverage: Contrary to expectations, this beverage doesn't have a chocolate taste.

This beverage is rich in antioxidants and flavonoids and is an excellent moisturizer. Cacao juice helps raise the serotonin level in the brain, acting as an anti-depressant that promotes an overall sense of well-being.

Cacao juice is also used as an ingredient in the industrial process of tropical beverages such as pineapple, orange, and other tropical fruits (KYB Anvoh, Bi, & Gnakri, 2009). Some studies have shown that cacao juice added with juice fruits (especially pineapple) increases the blend´s nutritional compared with raw cacao juice. This increase in nutritional value will depend on the type and quality of fruit juice used in the mixture and its proportion (Afolabi, Ibitoye, & Agbaje, 2015).

Cacao juice has an important sugar content and naturally occurred strains of microorganisms like yeast (Saccharomyces cerevisiae), bacteria (like acetic acid bacteria), and lactic acid bacteria, which degrade sugars into alcohol and several secondary products like organic acids. Because of this, cacao juice has been successfully used as a raw material in the production of wine with an alcohol content of up to 12.5% (Koutoua Anvoh, Guéhi, Beugré, Kinimo, & Gnakri, 2010).

Figure 6. Cacao juice.

The current process has been developed to produce alcoholic beverages using selected yeasts at each step during alcoholic beverages production, obtaining products up to 17% of alcohol (Koutoua Anvoh et al., 2010).

A process that is naturally linked to wine production is vinegar production. Cacao juice is used as feedstock for vinegar production, another cacao sub-product of commercial importance but without nutritional value.

Cacao juice has been studied as a vehicle of probiotic microorganisms like Lactobacillus casei to produce probiotic juice. This probiotic juice shows different composition than raw cacao juice due to the fermentation process. During fermentation, glucose is the carbohydrate consumed; the ascorbic and citric acid content are maintained. The phenolic compound´s content increases from 40.00 ± 0.02 to 49.63 ± 0.04 mg/100 mL, and consequently, its antioxidant activity. (dos Santos Filho et al. 2019)(dos Santos Filho et al. 2019)

Cacao juice is also used in jelly fabrication, among other applications. Jam has been obtained using cacao bean juice as raw material. It has been seen that some of the original components of cacao remain on the final product, especially fat and ascorbic acid (Anvoh, Bi, and Gnakri 2009)(Anvoh, Bi, and Gnakri 2009).

Most cacao is not consumed raw; instead, it is subjected to a series of processes during its traditional or industrial transformation. Chocolate is the most popular and economically important cacao byproduct worldwide. The main steps in the fabrication of chocolate are described below.

Fermentation

The most common subsequent process after harvesting is fermentation. It is a crucial stage because, during this process, the characteristic flavors of chocolate are formed. Fermentation represents the first important step for developing chocolate flavor and taste since it produces aroma precursors (Di Mattia et al. 2017)(Di Mattia et al. 2017). During fermentation, the embryo dies, which prevents the seed from posterior germination. After opening the cacao pods, the seeds and part of its mucilage are placed in wooden boxes (Figure 7). Other techniques for cacao fermentation include the use of greenhouse yards.

Figure 7. Wooden boxes for drying cacao seeds.

During fermentation, both anaerobic and aerobic fermentation occur in ecological succession. Those reactions consume oxygen and raise the temperature above 52.1°C in the center of the pile by the 3rd day of fermentation (Gutiérrez 2017). The cacao pile is tumbled daily with a wooden shovel to avoid high temperatures and low oxygen levels.

In the first 24 hours, due to the seed´s acidic pH, and the high sugar content, yeast´s rapid growth can be observed. Yeasts are responsible for the pectinolysis of the cacao pulp from glucose, thus generating the maximum production of ethanol during this stage (Gutiérrez 2017). After this first stage, and due to the aeration process, the temperature rises 37°C, which leads to the replacement of the yeasts by lactic acid bacteria (LAB). These bacteria, under the newly established conditions, cause the oxidation of ethanol to lactic acid (Gutiérrez 2017). Subsequently and almost immediately, the acetic acid bacteria (AAB; maximum production of acetic acid) are activated and dominate the fermentation process. The AAB consume glucose, fructose, and citric acid and produce mannitol, lactic acid, and acetic acid, while the temperature rises to around 50°C (on the 3rd day). This process is known as curing (Gutiérrez 2017).

From a chemical point of view, fermentation can be divided into two stages: oxidation and hydrolysis. During the aerobic or oxidation phase of acetic fermentation, anthocyanins are degraded by glycosidase, and the pigments 3-ß-D galactosidylcyanidine and 3-α- L arabinocidycyanidine are hydrolyzed to sugars and cyanidins. Also, in this aerobic stage, proteins are hydrolyzed into amino acids and peptides (proteolysis).

During cocoa fermentation, polyphenols are subjected to biochemical modifications through oxidation and polymerization and binding with proteins (Redovniković et al. 2009). Polyphenols are chemically condensed into insoluble complex products that have little or no taste (Gutiérrez 2017). Fermentation is thus considered responsible for the decrease in the content of polyphenols and polyphenolic compounds. The level of polyphenol reduction is proportionate to the degree of fermentation (Di Mattia et al. 2017). The reduction levels have been reported from 20–40%, up to 80% (Di Mattia et al. 2017).

Nevertheless, an increase in the total antioxidant capacity was observed in cacao varieties after spontaneous fermentation (Di Mattia et al. 2017).

During fermentation, endogenous enzymes like carboxypeptidase and aspartic endoprotease are activated due to low pH (Barišić et al. 2019). These enzymes hydrolyze proteíns, and therefore the content of free amino acids increases.

Drying

Once the fermentation stage is finished, the grains have an approximate moisture content of 40%, so it becomes necessary to dry the nibs to make them biologically stable: microorganisms decrease their multiplication rate, making it easy for the storing and transporting process.

Drying is a crucial stage because it reduces the astringency, bitterness, and acidity of the beans. During drying, the seeds develop a brown coloration resulting from condensation reactions between quinones and proteins after the polyphenol´s enzymatic oxidation (Gutiérrez 2017, Barišić et al. 2019).

Drying ends when cacao has 7% to 8% humidity (Figure 8). Lower levels of moisture (less than 6%) yield very fragile beans that are difficult to process (Gutiérrez 2017). The drying speed is crucial since, when too fast, it can lead to the external cuticle drying too quickly and thus leave part of the moisture in the center of the grain, causing high acidity level that decreases its quality.

Figure 8. Fermented and dried cacao beans.

The most common form of cacao drying is sun-drying for approximately seven days; however, in some regions where the humidity is high, it is necessary to use heat exchangers to accelerate the drying process.

Some tests are carried out to evaluate the quality of the seeds after the drying process. The most common test is cutting the seed by half and observing the cacao grain´s color, texture, and general appearance. This process can be performed manually using a cutter (Figure 9) or a guillotine specially designed to allows the cut of multiple seeds in one go.

Figure 9. Test for evaluating the quality of dried and fermented cacao.

Cacao beans contain natural antioxidants and other process-derived antioxidants such as Maillard reaction products (MRPs) that form during high-temperature processing, like drying, roasting, and conching (Di Mattia et al. 2017).

The polyphenols content is substantially reduced during drying, mainly by enzymatic browning; this reduction is catalyzed by polyphenol oxidase (Redovniković et al. 2009, Barišić et al. 2019). Sun-drying reduces the polyphenol content to different extents: a reduction from 77 to 44%, has been reported (Di Mattia et al. 2017).

Roasting

Roasting is a significant step in cacao bean processing. During this step, and due to the temperature rise, the microorganisms are reduced, producing a more stable material. Besides, precursors originated during fermentation and drying stages are converted into compounds that give flavor, color, and texture to the final product. During roasting, the bean´s acidity decreases due to the loss of some volatile acids like acetic acid, while acids like oxalic, tartaric, succinic and lactic remain in the nibs. However, at the same time, as the positive changes in the nibs happen, other reactions lower the content of bioactive compounds such as polyphenols.

For the product to have the best sensory and nutritional characteristics, it is necessary to control the roasting variables such as time, temperature, airflow, particle size, moisture content of the roasted beans, among other variables.

The most common form of roasting is using a forced flow of hot air (convection roasting), for reaching temperatures in the range of 130°C and 150°C and for a time between 15 and 45 min (Żyżelewicz et al. 2016), to reach moisture contents of about 2 g/100g (Oracz, Nebesny, and Żyżelewicz 2015).

Polyphenols are thermolabile molecules, so during the roasting process, its concentration decreases. Rates of flavanol and total proanthocyanidin loss increase with increasing roasting temperatures (Ioannone et al. 2015). High-temperature-short time processes are better recommended (Oracz, Nebesny, and Żyżelewicz 2015, Ioannone et al. 2015).

In general, roasting causes small changes in polyphenolic concentration (Redovniković et al. 2009), contrary to the fermentation and drying stages.

Conching

In this stage, volatile polyphenols are lost due to evaporation. This reduction can be up to 80%. Also, short-chain fatty acids are evaporated during conching (Barišić et al. 2019).

Chocolate

Chocolate is one of the most popular products derived from cacao and is usually consumed for pleasure. Its production is a multi-step process: fermentation, drying, roasting, nib grinding, refining, conching, mixing with other ingredients, and finally, tempering. The content and type of health-related compounds present in each kind of chocolate depend on the final formulation, the proportion of cacao, and the type and amount of added ingredients like cacao butter, milk sugar, nuts, and aromatic spices, among others (Figure 10).

Figure 10. Cacao, chocolate, and some common ingredients for making chocolate.

Consuming dark chocolate (99% cacao) is the best way of acquiring cacao nutrients. Dark chocolate is one of the foods with the highest antioxidant activity. (measured as ORAC; Oxygen Radical Absorbance Capacity) (Figure 11).

Dark chocolate (90-70% cacao) consumption has been associated with remarkable benefits on health, like enhancing cognitive abilities (Socci et al. 2017) and contributing to preventing neurodegenerative diseases and aging (Martín and Ramos 2017, Orozco-Arbelaez et al. 2017). Chocolate is reported to having antioxidant capacity due to the reduction of the number of free radicals in the body, inhibiting low-density lipoprotein (LDL) oxidation, raising the high-density lipoprotein (HDL) level, and reducing the thrombotic tendency (Kelishadi 2010). Chocolate consumption inhibits the carcinogenic process of reactive oxygen species (ROS) (Kelishadi 2010, Martín and Ramos 2017).

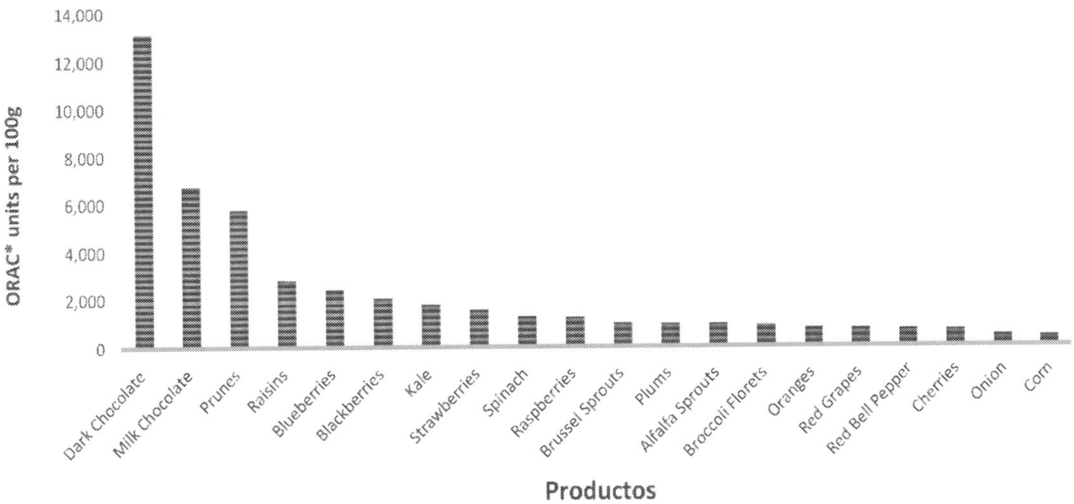

Figure 11. Antioxidant capacity, measured as ORAC* (Oxygen Radical Absorbance Capacity) Adapted of the information of (Kelishadi 2010)(Kelishadi 2010).

Chocolate consumption reduces anxiety due to its content of tryptophan and anandamide (Kelishadi 2010). It has been proved that moderate amounts of chocolate (≤6 servings/week) reduces the risk of coronary heart disease (CHD), stroke, and diabetes (Yuan et al. 2017, Gianfredi et al. 2018), especially in women (Gianfredi et al. 2018). The presence of methylxanthines in chocolate produces psychoactive effects, promoting more efficient thinking and exploring. Methylxabtines also regulate the immune response and are key players increasing the number of beneficial gut microbes and lowering the content of less beneficial (Franco, Oñatibia-Astibia, and Martínez-Pinilla 2013) among other benefits on health.

A Traditional Cacao Beverage

Since ancestral times, in Mesoamérica, cacao was widely consumed as a beverage for the elite class and was used in rituals; but also was an essential food. Cacao beverages were prepared mainly by grinding the dried fermented seeds and mixing it with cold water, ground chili, ground corn, and some flavorants (Crown and Hurst 2009). One of the most popular cacao beverages is tejate, which has its origins as a ceremonial beverage and is used to obtain the energy needed for hard fieldwork.

Tejate is a source of energy (2572 to 1999 kcal/100g), fat (2.82-20.21 g/100g), antioxidants, and methylxanthines (Franco, Oñatibia-Astibia, and Martínez-Pinilla 2013) (Figure 12).

Drinking traditional cacao beverages is an effective form of consuming nutrients from cacao and obtaining the health benefits associated with it due to the lower effect of the production process over the grain nutrients.

Figure 12. Cacao beverages.

Final Thoughts

Consuming chocolate is one of the most innocuous and pleasant habits (Franco, Oñatibia-Astibia, and Martínez-Pinilla 2013) that has significant benefits not only on well being but also in health. The degree of that benefit depends on several variables such as the level of consumption, type of cacao or byproduct, the consumer´s (normal, borderline, or with a disease), age, and even gender (for cardiovascular risk). The more processed the product, the lower content of nutrients; therefore, the consumption of cocoa nibs, traditional cocoa drinks, and dark chocolate (70%-90%) should be privileged over other products to obtain the expected benefits.

References

Anvoh, KYB, A Zoro Bi, and D Gnakri. 2009. "Production and characterization of juice from mucilage of cocoa beans and its transformation into marmalade." *Pakistan Journal of Nutrition* 8 (2):129-133.

Apgar, Joan L, Stanley M Tarka, and M Stanley. 1998. *Methylxanthine composition and consumption patterns of cocoa and chocolate products*: CRC Press, Boca Raton, FL.

Barišić, Veronika, Mirela Kopjar, Antun Jozinović, Ivana Flanjak, Đurđica Ačkar, Borislav Miličević, Drago Šubarić, Stela Jokić, and Jurislav Babić. 2019. "The Chemistry behind Chocolate Production." *Molecules* 24 (17):3163.

Bravo, Laura. 1998. "Polyphenols: chemistry, dietary sources, metabolism, and nutritional significance." *Nutrition reviews* 56 (11):317-333.

Cinquanta, Luciano, Cinzia Di Cesare, Remo Manoni, Angela Piano, Piero Roberti, and Giancarlo Salvatori. 2016. "Mineral essential elements for nutrition in different chocolate products." *International journal of food sciences and nutrition* 67 (7):773-778. doi: 10.1080/09637486.2016.1199664.

Crown, Patricia L, and W Jeffrey Hurst. 2009. "Evidence of cacao use in the Prehispanic American Southwest." *Proceedings of the National Academy of Sciences* 106 (7):2110-2113. doi: 10.1073/pnas.0812817106.

Decker, Eric A, Bingcan Chen, Atikorn Panya, and RJ Elias. 2010. "Understanding antioxidant mechanisms in preventing oxidation in foods." In *Oxidation in foods and beverages and antioxidant applications*, 225-248. Elsevier.

Di Mattia, Carla D, Giampiero Sacchetti, Dino Mastrocola, and Mauro Serafini. 2017. "From cocoa to chocolate: The impact of processing on in vitro antioxidant activity and the effects of chocolate on antioxidant markers in vivo." *Frontiers in immunology* 8:1207. doi: 10.3389/fimmu.2017.01207.

dos Santos Filho, Antonio Luiz, Hildeane Veloso Freitas, Sueli Rodrigues, Virgínia Kelly Gonçalves Abreu, Tatiana de Oliveira Lemos, Wesley Faria Gomes, Narendra Narain, and Ana Lúcia Fernandes Pereira. 2019. "Production and stability of probiotic cocoa juice with sucralose as sugar substitute during refrigerated storage." *LWT* 99:371-378. doi: 10.1016/j.lwt.2018.10.007.

Duncan, Susan E, and Hao-Hsun Chang. 2012. "Implications of light energy on food quality and packaging selection." In *Advances in food and nutrition research*, 25-73. Elsevier.

Fernández, Carmen Febles, Carmen Soto Febles, Alberto Saldaña Bernabeu, and Bárbara E García Triana. 2002. "Funciones de la vitamina E: Actualización." *Rev Cubana Estomatol* 39 (1):28-32.

Franco, Rafael, Ainhoa Oñatibia-Astibia, and Eva Martínez-Pinilla. 2013. "Health benefits of methylxanthines in cacao and chocolate." *Nutrients* 5 (10):4159-4173. doi: 10.3390/nu5104159.

Gianfredi, Vincenza, Tania Salvatori, Daniele Nucci, Milena Villarini, and Massimo Moretti. 2018. "Can chocolate consumption reduce cardio-cerebrovascular risk? A systematic review and meta-analysis." *Nutrition* 46:103-114. doi: 10.1016/j.nut.2017.09.006.

Guerrera, Mary P, Stella Lucia Volpe, and Jun James Mao. 2009. "Therapeutic uses of magnesium." *American family physician* 80 (2):157-162.

Gutiérrez, Tomy J. 2017. "State-of-the-art chocolate manufacture: a review." *Comprehensive Reviews in Food Science and Food Safety* 16 (6):1313-1344. doi: 10.1111/1541-4337.12301.

Ioannone, F, Carla Daniela Di Mattia, Miriam De Gregorio, Manuel Sergi, Mauro Serafini, and Giampiero Sacchetti. 2015. "Flavanols, proanthocyanidins and antioxidant activity changes during cocoa (Theobroma cacao L.) roasting as affected by temperature and time of processing." *Food chemistry* 174:256-262. doi: 10.1016/j.foodchem.2014.11.019.

Jalil, Abbe Maleyki Mhd, and Amin Ismail. 2008. "Polyphenols in cocoa and cocoa products: is there a link between antioxidant properties and health?" *Molecules* 13 (9):2190-2219. doi: 10.3390/molecules13092190.

Kelishadi, Roya. 2010. "Cacao to cocoa to chocolate: healthy food?" *ARYA Atherosclerosis* 1 (1).

Martín, María Ángeles, and Sonia Ramos. 2017. "Health beneficial effects of cocoa phenolic compounds: a mini-review." *Current Opinion in Food Science* 14:20-25.

Oracz, Joanna, Ewa Nebesny, and Dorota Żyżelewicz. 2015. "Changes in the flavan-3-ols, anthocyanins, and flavanols composition of cocoa beans of different Theobroma cacao L. groups affected by roasting conditions." *European Food Research and Technology* 241 (5):663-681.

Oracz, Joanna, Dorota Zyzelewicz, and Ewa Nebesny. 2015. "The content of polyphenolic compounds in cocoa beans (Theobroma cacao L.), depending on variety, growing region, and processing operations: a review." *Critical reviews in food science and nutrition* 55 (9):1176-1192. doi: 10.1080/10408398.2012.686934.

Orozco-Arbelaez, Edilberto, José Ramón Banegas, Fernando Rodríguez-Artalejo, and Esther López-García. 2017. "Consumo habitual de chocolate y estado cognitivo en los adultos mayores españoles." *Nutrición Hospitalaria* 34 (4):841-846. doi: 10.20960/nh.630.

Peláez, Pedro P, Inés Bardón, and Pedro Camasca. 2016. "Methylxanthine and catechin content of fresh and fermented cocoa beans, dried cocoa beans, and cocoa liquor." *Scientia Agropecuaria* 7 (4):355-365. doi: 10.17268/sci.agropecu.2016.04.01.

Redovniković, I Radojčić, Karmela Delonga, Slavica Mazor, Verica Dragović-Uzelac, Marija Carić, and Jasna Vorkapić-Furač. 2009. "Polyphenolic content and composition and antioxidative activity of different cocoa liquors." *Czech journal of food sciences* 27 (5):330-337.

Scalbert, Augustin, Ian T Johnson, and Mike Saltmarsh. 2005. "Polyphenols: antioxidants and beyond." *The American journal of clinical nutrition* 81 (1):215S-217S. doi: 10.1093/ajcn/81.1.215S.

Scalbert, Augustin, Claudine Manach, Christine Morand, Christian Rémésy, and Liliana Jiménez. 2005. "Dietary polyphenols and the prevention of diseases." *Critical reviews in food science and nutrition* 45 (4):287-306. doi: 10.1080/1040869059096.

Socci, Valentina, Daniela Tempesta, Giovambattista Desideri, Luigi De Gennaro, and Michele Ferrara. 2017. "Enhancing human cognition with cocoa flavonoids." *Frontiers in nutrition* 4:19. doi: 10.3389/fnut.2017.00019.

Solorzano, Rey Gaston Loor, Olivier Fouet, Arnaud Lemainque, Sylvana Pavek, Michel Boccara, Xavier Argout, Freddy Amores, Brigitte Courtois, Ange Marie Risterucci, and Claire Lanaud. 2012. "Insight into the wild origin, migration and domestication history of the fine flavour Nacional Theobroma cacao L. variety from Ecuador." *PloS ONE* 7 (11). doi: 10.1371/journal.pone.0048438.

Stern, Bonnie Ransom. 2010. "Essentiality and toxicity in copper health risk assessment: overview, update and regulatory considerations." *Journal of Toxicology and Environmental Health, Part A* 73 (2-3):114-127. doi: 10.1080/15287390903337100.

Tapas, Ahmad R, DM Sakarkar, and RB Kakde. 2008. "Flavonoids as nutraceuticals: a review." *Tropical journal of Pharmaceutical research* 7 (3):1089-1099. doi: DOI: 10.4314/tjpr.v7i3.14693.

Walsh, Carol T, Harold H Sandstead, Anan da S Prasad, Paul M Newberne, and Pamela J Fraker. 1994. "Zinc: health effects and research priorities for the 1990s." *Environmental health perspectives* 102 (suppl 2):5-46. doi: 10.1289/ehp.941025.

Yuan, Sheng, Xia Li, Yalei Jin, and Jinping Lu. 2017. "Chocolate consumption and risk of coronary heart disease, stroke, and diabetes: a meta-analysis of prospective studies." *Nutrients* 9 (7):688. doi: 10.3390/nu9070688.

Żyżelewicz, Dorota, Wiesława Krysiak, Joanna Oracz, Dorota Sosnowska, Grażyna Budryn, and Ewa Nebesny. 2016. "The influence of the roasting process conditions on the polyphenol content in cocoa beans, nibs and chocolates." *Food Research International* 89:918-929.

In: Super and Nutraceutical Foods
Editors: Sandra Teresita Martín-del-Campo et al.
ISBN: 978-1-53619-082-3

Chapter 3

CACTUS FRUITS: AN ANCIENT AND EMERGING CROP

***Leticia Hernández Galán*[1] *and Sandra T. Martín del Campo*[2,*]**
[1]Zentrela Inc. Hamilton, ON, Canada
[2]Tecnologico de Monterrey, Escuela de Ingeniería y Ciencias, Querétaro, Qro, Mexico

ABSTRACT

Alimentation is fundamental for human subsistence to satisfy biological needs. Cactus plants have been an important solution for human alimentation in arid and semiarid zones of America. Utilization and consumption of the plants and their fruits date from prehispanic times. The most important cactus fruits by production volume are prickly pear (cactus pear), pitaya (dragon fruit), berrycacti (garambullo), and xoconoxtle. These fruits are flavorsome and present a variety of colors. They aport good quality nutrients for human subsistence, but additionally, they apport bioactive compounds such as polyphenols, betalains, flavonoids, among others, that serve for the prevention and treatment of diseases. Cactus fruits are consumed mostly fresh, but they are used to manufacture products such as processed foods, medicaments, and additives. The cultivation of these crops is emerging as new solutions for human alimentation since this adaptability to arid regions brings a solution to water shortage.

Keywords: cactus fruits, prickly pear, pitaya, dragon fruit, berrycacti, xoconoxtle, bioactive compounds

INTRODUCTION

Cactus plants are endemic on the arid and semiarid zones of America, where Mexico hoses a higher amount of species. These plants have been used for several centuries before

* Corresponding Author's E-mail: smartinde@tec.mx.

America's colonization as food, medicinal, forage source, or construction. Spanish introduced cactus plants in Europe in the XVI century, and then they were introduced in other countries where the crops were well established.

Nowadays, cactus fruits are appreciated not only in America but in different international markets, mostly in Europe and Asia. The most distributed fruits in international markets are prickly pear and pitayas. But there is an important local market for other cactus fruits such as garambullo (Myrtillocactus), xoconoxtle, tunillo, among others.

These plants developed morphological and physiological adaptations to grow in adverse climatic conditions, such as high temperatures, poor soils, as well as dry and infertile lands. Even if the consumption of cactus plants and fruits by America's populations is not new, this crop is acquiring new popularity and value. Their adaptability to arid regions due to their Crassulacean Acid Metabolism and the low water amounts requirements makes them the perfect crop since one-third of the world's population challenges water shortage.

The Cactaceae family is composed of more than 1500 species, and some genus produces edible and juicy fruits with a wide variety of colors, flavors, as well as nutritional and nutraceutical characteristics. Fruits of about 83 cacti species are the most consumed by humans, where the main groups are *Opuntia*, *Hylocereus*, *Stenocereus*, *Myrtillocactus*, and *Maillaria*. (Casas & Barbera, 2002).

These fruits exhibit not only interesting organoleptic and nutritive properties but also nutraceutical functions due to different bioactive phytochemicals with diverse biological activities such as antioxidant, antidiabetic, anti-inflammatory, antimicrobial, among others (Corzo-Rios, Bautista-Ramírez, Gómez y Gómez, & Torres-Bustillos, 2017). Most of the cactus fruits produced are consumed fresh, but they are also employed to manufacture other products such as food products (jams, juices, candies, alcoholic beverages), ingredients (pigments, extracts, etc.), medicinal products (pills, drinks, powders, etc.).

PRICKLY PEAR

The origin of prickly pear (*Opuntia* spp), also called cactus pear, is in the arid or semiarid regions of America. There are vestiges of the ancient consumption of this fruit in Mexico and Peru (Abdel-Hameed, Nagaty, Salman, & Bazaid, 2014; Angeles Millones, 2014). In those countries, most of the prickly pear produced is in a wild noncontrolled form, and only in some regions is produced as a controlled crop.

Prickly pear o cactus pear is a fruit highly appreciated by consumers since they are flavorsome and show a wide variety of attractive colors (green, yellow, orange, red, purple, violet, etc.) depending on the varieties or cultivars. Mexico is the main producer and exporter of prickly pear (Albano et al., 2015; Barba et al., 2017). Nowadays, this crop has been established in other countries with arid, semiarid, or more tropical climates such as Italy, Saudi Arabia, Spain, India, South Africa, the United States of America, Peru, among others. Because of this geographical distributions, prickly pear could be found in the market almost during the whole year since in the north hemisphere maturation occurs from July to October while in the southern hemisphere is from January to April (Barreiro Perera, 1996).

Figure 1. Prickly pear maturation of fruits.

Maturation of this fruit occurs after flowering when pollination occurs, not when the fruit is produced (Figure 1 A and B). After flowering, when the fruit reaches the optimal maturity is harvested (Figure 1 C) and consumed fresh after peeling (Figure 1 D). Maturation time could variate depending on variety or cultivar as well as the geo-climatic conditions in the case of *Opuntia albicarpa* S. Var. Burrona cultured in central Mexico, Martín-Del-Campo et al., (2019) reported a period between 80 and 90 days after flowering to reach optimal maturity.

It is well known prickly pear exhibits an important nutritional value, but the interest on this fruit has increased in recent years since its nutraceutical benefits have been highlighted. These fruits are a good source of different bioactive compounds such as vitamin C, polyphenols, betacyanins, betalains, flavonoids, etc. These compounds confer important antioxidant activity, antiatherogenic properties, cardiovascular protective effects, antiulcerogenic properties, etc.

Prickly Pear Nutritional Composition

Fresh prickly pear or cactus pear present an important nutritional value not only as an energy source. This fruit is a good source of different macro- and micronutrients. The nutritional composition of this fruit has important variations depending on the cultivar, geoclimatic conditions, seasonality, among others (de Wit, Nel, Osthoff, & Labuschagne, 2010). The fresh fruit composition could also variate due to the postharvest storage temperature and time (Cruz-Bravo, Guzmán-Maldonado, Araiza-Herrera, & Zegbe, 2019; González González, Morales, Olivares Sáenz, Aranda Ruiz, & Gallegos Vázquez, 2001).

Table 1. Mexican prickly pear nutritional composition. Content for 100 g of the edible fraction. Data presented by the National Institute of Medical Sciences and Nutrition Salvador Zubirán (INCMNSZ, 2016)

Compositional Parameter	*Opuntia* sp.											*Opuntia hyptiacantha* Weber	*Opuntia robusta* Wendl
	General		Cardona		Cascarona	Chaveña	red	Mansa			Round	White	Tapona
		with seeds	with seeds	seedless	seedless	seedless	with seeds	with seeds	red, with seeds	seedless	with seeds	pressed	with seeds
Edible portion (%)	0.57	0.57	0.55	0.55	0.55	0.55	0.55	0.55	0.55	0.55	0.55	0.55	0.55
Energy (kJ)	186.00	201.00	132.00	131.00	143.00	88.00	135.00	91.00	106.00	73.00	110.00	863.00	134.00
Water (g)	83.84	77.30	88.10	91.00	88.00	90.80	82.60	87.80	85.70	89.40	85.60	20.80	86.30
Ether extract (g)	0.58	*	0.04	0.08	0.24	0.03	0.21	0.16	0.21	0.10	0.16	0.90	0.12
S. F. A.(g)[a]	0.07	*	*	*	*	*	*	*	*	*	*	*	*
M.F.A.(g)[a]	0.08	*	*	*	*	*	*	*	*	*	*	*	*
P.U.F.A.(g)[a]	0.21	*	*	*	*	*	*	*	*	*	*	*	*
Crude protein (g)	0.84	*	0.56	0.05	0.69	0.50	1.81	1.21	1,43	0.93	1.62	4.50	1.93
Carbohydrates (g)	10.24	10.10	8.15	8.43	8.32	5.32	6.82	4.58	5.21	3.78	5.42	49.60	4.55
Fiber (g)	4.65	10.03	2.85	0.24	0.25	0.45	8.16	5.55	6.75	0.19	6.70	21.20	6.60
Ca (mg)	56.00	90.00	49.00	25.00	29.00	29.00	32.00	40.00	35.00	50.00	21.00	340.00	*
P (mg)	32.00	31.00	*	31.00	8.00	8.00	33.00	39.00	39.00	26.00	29.00	67.00	25.00
Fe (mg)	1.24	1.07	2.64	1.25	2.38	2.72	2.30	0.45	0.61	2.57	1.71	8.70	0.65
Carotene (mg)	0.06	0.00	*	*	*	*	*	*	*	*	*	*	*
b-Carotene (mg)	0.05	*	0.06	0.00	0.03	0.01	0.00	0.02	0.03	0.00	0.02	0.00	0.02
Vit. B1 (mg)	0.01	0.02	0.02	0.00	0.00	0.00	0.01	0.01	0.02	0.00	0.01	0.03	0.04
Vit B2 (mg)	0.06	0.02	0.02	0.03	0.02	0.03	0.02	0.04	0.04	0..03	0.08	0.11	0.10
Niacin (mg)	0.46	0.29	0.20	0.24	0.19	0.12	0.44	0.55	0.30	0.30	0.91	2.25	0.44
Ascorbic acid (mg)	14.00	31.00	*	22.40	25.00	21.9	23.10	15.60	11.90	9.50	25.50	50.50	22.00

* Value not found.

[a]S.F.A.: Saturated Fatty Acids; M.F.A.: Monounsaturated Fatty Acids; P.U.F.A.: Poliunsaturated Fatty Acids.

Table 2. Different origin prickly pear nutritional composition. Content for 100 g

Compositional Parameter	Albano et al., (2015)		Allegra et al., (2015)	Abdel-Hameed et al., (2014)		Aparicio et al., (2017)	Barba et al., (2017)	Cerezal and Duarte (2005)	Chauhan, Sheth, Rathod, Suhagia and Maradia (2013)
	Opuntia ficus-indica (L.) Mill. (Italy) Juice		*O. ficus indica*, cv. Gialla (Italy)	*Opuntia ficus indica* Mill. (Saudi Arabia) Whole fruit		*Opuntia* (Mexico)	*Opuntia* sp.	*Opuntia ficus-indica* (L.) Miller (Chile)	*Opuntia elatior* Mill (India)
	Purple	Orange	Orange	Red	Yellow	18 wild varieties	Mean value		red
Edible portion (%)	NR	NR	NR	NR	NR	29.9-63.7	NR	NR	NR
Energy (kJ)							172		86
Water (%)						79.6-90.9	87.55	78.31	
TSS [a]	12.67	12.10	12.3-13.5			7.6-15.5		14.8	11.41
Ether extract (g)							0.51	0.49	
S. F. A.(g)[b]							0.067		
M.F.A.(g)[b]							0.075		
P.U.F.A.(g)[b]							0.213		
Crude protein (g)							0.73	0.95	
Carbohydrates (g)							9.57	10.29	
Glucose (g)	1.88	2.14		3.17	1.45				
Fructose (g)	0.78	1.04		3.048	0.513				
Fiber (g)							3.60	3.24	
Ca (mg)							56		50.24
P (mg)							24		
Mg (mg)							85		
Fe (mg)				226.23	205.15		0.30		0.32
Cu (mg)				227.88	204.52		0.08		0.16
Vit. B1 (mg)							0.014		
Vit B2 (mg)							0.06		
Niacin (mg)							0.46		
Ascorbic acid (mg)	36.6	30.2					14	15.99	

[a]TSS: Total soluble solids, Brix.

[b] S.F.A.: Saturated Fatty Acids; M.F.A.: Monounsaturated fatty acids; P.U.F.A.: Polyunsaturated Fatty Acids.

Table 1 shows the values of the nutritional composition of some Mexican cultivars of prickly pear reported by the National Institute of Medical Sciences and Nutrition Salvador Zubirán (INCMNSZ, 2016). The nutritional composition of different prickly pear varieties from different countries are shown in Table 2. Even wild varieties with less commercial value, due to the lower proportion of pulp, have a similar nutritional composition (Aparicio, Loza-Cornejo, Torres-Bernal, Velázquez-Placencia, & Arreola-Nava, 2017).

According to specifications of the Codex Alimentarius (CAC/GL 23-1997, 2013), for most of the prickly pear cultivars, it could be claimed a low energy apport, with the exception is the *Opuntia hyptiacantha* Weber (Table 1). Most of the energy is provided by carbohydrates since most of the cultivars' present low concentrations of fat and could be declared fat free. Additionally, this fruit is not a good source of protein.

The most abundant carbohydrates are monosaccharides and fiber. Glucose concentration is higher than fructose (Table 2). Prickly pear could be a good source of fiber since 150 g could apport between 32-57% of the recommended daily intake (RDI) (Medina, Rodríguez, & Romero, 2007). The most abundant fiber reported is pectin, followed by cellulose and hemicellulose (El Kossori, Villaume, El Boustan, Sauvaire, & Méjean, 1998; Stintzing, Schieber, & Carle, 2001). Additionally, prickly pear fruits are a source of microelements, such as Fe, Ca, Cu, among others. Prickly pear is a good source of Fe and vitamin C, since 100 g of fruit could apport 25%-100% and up to 50% of RDI, respectively. The microelements profile also varies according to cultivars and geographical origin (Dehbi, Hasib, Ouatmane, & Elbatal, 2014; INCMNSZ, 2016; Medina et al., 2007).

Even if the prickly pear is not a good source of fat, it presents a beneficial fatty acid profile. These fatty acids are present in the pulp but also in the seeds where higher concentrations are found. These compounds are important not only for their nutritive value as an energy source but also because they have different roles, such as modulators of physiological functions or for structural proposes. The most abundant are the monounsaturated (M.F.A.) and the polyunsaturated fatty acids (P.U.F.A.), and a small proportion of saturated fatty acids (S.F.A.). Andreu-Coll et al., (2019) reported for Spanish cultivars *Opuntia ficus-indica* (L.) Mill. the most abundant fatty acids were linoleic (29.2%), oleic (26.8%), palmitic (22.9%), and linolenic (12.2%) acids. Similar results were reported by Melgar et al., (2017) for Italian and Portuguese fruits. The fatty acid profile reported in the pulp similar to the fatty acid profile reported in *O. ficus indica* and *O. stricta* seed oil by Ennouri, Evelyne, Laurence and Hamadi (2005). The fatty acid profile could be affected by the harvesting time (maturity), as reported by Juhaimi et al., (2020) for *Opuntia ficus-barbarica* A. *Berger* from Turkey. Prickly pear is also a source of free amino acids comprised of all the essential amino acids (Stintzing et al., 2001). Concentration depends on the amino acid and the fruit variety: histidine (32-53 mg/L), isoleucine (22-39 g/L), leucine (up to 53 g/L), lysine (up to 18 g/L), methionine (32-77 g/L), phenylalanine 22-24 g/L), threonine (12-15 g/L), tryptophan (9-17 g/L), and valine (33-50 g/L).

Nutraceutical Benefits and Bioactive Compounds of Prickly Pear

Prickly pear or cactus fruit shows an important antioxidant capacity (Table 3). Antioxidant capacity varies between cultivars and geographical origin. For most of the cultivars, red fruits show higher antioxidant capacity than yellow or orange ones.

Table 3. Antioxidant capacity and bioactive compounds of prickly pear from different origin and cultivars. Composition content for 100 g of the edible fraction or 100 mL of juice

Parameter	**Albano et al., (2015)**		**Abdel-Hameed et al., (2014)**		**Barba et al., (2017)**	**Butera et al., (2002)**			**Cano et al., (2017)**		**Cayupán et al., (2011)**	**Chang, Hsieh and Yen (2008)**	**Chauhan et al., (2013)**
	Opuntia ficus-indica (L.) Mill. (Italy) Juice		*Opuntia ficus indica* Mill. (Saudi Arabia) Whole fruit		*Opuntia* sp.	*Opuntia ficus-indica.* (Italy) Juice			*Opuntia ficus-indica*, spp.) (Spain)		*Opuntia* sp. (Argentin)	*Opuntia dillenii* Haw (Taiwan) Fruit	*Opuntia elatior* Mill (India)
	Purple	Orange	Red	Yellow	Mean value	Yellow	Red	White	9 varieties	Verdal (orange)	7 varieties	Red	Red
Total Phenols (mg GAE)	89.2	69.8	1065.15	667.82					457.25	432.50	36-103	91.5	49.823
ABTS (mmol)	0.61	0.37											
ABTS µmol of Trolox equivalent						5.31	4.20	4.36	108.85	122.47		1.01	
TAC (mg ascorbic acid equivalent)*			735.35	204.31									
ORAC (mmol)	1.28	0.98							30.28	28.53		0.10	
DPPH (mg ascorbic acid equivalent)			81.90	391.90									
Reducing power activity (mg equivalent to ascorbic acid)			123.23	81.38									
Total Flavonoids (mg rutin equivalent)			159.49	80.35								29.2	
Total flavonols (mg rutin equivalent)			341.81	160.02									

*TAC: Total antioxidant capacity.

The antioxidant capacity depends additionally on the postharvest managing and harvesting time. Cruz-Bravo et al., (2019) reported that the antioxidant activity increased significantly for two Mexican varieties of prickly pear: *Opuntia ficus-indica* L. Mill cv. 'Roja Lisa' (Red) and *Opuntia* spp. cv. 'Amarilla Olorosa' (yellow) during storage at 24ºC (37% RH) or 10ºC (95% RH). Juhaimi et al., (2020) reported variations on Turkish *Opuntia ficus-barbarica* A. *Berger* harvested at different harvesting times (maturity) in the same orchard.

The antioxidant activity is associated not only with the vitamin C content, but also with the polyphenols, flavonoids, and flavonol content. Fresh prickly pear fruits have a high content of those compounds that depends on the variety and geographical origin. Abdel-Hameed et al., (2014) reported a higher concentration of those compounds in Saudi Arabian red fruits than in yellow fruits. Albano et al., (2015) reported significantly higher total phenol content in purple than in orange fruits and Allegra et al., (2015) for *O. ficus indica*, cv. Gialla. The most abundant polyphenols reported are rutin, gallic acid, catechin, and Kaempferol 3-O-b-D-glucoside (Abdel-Hameed et al., 2014; Barba et al., 2017; Cayupán, Ochoa, & Nazareno, 2011). Similar antioxidant results and lower total phenol content were reported for Brazilian prickly pear varieties (de Souza, Santana, de Macedo, de Brito, & Correia, 2015) and South African cultivars (de Wit, du Toit, Osthoff, & Hugo, 2019). The most abundant flavonoids reported are quercetin, isorhamnetin, luteolin, and kaempferol (Fernández-López, Almela, Obón, & Castellar, 2010; Kuti, 2004). The concentration of total phenols and total flavonoids increase during postharvest storage as reported by Cruz-Bravo et al., (2019). These authors also reported a significant decrease in condensed tannins during storage.

The protective effect against oxidative stress of prickly pear was evaluated by Tesoriere, Butera, Pintaudi, Allegra and Livrea (2004). These authors used healthy human volunteers and obtained an important reduction in oxidative damage to lipids plasma markers.

On the other hand, prickly pear fruits present a wide variety of colors in the pulp as in the peel. This variety of colors is given by different kinds of pigments where the most abundant are betacyanins and betaxanthins, but carotenoids could be found in some varieties. Generally, red, pink, and purple prickly pear show higher amounts of betacyanins (Albano et al., 2015; Butera et al., 2002; de Wit et al., 2019; Figueroa-Cares et al., 2010); while yellow and orange fruits are richer in betaxanthins (Albano et al., 2015; Butera et al., 2002). Nevertheless, Cano, Gómez-Maqueo, García-Cayuela and Welti-Chanes (2017) reported higher concentrations of betaxanthins than betacyanins in the fruit pulp of both Sanguinos (red) and Verdal (orange) varieties of *Opuntia ficus-indica*, spp. Similar results were reported by Fernández-López et al., (2010) for red-skinned *Opuntia* prickly pear.

Cejudo-Bastante, Chaalal, Louaileche, Parrado and Heredia (2014) reported important differences in betalain content in different varieties of *Opuntia ficus-indica* from Algeria; they identified 9 betaxanthins and 9 betacyanins. In this study, betaxanthins presented a higher concentration. Indicaxanthin was the most abundant betaxanthin (90% of total betalain area), betanin was the most abundant betacyanin (2% of total betalain area). The concentration of betalains could decrease during the postharvest storage, as reported by (Cruz-Bravo et al., 2019).

On the other hand, some studies identified carotenoid pigments in prickly pear. El Kharrassi et al., (2016) reported total carotenoid concentrations from 4.8-28 μg/L in juices obtained from Marocain *Opuntia ficus* indica and *Opuntia megacantha* fruits. Cano et al., (2017) reported a concentration of 255.93 mg/100 g fresh weight and 379.45 mg/100 g fresh weight Spanish Sanguinos (red) and Verdal (orange) varieties of *Opuntia ficus-indica*, spp., respectively. Additionally, these authors identified and quantified 10 different carotenoids in the fruit pulp. de Wit et al., (2019) reported carotenoids concentrations of 0.6-1.86 µg/g of pulp and 32.6-69.2 µg/g of the seed of South Africa prickly pear cultivars.

There are other important nutraceutical effects reported for prickly pear fruits. One is the neuroprotective effect reported by Dok-Go et al., (2003). These authors reported quercetin, (1)-dihydroquercetin, and quercetin 3-methyl ether isolated from Korean *Opuntia ficus-indica* var. saboten fruits inhibited neuronal injuries in cultured rat cortical cells.

On the other hand, the inhibition of Alpha-amylase and Alpha-glucosidase is associated with energy metabolism and homeostasis, as well as with the reduction of insulin requirements and glycemia. de Souza et al., (2015) evaluated Brazilian *Opuntia ficus-indica* fruits and reported reductions of 38-50% reduction of Alpha-amylase activity 53-70% reduction of Alpha-glucosidase inhibitory activity.

The hepatoprotective effect of prickly pear juice has been reported (Galati et al., 2005). These authors found when *O. ficus indica* fruit juice was given to rats, it promoted liver recovery and was able to prevent CCl_4 hepatotoxic activity and could help to restore the health state of the liver.

Another important biological activity reported is antimicrobial activity. Melgar et al., (2017) reported that resuspended dry pulp showed important antibacterial and antifungal activities against different microorganisms; these activities were even higher than reference antibiotics.

Processed Products from Prickly Pear

Since most of this fruit is consumed in the fresh state, one important process is focused on the storage at low temperatures, controlled temperature, or even under modified atmospheres. This process makes it possible not only to achieve new markets into the production countries but to reach international markets.

From the cactus fruits, prickly pear is one of the most stable. Nevertheless, studies have been performed to storage the fruit for longer periods. Harvesting methods avoiding fruit damage and cold storage made it possible to increase the storage time (González González et al., 2001; Ochoa-Velasco & Guerrero-Beltrán, 2013; Ochoa & Guerrero, 2012). Using only cold storage, Mexican prickly pear varieties 'Roja' (*Opuntia ficus indica* (L.) Miller) and "blanca Villanueva" (*Opuntia albicarpa*) maintains its antioxidant activity; nevertheless, fruit weight loss, texture, and color are affected by temperature. Storage at room temperature (24ºC) and 37% of RH and low temperature (10ºC) and 95% RH made it possible to store the fruit for 5 weeks to be transported (Cruz-Bravo et al., 2019). Under these conditions, the nutraceutical proprieties of the evaluated cultivars increased significantly.

Traditionally, some prickly pear products are produced locally using artisanal methods during the fruit harvesting season (few months by year). One of these products is 'Queso de tuna,' a candy made by the concentration of juice without the addition of other ingredients. Another product is called 'colonche,' 'coloche,' or 'nochoctli' that is a group of fermented beverages made by the spontaneous fermentation of different cactus fruits (Ojeda-Linares, Vallejo, Lappe-Oliveras, & Casas, 2020).

To reach new consumers, new products have been developed using processes focused on preserving or even increasing the physicochemical, nutritional, or nutraceutical properties

(Allegra et al., 2015; Gómez-Maqueo, García-Cayuela, Welti-Chanes, & Cano, 2019). Allegra et al., (2015) evaluated the suitability of offering the packed peeled fruits under modified atmospheres and kept at 4 °C. These authors found that depending on the harvest season and fruit ripening state, some physicochemical characteristics are modified depending on storage time; color parameters, fruit crunchiness, acidity, and polyphenol content decreased significantly. Gómez-Maqueo et al., (2019) evaluated the effect of high hydrostatic pressure (HHP) over prickly pear slices. These authors found that HHP treatment increased the in-vitro antioxidant and anti-inflammatory activities due to the increase of the bioactive compounds extractability.

Prickly pear juices are gaining popularity among consumers since they offer the nutritive and nutraceutical properties in a more practical consumption form. Pasteurization is a traditional method that, depending on the temperature and time used, could retain nutritional values but could affect the consumer's acceptance by affecting the sensorial attributes of the juice (Gurrieri et al., 2000). Purple prickly pear juices treated with thermoultrasonication retained higher antioxidant properties during 28 days of storage at 4°C compared with pasteurized juices (Cruz-Cansino et al., 2015). Moussa-Ayoub et al., (2011) increased the extraction yield of *O. macrorhiza* juices, without increasing the betacyanins content or decreasing the antioxidant activity, using proteolytic enzymes and a thermal treatment with microwave heating followed by the proteolytic enzymes' treatment. Zafra-Rojas et al., (2013) applied ultrasound treatments to purple cactus pear juice with good results. They reported that this treatment not only reduces the microbial count but preserves the juice antioxidant activity and quality; they did not report the effect on shelf life.

Dried prickly pear is another interesting product since it could be used as an ingredient to elaborate drinks, jams, ice creams, etc. Solar drying could be useful in the production areas where the climatic conditions for fruit growth are idea for this drying method. Freeze-drying is a more expensive method that made it possible to produce fruits with good rehydration properties without a significant loosing of visual properties (Martín-Del-Campo et al., 2019) but also to dry the pulp to preserve the stability of betalains (Viloria-Matos, Corbelli-Moreno, Moreno-Alvarez, & Belén C, 2002).

Other products include jams, marmalades, etc. Peeled prickly pear in syrup is a product that could be manufactured in rural zones using simple and artisanal methodologies using combined methods (Cerezal & Duarte, 2004).

Pitaya or Dragon Fruit

Pitaya, pitahaya, or dragon fruit is a tropical fruit originated from the tropical regions of America. Nowadays, this fruit is produced in America (North, Central, and South) as well as in Asia. This fruit has an oblong fruit with a pleasant flesh texture, and the pulp color goes from white to red and purple, and small edible black seeds. These fruits belong to the *Hylocereus* spp, *Selenicereus*, and *Stenocereus* species.

Hylocereus and *Selenicereus* are shade tolerant climbing cacti from tropical origin (Figure 2). These cacti are endemic to Mexico, Martinique, and Colombia; and cultivated in America: Nicaragua, Guatemala, Costa Rica, Peru, Venezuela, Bahamas, Bermuda, and the U.S.; Asia: China, India, Thailand, Taiwan, Malaysia, Philippines, Japan, Vietnam,

Cambodia, Indonesia, and Israel; Europe: Spain; Africa: Reunion Island; and Oceania: Australia, New Zealand (Ramírez-Rodríguez et al., 2020).

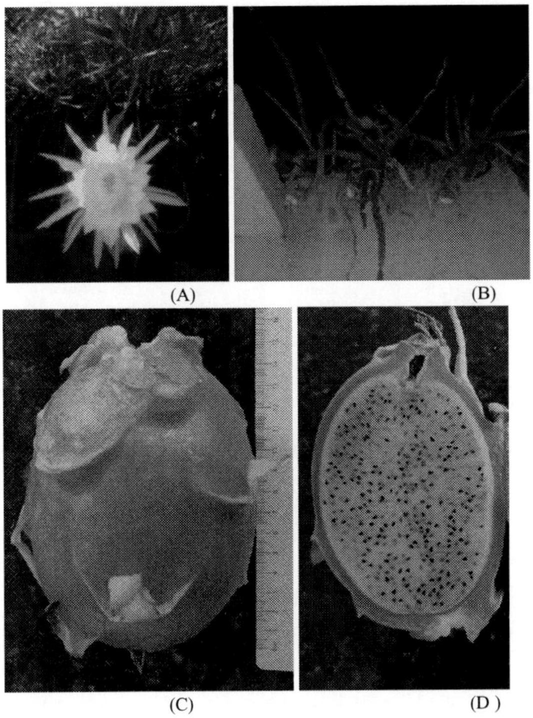

Figure 2. Pitaya *Selenicereus* sp. A) Flower, B) Cactus plant, C) Fruit, D) Flesh.

Figure 3. Pitaya *Stenocereus queretaroensis*. A) Columnar cactus, B) Fruits in cactus, C) Recollected fruits, D) Fruits in the market.

The most known and world distributed pitaya fruits belong to the *Hylocereus* sp species. This plant also receives the name of Moonflower, Lady of the night, or Quin of the night since the flowers bloom only at nighttime.

Stenocereus are columnar cacti growing in semiarid areas. The fruits are surrounded by large prickles (horns), and the flesh of the fruit presents colors such as white, red, orange, violet, and yellow (Figure 3). These cacti are distributed from the southwest U.S., Mexico, Colombia, Costa Rica, to Venezuela (Ramírez-Rodríguez et al., 2020). The species with commercial importance include *Stenocereus queretaroensis* (F. A.C. Weber) Buxbaum, *Stenocereus griseus* (Haworth) Buxbaum, *Stenocereus pruinosus* (Otto) Buxbaum, and *Stenocereus stellatus* (Pfeiffer) Riccobono (García-Cruz, Valle-Guadarrama, Salinas-Moreno, & Luna-Morales, 2016).

Pitaya Nutritional Composition

Pitaya is a fruit presenting not only an important nutritional value because of the macronutrients but also since it is a source of micronutrients and different phytochemicals. The fruit composition variates, according to species, genotypes, and origin. As other cactus fruits, the composition is affected by postharvest conditions.

Table 4 shows the nutritional composition of some pitaya cultivars from Mexico, as reported by the National Institute of Medical Sciences and Nutrition Salvador Zubirán (INCMNSZ, 2016). There are also reports about the nutritional value of pitaya fruits from other species, varieties, and countries (Table 4).

Pitaya is a good source of energy provided mainly by carbohydrates simples. Most of the species and varieties present a low-fat concentration, and some of them could be declared fat free according to the Codex Alimentarius specifications (CAC/GL 23-1997, 2013). Additionally, this fruit is not a good source of protein.

The most abundant carbohydrates are monosaccharides, such as glucose and fructose, and fiber. Esquivel, Stintzing and Carle (2007a), studied 5 *Hylocereus* genotypes from Costa Rica, and observed variations associated with the genotype, on the concentration of different carbohydrates such as glucose (49.14-103.95 g/L), fructose (19.2-28.97 g/L), sucrose (0.34-0.83 g/L), and pectin (0.64-1.36 mg/g). Beltrán-Orozco, Oliva-Coba, Gallardo-Velázquez and Osorio-Revilla (2009) reported variations in vitamin C content from 8.49 mg/100 g to 17.04 mg/100 g of fresh pulp for 4 types of *Stenocereus stellatus* Riccobono from Puebla Mexico.

Furthermore, pitaya fruits are a source of microelements such as potassium, magnesium, copper, among others. This fruit is a good source of magnesium, vitamins A and C, since 100 g of fruit could apport between 10-80%, 100%, and 30-100% of the IDR, respectively.

Even tough pitaya is not a good source of fat as other cactus fruits, its fatty acid profile present is beneficial for health. Fatty acids *Hylocereus* seed oil: Linolenic (mean 49.5%), linoleic (mean 22.7%), palmitic (mean 17%%); must abundant PUFA (mean 51%), MUFA (mean 26.4%) and SFA (mean 22.6%) (Ariffin et al., 2009; Choo, Koh, & Ling, 2016).

Table 4. Different origin cereus nutritional composition. Content for 100 g of fruit

Compositional Parameter	INCMNSZ (2016)					Cañar, Caetano and Bonilla (2014)	Khalili et al., (2006)	Nurul and Asmah (2014)		Pérez-Loredo, García-Ochoa and Barragán-Huerta (2016)
	Hylocereus ungatus B. et R.	*H. monacanthus* (Lem.) *H. undatus* (Haw.)				*Selenicereus megalanthus* (Colombia)	*Hylocereus* sp. (Malaysia)	*Hylocereus polyrhizus*		*Stenocereus stellatus* (Mexico)
	Fruit	Yellow	3 cultivars	Pink	Solferino	Yellow	Red	Red (Malaysia)	Red (australia)	4 colors
Edible portion (%)	55	45	45	45	45					
Energy (kJ)	228.00	205.00	179.00	189.00	239.00	162.12[c]		263.32*	147.91*	
Water (%)	82.3	85.90	87.05	84.50	83.40	84.8	87.30	85.05	89.98	85-86
TSS [a]										
Ether extract (g)	0.11	1.20	0.94	0.82	0.70	0.43	0.23	ND	ND	0.39-0.49
Crude protein (g)	1.44	1.10	1.60	1.25	0.80	0.65	0.16	1.45	0.41	1.08-1.30
Carbohydrates (g)	13.55	9.80	8.28	9.47	13.50	13.38		12.97	8.42	9.65-10.85
Sugars (g)		1.70	5.20		7.90		1.48			
Fiber (g)	2.10	1.40	1.29	3.26	0.90	0.77	10.10	2.65	ND	1.39-1.63
Ca (mg)	5	8.26	7.50	12.00	12.80		5.70	6.72	1.55	
P (mg)	15		34.00	25.00			23.00			
Fe (mg)	0.75	*	2.90	1.98			3.40	0.30	0.03	
Na (mg)		137.9	52.00		62.50		50.15	35.63	14.3	
K (mg)		*					56.96	158.29	437.35	
Mg (mg)		253.40	272.60		170.90		28.30			
Cu (mg)		27.77	36.60		21.90		0.03			
Mn (mg)		0.20	0.20		0.30			26.4	13.7	
Vit A (U.I.)		558.44	7400.79		6704.95					
Vit. B1 (mg)	0.11		0.04	0.06						
Vit B2 (mg)	0.13		0.04	0.05						
Niacin (mg)	0.37		0.22	0.26						
Ascorbic acid (mg)	25.80	7.34	13.70	9.50				0.24	0.30	261-411[c]

[a]TSS: Total soluble solids, °Brix.

[b]S.F.A.: Saturated Fatty Acids; M.F.A.: Monounsaturated fatty acids; P.U.F.A.: Polyunsaturated Fatty Acids.

[c]Calculated from reported values.

Table 5. Antioxidant capacity and bioactive compounds of pitaya from different origins and cultivars. Composition content for 100 g of the edible fraction or 100 mL of juice

Parameter	Esquivel, Stintzing and Carle (2007b)	Hua et al., (2018)	Nurul and Asmah (2014)	Rebecca, Boyce and Chandran (2010)	Beltrán-Orozco et al., (2009)	Pérez-Loredo et al., (2016)
	Hylocereus	*H. monacanthus* (Lem.) *H. undatus* (Haw.)	*Hylocereus polyrhizus*	*Hylocereus polyrhizus*	*Stenocereus stellatus* Riccobono	*Stenocereus stellatus*
	5 genotypes	3 cultivars	2 regions	red	4 types	4 colors
Total Phenols (mg GAE)	9.2-13-3	40-140	70-72	86	1384-2395*	372-579*
ABTS µmol of Trolox equivalent		2000-5000*			1100-1700	5720-12869*
TAC (mg ascorbic acid equivalent)[a]	24.5-36.1					
Total Flavonoids (mg)		60-190	41-49	230		

[a]TAC: Total antioxidant capacity.

*Calculated from the reference to the units on the table.

Pitaya Nutraceutical Benefits and Bioactive Compounds

Pitaya fruits present important antioxidant capacity (Table 5). This capacity variates depending on the species, varieties, and geographical origin. This antioxidant capacity could be associated with the fruit vitamin C content as well as with the polyphenols and flavonoids present in these fruits.

Pitayas present a variety of colors such as purple, red, orange, yellow, and white. These colors are due to pigments whose concentration variates depending on the species, varieties, and cultivars. The proportion between betalains and betacyanins is different depending not only on the species but also according to the flesh colors.

Ding, Chew, Abdul Aziz, Lai and Abdullah (2009) reported a total concentration of betacyanins of 11.703 mg/mL in red-fleshed Malaysian *Hylocereus polyrhizus*, fruits. The most abundant was Betanin (8.057 mg/mL), followed by Isobetanin (3.646 mg/mL). Meanwhile, Esquivel et al., (2007b), reported a total betalains concentration of 45.2-75.1 mg/100 mL in purple pitaya (*Hylocereus* sp.) genotypes from Costa Rica, where betacyanins represented between 70-78%. Hua et al., (2018) reported 1-6 mg of betacyanins/100 g fresh pulp and 0.6-2.1 mg of betaxanthin/100 g fresh pulp on new Chinese cultivars of hybrid offspring of *H. monacanthus* (Lem.) and *H. undatus* (Haw.)

Stenocereus species also present betalains as principal pigments; nevertheless, the proportion between betaxanthins and betacyanins variates according to the pitaya varieties. García-Cruz, Salinas-Moreno and Valle-Guadarrama (2012), for *Stenocereus griseus* H. from Mexico, reported a higher concentration of betacyanins in red-fleshed fruits (betacyanins: 199.6 mg/100 g dry fruit and betaxanthins: 147.61 dry fruit) and betaxanthins for orange-fleshed ones (betacyanins: 37.6 mg/100 g dry fruit and betaxanthins: 177.37 dry fruit). García-Cruz, Dueñas, Santos-Buelgas, Valle-Guadarrama and Salinas-Moreno (2017) observed higher betaxanthins content than betacyanins in *S. pruinosus* (red-fleshed and orange-fleshed) and *S. stellatus* (red-fleshed) from Mexico. They reported total betaxanthin concentration of 17706-21681 μg/g of dry fruit where the most abundant was Indicaxanthin, while betacyanins concentrations were from 162-5423 μg/g of dry fruit. Similar results were reported by Pérez-Loredo et al., (2016) for yellow, purple, and red *S. stellatus* fruits; betalains were not detected in white fruits.

Pitaya's betacyanins have been associated with protection against dieted-induced obesity. Song et al., (2016) suggested that pitaya betacyanins possess a modulatory effect on gut microbiota and an improvement of the inflammatory status that helps to control obesity in mice.

Cardioprotective effects and anti-insulin resistant activity have been reported for red pitaya (*Hylocereus* sp.). Mohd Adzim Khalili et al., (2009) evaluated the effect of Malaysian red pitaya on hypercholesterolemic rats. Rats were feed with a diet supplemented with freeze-dried pitaya, and after 11 weeks, they observed significant and dose-dependent reduction of blood cholesterol, triglycerides, and LDL, as well as a significant increase of HDL. Omidizadeh et al., (2011) reported that cardioprotective effects were observed only when rats' diet was supplemented with fresh fruit since the no-significant effect was observed for diets supplemented with cooked fruit (95°C, 30 min; or 105°C, 60 min). On the other hand, Omidizadeh et al., (2014) reported that anti-insulin resistant activity was observed when the rats' diet was supplemented with fresh fruit and fruit cooked at 95°C for 30 min. Still, the activity was not significant with a strong cooking procedure (105°C, 60 min). These authors attributed the cardioprotective effects to polyphenols and antioxidant content on fresh fruit, and the anti-insulin resistant activity to the antioxidant and soluble dietary fiber contents.

Processed Products from Pitayas

Most of the pitaya produced is sell and consumed as a fresh fruit due to its pleasant flavor. Asia is an important marked for fresh pitaya, even if this is a new fruit on those markets. In countries such as Thailand, Vietnam, Indonesia, or China, the culture of the pitaya has increased in the last 15 years. In China, during the New Year celebration, the consumption of this fruit increases significantly since they believe pitaya brings good luck (Damar Jaya, 2018).

Fresh fruit preservation has been an important issue. Fruits from the *Hylocereus* spp, and *Selenicereus* species reach international markets since they show higher resistance and bigger shelf life. *Stenocereus* fruits are highly perishable, so they normally reach only local markets, due to the elimination of the long thorns after harvesting. The fruit epidermis is sensitive to abrasion damages, and elimination of thorns make the fruit more sensitive to cracking (Quiroz-González, García-Mateos, Corrales-García, & Colinas-León, 2018). One important issue in postharvest managing is storage temperature since this fruit suffers chilling damages when it is stored below 12ºC. To reach more diverse markets, fresh fruit must be preserved controlling storage conditions such as temperature, atmospheres, and time.

Pitayas from *S. pruinosus* (red and orange flesh) and *S. stellatus* (red and white flesh) from Mexico could be preserved 6 days at 24ºC and 90% of relative humidity after harvesting not only with good organoleptic characteristics but also preserving the biological activities (García-Cruz et al., 2016).

Pitaya is a non-climacteric fruit that could act as a climacteric fruit after only 3 days after harvesting and fruits show important enzymatic activities leading to fruit damage in storage (Dueñas Gómez, Narváez Cuenca, & Restrepo Sánchez, 2008; Dueñas Gómez, Narváez Cuenca, & Restrepo Sánchez, 2009). A thermic shock at 25ºC during 24 h followed by storage at 2ºC (15 days), and storage at room temperature applied to *Acanthocereus pitajaya* fruits showed in important preservation of total acidity and total soluble solids content, and fruit hardness, as well as a reduction in weight loss compared with room temperature and 2ºC storage (Dueñas Gómez et al., 2009). Additionally, it was observed an important effect on the polyphenol oxidase, catalase, and peroxidase enzymatic activity after thermal shock treatment (Dueñas Gómez et al., 2008). On the other hand, García and Robayo (2008) evaluated different storage temperatures and polymeric materials to increase the shelf life of *Selinicereus megalanthus* Shuman from Colombia. They reported the principal advantage using polymeric materials as packaging reduces fruit dehydration. While de Freitas and Mitcham (2013), evaluating different storage temperatures for *Hylocereus undatus* from California, found that at 5ºC with perforated plastic bags gave good results maintaining fruit firmness, titratable acidity, and a weight loss reduction keeping a good visual appearance.

Fresh pitaya fruit, already peeled and cut, is also an interesting product to reach new consumers since the size of some varieties could be relatively large. Li et al., (2017) describe the effect of cutting styles on fruit quality throughout 4 d storage at 15ºC in polypropylene containers. They observed that the cutting style did not produce an important effect on the organoleptic quality of the fruit. This cutting induced a little deterioration on total soluble solids, total acidity, vitamin C content, as well as, an increase in b* value du to oxidation attributed to total phenolics making the fruit unacceptable after 4 d storage. On the other hand, these authors mentioned cutting style could induce the production of reactive oxygen species, increase the antioxidant activity, and the accumulation of phenolics.

Pitaya juices are also attractive products not only for the direct health benefices but also to be added to other products. Pigmented juices have high betalain content and could be used to add color to other kinds of fruit juices. Herbach, Maier, Stintzing and Carle (2007) evaluated betalain stability of the clarified juice of purple pitaya (*Hylocereus polyrhizus* [Weber] Britton & Rose) from Israel. On the other hand, since pitaya juices present a high viscosity that could be undesirable to consumers, different methods have been applied to clarify red and white pitaya juices (Nur 'Aliaa, Siti Mazlina, & Taip, 2011; Nur 'Aliaa, Siti Mazlina, Taip, & Liew Abdullah, 2010).

Additionally, emerging technologies have been applied to pitaya juices to decrease the microbial charge while keeping the biological activities instead of the traditional pasteurization methods. Ochoa-Velasco and Guerrero-Beltrán (2012) treated *Stenocereus griseus* juice from Mexico with UV-C light. Treatment of juices with the UV-C light during 25 min at 30.33 mL/s flow, and 4°C (juice temperature), made it possible a reduction of 2.11 log reduction of aerobic mesophyll bacteria and 1.14 log reduction of molds plus yeasts maintaining the juice color; nevertheless, some parameters suffered a reduction such as phenolic compounds (13.27% reduction), antioxidant activity (3.55% reduction), and betalains (26.02% reduction).

Garambullo or Berrycacti

Myrtillocactus species are a type of columnar cactus of 4-5 m high (Figure 4A), with a chandelier growth that posses greenish-white flowers of 15–30 mm (Figure 4B), (Pérez-González, 1999) that develop into unique edible fruits, known as berrycacti (Figure 4 C and D), that can be eaten fresh or processed as artisanal products. Thanks to its Crassulacean Acid Metabolism, *Myrtillocactus* grows widely in arid or semiarid regions where agriculture is not suitable.

The fruit of *Myrtillocactus* is locally called "garambullo", receive its name due to the appearance of their fruits that resemble a myrtle (Greek myrtillo, meaning small, myrtle-like). These fruits, also known as berrycatus, have been compared to blueberries or whortleberries because of their size and color (Hernández-López, Vaillant, Reynoso-Camacho, & Guzman-Maldonado, 2008; Reynoso-Camacho, Martinez-Samayoa, Ramos-Gomez, Guzmán, & Salgado, 2015), and their taste has been described as similar to the kiwi fruit.

Myrtillocactus production level is significant, due to their ability to grow without the need for any special agronomic care. However, the consumption of its fruit is very limited despite the recent interest in its potential as nutraceutical food (González Nava, 2010; Santiago-Mora, Cardador-Martinez, Tellez-Perez, Montejano-Gaitan, & Martin del Campo, 2017).

Nowadays they are four identified species of *Myrtillocactus*: *M. cochal, M. geometrizans M. schenckii,* and *M. eichlamii*. The former three are endemic from Mexico, while the fourth one is found in Guatemala, as shown in Figure 5. (Hernández-López et al., 2008).

Figure 4. Mirtillocactus geometricans.

The harvest period of Berrycacti is usually from May through September (Hernández-López et al., 2008) and local collectors collect the fruit ripe (purple) and overripe (deep purple) (Herrera-Hernández, Guevara-Lara, Reynoso-Camacho, & Guzmán-Maldonado, 2011) to be eaten fresh or processed in artisanal products like ice cream, mixed with aguardiente to prepare spirits, jam, sun-dried raisin-like product and consumed by the communities that are close to where they are harvested (Arias, 2010; Pérez-Negrón, Dávila, & Casas, 2014; Vazquez-Cruz et al., 2012).

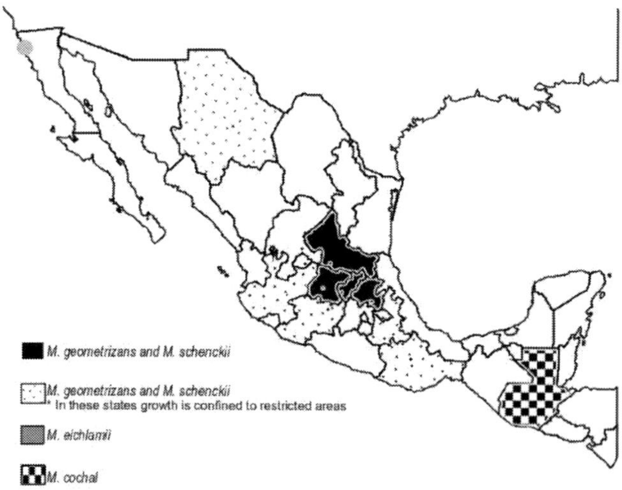

Figure 5. Distribution of the Myrtillocactus species.

Berrycacti Nutritional Composition

The fresh berrycacti has an average weight of 1.3 – 1.4 g and a pH of 3.5-3.8 (González Nava, 2010). Several authors have measured the proximate composition of berrycacti showing that this composition depends on the variety, the grade of maturation and state (but not locality) where the fruit was collected (Guzmán-Maldonado et al., 2010a; Herrera-Hernández et al., 2011; Herrera Hernández, 2008; Topete Viniegra, 2014). There is a lack of commercial indicators of the ripening stages of berrycacti; therefore the assessment of the ripening degree is based on the experience of the collectors, who use the color of the fruit as the accepted parameter, being a red-purple color characteristic of an early full maturity stage (Guzmán-Maldonado et al., 2010a; Herrera-Hernández et al., 2011).

Berrycacti has excellent nutritional characteristics, as shown in Table 6. This fruit has low carbohydrates, protein, and fat concentrations. Still, it has an interesting content of dietary fiber and vitamin C. Also, berrycacti is also a source of minerals and vitamins, It also represents a source of iron and copper (3.8 and 0.29 mg/kg FW), characteristic important for the rural population that lives mostly on diets based in local plants, risking the children with deficient growth and development as suggested by Guzmán-Maldonado et al., (2010a).

Table 6. Berrycacti nutritional composition. Content for 100 g of the edible fraction

Compositional Parameter	**INCMNSZ (2016)**	**Herrera Hernández (2008)**	**Guzmán-Maldonado et al., (2010a)**		**Topete Viniegra (2014)**
	Mirtillocactus geometrizans	*Mirtillocactus geometrizans*	*M. geometrizans*	*M. shenckii*	*Mirtillocactus geometrizans*
	fruit	Fruit[a]	2 regions	2 regions	4 regions
Edible portion (%)	80				
Water (g)	79.6		83.8-84.4	83.4-84.2	
Ether extract (g)	1.03	3.8	0.61-0.66	0.63-0.69	0.03-0.22[a]
Crude protein (g)	2.12	3.84	0.71-0.87	0.79	3.46-5.13[a]
Carbohydrates (g)		53.18	10.38-10.56	10.30-10.55	
Fiber (g)		35.6	3.49-3.69	3.49-3.61	0.46-0.49[a]
Ca (mg)	44		0.47-0.55	0.43	
P (mg)	23		0.18	0.15-0.19	
Fe (mg)					
Mg (mg)			0.26-0.35	0.32	
K (mg)			1.94-2.09	2.31	
Fe (mg)			0.37-0.41	0.35-0.40	
Mn (mg)			0.53	0.59-0.69	
b-Carotene (mg)	0.08				
Vit. B1 (mg)	0.04				
Vit B2 (mg)	0.03				
Niacin (mg)	0.48				
Ascorbic acid (mg)	32.5	195.5	22.6-26.4	47.6-49.7	

* Value not found.

[a] Calculated from reference data.

Berrycacti Nutraceutical Benefits and Bioactive Compounds

Several works have shown that berrycacti is an important source of bioactive compounds with a potential effect on human health. Most of these compounds include flavonoids, phenolic acids, betalains, and phytosterols. Its concentration is variable according to the ripening time of berrycacti, varieties, and geographical origin.

The property of *M. geometizans* that has been most widely described is its antioxidant capacity due to the presence of betalains, vitamin C, and phytosterols.

Gordon (2012) reported an antioxidant capacity of 6.98 9.67 mg/100 g of dry matter for *M. geometizans* at different ripening stages. Meanwhile, Herrera-Hernández et al., (2011) reported antioxidant activities of 748.4-2291.7 μmol Trolox equivalents/kg of berrycacti depending on the ripening stage. They stated that betacyanins and total phenols showed the highest antioxidant capacity in *M. geometizans*. Topete Viniegra (2014) found the variability of the antioxidant activity between *M. geometizans* harvested in different localities from Queretaro, Mexico. These authors reported values from 816-903 TEAC/100 mg of lyophilized fruit.

Polyphenols are important bioactive compounds found in berrycacti. Guzmán-Maldonado et al., (2010a) reported variation in polyphenol content associated with the berrycacti variety and the production zone. They reported concentrations of 3.41-3.78 mg/100 g of *M. geometrizans* fruit from 2 regions and 3.36-3.98 mg/100 g of *M. shenckii* fruit from 2 regions. Correa-Betanzo, Jacob, Perez-Perez and Paliyath (2011) reported a total polyphenol content of 3534.1 mg/100 g of dry berrycacti fruit. Herrera-Hernández et al., (2011) reported total phenols concentrations of 1850-10021 mg/kg of fresh weight depending on the maturation stage. They also said that the most abundant polyphenols in berrycacti are protocatechuic and gallic acids (41-53 mg/kg of fruit and 27-37 mg/kg of fruit, respectively). These authors also reported other bioactive compounds such as quercetin (155- 286 mg/kg of fruit), epigallocatechin gallate (13-27 mg/kg of fruit), and vanillin (5.2 mg/kg of ripe fruit). Meanwhile, Gordon (2012) reported quercetin 3-O-rutinoside and kaempferol 3-O-rutinoside were the most abundant phenolic compounds in *Myrtillocactus geometricans*.

Grambullo shows a unique purple color that variates according to the ripening stage. As in other cactus fruits, the color is given by different betalain pigments. The principal pigments reported for *M. geometrizans* and *M. shenckii* are betacyanins, where betaxanthins represent only a small fraction (Correa-Betanzo et al., 2011; Guzmán-Maldonado et al., 2010a; Herrera-Hernández et al., 2011).

The richness of berrycacti in bioactive compounds makes it an interesting target for the functional food market. It has been reported that the consumption of both berrycacti juice and its bioactive compounds is associated with functional and medicinal effects as anti-inflammatory, hypocholesterolemic, anticancer, antimicrobial, and antidiabetic (hypoglycemic).

Reynoso-Camacho et al., (2015) showed that the juice of berrycacti (*M. geometrizans*) had a hypoglycemic and antioxidant effect on Wistar rats. The authors evaluated the effect of the berrycacti juice (2 or 4 g/kg) on glucose levels and renal damage in diabetic animals. After four weeks of treatment, circulating glucose, triglycerides, and total cholesterol diminished. Besides, the authors observed an improvement in renal function of diabetic rats caused by a restoration of glutathione and glutathione S-transferase in the kidney. These observations supported previous hypoglycemic and anti-inflammatory effects of berrycacti juice observed by Martínez Samayoa (2008) and Topete Viniegra (2014) on the same animal model.

Processed Products from Berrycacti

The consumption of fresh berrycacti is mostly limited to the collector's family. The commercialization of this fruit is very scarce because of the extremely short shelf-life, which is approximate of 2 days at 25°C, and 5 days at 4°C (Hernández-López et al., 2008), because of its high content in carbohydrates, mature berrycacti can easily ferment (Topete Viniegra, 2014); thus the only way that collectors have found to preserve the fruit is freezing it after collected (Herrera-Hernández et al., 2011). Hence some authors have suggested the importance of finding postharvest technologies that could extend the shelf-life of berrycacti.

Correa-Betanzo et al., (2011) analyzed the use of an edible coating to protect the nutritional quality of berrycacti during its storage. These authors treated berrycacti with sodium caseinate based edible coating, plasticized with sorbitol and glycerol, and stored the fruit at 5°C for 15 days. After 7 days, the content of polyphenols decreased, but the betalains remain stable during all the storage time. In a similar work, López-Palestina et al., (2018) used oily tomato extract as part of a gelatin coating to increase the shelf-life of berrycacti while storage at 5 ± 1°C, founding that this coating was able to delay changes in sensorial characteristics and maintain bioactive compounds for up to 15 days.

In a different approach, Santiago-Mora et al., (2017) suggest the use of two different drying methods, freeze-drying (FD) and Instant Controlled Pressure Drop (DIC) as options to preserve berrycacti. In their work, authors found that both methods are capable of preserving not only the nutrimental characteristics of berrycacti but also its functional properties expressed by antioxidant capacity and content of bioactive compounds.

Meanwhile, Santiago-Mora et al., (2017), also analyzed the effect of drying methods (to increase shelf-life) on the antioxidant activity of *M. geometizans*. Authors used fruit at different ripening stages, unripe, ripe, and overripe, as well as ripe fruit dried by two different methods, freeze-drying (FD) and Instant Controlled Pressure. Antioxidant activity increased from unripe (7215.3 μM Teq/100g) to ripe fruit (8334 μM Teq/100g) but decreased in overripe fruit (9110 μM Teq/100g). Based on these results, the ripe fruit was selected to test the drying procedures. The authors concluded that these drying methods were able to maintain the bioactive compounds responsible for the antioxidant activity in ripe *M. geometizans*.

XOCONOSTLE

The word xoconostle derives from the nahuatl xoco = acid, noxtle = pear-like (Arias, 2010; Gallegos-Vázquez, Scheinvar, Núñez-Colín, & Mondragón-Jacobo, 2012) and as its name indicates, is a type of cactus pear characterized by its acidic flavour and low pH of 2.8 – 3.2 (Contreras, Jaimez, Castañeda, Añorve, & Villanueva, 2011). Like prickly pear, xoconostle grows in the arid and semiarid regions being a highly resistant fruit that after reaching maturity, can remain attached to the nopal cladodes for several months, without deterioration (Gallegos-Vázquez et al., 2012).

Xoconostle is a small round shape fruit with a diameter of 4-5 cm (Figure 6A). Its color varies between pale green, pink, or red, depending on the variety (Hernández-Fuentes et al., 2015). According to Gallegos-Vázquez et al., (2012), there are ten species of xoconostle fruit, nine from the Opuntia genus, and one from the genus Cylindropuntia (*Cylindropuntia imbricata* DC). From the Opuntia genus, there are four species cultivated for commercial purposes: *O. leucotricha*, *O. joconostle*, *O. matudae*, and *O. oligacantha*; and five wild species (with agronomical potential): *O. heliabravoana*, *O spinulifera*. *O. zamudioi*, *O. elizondoana* y *O. durangensis* (Eguiarte Fruns, 2011). However, phenotypic plasticity, interspecific hybridization, caused by bee pollinating between other factors, makes the delimitation of xoconostle species difficult (Gallegos 2012, LunaValdez 2016). Still, multiple efforts have been made to elucidate the different varieties of xoconostle and the relationship between them (de Luna-Valadez, Gallegos-Vázquez, Valdez-Cepeda, & Scheinvar, 2016; Gallegos-Vázquez et al., 2012; Valadez-Moctezuma, Ortiz-Vásquez, & Samah, 2014).

Xoconostle fruits are recognized by their edible mesocarp, which is usually 1–1.3 mm thick (Figure 6B). Its pericarp is 1–2 mm thick and characterized by short, flexible thorns (Arias-Rico et al., 2020), when the fruit reaches maturity, white spots appear and the fruit is ready to be harvested. The endocarp is a mucilaginous part that contains small brown seeds and, together with the peel, is usually discarded before consumption (Guzmán-Maldonado et al., 2010b; Morales, Barros, Ramírez-Moreno, Santos-Buelga, & Ferreira, 2015).

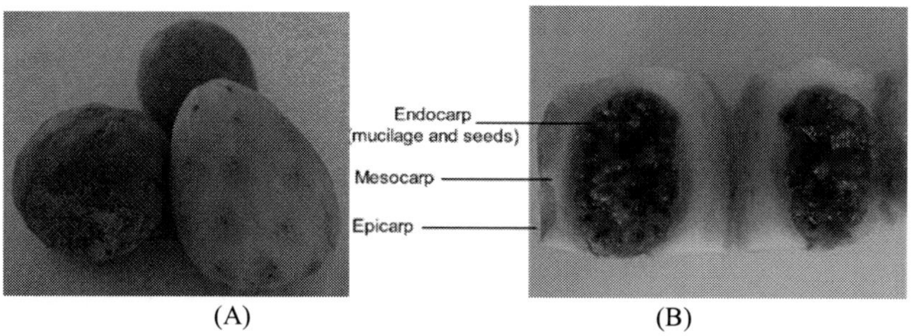

Figure 6. Xoconoxtle. A) Whole fruit, B) Open fruit.

Xoconostle has existed in Mexican cuisine since pre-Columbian times (López Martínez, García Mateos, Gallegos Vázquez, & Sahagún Castellanos, 2015), its traditional use it's mostly as spice or condiment in Mexican cuisine. It is also processed in products as jams, that are highly appreciated by its attractive color; candies, jellies, liquors, and syrups (Sáenz et al., 2013).

Xoconostle is currently market in Mexico, canned in syrup. Its acidity provides it with natural protection that inhibits the growth of harmful microorganisms and allows the canning process to take place at lower temperatures maintaining the sensorial quality of the fruit. (Inglese et al., 2018). Meanwhile, internationally the use of xoconostle for the production of powders and spicy sauces has attracted attention (Guzmán-Maldonado et al., 2010b).

However, xoconostle's consumption is still very low when compared to other vegetables like tomatoes or potatoes (0.67 g against 24.67 and 16.64 g fresh matter of edible portion/day/person) (Hervert-Hernández, García, Rosado, & Goñi, 2011), mostly due to its acidity and the lack of knowledge about its nutritional composition and potential biological activity.

Xoconoxtle Nutritional Composition

There are slight differences in nutritional composition according to the xoconostle variety and/or genotype, however until now only a few varieties have been studied been *O. joconostle* and *O. matudae* the most characterized because of its higher commercialization (Arias-Rico et al., 2020; Garcia-Saavedra et al., 2019; Hernández-Fuentes et al., 2015; López Martínez et al., 2015; Morales, Ramírez-Moreno, Sanchez-Mata, Carvalho, & Ferreira, 2012). Recent studies have also reported differences in nutrients between fruits from the same genotype and variety, but collected in different regions as observed by Guzmán-Maldonado et al., (2010b), In a different approach, Contreras et al., (2011), found no significant differences in chemical compositions between fruits from different locations, but the flavor profile was different between samples, probably due to differences in the maturity stage and chemical composition of the soil from the region.

The distribution of nutrients also varies within the structure of the fruit, and some important nutrients like protein and fat are higher in the by-products than in the edible part (Guzmán-Maldonado et al., 2010b; Morales et al., 2012). In general, the content of carbohydrates in xoconostle is low compared to other fruits (Table 7), which contributes to the sour taste of the fruit (Guzmán-Maldonado et al., 2010b).

Xoconostle has very low-fat content, which is mostly concentred in the seeds. Its fatty acid profile is rich in unsaturated fatty acids, which are associated with a reduced risk of developing cardiovascular, inflammatory, and autoimmune diseases (Morales et al., 2012). Morales et al., (2012) reported the most abundant fatty acids in *O. joconostle* cv. Cuaresmeño where S.F.A. (54.43%), followed by P.U.F.A. (41.41%); while for *O. matudae* cv. Rosa, the most abundant where P.U.F.A. (71.95%), followed by S.F.A. (25%).

Content of vitamin C is variable between xoconostle species (Arias-Rico et al., 2020; López Martínez et al., 2015; Morales et al., 2012), but in general, this fruit is rich in ascorbic acid, a potent water-soluble antioxidant, that protects lipids, proteins, and membranes from oxidative damage.

Xoconoxtle Nutraceutical Benefits and Bioactive Compounds

There has been a recent interest in the use of xoconostle and its by-products as sources of compounds with potential biological activity. The most important bioactive compounds in any cactus pear fruit are vitamin C, tocopherols, phenolics, and betalains (Table 8). All of these have potent antioxidant properties and protect the body against oxidative stress (de Wit et al., 2019).

Tocopherol in its 4 isoforms, (α, β, γ, and δ-tocopherol) have been found in different xoconostle varieties. The total content is significantly higher in the seeds than in the pulp, making seeds an interesting source of antioxidant compounds (Morales et al., 2012).

Xoconostle has also probed to be rich in phenolic acids and flavonoids (Table 8) distributed in the different structures of the fruit, where the mesocarp is an important contributor to the total phenolic content (Guzmán-Maldonado et al., 2010b; Morales, Barros, Ramírez-Moreno, Santos-Buelga, & Ferreira, 2014; Morales et al., 2015; Morales et al., 2012; Osorio-Esquivel, Alicia Ortiz, Álvarez, Dorantes-Álvarez, & Giusti, 2011). Xoconostle's phenolic profiles include gallic, vanillic, 4-hydroxybenzoic, syringic, caffeic, protocatechuic acids as well as the flavonoids catechin, epicatechin, vanillin, kaempferol and other isorhamnetin derivatives (Cortez-García, Ortiz-Moreno, Zepeda-Vallejo, & Necoechea-Mondragón, 2015; Garcia-Saavedra et al., 2019; Guzmán-Maldonado et al., 2010b; Morales et al., 2014, 2015; Osorio-Esquivel et al., 2011). However, this profile could be modified during the processing of xoconostle (Cortez-García et al., 2015; Dávila-Hernández, Sánchez-Pardo, Gutiérrez-López, Necoechea-Mondragon, & Ortiz-Moreno, 2019).

Phenolic content in xoconostle might be the reason why the mature fruit is capable of remaining in the plant for several months since these compounds protect against ultraviolet rays and pathogenic microorganisms (Osorio-Esquivel et al., 2011). Incorporation of phenolic compounds in the human diet contributes to conserve the quality of the food and reduce the risk of developing some diseases related to oxidative damage, like cancer (Chougui 2013).

Different studies have highlighted the importance of betalains as contributors to the antioxidant capacity in xoconostle (Guzmán-Maldonado et al., 2010b; Hernández-Fuentes et al., 2015; López Martínez et al., 2015). Betalains are natural pigments responsible for the characteristic color of xoconostle (betacyanins that provide a reddish to violet color, and betaxanthins that are yellow to orange pigments) (García-Cruz et al., 2012). Therefore is not surprising that its content is variable between xoconostle varieties (Hernández-Fuentes et al., 2015; López Martínez et al., 2015) and within the different structures of the fruit (Guzmán-Maldonado et al., 2010b).

Thus, xoconostle provides an important antioxidant activity that is not only associated with one type of compound but to a synergic effect due to the levels found of total betalains, phenolic compounds, flavonoids, and vitamin C.

Processed Products from Xoconoxtle

Because of its biological properties, there has been an increased interest in the development of new technologies and products that help exploit and preserve the nutraceutical properties of xoconostle. Xoconostle juice is a colorful substance with nutritional and functional properties that needs to increase its market value. The use of a hollow fiber membrane removes turbid components, enhancing the color and producing a clear juice with high antioxidant activity, by its content in phenolic compounds and betalains, that could be a good alternative for the commercialization of the fruit (Castro-Muñoz et al., 2018). Like every other fruit, xoconostle deteriorates with time; therefore, there is the need to find a way to preserve the properties of the fruit, and dehydration is the preferred method. Álvarez-Castro, Corrales-Garcia and Hernandez-Montes (2014) worked in the production of a snack using freeze-dried xoconostle sweetened with neonatame as an artificial sweetener (non-caloric) to improve the characteristic sour-flavor of xoconostle. The authors found that the final product had improved sensorial characteristics, but lower antioxidant activity due to a decreased in the total phenolic content. Studies have been made on how cooking could affect xoconostle's nutraceutical properties (Cortez-García et al., 2015). The use of microwave heating has reached attention as an enhancer during compounds extraction. Dávila-Hernández et al., (2019) used microwave pretreatment to improved the extraction yield of the antioxidant compounds from xoconostle's endocarp. The authors found that pretreatment increased the concentration of betalains, ascorbic acid, phenolic compounds, and flavonoids, that was traduced in increased antioxidant activity.

Table 7. Nutritional composition of different species and genotypes of xoconostle. Composition content for 100 g of the edible fraction or whole fruit

Compositional Parameter	INCMNSZ (2016)	Morales et al., (2012)		Arias-Rico et al., (2020)		Guzmán-Maldonado et al., (2010)	López Martínez et al., (2015)							
	O. imbricata D.C/ *Opuntia* sp.	*O. joconostle* cv. *Cuaresmeño*	*O. matudae* cv. *Rosa*	*O. joconostle* F.A.C. Weber	*O. matudae* Scheinvar	*O. matudae*	*O. matudae* Scheinvar	*O. duranguensis* Britton & Rose	*O. joconostle* Weber	*O. sainaltense*	*O. tezontepecana* Gallegos & Scheinvar	*O. leucotricha* Salm-Dick x *O. joconostle*	*O. leiascheinvaria* na Martínez &	*O. oligacantha*
	General			2 cultivars	*Rosa*	*Cuaresmeño*	Different genotypes	Different genotypes	Different genotypes	Chaveñito	Invierno	Cafe	Matizado	Borrego
	Mesocarp	Mesocarp	Mesocarp	Total fruit (dry matter)	Total fruit (dry matter)	Mesocarp (dry weight)	Mesocarp (dry weight)	Mesocarp (dry weight)	Mesocarp (dry weight)	Mesocarp (dry weight)	Mesocarp (dry weight)	Mesocarp (dry weight)	Mesocarp (dry weight)	Mesocarp (dry weight)
Edible portion (%)	0.55	0.70	0.56			0.58-0.63								
Energy (kJ)	92.00													
Water (g)	86.30	93.24	94.11	7.2	8.2		91.16-93.12	89.94-93.84	93.39-93.8	91.31	91.55	92.01	90.98	91.85
Ether extract (g)	0.36	*				5.05-5.28[c]								
Crude protein (g)	0.09	0.66	0.56	4.0	4.8	1.63–1.78[c]	0.16-0.39	0.17-0.34	0.14-0.28	0.26	0.29	0.25	0.25	0.24
Carbohydrates (g)	5.20	3.69[a]	3.93[a]			74.41-75.23[c]	5.08-6.55	5.02-7.25	4.64-4.99	6.68	6.12	5.53	7.08	6.41
Fat		0.03	0.04											
Ash (g)		0.07	0.11											
Fiber (g)	4.55	2.31	1.74	30.8	36.8	8.51-10.63[b]	0.53-1.04	0.31-1.03	0.45-0.49	0.66	0.72	0.71	0.69	0.42
Ca (mg)	128.50					1410-1960								
Fe (mg)	0.17					1.3-1.6								
Ascorbic acid (mg)	21.80	20.63	31.67	169.4	385.5	200.3-229.2	6.83-13.46	4.4-11.5	6.8-8.59	9.56	11.66	9.62	10.18	9.56

[a] Result refers to the total available carbohydrates, calculated as the difference of moisture, protein, fat, ash, and fiber values.

[b] Results were expressed as % dry weight basis.

Table 8. Antioxidant capacity and bioactive compounds of xoconostle from different species and genotypes. Composition content for 100 g of the edible fraction or whole fruit

Compositional Parameter	Cruz-Bravo et al., (2019)		Arias-Rico et al., (2020)		Guzmán-Maldonado et al., (2010)	López Martínez et al., (2015)							
	O. joconostle cv. Cuaresmeño	*O. matudae cv. Rosa*	*O. joconostle F.A.C.* Weber	*O. matudae* Scheinvar	*O. matudae*	*O. matudae* Scheinvar	*O. duranguensis* Britton & Rose	*O. joconostle* Weber	*O. sainaltense* Scheinvar	*O. tezontepecana* Gallegos & Scheinvar	*O. leucotricha Salm-Dick x O. joconostle* Weber	*O. leiascheinvariana* Martínez & Gallegos	*O. oligacantha*
			2 cultivares	*Rosa*	*Cuaresmeño*	Different genotypes	Different genotypes	Different genotypes	Chaveñito	Invierno	Cafe	Matizado	Borrego
	Mesocarp	Mesocarp	Total fruit (dry matter)	Total fruit (dry matter)	Mesocarp (dry weight)	Mesocarp (dry weight)	Mesocarp (dry weight)	Mesocarp (dry weight)	Mesocarp (dry weight)	Mesocarp (dry weight)	Mesocarp (dry weight)	Mesocarp (dry weight)	Mesocarp (dry weight)
Total Phenols (mg GAE)	38.57[a]	33.71[a]	1415.2 - 1752.4	1068.5-1438.5	919-986	141.53-224.32	176.86-189.04	151.2-161.2	231.37±3.14	140.48±3.05	132.83±2.88	190.1±3.37	196.62±2.94
ABTS (mmol)	*	*	*	*	6.87	*	*	*	*	*	*	*	*
ABTS µmol of Trolox equivalent	*	*	3261.4	1348.1	*	*	*	*	*	*	*	*	*
β-carotene bleaching inhibition	0.32[b]	0.02[b]	*	*	*	*	*	*	*	*	*	*	*
DPPH scavenging activity (mg QE)	5.14[b]	>16[b]	3318.78	1753.5	*	40.78-96.77 (%) 214.37-313.49	83.79-96.9 (%) 290.51-313.71	62.45-89.05 (%) 252.73-299.82	95.88 (%) 311.91	49.26% 229.39	93.78 308.18	90.07(%) 301.61	64.1(%) 255.65

Compositional Parameter	Cruz-Bravo et al., (2019)		Arias-Rico et al., (2020)		Guzmán-Maldonado et al., (2010)	López Martínez et al., (2015)							
Reducing power activity	3.16[b]	5.35[b]	*	*	*	*	*	*	*	*	*	*	*
Total flavonoids (mg)	3.93[c]	0.86 [c]	*	*	*	2.18-4.24	1.98-2.77	2.23-2.93	3.59	3.63	2.26	4.97	4.77
Total tocopherols (mg)	0.22	0.14	*	*	*	*	*	*	*	*	*	*	*

[a] Result refers to the total available carbohydrates, calculated as the difference of moisture, protein, fat, ash and fiber values.

[b] Results were expressed as % of total fatty acids.

[c] Results were expressed as % dry weight basis.

Other authors have proposed the use of encapsulation to preserve the bioactive compounds of xoconostle. Gorostiola Espinosa (2018), obtained xoconostle powder by spray-dry and then encapsulated the powder with nopal mucilage. The authors tested the hypoglycemic activity of both xoconostle powder and the encapsulated juice, finding that xoconostle powder has activity equivalent to commercial drugs that maintain while encapsulated.

Conclusion

There are other eatable cactus fruits produced in different arid regions. Normally they are produced and consummed only locally. Must of these cactus fruits cannot be produced in higher ammounts, others have a high perishability.

There are only few studies about these cactus fruits since they do not have a commercial importance, neverthess, they apport a source of nutrients and bioactive components to local populations.

Interesting examples are the sahuaro, borrachitos, and chilitos de biznaga, among others. The *Carnegiea gigantea* (sahuaro or saguaro) is the tallest columnar cati, from Sonora desert. Their growth is really lent, and thakes several years before having fruits. Their fruits are 6-9 cm long, red, and with a high ammount of small seeds. Fruits of *Ferocactus haematacanthus* called little drunk man (borrachitos) these fruits are collected riped or nearly so since they are sour when unriped. Some species of *Mammillaria* produce fruits called chili-peppers from cacti (chilitos de binznaga). Those are small (about 12 mm length), red to violed sweet fruits. *Echinocereus stramineus* produces small strawberry-flavoured fruits called strawberry pitaya or spiny strawberry cactus.

References

Abdel-Hameed, E. S. S., Nagaty, M. A., Salman, M. S., & Bazaid, S. A. (2014). Phytochemicals, nutritionals and antioxidant properties of two prickly pear cactus cultivars (*Opuntia ficus indica* Mill.) growing in Taif, KSA. *Food Chemistry, 160*, 31-38. doi: 10.1016/j.foodchem.2014.03.060.

Albano, C., Negro, C., Tommasi, N., Gerardi, C., Mita, G., Miceli, A., De Bellis, L., & Blando, F. (2015). Betalains, phenols and antioxidant capacity in cactus pear [*Opuntia ficus-indica* (L.) Mill.] fruits from Apulia (South Italy) genotypes. *Antioxidants, 4*(2), 269-280. doi: 10.3390/antiox4020269.

Allegra, A., Sortino, G., Miciletta, G., Riotto, M., Fasciana, T., & Inglese, P. (2015). The influence of harvest period and fruit ripeness at harvest on minimally processed cactus pears (*Opuntia ficus-indica* L. Mill.) stored under passive atmosphere. *Postharvest Biology and Technology, 104*, 57-62. doi: 10.1016/j.postharvbio.2015.03.001.

Álvarez-Castro, N. V., Corrales-Garcia, J., & Hernandez-Montes, A. (2014). Development of a snack from xoconostle (*Opuntia matudae* Scheinvar) sweetened with neotame and its antioxidant capacity. *Journal of the Professional Association for Cactus Development, 16*, 15-31.

Andreu-Coll, L., Cano-Lamadrid, M., Sendra, E., Carbonell-Barrachina, Á., Legua, P., & Hernández, F. (2019). Fatty acid profile of fruits (pulp and peel) and cladodes (young and old) of prickly pear [*Opuntia ficus-indica* (L.) Mill.] from six Spanish cultivars. *Journal of Food Composition and Analysis, 84*, 103294. doi: 10.1016/j.jfca.2019.103294.

Angeles Millones, J. (2014). *Estudio técnico económico para la elaboración de una bebida gasificada a base de la tuna fruta (Opuntia ficus) en la región de Ayacucho* [*Technical economic study for the elaboration of a carbonated drink based on prickly pear (Opuntia ficus) in the Ayacucho region.*].

Aparicio, X., Loza-Cornejo, S., Torres-Bernal, M. G., Velázquez-Placencia, N. J., & Arreola-Nava, H. J. (2017). Physicochemical characteristics of fruits from wild *Opuntia* species from two semiarid regions of Jalisco, Mexico. *Polibotanica, 43*, 219-244. doi: 10.18387/polibotanica.43.10.

Arias-Rico, J., Cruz-Cansino, N. d. S., Cámara-Hurtado, M., López-Froilán, R., Pérez-Rodríguez, M. L., Sánchez-Mata, M. d. C., Jaramillo-Morales, O. A., Barrera-Gálvez, R., & Ramírez-Moreno, E. (2020). Study of xoconostle (*Opuntia* spp.) powder as source of dietary fiber and antioxidants. *Foods 9*(4), 403. doi: 10.3390/foods9040403.

Arias, S. (2010). *Diversidad de Cactáceas y su aprovechamiento en México* [*Cactus diversity and its use in Mexico*] (pp. 27): Instituto de Biología UNAM.

Ariffin, A. A., Bakar, J., Tan, C. P., Rahman, R. A., Karim, R., & Loi, C. C. (2009). Essential fatty acids of pitaya (dragon fruit) seed oil. *Food Chemistry, 114*(2), 561-564. doi: 10.1016/j.foodchem.2008.09.108.

Barba, F. J., Putnik, P., Bursać Kovačević, D., Poojary, M. M., Roohinejad, S., Lorenzo, J. M., & Koubaa, M. (2017). Impact of conventional and non-conventional processing on prickly pear (*Opuntia* spp.) and their derived products: From preservation of beverages to valorization of by-products. *Trends in Food Science & Technology, 67*, 260-270. doi: 10.1016/j.tifs.2017.07.012.

Barreiro Perera, M. (1996, 1997). De nuestra cosecha [Of our harvest]. *Claridades Agropecuarias,* 3-28.

Beltrán-Orozco, M. C., Oliva-Coba, T. G., Gallardo-Velázquez, T., & Osorio-Revilla, G. (2009). Ascorbic acid, phenolic content, and antioxidant capacity of red, cherry, yellow and white types of pitaya cactus fruit (*Stenocereus stellatus* Riccobono). *Agrociencia, 43*(2), 153-161.

Butera, D., Tesoriere, L., Di Gaudio, F., Bongiorno, A., Allegra, M., Pintaudi, A. M., Kohen, R., & Livrea, M. A. (2002). Antioxidant activities of Sicilian prickly pear (*Opuntia ficus indica*) fruit extracts and reducing properties of its betalains: betanin and indicaxanthin. *Journal of Agricultural and Food Chemistry, 50*(23), 6895-6901. doi: 10.1021/jf025696p.

CAC/GL 23-1997. (2013). Guidelines for use of nutrition and health claims (pp. 8): Codex Alimentarius.

Cano, M. P., Gómez-Maqueo, A., García-Cayuela, T., & Welti-Chanes, J. (2017). Characterization of carotenoid profile of Spanish Sanguinos and Verdal prickly pear (*Opuntia ficus-indica,* spp.) tissues. *Food Chemistry, 237*, 612-622. doi: 10.1016/j.foodchem.2017.05.135.

Cañar, D., Caetano, C., & Bonilla, M. (2014). Caracterización fisicoquímica y proximal del fruto de pitahaya amarilla [*Selenicereus megalanthus* (k. schum. ex vaupel) moran] cultivada en Colombia [Physicochemical and proximal characterization of yellow pitahaya fruit [*Selenicereus megalanthus* (k. schum. ex vaupel) moran] cultivated in Colombia]. *Revista Agronomía, 22*(1), 77-87.

Casas, A., & Barbera, G. (2002). Mesoamerican domestication and diffusion. In P. S. Nobel (Ed.), *Cacti: Biology and uses* (pp. 143-162).

Castro-Muñoz, R., Fíla, V., Barragán-Huerta, B. E., Yáñez-Fernández, J., Piña-Rosas, J. A., & Arboleda-Mejía, J. (2018). Processing of Xoconostle fruit (*Opuntia joconostle*) juice for improving its commercialization using membrane filtration. *Journal of Food Processing and Preservation, 42*(1), e13394. doi: 10.1111/jfpp.13394.

Cayupán, Y. S. C., Ochoa, M. J., & Nazareno, M. A. (2011). Health-promoting substances and antioxidant properties of *Opuntia* sp. fruits. Changes in bioactive-compound contents during ripening process. *Food Chemistry, 126*(2), 514-519. doi: 10.1016/j.foodchem.2010.11.033.

Cejudo-Bastante, M. J., Chaalal, M., Louaileche, H., Parrado, J., & Heredia, F. J. (2014). Betalain profile, phenolic content, and color characterization of different parts and varieties of *Opuntia ficus-indica. J Agric Food Chem, 62*(33), 8491-8499. doi: 10.1021/jf502465g.

Cerezal, P., & Duarte, G. (2004). Sensory influence of chemical additives in peeled cactus pears (*Opuntia ficusindica* (L.) Miller) in syrup conserved by combined methods. *Journal of the Professional Association for Cactus Development, 6*, 102-119.

Cerezal, P., & Duarte, G. (2005). Some characteristics of cactus pear (*Opuntia ficus-indica* (L.) Miller) harvested in the Andean altiplane of Region 2 of Chile. *Journal of the Professional Association for Cactus Development*, 34-60.

Chang, S. F., Hsieh, C. L., & Yen, G. C. (2008). The protective effect of *Opuntia dillenii* Haw fruit against low-density lipoprotein peroxidation and its active compounds. *Food chemistry, 106*(2), 569-575. doi: 10.1016/j.foodchem.2007.06.017.

Chauhan, S. P., Sheth, N., Rathod, I., Suhagia, B., & Maradia, R. (2013). Phytochemical screening of fruits of *Opuntia elatior* Mill. *American Journal of PharmTech Research, 3*, 1-6.

Choo, J. C., Koh, R. Y., & Ling, A. P. K. (2016). Medicinal properties of pitaya: a review. *Spatula DD, 6*(2), 69-76. doi: 10.5455/spatula.20160413015353.

Contreras, L., Jaimez, O., Castañeda, O., Añorve, M., & Villanueva, R. (2011). Sensory profile and chemical composition of *Opuntia joconostle* from Hidalgo, Mexico. *Stored Prod. Postharvest Res, 22*, 37-39.

Correa-Betanzo, J., Jacob, J. K., Perez-Perez, C., & Paliyath, G. (2011). Effect of a sodium caseinate edible coating on berry cactus fruit (*Myrtillocactus geometrizans*) phytochemicals. *Food Research International, 44*(7), 1897-1904. doi: 10.1016/j.foodres.2010.10.053.

Cortez-García, R. M., Ortiz-Moreno, A., Zepeda-Vallejo, L. G., & Necoechea-Mondragón, H. (2015). Effects of cooking methods on phenolic compounds in Xoconostle (*Opuntia joconostle*). *Plant Foods for Human Nutrition, 70*(1), 85-90. doi: 10.1007/s11130-014-0465-2.

Corzo-Rios, L. J., Bautista-Ramírez, M. E., Gómez y Gómez, Y. d. l. M., & Torres-Bustillos, L. G. (2017). Frutas de cactáceas: Compuestos bioactivos y sus propiedades nutracéuticas [Cactaceae fruits: Bioactive compounds and their nutraceutical properties]. In M. E. Ramírez Ortíz (Ed.), *Propiedades Funcionales de Hoy* (pp. 35-65): OmniaScience Monographs.

Cruz-Bravo, R. K., Guzmán-Maldonado, S. H., Araiza-Herrera, H. A., & Zegbe, J. A. (2019). Storage alters physicochemical characteristics, bioactive compounds and antioxidant capacity of cactus pear fruit. *Postharvest Biology and Technology, 150*, 105-111. doi: 10.1016/j.postharvbio.2019.01.001.

Cruz-Cansino, N. d. S., Ramírez-Moreno, E., León-Rivera, J. E., Delgado-Olivares, L., Alanís-García, E., Ariza-Ortega, J. A., Manríquez-Torres, J. d. J., & Jaramillo-Bustos, D. P. (2015). Shelf life, physicochemical, microbiological and antioxidant properties of purple cactus pear (*Opuntia ficus indica*) juice after thermoultrasound treatment. *Ultrasonics Sonochemistry, 27*, 277-286. doi: 10.1016/j.ultsonch.2015.05.040.

Damar Jaya, I. K. (2018). Morphology and physiology of pitaya and it future prospects in Indonesia. *CROP AGRO, Jurnal Ilmiah Budidaya*(1), 44-50.

Dávila-Hernández, G., Sánchez-Pardo, M., Gutiérrez-López, G., Necoechea-Mondragon, H., & Ortiz-Moreno, A. (2019). Effect of microwave pretreatment on bioactive compounds extraction from xoconostle (*Opuntia joconostle*) by-products. *Revista Mexicana de Ingeniería Química, 18*(1), 191-204. doi: 10.24275/uam/izt/dcbi/revmexingquim/2019v18n1/Davila.

de Freitas, S. T., & Mitcham, E. J. (2013). Quality of pitaya fruit (*Hylocereus undatus*) as influenced by storage temperature and packaging. *Scientia Agricola, 70*, 257-262.

de Luna-Valadez, J., Gallegos-Vázquez, C., Valdez-Cepeda, R., & Scheinvar, L. (2016). Morphometric characterization of wild cultivars of xoconostle (*Opuntia* spp.) from Zacatecas, México. *AGRO Productividad, 9*(4), 19-27.

de Souza, R., Santana, M. S., de Macedo, E. S., de Brito, E., & Correia, R. P. (2015). Physicochemical, bioactive and functional evaluation of the exotic fruits *Opuntia ficus-indica* and *Pilosocereus pachycladus* Ritter from the Brazilian caatinga. *Journal of Food Science and Technology*, 1-8. doi: 10.1007/s13197-015-1821-4.

de Wit, M., du Toit, A., Osthoff, G., & Hugo, A. (2019). Cactus pear antioxidants: a comparison between fruit pulp, fruit peel, fruit seeds and cladodes of eight different cactus pear cultivars (*Opuntia ficus-indica* and *Opuntia robusta*). *Journal of Food Measurement and Characterization, 13*(3), 2347-2356. doi: 10.1007/s11694-019-00154-z.

de Wit, M., Nel, P., Osthoff, G., & Labuschagne, M. T. (2010). The effect of variety and location on cactus pear (*Opuntia ficus-indica*) fruit quality. *Plant Foods for Human Nutrition, 65*(2), 136-145. doi: 10.1007/s11130-010-0163-7.

Dehbi, F., Hasib, A., Ouatmane, A., & Elbatal, H. (2014). Physicochemical characteristics of Moroccan prickly pear juice (*Opuntia ficus indica* L.). *International Journal of Emerging Technology and Advanced Engineering, Volume 4*, 300-306.

Ding, P., Chew, M. K., Abdul Aziz, S., Lai, O. M., & Abdullah, J. O. (2009). Red-fleshed pitaya (*Hylocereus polyrhizus)* fruit colour and betacyanin content depend on maturity. *International Food Research Journal, 16*, 233-242.

Dok-Go, H., Lee, K. H., Kim, H. J., Lee, E. H., Lee, J., Song, Y. S., Lee, Y. H., Jin, C., Lee, Y. S., & Cho, J. (2003). Neuroprotective effects of antioxidative flavonoids, quercetin, (+)-dihydroquercetin and quercetin 3-methyl ether, isolated from *Opuntia ficus-indica* var. saboten. *Brain research, 965*(1-2), 130-136. doi: 10.1016/S0006-8993(02)04150-1.

Dueñas Gómez, Y. M., Narváez Cuenca, C. E., & Restrepo Sánchez, L. P. (2008). Inhibition of Chillyng Injury in Pitaya Amarilla (*Acanthocereus pitajaya*) Fruit by Heat Shock: Catalese, Peroxidase and Polyphenoloxidase. *Acta Biologica Colombiana, 13*(1), 95-106.

Dueñas Gómez, Y. M., Narváez Cuenca, C. E., & Restrepo Sánchez, L. P. (2009). Heat shock improves refrigerated storage performance of yellow pitaya. *Agronomía Colombiana, 27*(1), 105-110.

Eguiarte Fruns, L. E. (2011). El futuro del nopal y sus frutos, xoconostles y tunas [The future of opuntina and its fruits, xoconoxtles and prickly pears]. *Cactáceas y Suculentas Mexicanas, 51*(1), 22-28.

El Kharrassi, Y., Mazri, M. A., Benyahia, H., Benaouda, H., Nasser, B., & El Mzouri, E. H. (2016). Fruit and juice characteristics of 30 accessions of two cactus pear species (Opuntia ficus indica and Opuntia megacantha) from different regions of Morocco. *LWT - Food Science and Technology, 65*, 610-617. doi: 10.1016/j.lwt.2015.08.044.

El Kossori, R. L., Villaume, C., El Boustan, E., Sauvaire, Y., & Méjean, L. (1998). Composition of pulp, skin, and seeds of prickly pears fruit (*Opuntia ficus-indica* sp.). *Plant Foods for Human Nutrition,* 8.

Ennouri, M., Evelyne, B., Laurence, M., & Hamadi, A. (2005). Fatty acid composition and rheological behaviour of prickly pear seed oils. *Food Chemistry, 93*(3), 431-437. doi: 10.1016/j.foodchem.2004.10.020.

Esquivel, P., Stintzing, F. C., & Carle, R. (2007a). Comparison of morphological and chemical fruit traits from different pitaya genotypes (*Hylocereus* sp.) grown in Costa Rica. *Journal of Applied Botany and Food Quality, 81*(1), 7.

Esquivel, P., Stintzing, F. C., & Carle, R. (2007b). Phenolic compound profiles and their corresponding antioxidant capacity of purple pitaya (*Hylocereus* sp.) genotypes. *Zeitschrift für Naturforschung C, 62*(9-10), 636-644. doi: 10.1515/znc-2007-9-1003.

Fernández-López, J. A., Almela, L., Obón, J. M., & Castellar, R. (2010). Determination of Antioxidant Constituents in Cactus Pear Fruits. *Plant Foods for Human Nutrition, 65*(3), 253-259. doi: 10.1007/s11130-010-0189-x.

Figueroa-Cares, I., Martínez-Damián, M. T., Rodríguez-Pérez, E., Colinas-León, M. T., Valle-Guadarrama, S., Ramírez-Ramírez, S., & Gallegos-Vázquez, C. (2010). Pigments content, other compounds and antioxidant capacity in 12 cactus pear cultivars (*Opuntia* spp.) from Mexico. *Agrociencia, 44*, 763-771.

Galati, E. M., Mondello, M. R., Lauriano, E. R., Taviano, M. F., Galluzzo, M., & Miceli, N. (2005). *Opuntia ficus indica* (L.) Mill. fruit juice protects liver from carbon tetrachloride-induced injury. *Phytotherapy Research, 19*(9), 796-800. doi: 10.1002/ptr.1741.

Gallegos-Vázquez, C., Scheinvar, L., Núñez-Colín, C. A., & Mondragón-Jacobo, C. (2012). Morphological diversity of xoconostles (*Opuntia* spp.) or acidic cactus pears: a Mexican contribution to functional foods. *Fruits, 67*(2), 109-120. doi: 10.1051/fruits/2012001.

García-Cruz, L., Dueñas, M., Santos-Buelgas, C., Valle-Guadarrama, S., & Salinas-Moreno, Y. (2017). Betalains and phenolic compounds profiling and antioxidant capacity of pitaya (*Stenocereus* spp.) fruit from two species (*S. Pruinosus* and *S. stellatus*). *Food Chemistry, 234*, 111-118. doi: 10.1016/j.foodchem.2017.04.174.

García-Cruz, L., Salinas-Moreno, Y., & Valle-Guadarrama, S. (2012). Betalains, phenolic compounds and antioxidant activity in pitaya de mayo (*Stenocereus griseus* H.). *Revista fitotecnia mexicana, 35*, 01-05.

García-Cruz, L., Valle-Guadarrama, S., Salinas-Moreno, Y., & Luna-Morales, C. d. C. (2016). Postharvest quality, soluble phenols, betalains content, and antioxidant activity of *Stenocereus pruinosus* and *Stenocereus stellatus* fruit. *Postharvest Biology and Technology, 111*, 69-76. doi: 10.1016/j.postharvbio.2015.07.004.

Garcia-Saavedra, N. M., Barros, L., Reis, F. S., Roriz, C. L., Alves, M. J., García-Hernandez, M., Pérez-Rodríguez, M. L., Sánchez-Mata, M. d. C., Ramírez-Moreno, E., & Ferreira, I. C. F. R. (2019). Chemical characterization and biological activities of two varieties of xoconostle fruits *Opuntia joconostle* F.A.C. Weber ex Diguet and *Opuntia matudae* Scheinvar. *Food & Function, 10*(6), 3181-3187. doi: 10.1039/C9FO00737G.

García, M. C., & Robayo, P. (2008). Evaluación del uso de atmósferas modificadas pasivas y temperaturas bajas en la conservación de pitaya amarilla (*Selinicereus megalanthus* Shuman) [Evaluation of the use of passive modified atmospheres and low temperatures on the conservation of yellow pitahaya]. *Ciencia y Tecnología Agropecuaria, 9*(1), 30-39.

Gómez-Maqueo, A., García-Cayuela, T., Welti-Chanes, J., & Cano, M. P. (2019). Enhancement of anti-inflammatory and antioxidant activities of prickly pear fruits by high hydrostatic pressure: A chemical and microstructural approach. *Innovative Food Science & Emerging Technologies, 54*, 132-142. doi: 10.1016/j.ifset.2019.04.002

González González, R., Morales, O., Olivares Sáenz, E., Aranda Ruiz, J., & Gallegos Vázquez, C. (2001). Conservación de una variedad de tuna (Burrona) bajo diferentes manejos poscosecha [Conservation of a variety of prickly pear (Burrona) under different post-harvest management]. *Ciencia UANL, 4*(3).

González Nava, C. (2010). *Caracterización físico química del fruto de garambullo (Myrtillocactus geometrizans).* [*Physical chemical characterization of the garambullo fruit (Myrtillocactus geometrizans).*]. (Maestría en Ciencias-Recursos Bióticos), Universidad Autónoma de Querétaro, Santiago de Querétaro, Querétaro.

Gordon, A. (2012). *Bioactive Compounds in Underutilized Tropical Fruits from Latin America.* (Doctorat), Universitäts-und Landesbibliothek Bonn.

Gorostiola Espinosa, H. (2018). *Efecto del secado por aspersión en la actividad hipoglucemiante de jugo de xoconostle (Opuntia joconostle)* [*Effect of spray drying on hypoglycemic activity of xoconostle juice (Opuntia joconostle)*]. Instituto Politécnico Nacional.

Gurrieri, S., Miceli, L., Lanza, C. M., Tomaselli, F., Bonomo, R. P., & Rizzarelli, E. (2000). Chemical characterization of Sicilian prickly pear (*Opuntia ficus indica*) and perspectives for the storage of its juice. *Journal of Agricultural and Food Chemistry, 48*(11), 5424-5431. doi: 10.1021/jf9907844.

Guzmán-Maldonado, S. H., Herrera-Hernández, G., Hernández-López, D., Reynoso-Camacho, R., Guzmán-Tovar, A., Vaillant, F., & Brat, P. (2010a). Physicochemical, nutritional and functional characteristics of two underutilised fruit cactus species (*Myrtillocactus*) produced in central Mexico. *Food Chemistry, 121*(2), 381-386. doi: 10.1016/j.foodchem.2009.12.039.

Guzmán-Maldonado, S. H., Morales-Montelongo, A. L., Mondragón-Jacobo, C., Herrera-Hernández, G., Guevara-Lara, F., & Reynoso-Camacho, R. (2010b). Physicochemical, nutritional, and functional characterization of fruits xoconostle (*Opuntia matudae*) pears from Central-México Region. *75*(6), C485-C492. doi: 10.1111/j.1750-3841.2010.01679.x.

Herbach, K. M., Maier, C., Stintzing, F. C., & Carle, R. (2007). Effects of processing and storage on juice colour and betacyanin stability of purple pitaya (*Hylocereus polyrhizus*) juice. *European Food Research and Technology, 224*(5), 649-658. doi: 10.1007/s00217-006-0354-5.

Hernández-Fuentes, A. D., Trapala-Islas, A., Gallegos-Vásquez, C., Campos-Montiel, R. G., Pinedo-Espinoza, J. M., & Guzmán-Maldonado, S. H. (2015). Physicochemical variability and nutritional and functional characteristics of xoconostles (*Opuntia* spp.) accessions from Mexico. *Fruits, 70*(2), 109-116. doi: 10.1051/fruits/2015002.

Hernández-López, D., Vaillant, F., Reynoso-Camacho, R., & Guzman-Maldonado, S.-H. (2008). *Myrtillocactus* (*cactaceae*): botanical, agronomic, physicochemical and chemical characteristics of fruits. *Fruits, 63*(05), 269-276. doi: doi: 10.1051/fruits:2008042.

Herrera-Hernández, M. G., Guevara-Lara, F., Reynoso-Camacho, R., & Guzmán-Maldonado, S. H. (2011). Effects of maturity stage and storage on cactus berry (*Myrtillocactus geometrizans*) phenolics, vitamin C, betalains and their antioxidant properties. *Food Chemistry, 129*(4), 1744-1750. doi: 10.1016/j.foodchem.2011.06.042.

Herrera Hernández, M. G. (2008). *Efecto del estado de madurez y el almacenamiento sobre la calidad funcional del garambullo (Myrtillocactus geometrizans).* [*Effect of maturity status and storage on the functional quality of garambullo (Myrtillocactus geometrizans).*]. (Maestrí en Ciencia y Tecnología de Alimentos), Universidad Autónoma de Querétaro, Santiago de Querétaro, Querétaro.

Hervert-Hernández, D., García, O. P., Rosado, J. L., & Goñi, I. (2011). The contribution of fruits and vegetables to dietary intake of polyphenols and antioxidant capacity in a Mexican rural diet: Importance of fruit and vegetable variety. *Food Research International, 44*(5), 1182-1189. doi: 10.1016/j.foodres.2010.09.021.

Hua, Q., Chen, C., Tel Zur, N., Wang, H., Wu, J., Chen, J., Zhang, Z., Zhao, J., Hu, G., & Qin, Y. (2018). Metabolomic characterization of pitaya fruit from three red-skinned cultivars with different pulp colors. *Plant Physiology and Biochemistry, 126*, 117-125. doi: 10.1016/j.plaphy.2018.02.027

INCMNSZ. (2016). *Tables of composition of foods and food products (condensed version 2015)* (J. Morales de León, H. Bourges Rodríguez & M. Camacho Parra Eds.): Instituto Nacional de Ciencias Médicas y Nutrición Salvador Zubirán: México City, México.

Inglese, P., Mondragon, C., Nefzaoui, A., Saenz, C., Taguchi, M., Makkar, H., & Louhaichi, M. (2018). *Ecologia del cultivo, manejo y usos del nopal* [*Ecology of the cultivation, management and uses of the nopal*]. Rome, Italy: Food and Agriculture Organization of the United Nations (FAO).

Juhaimi, F. A., Ghafoor, K., Uslu, N., Mohamed Ahmed, I. A., Babiker, E. E., Özcan, M. M., & Fadimu, G. J. (2020). The effect of harvest times on bioactive properties and fatty acid compositions of prickly pear (*Opuntia ficus-barbarica* A. Berger) fruits. *Food Chemistry, 303*, 125387. doi: 10.1016/j.foodchem.2019.125387.

Khalili, R. M. A., Norhayati, A., Rokiah, M., Asmah, R., Nasir, M. M., & Muskinah, M. S. (2006). Proximate composition and selected mineral determination in organically grown red pitaya (*Hylocereus* sp.). *Journal of Tropical Agriculture and Food Science, 34*(2), 269.

Kuti, J. O. (2004). Antioxidant compounds from four *Opuntia* cactus pear fruit varieties. *Food chemistry, 85*(4), 527-533. doi: 10.1016/S0308-8146(03)00184-5.

Li, X., Long, Q., Gao, F., Han, C., Jin, P., & Zheng, Y. (2017). Effect of cutting styles on quality and antioxidant activity in fresh-cut pitaya fruit. *Postharvest Biology and Technology, 124*, 1-7. doi: 10.1016/j.postharvbio.2016.09.009.

López-Palestina, C. U., Aguirre-Mancilla, C. L., Raya-Pérez, J. C., Ramírez-Pimentel, J. G., Gutiérrez-Tlahque, J., & Hernández-Fuentes, A. D. (2018). The effect of an edible coating with tomato oily extract on the physicochemical and antioxidant properties of garambullo (*Myrtillocactus geometrizans*) fruits. *Agronomy, 8*(11), 248. doi: 10.3390/agronomy8110248.

López Martínez, C. R., García Mateos, R., Gallegos Vázquez, C., & Sahagún Castellanos, J. (2015). Antioxidant components and nutritional quality of 15 genotypes of Xoconostle (*Opuntia* spp.). *Journal of the Professional Association for Cactus Development, 17*, 33-49.

Martín-Del-Campo, S. T., Ramírez-Anaya, J. d. P., Pimienta-Barrios, E., Castañeda-Saucedo, M. C., Samaniego-Sánchez, C., & Gómez-Hernández, H. E. (2019). Identifying non-destructive growth and maturity indexes of Prickly pear (*Opuntia albicarpa* S. Var. Burrona) and evaluation of freeze-drying conditions. *CyTA - Journal of Food, 17*(1), 917-925. doi: 10.1080/19476337.2019.1676315.

Martínez Samayoa, P. M. (2008). *Análisis molecular de la capacidad antioxidante del fruto de garambullo (Myrtillocactus geometrizans) en ratas diabéticas.* [*Molecular analysis of the antioxidant capacity of garambullo fruit (Myrtillocactus geometrizans) in diabetic rats.*]. (Tesis para obtener el grado de maestro en ciencia y tecnología de los alimentos.), Universidad Autónoma de Querétaro, Querétaro, México, Santiago de Querétaro, Querétaro.

Medina, E. D., Rodríguez, E. R., & Romero, C. D. (2007). Chemical characterization of *Opuntia dillenii* and *Opuntia ficus indica* fruits. *Food chemistry, 103*(1), 38-45. doi: 10.1016/j.foodchem.2006.06.064.

Melgar, B., Pereira, E., Oliveira, M. B. P. P., Garcia-Castello, E. M., Rodriguez-Lopez, A. D., Sokovic, M., Barros, L., & Ferreira, I. C. F. R. (2017). Extensive profiling of three varieties of *Opuntia* spp. fruit for innovative food ingredients. *Food Research International, 101*, 259-265. doi: 10.1016/j.foodres.2017.09.024.

Mohd Adzim Khalili, R., Norhayati, A., Rokiah, M., Asmah, R., Siti Muskinah, M., & Abdul Manaf, A. (2009). Hypocholesterolemic effect of red pitaya (*Hylocereus* sp.) on hypercholesterolemia induced rats. *International Food Research Journal, 16*(3).

Morales, P., Barros, L., Ramírez-Moreno, E., Santos-Buelga, C., & Ferreira, I. C. F. R. (2014). Exploring xoconostle by-products as sources of bioactive compounds. *Food Research International, 65*, 437-444. doi: 10.1016/j.foodres.2014.05.067.

Morales, P., Barros, L., Ramírez-Moreno, E., Santos-Buelga, C., & Ferreira, I. C. F. R. (2015). Xoconostle fruit (*Opuntia matudae* Scheinvar cv. Rosa) by-products as potential functional ingredients. *Food Chemistry, 185*, 289-297. doi: 10.1016/j.foodchem.2015.04.012.

Morales, P., Ramírez-Moreno, E., Sanchez-Mata, M. d. C., Carvalho, A. M., & Ferreira, I. C. F. R. (2012). Nutritional and antioxidant properties of pulp and seeds of two xoconostle cultivars (*Opuntia joconostle* F.A.C. Weber ex Diguet and *Opuntia matudae* Scheinvar) of high consumption in Mexico. *Food Research International, 46*(1), 279-285. doi: 10.1016/j.foodres.2011.12.031.

Moussa-Ayoub, T. E., Jaeger, H., Knorr, D., El-Samahy, S., Rohn, S., & Kroh, L. W. (2011). Impact of traditional and innovative technologies on some characteristics and bioactive compounds of *Opuntia macrorhiza* juice. *Procedia Food Science, 1*, 1410-1416. doi: 10.1016/j.profoo.2011.09.209.

Nur 'Aliaa, A. R., Siti Mazlina, M. K., & Taip, F. S. (2011). Effects of commercial pectinases application on selected properties of red pitaya juice. *Journal of Food Process Engineering, 34*(5), 1523-1534. doi: 10.1111/j.1745-4530.2009.00388.x.

Nur 'Aliaa, A. R., Siti Mazlina, M. K., Taip, F. S., & Liew Abdullah, A. G. (2010). Response surface optimization for clarification of white pitaya juice using a commercial enzyme. *Journal of Food Process Engineering, 33*(2), 333-347. doi: 10.1111/j.1745-4530.2008.00277.x.

Nurul, S., & Asmah, R. (2014). Variability in nutritional composition and phytochemical properties of red pitaya (*Hylocereus polyrhizus)* from Malaysia and Australia. *International Food Research Journal, 21*(4).

Ochoa-Velasco, C. E., & Guerrero-Beltrán, J. Á. (2012). Ultraviolet-C light effect on pitaya (*Stenocereus griseus*) juice. *Journal of Food Research, 1*(2), 60. doi: 10.5539/jfr.v1n2p60.

Ochoa-Velasco, C. E., & Guerrero-Beltrán, J. Á. (2013). Effects of the storage temperature on quality characteristics of green prickly pear (*Opuntia albicarpa*). *Revista Iberoamericana de Tecnología Postcosecha, 14*(2), 149-161.

Ochoa, C. E., & Guerrero, J. A. (2012). Efecto del almacenamiento a diferentes temperaturas sobre la calidad de tuna roja (*Opuntia ficus indica* (L.) Miller) [Effect of storage at different temperatures on the quality of red prickly pear (*Opuntia ficus indica* (L.) Miller)]. *Información tecnológica, 23*, 117-128. doi: 10.4067/S0718-07642 012000100013.

Ojeda-Linares, C. I., Vallejo, M., Lappe-Oliveras, P., & Casas, A. (2020). Traditional management of microorganisms in fermented beverages from cactus fruits in Mexico: an ethnobiological approach. *Journal of Ethnobiology and Ethnomedicine, 16*(1), 1. doi: 10.1186/s13002-019-0351-y.

Omidizadeh, A., Yusof, R. M., Ismail, A., Roohinejad, S., Nateghi, L., Zuki, M., & Bakar, A. (2011). Cardioprotective compounds of red pitaya (*Hylocereus polyrhizus*) fruit. *Journal of Food, Agriculture & Environment, 9*(3-4), 152-156.

Omidizadeh, A., Yusof, R. M., Roohinejad, S., Ismail, A., Bakar, M. Z. A., & Bekhit, A. E.-D. A. (2014). Anti-diabetic activity of red pitaya (*Hylocereus polyrhizus*) fruit. *RSC Advances, 4*(108), 62978-62986. doi: 10.1039/C4RA10789F.

Osorio-Esquivel, O., Alicia Ortiz, M., Álvarez, V. B., Dorantes-Álvarez, L., & Giusti, M. M. (2011). Phenolics, betacyanins and antioxidant activity in *Opuntia joconostle* fruits. *Food Research International, 44*(7), 2160-2168. doi: 10.1016/j.foodres.2011.02.011.

Pérez-González, S. (1999). *Estudio etnobotánico, ecológico y de usos potenciales del garambullo (Myrtillocactus geometrizans) como base para su domesticación y cultivo.* [*Ethnobotanical, ecological and potential uses of the garambullo (Myrtillocactus geometrizans) as a basis for domestication and cultivation*]. Querétaro, Querétaro: Consejo Nacional de Ciencia y Tecnología.

Pérez-Loredo, M. G., García-Ochoa, F., & Barragán-Huerta, B. E. (2016). Comparative analysis of betalain content in *Stenocereus sellatus* fruits and other cactus fruits using principal component analysis. *International Journal of Food Properties, 19*(2), 326-338. doi: 10.1080/10942912.2015.1022259.

Pérez-Negrón, E., Dávila, P., & Casas, A. (2014). Use of columnar cacti in the Tehuacán Valley, Mexico: perspectives for sustainable management of non-timber forest products. *Journal of ethnobiology and ethnomedicine, 10*(1), 79. doi: 10.1186/1746-4269-10-79.

Quiroz-González, B., García-Mateos, M., Corrales-García, J., & Colinas-León, M. (2018). Pitaya (*Stenocereus* spp.): An under-utilized fruit. *Journal of the Professional Association for Cactus Development, 20*, 82-100.

Ramírez-Rodríguez, Y., Martínez-Huélamo, M., Pedraza-Chaverri, J., Ramírez, V., Martínez-Tagüeña, N., & Trujillo, J. (2020). Ethnobotanical, nutritional and medicinal properties of Mexican drylands Cactaceae Fruits: Recent findings and research opportunities. *Food Chemistry, 312*, 126073. doi: 10.1016/j.foodchem.2019.126073.

Rebecca, O., Boyce, A., & Chandran, S. (2010). Pigment identification and antioxidant properties of red dragon fruit (*Hylocereus polyrhizus*). *African Journal of Biotechnology, 9*(10), 1450-1454.

Reynoso-Camacho, R., Martinez-Samayoa, P., Ramos-Gomez, M., Guzmán, H., & Salgado, L. M. (2015). Antidiabetic and renal protective properties of berrycactus fruit (*Myrtillocactus geometrizans*). *Journal of medicinal food, 00*(0), 1-7. doi: 10.1089/jmf.2013.0164.

Sáenz, C., Berger, H., Rodríguez-Félix, A., Galleti, L., Corrales García, J., & Sepúlveda, E. (2013). *Agro-industrial utilization of cactus pear.* Rome: Food and Agriculture Organization of the united Nations.

Santiago-Mora, P. D., Cardador-Martinez, A., Tellez-Perez, C., Montejano-Gaitan, J. G., & Martin del Campo, S. T. (2017). In-Vitro antioxidant capacity and bioactive compounds preservation post-drying on berrycacti (*Myrtillocactus geometrizans*). *Journal of Food Research, 6*(4), 121-133. doi: 10.5539/jfr.v6n4p121.

Song, H., Chu, Q., Yan, F., Yang, Y., Han, W., & Zheng, X. (2016). Red pitaya betacyanins protects from diet-induced obesity, liver steatosis and insulin resistance in association with modulation of gut microbiota in mice. *Journal of Gastroenterology and Hepatology, 31*(8), 1462-1469. doi: 10.1111/jgh.13278.

Stintzing, F. C., Schieber, A., & Carle, R. (2001). Phytochemical and nutritional significance of cactus pear. *European Food Research and Technology, 212*(4), 396-407. doi: 10.1007/s002170000219.

Tesoriere, L., Butera, D., Pintaudi, A. M., Allegra, M., & Livrea, M. A. (2004). Supplementation with cactus pear (*Opuntia ficus-indica*) fruit decreases oxidative stress in healthy humans: a comparative study with vitamin C. *American Journal of Clinical Nutrition, 80*(2), 391-395. doi: 10.1093/ajcn/80.2.391.

Topete Viniegra, R. (2014). *Caracterización química y evaluación del efecto hipoglucemiante y antioxidante del fruto de garambullo (Myrtillocactus geometrizans).* [*Chemical characterization and evaluation of the hypoglycemic and antioxidant effect of garambullo fruit (Myrtillocactus geometrizans)*]. (Mastría en Ciencia y Tecnología de los Alimentos), Universidad Autónoma de Querétaro, Santiago de Querétaro, Querétaro.

Valadez-Moctezuma, E., Ortiz-Vásquez, Q., & Samah, S. (2014). Molccular based assessment of genetic diversity of xoconostle accessions (*Opuntia* spp.). *African Journal of Biotechnology, 13(2). 13*(2), 202-210. doi: 10.5897/AJB2013.13350.

Vazquez-Cruz, A. M., Jimenez-Garcia, S. N., Torres-Pacheco, I., Horacio Guzman-Maldonado, S., Guevara-Gonzalez, R. G., & Miranda-Lopez, R. (2012). Effect of maturity stage and storage on flavor compounds and sensory description of berrycactus (*Myrtillocactus geometrizans*). *Journal of Food Science, 77*(4), C366-C373. doi: 10.1111/j.1750-3841.2012.02621.x.

Viloria-Matos, A., Corbelli-Moreno, C., Moreno-Alvarez, M. J., & Belén C, D. R. (2002). Estabilidad de betalainas en pulpa de tuna (*Opuntia boldinghii* Br. et R.) sometidas a un proceso de liofilización [Stability of betalains in prickly pear pulp (*Opuntia boldinghii* Br. Et R.) subjected to a lyophilization process]. *Revista de la Facultad de Agronomía, 19*, 324-331. [*Journal of the Faculty of Agronomy*]

Zafra-Rojas, Q. Y., Cruz-Cansino, N., Ramírez-Moreno, E., Delgado-Olivares, L., Villanueva-Sánchez, J., & Alanís-García, E. (2013). Effects of ultrasound treatment in purple cactus pear (*Opuntia ficus-indica*) juice. *Ultrasonics Sonochemistry, 20*(5), 1283-1288. doi: 10.1016/j.ultsonch.2013.01.021.

In: Super and Nutraceutical Foods
Editors: Sandra Teresita Martín-del-Campo et al.
ISBN: 978-1-53619-082-3

Chapter 4

POMEGRANATE (*PUNICA GRANATUM*): BIOFUNCTIONAL COMPOUNDS AND THERAPEUTIC PROPERTIES

Ma Claudia Castañeda-Saucedo[1,*], Ernesto Tapia Campos[2,*], Jessica del Pilar Ramirez-Anaya[3], Adriana Delgado-Alvarado[4], Adriana Cavazos Garduño[5] and Julio C. Serrano-Niño[5]

[1]Departamento Ciencias del Naturaleza
[2]Unidad de Biotecnologia Vegetal, Centro de Investigación y Asistencia en Tecnología y diseño del Estado de Jalisco (CIATEJ), Guadalajara, México
[3]Departamento de Ciencias Computacionales e Innovación Tecnólogica, CUSUR-Universidad de Guadalajara, Ciudad Guzmán, México
[4]Campus Puebla, Colegio de Postgraduados, Puebla, México.
[5]Departamento de Farmacobiologia, CUCEI-Universidad de Guadalajara, Guadalajara, México

ABSTRACT

Pomegranate, *Punica granatum*, is considered a superfood worldwide. It has gained popularity all over the world as a functional nutraceutical food. Pomegranate has attracted scientific, medicinal, and nutritional interest because of its multifunctionality in human diets and its organoleptic properties. Most of the therapeutic effects of pomegranate are attributed to its primary and secondary metabolites, such as organic acids (citric, mallic, oxalic, succinic, tartaric, ascorbic), fatty acids (linoleic, palmitic and oleic acids), sugars, proteins, lipids, amino acids and minerals (Cu, Fe, Zn, Mg, P, Na, Ca, K), polyphenols, flavonoids (anthocyanins, flavonols, flavone, flavanone, flavanols) and non-flavonoid polyphenols (phenolic acids, hydrolysable tannins, etc.). The metabolites are synthesized in all parts of the fruit: peel (ellagitannins, flavonoids, anthocyanins), arils (ellagitannins. flavonoids, anthocyanins), seeds (fatty acids, lipids, punicic acid), membranous walls (mostly ellagitannins). These compounds are associated

* Corresponding Author's E-mail: claudia.saucedo@cusur.udg.mx., etapia@ciatej.mx

with a reduction of risk of contracting degenerative and cardiovascular diseases, cancer, diabetes, inflammation, ulcers, microbial infections, diarrhea, helminth infections, hemorrhages, respiratory diseases, colitis, menorrhagia, oxyuriasis, headaches, diuretics, acne, allergic dermatitis, among others. The purpose of this chapter of the book is to present a brief review of pomegranate's biofunctional compounds of primary and secondary metabolism and their importance in human health.

Keywords: polyphenols, antioxidant activity, total phenols

INTRODUCTION

Pomegranate (*Punica granatum* L.) is a perennial crop (Qin et al. 2017, 1108), native to Iran (Tavallali et al. 2020, 1). Today, it is distributed all over the world (Bar-Ya'akov et al. 2019, 1) since it can adapt to a wide variety of agroclimatic conditions (Qin et al. 2017, 1108). The FAO has declared it to be an emergent, developing crop due to its capacity to adapt to all types of soils and climates and its tolerance to drought, salinity, ferric chlorosis and active limestone. Due to the previously mentioned characteristics, the pomegranate can be grown where other crops cannot tolerate (Mira 2010, 252). The pomegranate is considered a lesser fruit because it is not consumed in large quantities as are apples, bananas, grapes, or citruses, but it is very important in a cultural sense and in its therapeutic use (Bar-Ya'akov et al. 2019, 1). Pomegranates are consumed fresh, in beverages and food products such as jams and jellies. Pomegranate extracts are used as ingredients in herbal medicines and dietary supplements (Elfalleh et al. 2011, C1). Moreover, it has currently awoken much interest in terms of science (Bar-Ya'akov et al. 2019, 1), medicine (Attanayake et al. 2018, 8761), and nutrition because of its multifunctionality in human diets (Viswanath et al. 2019, 172).

The functional properties of the fruit are attributed to metabolites of secondary metabolism (Qin et al. 2017, 1116-1118), mainly polyphenols, including ellagic acid, ellagitannins (punicalagin), punicic acid, anthocyanins, flavonols, flavan-3-ols and flavones (Pareek et al. 2015, 2362). The Secondary metabolites are compounds associated with reduction of risk of contracting diseases (Viswanath et al. 2019, 174) such as cancer, inflammation and cardiovascular and degenerative diseases (Fischer et al. 2011, 807), acidosis, dysentery, microbial infections, diarrhea, helminth infections, hemorrhages, and respiratory diseases (Viswanath et al. 2019, 172). Estrogenic compounds, estrone and estradiol have been found in the seeds. The dry pericarp and juice have been used to treat colic, colitis, menorrhagia, oxyuriasis, headaches, diuretics, acne, allergic dermatitis (Viswanath et al. 2019, 172) and Alzheimer's disease (Morzelle 2016, 15). In food, encapsulated extracts of pomegranate peel have been used as functional compounds added to cookies without affecting sensorial quality, although many phenolic compounds are degraded in the cooking process (Kaderides et al. 2020, 1). It has also been reported that pomegranate syrup acts in synergy with the antibiotic's ampicillin and cefotaxime against *Escherichia coli*, a bacterium that is resistant to these antibiotics, but the synergic action of pomegranate syrup with the antibiotics was shown to act against this pathogen (Al-Hisnawi et al. 2019, 1, 5, 6). The scientific literature undoubtedly demonstrates the great importance of the pomegranate fruit since Biblical times, and today, an infinity of applications of its compounds is reported in the cosmetic, food, and medicinal industries. However, pomegranate is a widely adapted crop mainly due to its high heterozygosis and the number of genotypes and cultivars that exist, and many of the compounds studied are highly affected by the environment. Studies report a broad variation in results. Nevertheless, this review will serve as a starting point from which knowledge of pomegranate and its compounds can continue to broaden.

IMPORTANCE OF POMEGRANATE

Worldwide, pomegranate is considered a superfood (Puneeth and Chandra 2020, 9). In recent decades, it has gained popularity throughout the world as functional and nutraceutical food (Mphahlele et al. 2014, 36; Erkan and Dogan 2020, 477). The genus *Punica* belongs to the Punicaceae family, which includes two species: *Punica granatum* L. and *P. protopunica* Balf. f. (Khadivi et al. 2020, 1). Pomegranate is a source of bioactive components that make it highly nutritive (Khadivi et al. 2020, 10). Most of the therapeutic effects of the pomegranate are attributed to its primary and secondary metabolites (Bar-Ya'akov et al. 2019, 1): sugars (glucose, fructose), organic acids (citric, malic, oxalic, succinic, tartaric), vitamin C or ascorbic acid (Hasnaoui et al. 2011, 378; Mphahlele et al. 2014, 41; Kyriacou et al. 2020, 12), fatty acids (linoleic, palmitic and oleic acids) flavonoids (El-Hadary and Ramadan 2019, 4) anthocyanins (Fischer et al. 2011, 811), phenolic acids (chlorogenic, caffeic, syringic, sinapic, *p*-coumaric, ferulic, ellagic, gallic and cinnamic acid) (Singh et al. 2018, 75).

These compounds have been reported to be present in all fruit parts. Punicic acid has been reported in seeds (Ustun-Argon 2019, 1347, 1349, 1350), as well as glutamate, arginine and aspartate, fatty acids and lipids, and membranes contain ellagitannins (Bar-Ya'akov et al. 2019, 15). The juice contains glutamine, serine, aspartate, while the peel and arils contain ellagitannins, flavonoids, and anthocyanins (Bar-Ya'akov et al. 2019, 15).

The conditions in which pomegranate is produced strongly affect fruit development and its physical and chemical characteristics (Tapia-Campos et al. 2016, 163, Mphahlele et al. 2014, 36). Likewise, accumulation of biofunctional compounds is also affected by production conditions, and therefore it is important to consider this factor so that functional quality is not lost (Zeynalova et al. 2019, 45). Recent studies report that pomegranate cultivation in dry climates maximizes the production of biochemical compounds that are beneficial to health; this is associated with the expression of genes related to the synthesis of these compounds (Attanayake et al. 2018, 8766-8767).

Primary Metabolites

Carbohydrates

The pomegranate fruit consists of peel and membranes (41-46%), arils (54-59%), and seeds (15-21%) (Figure 1) (Tapia-Campos et al. 2016, 165). The juice is 84.57% water, 14.1% carbohydrates, and 0.33% ash (Al-Maiman and Ahmad 2002, 437). Pomegranate is consumed mainly fresh or use it to prepare juice or extracts of different parts of this. Alcaraz-Mármol et al. (2017, 152) reported 15.5- 15.9 g/100 mL total sugars on average of 20 Spanish cultivars. In pomegranate from Iran, reducing sugars, fructose, and glucose were determined in concentrations of 1.72-5.81, 0.67-1.28 and 1.14-5.81 mg/100g, respectively; that is, there is four times more glucose than fructose (Nikdel et al. 2016, 277). In contrast, in pomegranate from Cyprus, Kyriacou et al. (2020, 7, 10) reported very similar quantities of glucose and fructose, 64-82 and 65-83 g/L, and very little sucrose (0.2 gr/L). Carbohydrates have also been reported in pomegranate leaves, 19.6% (Sreedevi et al. 2017, 16).

Minerals

The minerals reported in pomegranate juice are Cu, Fe, Zn, Mg, P, Na, Ca, and K, in quantities of 0.07, 2.21, 0.3, 5.13, 6.25, 72.1, 24.5, and 333 mg/100 g, respectively. The amounts of these minerals in seeds are higher, except for Cu, Fe, and Zn. The values reported for seeds are 0.04, 1.88, 1.26, 11.9, 7.49, 95.7, 59.3, 243 mg/100 g of Cu, Fe, Zn, Mg, P, Na, Ca and K, respectively (Al-Maiman and Ahmad, 2002, 440).

Figure 1. Pomegranate fruits, arils, peel, and seeds.

Organic Acids

The main organic acids reported in pomegranate fruits are 44.7% citric acid, 39.1% malic acid and 15.8% succinic acid (Kyriacou et al. 2020, 7), as well as oxalic and tartaric acids where the percentages ranged from 0.03-0.65 and 0.0-0.18%, respectively (Hasnaoui et al. 2011, 378). In seeds of pomegranate from Saudi Arabia, ascorbic acid was quantified at 0.2% (Al-Maiman and Ahmad 2002, 439), and Mphahlele et al. (2014, 41) reported in pomegranate juice to ascorbic acid in concentrations of 92.55-114 μg AAE/mL.

Proteins

There is little data on proteins in pomegranate fruits. Most studies do not report specific functions except for reserve proteins. In general, the percentage of total protein in juice is less than 1%. However, the percentage of total protein in seed varies from 4.1 to 16.9% (Bar-Ya'akov et al. 2019, 7). This was confirmed by Al-Maiman and Ahmad (2002, 439) in pomegranate from Saudi Arabia, reporting 1.05% and 4.06% total protein in juice and seed, respectively. Reserve proteins that have been reported in seeds are globulins, albumins, glutelins, and prolamins (Bar-Ya'akov et al. 2019, 7). Total proteins (11.8%) have also been reported in leaves (Sreedevi et al. 2017, 16).

Lipids and Fatty Acids in Pomegranate

Fat in pomegranate seeds has been reported at 0.15% (Al-Maiman and Ahmad 2002, 439). Oil from pomegranate seed has become more popular in the cosmetics industry because of its high content of punicic acid (omega 5) and polyunsaturated fatty acids, which make pomegranate seed oil unique. The concentration of punicic acid in pomegranate seed is 67.16-76.5%; it also contains 2.27-2.57% palmitic acid, 1.62-1.97% stearic acid, 4.2-4.72%, oleic acid, 4.72-45.84% linoleic acid and 0.24-6.8% eicosapentaenoic acid (Ustun-Argon 2019, 1350). Other acids and lipids that have been reported include caproic acid, capric acid, lauric acid, myristic acid, caprylic acid, myristoleic acid, α-linolenic acid, γ-linolenic acid, α-eleostearic acid, β-eleostearic acid, catalpic acid, arachidic acid, gadoleic acid, behenic acid, palmitoleic acid, nervonic acid, 1-O-9E,11Z,13E-octadecatrienoyl glycerol, 1-O-Isopentyl-3-O-octadec-2-enoyl glycerol, Tri-O-punicylglycerol, 11Z, 13E-trienylglycerol, Di-O-punicyl-O-octadeca-8Z, and N-palmitoyl cerebroside (Wu and Tian, 2017, 3). Lipids have also been obtained from leaves, 0.96% (Sreedevi et al. 2017, 16).

Amino Acids

In pomegranate juice from China, Li et al. (2017, 5) detected glutamate, aspartate, pyruvate, and serine. Bar-Ya'akov et al. (2019, 15) reported other amino acids in pomegranate such as alanine, asparagine, lysine, arginine, valine, phenylalanine, threonine, histidine, tyrosine, methionine, cysteine, proline, isoleucine, tryptophan, leucine, and glycine.

Secondary Metabolites

Pomegranates are rich in secondary metabolites, such as polyphenols, a group of phytochemicals (Figure 1) that are not identified as essential nutrients but have been attributed with therapeutic properties for those who consume it habitually (Vázquez-Flores 2012, 85). Polyphenols are a group of secondary metabolites synthesized by plants and are considered important from a chemical perspective because they all exhibit antioxidant properties (Olivas-Aguirre et al. 2015, 55). They constitute one of the groups of micronutrients present in the plant kingdom (Porras-Loaiza and López-Malo 2009, 121). More than 8000 compounds have been identified (Altemimi et al. 2017, 1). The secondary metabolites are important in the pomegranate's organoleptic properties, colour and astringency (Fischer et al. 2011, 808). The secondary metabolites are biosynthesized by plants in peel (Attanayake et al. 2018, 8764), juice (Kyriacou et al. 2020, 1), arils (Attanayake et al. 2018, 8764), seeds, flowers and leaves (Elfalleh 2012, 4727), and other parts (Figure 2). Structurally, they contain at least one phenol group, one aromatic ring bonded to at least one hydroxyl functional group (Vázquez-Flores 2012, 85; Olivas-Aguirre et al. 2015, 56).

Outstanding among the polyphenols are flavonoids (El-Hadary and Ramadan 2019, 4) and non-flavonoids (Vázquez-Flores 2012, 85). A highly important metabolite among the flavonoids are anthocyanins (Fischer et al. 2011, 820), as well as flavonols, flavone, flavanones, isoflavones, and flavanols. In the classification of non-flavonoids, phenolic acids, hydrolyzable tannins, benzoquinones, acetophenones, hydroxycinnamic acid, naphthoquinones, and stilbenes phenolic acids (chlorogenic, caffeic, syringic, sinapic, *p*-coumaric, ferulic, ellagic, gallic and cinnamic acid have been reported (Figure 2 y 3) (Singh et al. 2018, 75), among others. In pomegranate leaves, other metabolites from secondary metabolism have been reported, such as alkaloids (3.5%) (Sreedevi et al. 2017, 16). Because of all these compounds present in pomegranate, it can be considered a source of natural antioxidants for the development of functional products (Alcaraz-Mármol et al. 2017, 152, 159).

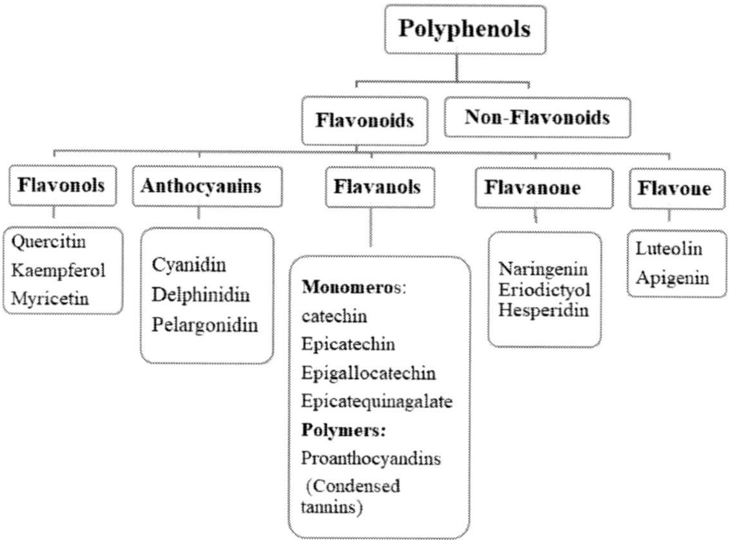

Figure 2. Flavonoids compounds in pomegranate.

Table 1. Total phenolic compounds content (TPC) in pomegranate fruit from different countries

Country	Tissue	TPC	Source
Sri Lanka	Peel	28.54-36.39 mgGAE/g	Attanayake et al. 2018, 8764
Mexico	Peel	226.3-303.3 mgGAE/g	Gómez 2018, 33
Egypt	Peel	188.9 ± 0.11 mgGAE/g	El-Hadary and Ramadan 2019, 4
Tunisia	Peel	53.6-85.6 mgGAE/g	Elfalleh 2012, 4727
Germany	Peel	101856 mg/kg	Fischer et al. 2011, 817
Iran	Peel	276-413 mgGAE/g	Derakhshan et al. 2018, 109
Cyprus	Juice	384-896 mg/L	Kyriacou et al. 2020, 11
Germany	Juice	20152-5186 mg/L	Fischer et al. 2011, 817
Iran	Juice	1.8-6.4 mgGAE/100 mL	Nikdel et al. 2016, 283
Africa	Juice	159-367 mgGAE/100 mL	Mphahlele et al. 2014, 40
Italy	Juice	424-742 mgGAE/100 mL	Stefano et al. 2019, 1040
Iran	Juice	12.4-23.8 mgGAE/g	Derakhshan et al. 2018, 109
Sri Lanka	Arils	4.69-6.37 mgGAE/g	Attanayake et al. 2018, 8764
Iran	Pomace	281-291 mgGAE/g	Tavallali et al. 2020, 8
Tunisia	Leaves	9.85-14.78 mgGAE/g	Elfalleh 2012, 4727
India	Leaves	117-394 mg CGE/g	Sreedevi et al. 2017, 16
Tunisia	Flowers	42.7-66.29 mgGAE/g	Elfalleh 2012, 4727
Saudi Arabia	Seeds	1.90-3.65 mgGAE/g	Al-Maiman and Ahmad 2002, 439
Tunisia	Seeds	7.94-11.84 mgGAE/g	Elfalleh 2012, 4727
Iran	Seeds	72-73 mgGAE/g	Derakhshan et al. 2018, 109

GAE = Gallic acid equivalent; CGE = Chlorogenic acid equivalent.

There is a strong interest in investigating polyphenols because of their antioxidant properties, their participation in the sensorial processes of food, and their applications to benefit human health (Porras-Loaiza and López-Malo 2009, 121). The total phenol content in pomegranate has been documented in the peel of pomegranate from Sri Lanka with 28.54-36.39 mgGAE/g (Attanayake et al. 2018, 8764), Mexico with 226.3-303.3 mgGAE/g (Gómez 2018, 33), Egypt with 188.9 ± 0.11 mgGAE/g (El-Hadary and Ramadan 2019, 1), Tunisia with 53.6-85.6 mgGAE/g (Elfalleh 2012, 4727), Germany with 101856 mg/kg (Fischer et al. 2011, 817) and Iran with 276-413 mgGAE/g (Derakhshan et al. 2018, 109). They have also been reported in arils (4.69-6.37 mgGAE/g) of pomegranate from Sri Lanka (Attanayake et al. 2018, 8764), in leaves of pomegranate from Tunisia (Elfalleh 2012, 4727) and India (Sreedevi et al. 2017, 16), and in flowers, phenol content has been reported to be 42.7-66.29 mgGAE/g (Elfalleh 2012, 4727) (Table 1).

Flavonoids

Flavonoids act as plugs, capturing free radicals to generate flavinic radicals, which are much less reactive (Martín 2018, 94). They are compounds of low molecular weight that share a common diphenyl-pyran (C_6-C_3-C_6) skeleton composed of two phenyl rings (A and B) linked by a C ring of heterocyclic pyran. The main subgroups of flavonoid compounds are flavonols, flavones, flavanones (dihydroflavones), isoflavones, anthocyanidins, and flavanols (Quiñones 2012, 79). They constitute a large group of chemical substances considered to be plant secondary metabolites (Porras-Loaiza and López-Malo 2009, 121). Recent studies have documented total flavonoid contents of pomegranate in several parts of the world. In Iran, the total flavonoid content was reported to be 14.2 and 85.1 mg quercetin equivalent/g extracted with 80% methanol and water at 60°C, respectively (Tavallali et al. 2020, 7). In Africa, concentrations of 97.1-458.4 mg catechin equivalent/100 mL of juice were found (Mphahlele et al. 2014, 40). Tunisia has reported total flavonoids of 3.30-6.79, 12.77-26.08, 21.45-72.45, and 21.03-51.52 mg rutin equivalent/g dry weight in seeds, leaves, flowers, and peel, respectively (Elfalleh 2012, 4727). Also, in pomegranate from Iran, total flavonoids ranged from 36-54, 7-38, and 1.8-8.7 mg rutin equivalent/g in peel, seed and juice, respectively (Derakhshan et al. 2018, 109).

In peel, individually, the following flavonoids have been reported: apigenin, rhamnetin, acacetin 7-O-rutinoside, hesperetin, naringenin, acacetin 7 neo hesperside, quercetin, laem-3-(2-p-comaroyl) glucoside, quercetin, apigenin 7-O- glucoside, apigenin.7-O-neohespiroside, kaempferol, kamp.3,7-di rhamoside, quercetrin-3-O-glucoside, hesperidin rutin, luteo-7-glucoside, naringin, apigenin-6-rhamnose 8-galactoside and apigenin-6-arbinose 8-galactoside (El-Hadary and Ramadan 2019, 5), as well as dihydroquercetin (taxifolin) (Mphahlele et al. 2014, 39).

Anthocyanins

Anthocyanins are a relatively small group of pigments in the flavonoid family (Pervaiz et al. 2017, 1). They are derived from phenols (Bar-Ya'akov et al. 2019, 2), are soluble in water and are responsible for the bright red color of the juice (Elfalleh et al. 2011, C1) and the antioxidant capacity of the pomegranate (Pervaiz et al. 2017, 1; Kyriacou et al. 2020, 7, 13). It is abundant in pomegranate fruits, flower tissues, and leaves (Bar-Ya'akov et al. 2019, 11).

Table 2. Anthocyanins reported in different pomegranate (*P. granatum*) fruit parts

Anthocyanins	peel	aril	juice	Reference
Cyanidin	″◊	″◊	″◊	″Bar-Ya'akov et al. 2019, 15 ; ◊Yaranga et al. 2019, 35
Cyanidin 3,5-diglucoside	x	*	°+x∞	+Fischer et al. 2011, 819 ; ∞Kyriacou et al. 2020, 7 °Stefano et al. 2019, 1041 ; xZeynalova et al. 2019, 44 ; *Russo et al. 2018, 3513
Cyanidin pentoside-hexoside			+	+Fischer et al. 2011, 819
Cyanidin 3-diglucoside		*	+*	+Fischer et al. 2011, 819 ; *Russo et al. 2018, 3513
Cyanidin pentoside			+	+Fischer et al. 2011, 819
Cyanidin-3-O-glucoside			°	°Stefano et al. 2019, 1041
Cyanidin -3,5-glucoside	x		x	xZeynalova et al. 2019, 44
Cyanidin 3-rutinoside	+		+	+Fischer et al. 2011, 819
Delphinidin		″	″	″Bar-Ya'akov et al. 2019, 15
Delphinidin 3,5-diglucoside	x	*	°x +*	+Fischer et al. 2011, 819 ; xZeynalova et al. 2019, 44 °Stefano et al. 2019, 1041 ; *Russo et al. 2018, 3513
Delphinidin-3,5-glucoside		*	°x *	°Stefano et al. 2019, 1041 xZeynalova et al. 2019, 44 *Russo et al. 2018, 3513
Delphinidin 3-diglucoside			+	+Fischer et al. 2011, 819
Pelargonidin	″	″	″	″Bar-Ya'akov et al. 2019, 15
Pelargonidin 3,5-diglucoside	x	*	°+ x *	+Fischer et al. 2011, 819 ; °Stefano et al. 2019, 1041 ; xZeynalova et al. 2019, 44 ; *Russo et al. 2018, 3513
Pelargonidin 3,5-glucoside	x	*	+ x *	x Zeynalova et al. 2019, 44 ; *Russo et al. 2018, 3513
Pelargonidin 3-diglucoside	+		+	+Fischer et al. 2011, 819
Pelargonidin-3-O-glucoside			°	°Stefano et al. 2019, 1041

Anthocyanin synthesis is regulated by edaphoclimatic factors, low temperatures, high precipitation and availability of soil moisture. A high quantity of organic matter is associated with an increase in anthocyanins in pomegranate fruit development (Attanayake et al. 2018, 8769). Anthocyanins have been found in the peel, juice, and aril. The identified anthocyanins in the fruit included cyanidin (red), delphinidin (purple) pelargonidin (orange) and their derivatives (Table 2) (Fischer et al. 2011, 819); Russo et al. 2018, 3513; Kyriacou et al. 2020, 7). In pomegranate from Tunisia, total anthocyanins in peel and juice were reported at 73.9 and 39.19 mg cyanidin-3-glucoside equivalent/g dry weight, respectively (Elfalleh et al. 2011, C3) and in seeds, leaves, flowers 19.62-40.84, 40.91-89.81, 80.20-168.91 mg cyanidin-3-glucoside equivalent/g dry weight, respectively (Elfalleh 2012, 4727). Pomegranate from Cyprus present concentrations of 1.2-191 mg cyanidin-3-glucoside equivalent/L in juice (Kyriacou et al. 2020, 12). Also, Yaranga et al. (2019, 35) reported 0.13359, 0.6345, 9.6186 mg cyanidin/L in juice, seeds, and peel, respectively.

Recently, a study of pomegranate juice conducted in Italy reported 34.68–97.64 mg/L cyanidin-3,5-O-diglucoside, 9.92-40.29 mg/L pelargonidin-3,5-O-diglucoside, 6.28-26.00 mg/L cyanidin-3-O-glucoside, 4.99-20.67 mg/L delphinidin-3,5-O-diglucoside, 1.75-8.84 mg/L pelargonidin-3-O-glucoside, and 0.63-2.51 mg/L delphinidin-3-O-diglucoside (Stefano et al. 2019, 1041). Likewise, in pomegranate from the Azerbaijan region delphinidin-3,5-diglucoside, cyanidin-3,5-diglucoside, delphinidin 3-glucoside, pelargonidin-

3,5-diglucoside, cyanidin-3-glucoside, and pelargonidin-3-glucoside, in percentages of 20.8, 34.65, 13.19, 1.67, 27.48 and 1.82% and 3.85, 14.27, 2.97, 25.9, 18.79 and 32.9%, respectively, were reported in juice and exocarp, respectively. Cyanidin-3,5-diglucoside makes up the highest percentage of anthocyanins in juice, while in the exocarp, it is pelargonidin-3-glucoside (Zeynalova et al. 2019, 40). The red purple color of the juice is attributed the anthocyanins like delphinidin, cyanidin, and pelargonidin 3-glucosides and 3,5-diglucosides. Also, the anthocyanins are responsible for with properties related to elimination of free radicals and are markers of juice quality control (Stefano et al. 2019, 1044).

Flavonols

Flavonols are characterized by a keto group in the C_4 carbon and an unsaturation between the C_2 and C_3 carbons. Moreover, it possesses an additional hydroxyl group in the C_3 carbon. A study conducted in Iran reported 25-45 mg rutin/g total flavonols in pomegranate peel, 3.4-22 mg rutin/g in seed, and 1.5-2 mg rutin/g in juice (Derakhshan et al. 2018, 109). The most abundant flavonoid is quercetin, the most representative of flavonols (Quiñones, 2012, 78, 79). Yaranga et al. (2019, 34) reported quercetin concentrations in pomegranate of 12.11% in juice, 15.5% in seed, and 95.87% in the peel. In the same way, a study in China reported that the concentrations of quercetin, kaempferol, and myricetin vary depending on the stage of fruit development and of the cultivar, reporting kaempferol in concentrations of 31.98 to 137.15 mg/100 g fresh weight (fw), quercetin 33.25-52.24 mg/100 g fw, while myricetin was 10-185.27 mg/100 g (Zhao et al. 2014, 1654, 1655). Also, in pomegranate from Egypt kaempferol was reported in concentrations of 0.106 mg/g in peel, while in peel the concentration was 0.106 mg g^{-1} (El-Hadary and Ramadan 2019, 5). In contrast, in juice of pomegranate from Africa was reported kaempferol in concentrations of 0.32-4.05 mg/L (Mphahlele et al. 2014, 39). Also, in juice of pomegranate from Italy was reported kaempferol in concentrations of 36.0 mg/L (Stefano et al. 2019, 1043). In pomegranate from the regions of Azerbaijan, quercetin was reported in concentrations of 0.44 mg/kg in exocarp and 3.05 mg/kg in juice (Zeynalova et al. 2019, 44). Likewise, in pomegranate from Italy 0.82-10.35 mg/L quercetin 3-O- hexoside was reported (Stefano et al. 2019, 1043). It has been reported that quercetin can chelate transition metallic ions, such as iron or copper, thus preventing the formation of reactive species of oxygen (Martín 2018, 94).

Flavone

Flavones possess a keto group in the C_4 carbon and an unsaturation between the C_2 and C_3 carbons. They are the least abundant flavonoids in foods (Quiñones, 2012, 79). Flavones that have been reported are apigenin (El-Hadary and Ramadan 2019, 5) and luteolin (Zhao et al. 2014, 1649, 1652-3). A study conducted in pomegranate from China reported that the level of luteolin (3.5-8 mg/100 g fw) was higher than that of apigenin (2.5-4.5 mg/100 g fw) in all of the fruit development stages of the evaluated cultivars (Zhao et al. 2014, 1654).

Flavanone

Flavanones are analogs of flavones, with a saturated C ring. They glycosylate mainly by attaching with a disaccharide in the C_7 carbon. They constitute a minor group in edible parts. Up to five times more is found in membranes than in edible parts (Quiñones 2012, 79). The flavonone reported in pomegranate are naringenin, eriodictyol and hesperidin. The naringenin have been found in concentrations of 0.089 mg/ g^{-1} in peel of pomegranate (El-Hadary and Ramadan 2019, 5). Eriodyctiol 7-o-ßglucoside was reported in concentration of 1.77-3.48 mg/L in juice of pomegranate from Africa (Mphahlele et al. 2014, 39), as well as hesperidin 0.19-0.55 mg/g in the pomace (Tavallali et al. 2020, 4).

Flavanols

Flavanols possess a saturated C ring and a hydroxyl group in the C_3 carbon. They can appear as monomers or polymers with different degrees of polymerization (Quiñones 2012). The main dimers and polymers are condensed or non-hydrolyzable tannins; the monomers are catechin, epicatechin, epigallocatechin y epigallocatechin gallate. Condensed tannins are represented by procyanidins, proanthocyanidins, prodelphinidins, and profisetinidins (Olivas-Aguirre et al. 2015, 56, 57). Proanthocyanidins are between 0.1-7% of the total polyphenols in pomegranate. They are composed of a very rich mixture of flavan-3-ols: catechin, epicatechin, gallocatechin, epigallocatechin, epigallocatechin gallate, and epicatechin gallate. A study conducted in Spain reported different concentrations of proanthocyanidins in peel (1840.5 µg/g), seed (316.2 µg/g), membranes (51.4 µg/g) and the edible portion (arils) (34.4 µg/g) (Díaz-Mula et al. 2019, 9160). Several authors have reported catechin in the juice of pomegranate from Africa (15.22-29.01 mg/L) (Mphahlele et al. 2014, 39), Iran (1.33-3.57 mg/g) (Tavallali et al. 2020, 4) and Azerbaijan (0.22 mg/kg) (Zeynalova et al. 2019, 44). Also, catechin has been reported in pomegranate peel from Egypt (3.275 mg/g) (El-Hadary and Ramadan 2019, 5) and the region of Azerbaijan (2.04 mg/kg) (Zeynalova et al. 2019, 44). Mphahlele et al. (2014, 39) reported 8.82-28.72 mg/L epicatechin in the juice of pomegranate from Africa, and Zeynalova et al. (2019, 44) reported 67.23 mg/kg in juice and 2.39 mg/kg in exocarp of pomegranate from Azerbaijan.

Non-Flavonoid Polyphenols

Non-flavonoid polyphenols include the simplest of polyphenol molecules, such as phenolic acids, with chemical structures of six carbons (C_6) bonded or not to skeletons of two or up to four carbons (C_6-C_4). The most complex are those that in their skeleton have their C_6 portion attached to C_2 portions and another C_6 ring, such as in the case of stilbenes, gallotannins, or ellagitannins (Vázquez-Flores 2012, 85). The subgroups into which non-flavonoids in pomegranate are divided are simple phenols and hydrolyzable tannins (Figure 3) (Vázquez-Flores 2012, 85).

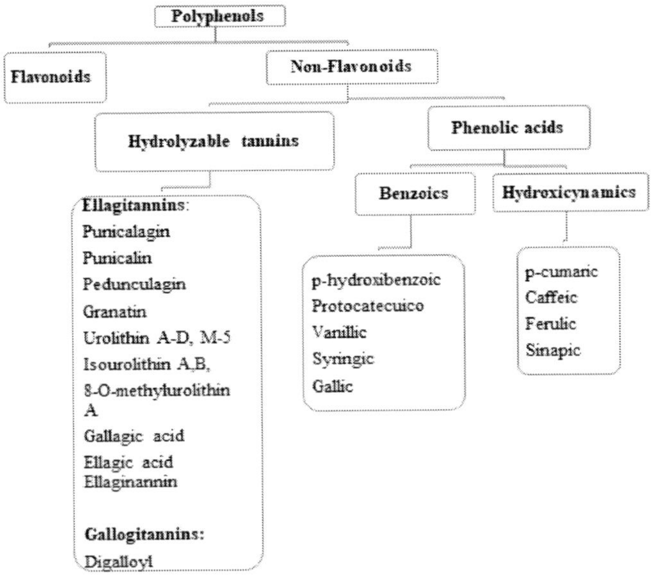

Figure 3. Non-Flavonoids compounds in pomegranate.

Phenolic Acids

Phenolic acids are divided into two subgroups, benzoics acid and hydroxycinnamic acid and derivatives.

Table 3. Phenolic acids reported in pomegranate

Acid	Country	Tissue	Quantity	Source
Gallic	Tunisian	Peel	123.79 ± 9.56 mg/100 g	Elfalleh et al. 2011, C4
Gallic	Tunisian	Pulp	17.50 ± 2.62 mg/100 g	Elfalleh et al. 2011, C4
Gallic	Egyptian	Peel	2.5 mg/g	El-Hadary and Ramadan 2019, 5
Gallic	Iran	Juice	11.25 mg/g	Tavallali et al. 2020, 4
Gallic	Africa	Juice	17.9-39.3 mg/L	Mphahlele et al. 2014, 40
Gallic	Italy	Juice	2.47-53.2 mg/L	Stefano et al. 2019, 1043
Ellagic	Tunisian	Peel	35.89 ± 2.42 mg/100 g	Elfalleh et al. 2011, C4
Ellagic	Tunisian	Pulp	23.10 ± 0.44 mg/100 g	Elfalleh et al. 2011, C4
Ellagic	Egyptian	Peel	12.561 mg/g	El-Hadary and Ramadan 2019, 5
Ellagic	Iran	Juice	0.68 mg/g	Tavallali et al. 2020, 4
Caffeic	Tunisian	Peel	20.56 ± 1.54 mg/100 g	Elfalleh et al. 2011, C4
Caffeic	Tunisian	Pulp	13.52 ± 0.72 mg/100 g	Elfalleh et al. 2011, C4
Caffeic	Egyptian	Peel	0.458 mg/g	El-Hadary and Ramadan 2019, 5
p-Coumaric	Tunisian	Peel	4.48 ± 0.39 mg/100 g	Elfalleh et al. 2011, C4
p-Coumaric	Tunisian	Pulp	11.45 ± 0.80 mg/100 g	Elfalleh et al. 2011, C4
p-Coumaric	Egyptian	Peel	0.086 mg/g	El-Hadary and Ramadan 2019, 5
p-Coumaric	Iran	Pomace	0.40 mg/g	Tavallali et al. 2020, 4
Chlorogenic	Iran	Pomace	0.07-0.12 mg/g	Tavallali et al. 2020, 4
Chlorogenic	Egyptian	Peel	1.562 mg/g	El-Hadary and Ramadan 2019, 5
Ferulic	Egyptian	Peel	0.492 mg/g	El-Hadary and Ramadan 2019, 5
Iso-ferulic	Egyptian	Peel	0.117 mg/g	El-Hadary and Ramadan, 2019, 5
α-coumaric	Egyptian	Peel	0.268 mg/g	El-Hadary and Ramadan 2019, 5
Coumarin	Egyptian	Peel	0.912 mg/g	El-Hadary and Ramadan 2019, 5
Cinnamic	Egyptian	Peel	2.50 mg/g	El-Hadary and Ramadan 2019, 5
Protocatechuic	Africa	Juice	1.55-3.13 mg/L	Mphahlele et al. 2014, 40
Vanillic acid	Italy	Juice	44.4-124 mg/L	Stefano et al. 2019, 1043.

Benzoic acids include p-hydroxybenzoic, protocatechuic, vanillic, syringic, and gallic acids. In the hydroxycinnamic acid group is *p*-coumaric, caffeic, ferulic, and sinapic acids. Several authors have reported phenolic acids in pomegranate peel, pulp, juice, and pomace. Other phenolic acids reported in pomegranate are gallic, ellagic, caffeic, *p*-coumaric, chlorogenic, ferulic, iso-ferulic, α-coumaric, cinnamic, and protocatechuic acids. These compounds have been found in pomegranate peel, pulp, and pomace (Table 3) (Elfalleh et al. 2011, C4; Mphahlele et al. 2014, 40; El-Hadary and Ramadan 2019, 5; Tavallali et al. 2020, 4). Gallic, ellagic and caffeic acids are found in a larger proportion in the peel than in the pulp or arils (Elfalleh et al. 2011, C4). The variability of data among authors is wide. However, this could be due to the methods of extraction, quantification, production system, and production environment.

Hydrolyzable Tannins

Tannins are polyphenolic compounds defined as secondary metabolites derived from plants. Tannins are classified into four groups: condensed tannins or non-hydrosoluble tannins called proanthocyanidins (flavonoid origin), hydrolysable tannins (of non-flavonoid origin), phlorotannins (derived from algae and coffee) and complex tannins (Olivas-Aguirre et al. 2015, 56). Pomegranates are rich in hydrosoluble tannins (Wu and Tian 2017, 2).

Hydrolyzed tannins are units of gallic or ellagic acid joined to carbohydrates (Vázquez-Flores 2012, 86). They have an important role in defense against herbivores and in promoting benefits to human health (Ono et al. 2016, 1). Hydrolyzed tannins are the most studied phytochemicals in pomegranate. They can be grouped into ellagitannins and gallotannins (Wu and Tian 2017, 3). Gallotannins and ellagitannins are derived from the esterification of non-flavonoid polyphenolic compounds, such as gallic acid and ellagic acid, respectively (Vázquez-Flores 2012, 86). Gallic acid is a precursor of ellagic acid and punicalagin, which have been identified in pomegranate tissues (Qin et al. 2017, 1116-1118). More than 60 ellagitannins have been identified by mass spectrometry and nuclear magnetic resonance (Wu and Tian 2017, 3).

Pomegranate is a rich source of many phenolic compounds, including ellagitannins (punicalagin, punicalin, pedunculagin, granatin, gallagic acid, ellagic acid, ellagitannin, urolithin A-D, M-5, isourolithin A and B, 8-O-methylurolithin A) and gallotannins (digalloyl) (Wu and Tian 2017, 2). Other authors have reported ellagitannins in peel (71.9 mg/g), membranes (56.1 mg/g), seeds (4.4 mg/g), arils (2.7 mg/g) and juice (2.5 g/L), 97%, 99.9%, 69, 35 and 19%, respectively, of the total phenolic compounds (Díaz-Mula et al. 2019, 9165, 9166). Punicalagin is the polyphenol with the highest molecular weight and it is the most abundant, making up 80-85% of the all the polyphenols extracted from the peel, while gallic acid is 1.3% (Seeram et al. 2005a, 49). Pedunculagin and granadin have also been reported in pomegranate juice in concentrations of 11.41-26.65 and 1.24-13.65 mg/L, respectively (Stefano et al. 2019, 1042). Recent studies have quantified gallotannins as digalloyl hexoside in concentrations of 6.09-40.88 mg/L pomegranate juice (Stefano et al. 2019, 1042).

Ellagitannins exhibit potent antioxidant, antiatherosclerotic and anti-cancer activities (Seeram et al. 2005a, 49). Moreover, urolothin, a compound derived from ellagitannins, has shown the potential to prevent breast cancers that are sensitive to estrogen (Adams et al. 2010, 108). A study that compared the effect of pomegranate juice to isolated compounds, such as total polyphenols, punicalagin and ellagic acid, found greater bioactivity in juice, illustrating the synergic action of all the compounds contained in the juice (Adams et al.

2010, 112). At concentrations of 100 μg/mL, ellagic acid and punicalagin, and total phenols produced apoptosis of HT-29 cancerous colon cells (Seeram et al. 2005b, 360). Another study reported that punicalagin prevents cardiac metabolic disorders induced by a diet high in fats (Cao et al. 2015, 1).

Antioxidant Activity

Antioxidants are a group of substances, such as vitamins, minerals, enzymes, natural pigments, and phenolic compounds (Salinas-Flores et al. 2019, 34).

Table 4. Antioxidant activity in peel, juice, aril, seed, leaf, and flower of pomegranate from different parts of the world quantified with the methods FRAP, ORAC, DPPH, and DPPH

Place	Tissue	Method	Antioxidant activity	Reference
Germany	Peel	TEAC	1361.9TE mM/kg	Fischer et al. 2011, 817
Germany	Peel	FRAP	589.1TE mM/kg	Fischer et al. 2011, 817
Tunisian	Peel	FRAP	25.63 ± 3.67TE mM/100 g	Elfalleh et al. 201, C5
Tunisian	Pee l	ORAC	22.08 ± 2.73TE mM/100 g	Elfalleh et al. 2011, C5
Sri Lanka	Peel	DPPH	22.96-45.12%	Attanayake et al. 2018, 8764
Egypt	Peel	ABTS	90.92%	El-Hadary and Ramadan 2019, 4
Egypt	Peel	DPPH	93.97%	El-Hadary and Ramadan 2019, 4
Tunisian	Peel	DPPH	3.9-11.5(EC50) μg/mL	Elfalleh et al. 2012, 4728
Tunisian	Peel	FRAP	3.8-7.5TE mM/100 g	Elfalleh et al. 2012, 4728
Iran	Peel	--	45-58%	Derakhshan et al. 2018, 109
Cyprus	Juice	DPPH	4.1-8.0 AE mM	Kyriacou et al. 2020, 12
Germany	Juice	TEAC	21.7-76.4 TE mM/L	Fischer et al. 2011, 817
Germany	Juice	FRAP	9.1-31.5 TE mM/L	Fischer et al. 2011, 817
Africa	Juice	DPPH	690.8-775 μM AE/mL	Mphahlele et al. 2014, 42
Tunisian	Juice	FRAP	7.24 ± 1.24 TE mM/L	Elfalleh et al. 2011, C5
Iran	Juice	DPPH	79.54-86.77%	Nikdel et al. 2016, 283
Italy	Juice	ABTS	36.73-221.5 μM TE/100 mL	Stefano et al. 2019, 1038
Tunisian	Juice	ORAC	7.00 ± 1.52 TE mM/L	Elfalleh et al. 2011, C5
Sri Lanka	Aril	DPPH	14.60-25.11%	Attanayake et al. 2018, 8764
European U.	Arils	ABTS	1.55-4.24TE mM/kg	Alcaraz-Mármol et al. 2017, 157
European U.	Arils	DPPH	6.39-7.74TE mM/ kg	Alcaraz-Mármol et al. 2017, 157
Tunisian	Oil	FRAP	8.17 ± 1.90 TE mM/100 g	Elfalleh et al. 2011, C5
Tunisian	Oil	ORAC	6.43 ± 2.03 TE mM/100 g	Elfalleh et al. 2011, C5
Turkey	Seed oil	DPPH	15.2-49.01%	Ustun-Argon 2019, 1350
Iran	Pomace	DPPH	200-250 μg/mL	Tavallali et al. 2020, 8
Tunisian	Seed	DPPH	21-45 EC50 μg/mL	Elfalleh et al. 2012, 4728
Tunisian	Seed	FRAP	0.76-1.1 TE mM/100 g	Elfalleh et al. 2012, 4728
Iran	Peel		26-54%	Derakhshan et al. 2018, 109
Tunisian	Leaves	DPPH	11.4-26.6 EC50 μg/mL	Elfalleh et al. 2012, 4728
Tunisian	Leaves	FRAP	1.81-4.2 TE mM/100 g	Elfalleh et al. 2012, 4728
Tunisian	Flowers	DPPH	4.5-13.9 EC50 μg/mL	Elfalleh et al. 2012, 4728
Tunisian	Flowers	FRAP	4.1-6.4 TE mM/100 g	Elfalleh et al. 2012, 4728
Italy	Juice	ABTS	36.7-221.5 μmol TE/100 mL	Stefano et al. 2019, 1038
Iran	Juice		9-10%	Derakhshan et al. 2018, 109

AE = Ascorbate equivalent, TE = Trolox equivalent.

Antioxidant activity refers to the ability to inhibit oxidative degradation due to its ability to react with free radicals (Londoño 2012, 129) by donating hydrogen atoms, electrons, or metallic cations that act as chelating agents. A free radical is a chemical species that contains an unpaired electron, making it highly reactive. Once the free radical subtracts what it needs, the stable molecule that ceded the electron becomes a free radical and initiates a chain reaction that destroys cells, greatly damaging molecules, cell membranes, and tissues (Avello and Suwalsky 2006, 162).

The antioxidants are obtained from fruits or agro-industrial residues that are capable of reacting with free radicals to prevent cell damage, and pomegranate is an excellent option because of its antioxidant capacity (Fischer et al. 2011, 807. 817; Elfalleh et al. 2011, C5-C6; Mphahlele et al. 2014, 42; Nikdel et al. 2016, 283; Alcaraz-Mármol et al. 2017, 157; Attanayake et al. 2018, 8764; Derakhshan et al. 2018, 109; El-Hadary and Ramadan 2019, 4,8; Stefano et al. 2019, 1041; Ustun-Argon 2019, 1350; Kyriacou et al. 2020, 12; Tavallali et al. 2020, 1, 8) mainly due to it high polyphenol contents (Adams et al. 2010, 108), specifically punicalagins, punicalins, gallagic acid, and ellagic acid (Johanningsmeier and Harris 2011, 181). Pomegranate juice has been reported to possess a 3-fold higher antioxidant activity than that of red wine or green tea (Gil et al. 2000, 4581), and 2-, 6- and 8-fold higher levels than those detected in grape/cranberry, grapefruit, and orange juice, respectively.

Numerous studies have documented pomegranate's antioxidant activity using different methods in juice, peel, seeds, leaves, seed oil, pomace, and flowers grown in different parts of the world (Table 4). Alcaraz-Mármol et al. (2017, 157) reported antioxidant activity in DPPH values of 6.39- 7.74 mM TE/kg fresh weight, while in ABTS values, it was 1.55 a 4.24 mM TE/kg. Elfalleh et al. (2011, C5) also quantified antioxidant activity in pomegranate peel, finding 25.63 ± 3.67 mM 100g and 22.08 ± 2.73TE mM/100 g by FRAP and ORAC, respectively. Antioxidant activity is generally higher in the peel than in juice or seed (Derakhshan et al. 2018, 109).

Therapeutic Uses of Pomegranate

Pomegranate has been used therapeutically since ancestral times. Today, it has gained much popularity as a functional and nutraceutical food (Johanningsmeier and Harris, 2011, 181) due to its antioxidant capacity (Derakhshan et al. 2018, 109; El-Hadary and Ramadan 2019, 4; Kyriacou et al. 2020, 12), reflected in inhibition of oxidative stress caused by free radicals (Londoño 2012, 129). Many disorders are related to oxidative stress, which is an unusually high level of oxidation that can cause damage to vital biomolecules, including proteins and DNA, and increase the risk of disease (Johanningsmeier and Harris 2011, 183). For this reason, compounds contained in pomegranate are seen as an option in the prevention of chronic diseases (Johanningsmeier and Harris, 2011, 189), metabolic disorders Ahmed et al. 2019), cancer (Jha et al. 2018; Toda et al. 2020), and diabetes (Tang et al. 2018, 1), among others.

Cancer

Cancer has become one of the most lethal diseases worldwide (Bassiri-Jahromi, 2018, 1). Therapeutic use of extracts from pomegranate leaves, peel, seed oil, juices, and arils concerning cancer has been widely documented (Table 5). Studies related to prevention of cancer in the colon (Waly et al. 2012, 4051), liver (Costantini et al. 2014, 8644), breast (Kim et al. 2002, 203; Dikmen et al. 2011, 1638; Ganeshkumar et al. 2013, 208; Jha et al. 2018;

Pinnamaneni, 2020, 303), prostate (Deng et al. 2017, 976), lung (Li et al. 2016, 227; Toda et al. 2020, 757), cervix, and kidney and nasopharyngeal carcinoma (Yusefi et al. 2020) have been extensively documented.

The therapeutic effect in pomegranate is due to phenolic compounds that prevent the proliferation of cancer stem cells (Dai et al. 2010, 1087), inhibit cell growth (Kim et al. 2002, 203; Ganeshkumar et al. 2013, 208), reduce oxidative stress (Waly et al. 2012, 4050), suppress microsomal prostaglandin E synthase-1 expression (Toda et al. 2020, 757), block endogenous active estrogen biosynthesis (Kim et al. 2002, 203), decrease viability of cancer cells (Pinnamaneni 2020, 303), induce apoptosis (Li et al. 2016, 227; Deng et al. 2017, 976; Toda et al. 2020, 757), and impair metastasis (Deng et al. 2017, 976).

Table 5. Effects of pomegranate metabolites on cancer

Disease	Method	Tissue	Reference	Observations
Colon cancer	*In vivo*	fruit peel extract	Waly et al. 2012, 4050	Reduction of oxidative stress and improving the redox status of colonic cells
Colon, liver, breast and prostate cancer	*In vitro*	Seed oil	Costantini et al. 2014, 8644	Anti-inflammatory effects
Prostate cancer	*In vitro*	Peel extract	Deng et al. 2017, 976	Apoptosis and impairs metastasis
Lung cancer	*In vitro*	Leaves extract	Li et al. 2016, 227	Induces apoptosis
Mouse mammary gland cancer	*In vitro*	Whole fruit	Dai et al. 2010, 1087	Prevents proliferation of cancer stem cells
Lung cancer	*In vitro*	Ellagitannins granatin a and granatin b isolated from pomegranate leaves	Toda et al. 2020, 757	Suppresses microsomal prostaglandin E synthase-1 expression, apoptosis
Colon, breast, cervical, lung, kidney cancer and nasopharyngeal carcinoma (NPC)	Iron Oxide Nanoparticles *in vitro*	Fruit peel extract	Yusefi et al. 2020	Anti-cancer therapeutic agents only against NPC cells
Breast and colon cancer	*In vitro*	Peel extract	Pinnamaneni 2020, 303	Decreased viability of cancer cells
Breast cancer	*In vivo*	Fruit peel extract	Dikmen et al. 2011, 1638	Induces apoptosis
Breast cancer	Gold nano particles *in vitro*	Fruit peel extract	Ganeshkumar et al. 2013, 208	Cell growth inhibition
Breast cancer	Nano particles *in vitro*	Arils	Jha et al. 2018	
Breast cancer	*In vitro*	Fermented juice, pericarp extract, seed oil	Kim et al. 2002, 203	Blocks active endogenous estrogen biosynthesis, Cell growth inhibition

Metabolic and Cardiovascular Disease

Pomegranate extracts have also been tested in treating cardiovascular and metabolic disorders, acne, and as an antibacterial, among others (Table 6) (Lei et al. 2007, 1023; Hossin 2009, 1251; Karodi et al. 2013, 160; Haghighian et al. 2016, 676; Durgadevi et al. 2018, 173; Khabeer et al. 2019; Rao and Krishnamurthy 2019, 673; Ahmed et al. 2019; Dițescu et al. 2019, 722).

Metabolic syndrome is a complex metabolic disorder caused by an accumulation of factors that include hypertension dyslipidemia, resistance to insulin, obesity, and glucose intolerance (Hashemi 2017, 227), all of which are a heavy load for the public health system. For this reason, several authors have conducted research with pomegranate extracts. One study with rats that had metabolic syndrome tested the ingestion of 500 and 750 mg of extract in tablet form with 40% ellagic acid and 40% polyphenols/kg body weight. The study found a reduction in total cholesterol (TC), triacylglycerol (TG), high-density lipoprotein cholesterol (HDL-C), low-density lipoprotein-cholesterol, and uric acid (Hashemi 2017, 227).

Table 6. Metabolic and cardiovascular disorders treated experimentally with pomegranate *in vivo* and *in vitro*

Disease	Method	Tissue	Reference	Observations
Obesity	*In vivo*, rats	Peel powder and extract	Hossin, 2009, 1251	Lipid peroxidation, antiatherogenic effects
Dyslipidemia, cardiovascular risk.	*In vivo*, women	peel extract	Haghighian et al. 2016, 676	Decrease in Systolic Blood Pressure (SBP) levels and hs-CRP, energy intake, and weight changes.
Obesity	*In vivo*, mice	leaf extract	Lei et al. 2007, 1023	Decrease in body weight, TC, TG, glucose levels and TC/HDL-C ratio.
Obesity	*In vitro*, human	Lipase inhibitor SHAMstat3pg from pomegranate seed oil	Khabeer et al. 2019	Reduce the differentiation of preadipocytes to adipocytes, anti-obesity gene expression.
Liver dysfunction	*In vivo*, rats	Leaf extract	Rao and Krishnamurthy 2019, 673	Liver alterations were restored more effectively using P. granatum leaf extract plus gallic acid
Metabolic syndrome	*In vitro*, enzyme inhibition	Rind methanol extract	Ahmed et al. 2019	Inhibition of metabolism-related enzymes: α-amylase, tyrosinase, and hyaluronidase
Cardiovascular disease	*In vitro*, human vascular endothelial cell	Seed oil and peel extract	Dițescu et al. 2019, 722	Antioxidant effect in vascular endothelium level, pro-inflammatory and pro-oxidative conditions
Acne	*In vitro*	Powder fruit rind extract	Karodi et al. 2013, 160	Effective anti acne
Antibacterial	*In vitro*	Leaves extract	Durgadevi et al. 2018, 173	Aspergillus niger, Candida albicans and Trichoderma viride were prevented by Punica granatum extract

Other studies on obesity also found a decrease in body weight, TC, TG, glucose and TC/HDL-C after five weeks of treatment (Lei et al. 2007, 1023), as well as a reduction in differentiation of preadipocytes to adipocytes, anti-obesity gene expression (Khabeer et al.

2019) and a reduction in α-amylase, tyrosinase and hyaluronidase, which are enzymes related to metabolism (Ahmed et al. 2019).

Antimicrobial Activity of Pomegranate

The popularity of pomegranate has increased tremendously, especially in the last decade because of the proved antimicrobial, anti-viral, anti-cancer, potent anti-oxidant, and anti-mutagenic effects of the fruit (Çam and Hısıl 2010, 878). Pomegranate juice have shown to have antimicrobial activity against diverse pathogenic bacterias such as *Escherichia coli*, *Salmonella enteritidis*, *Staphylococcus aureus*, *Bacillus subtilis*, *Klebsiella pneumoniae*, *Pseudomonas aeruginosa*, *Yersinia enterocolitica*, *Cryptococcus neoformans*, *Staphylococcus aureus*, *Listeria monocytogenes* and *Clostridia species*, yeast and molds such as *Candida albicans* and *Aspergillus fumigatus*. The resistance of microorganisms against the antibiotics commonly used to treat oral infections, the increasing number of oral pathologies, and the lack of medications without side effects require identifying new effective strategies against oral pathogens, on this regard, fresh pomegranate juice have shown antimicrobial activity in clinical isolates of multi drug-resistant *Staphylococcus epidermidis* strains isolated from ocular infections. A minimum inhibitory concentration (MIC) of 20% was observed for fresh pomegranate juice, which does not appear to indicate a strong antimicrobial effect. However, it should be taken into account that undiluted FPJ contains over 80% water. Therefore, a higher volume is necessary to produce a strong antimicrobial effect, unless concentrate or extracts are obtained (Betanzos-Cabrera et al. 2015, 27620, 2602, 186).

The antimicrobial activity has been attributed to antioxidant polyphenols, including ellagitannins (hydrolyzable tannins) and anthocyanins (condensed tannins) contained in pomegranate juice (Cowan 1999, 564). Phenolic contents vary among different cultivars of fruits and vegetables, as well as within different tissues, in this sense, antimicrobial activity of pomegranate juice has been evaluated among different pomegranate cultivars, novel Greek pomegranate cultivars "Persephone" and "Porphiroyeneti," compared with the most widespread "Wonderful" cultivar. All tested juices displayed good antibacterial activity with minimal inhibitory concentration ranging from 0.05 to 0.20 mg/mL and a minimal bactericidal concentration (MBC) in the range of 0.10-0.40 mg/mL. It should be mentioned that the best activity was obtained against L. monocytogenes, whereas *M. flavus* and *E. coli* were the most resistant to the tested samples. The highest antibacterial activity was observed for "Persephone," whereas "Wonderful" showed the lowest activity.

Similar to antibacterial activity results, it was observed that all samples possessed good antifungal activity against tested fungi with MIC values between 0.03 and 0.40 mg/mL and MFC between 0.05 and 0.80 mg/mL. T. viride was the most sensitive microfungi tested, whereas A. fumigatus was the most resistant species for the tested samples.

The highest antifungal activity was observed for "Wonderful", whereas "Persephone" showed the lowest activity. All the examined samples exhibited equal or higher antifungal activity than commercial antifungals such as bifonazole and ketoconazole, apart from A. *fumigatus*, where lower activities were observed (Lantzouraki et al. 2015).

In accordance with the above-mentioned results, previous studies showed that pomegranate juices and fruit peels had significant antimicrobial activity, which is due to the ellagitannins and flavonoids present in these and that the differences among antimicrobial and antifungal activities may be due to the differences in antimicrobial and antifungal compounds contents in cultivars evaluated (Altieri et al. 2019, 186).

It has been reported that pomegranate juices exhibited notable antibacterial and antifungal activity due to the ellagitannins and flavonoids present in it. Several compounds responsible for the antimicrobial action of each pomegranate beverage, depending on their abundance. One example is tannins (abundant in pomegranate), which are considered to be toxic to microorganisms (Viuda-Martos et al. 2010, 635). Their hydrophilic part interacts with the polar region of the membrane, whereas the hydrophobic part is immersed in the non-polar inner region of the bacterial membrane, which can cause instability of the membrane, thus affecting the transport of substrates into the cell (Cristani et al. 2007, 6300).

Another suggestion is that phenolic compounds cause toxicity through reactions with sulfhydryl groups or through more non-specific interactions with proteins leading to loss of function (Naz et al. 2007, 341).

On the other hand, phenols may also render substrates unavailable to microorganisms or interfere with bacterial protein secretions (Seeram et al. 2008, 360). Additionally, high-performance liquid chromatography analyses have shown that the major phenolic compounds in pomegranate are gallic acid and ellagic acid, in addition to punicalagin as a major ellagitannin. Thus, it is reasonable to consider that these compounds may be responsible for the antimicrobial activity, as suggested by Ahn et al. (1998, 439). However, authors agree that it might be possible that other unidentified compounds contribute to this activity.

Based on antimicrobial activity results, pomegranate juice could provide an alternative natural antibacterial treatment, although further studies should be conducted on a wide variety of bacteria and direct relationships established between antimicrobial activity and polyphenols.

Pomegranate peel represents about 40–50% of the total fruit weight it is produced as byproduct in huge amounts by the food industry, which could represent an important source of bioactive compounds, including hydrolysable tannins, include hexahydroxydiphenic acid and its derivatives, ellagic acid and its derivatives, punicalin, and punicalagin (Li et al. 2006, 254; Çam and Hısıl 2010, 878; Fischer et al. 2011, 807; Hasnaoui et al. 2014, 196; Gullon et al. 2016, 94).

Pomegranate peels have been traditionally used as natural remedies against infectious diseases in different countries and, for various diseases, for example, in traditional Chinese medicine, peels are used as an astringent and anti-inflammatory agent and applied in the treatment of infections, traumatic hemorrhage and diarrhea and infection of the gastrointestinal tract. In countries like India, Tunisia, and Guatemala, water decoction of dried pomegranate peels is employed as a cure for aphthae and ulcers (Lansky and Newman 2006, 177).

Pomegranate peels have been found to have effect against many pathogenic bacterial species, including inhibition of formation of biofilms (Meléndez and Capriles 2006, 272; Al-Zoreky 2009, 244; Dell´Agli et al. 2009, 279; Bakkiyaraj et al. 2013, 929), antiplasmodial activity, and effects against Entamoeba histolytica and Giardia lamblia, part of these properties have been attributed to the high content of tannins, such as anthocyanins and polyphenolic compounds mainly punicalagin. The main hypothesis is that the hydrophilic part of the tannin chemical structure interacts with the polar regions of the membrane, whereas the hydrophobic part is immersed in a non-polar inner section of the bacterial membrane; this may induce membrane instability, thus affecting the transport of substrates into the cell, another hypothesis is that tannins forms complex with enzymes changes their structural conformation, thereby inhibiting enzymatic activity. The formation of complexes with cell wall proteins decreases cell wall permeability and reduces the transport of substrates into the cell. In addition, tannins decrease metal ion availability to bacteria when forming stable complexes with these metal ions. Subsequently, metal depletion may adversely affect the activity of metalloenzymes in microbial cells (Goel et al. 2005, 497). Otherwise, it has been suggested that phenolic toxicity via reactions with sulfhydryl groups or through more non-specific interactions with proteins leading to a loss of function in microbial cells (Cristani et al. 2007, 6300; Naz et al. 2007, 341).

Antimicrobial activity of pomegranate peels seems to be specific of extraction method, in different researches water and ethanol extracts of pomegranate fruit peels showed anti-Aeromonas caviae activity with more effectiveness in water extracted samples, in contrast only water-methanol extract of peels have marked inhibition and the water extract was inactive against eleven microorganisms tested, such as *S. aureus* (2 strains), *B. subtilis, E. coli, Listeria monocytogenes, Pseudomonas aeruginosa, Klebsiella pneumoniae, Yersinia enterocolitica, Candida utilis, Saccharomyces cerevisiae*, and *Aspergillus*.

Differences in the antimicrobial activity of peel among different studies made, could be partially explained by variations in phenolic content in the peel and a sensitivity of assessed bacteria to these compounds in the different studies made it.

Pomegranate has showed to have effect inhibited communication between two bacterial strains, this kind of communication is knowing as Quorum sensing, Quorum sensing is an intercellular signaling mechanism used by bacteria to communicate as a colony about critical survival issues such as availability of nutrients, defense against other microorganisms, virulence, and biofilm formation. Bacteria produce and detect signaling molecules important for pathogenic bacteria during infection of a host to coordinate their virulence to escape the immune response of the host and establish a successful infection. In this order, pomegranate inhibited quorum-sensing signals in two bacterial strains, interfering with purple pigment production and bacterial swarming motility in Chromobacterium violaceum and P. aeruginosa, respectively (Zahin et al. 2010, 1219; Koh and Tham 2011, 144). The inhibition of these particular processes may be due to direct or indirect interference on quorum-sensing by pomegranate polyphenols or the interactive effect of different compounds present in the extract.

CONCLUSION

Today, pomegranate has gained much popularity as an important source of bioactive compounds, such as flavonoids (anthocyanins, flavonols, flavone, flavanone, flavanols) and non-flavonoids (phenolic acids, hydrolysable tannins, etc.) that are present in the fruit, peel, arils, seed oil, and leaves. Antioxidant activity is attributed to the polyphenols, which have been related to treatment or prevention of metabolic and cardiovascular disorders, cancer, and acne. Pomegranate has also exhibited anti-bacterial and anti-fungal properties. In sum, it has great potential for therapeutic use to benefit human health. However, it is necessary to conduct more in-depth studies on the therapeutic effects of pomegranate.

REFERENCES

Adams, Lynn S., Yanjun Zhang, Navindra P. Seeram, David Heber, and Dhiuan Chen. 2010. "Pomegranate Ellagitannin–Derived Compounds Exhibit Antiproliferative and Antiaromatase Activity in Breast Cancer Cells *in vitro*." *Cancer Prevention Research*, *3*(1), 108-113. https://doi.org/10.1158/1940-6207.CAPR-08-0225.

Ahmed, H. Mohamed, Hala M. Aldesouki and Farid A. Badria. 2019. "Effect of phenolic compounds from the rind of *Punica granatum* on the activity of three metabolism-related enzymes." *Biotechnology and Applied Biochemistry*. Accessed: March 17, 2020. doi:10.1002/bab.1866.

Ahn, Y. J., C. O. Lee, J. H. Kweon, J. W. Ahn, and J. H. Park. 1998. "Growth inhibitory effects of Galla Rhois-derived tannins on intestinal bacteria." *Journal of Applied Microbiology*, 84: 439_43.

Alcaraz-Mármol, Francisco, Nallely Nuncio-Jáuregui, Francisco García-Sánchez, Juan José Martínez-Nicolás, and Francisca Hernández. 2017. "Characterization of twenty pomegranates (*Punica granatum* L.) cultivars grown in Spain: Aptitudes for fresh consumption and processing." *Scientia Horticulturae 219*, 152-160. https://doi.org/10.1016/j.scienta.2017.03.008.

Al-Hisnawi, Ali A., Jassim M. Mustafa, and Yass K. Yasser. 2019. "The antimicrobial activity synergism between eucalyptus honey, pomegranate, date and antibiotics on Escherichia coli causing diarrhea in children." In AIP Conference Proceedings. 2144(1):040003, 1-9. AIP Publishing LLC. https://doi.org/10.1063/1.5123104.

Al-Maiman, Salah A., and Dilshad Ahmad. 2002. "Changes in physical and chemical properties during pomegranate (*Punica granatum* L.) fruit maturation." *Food Chemistry 76*(4), 437-441. https://doi.org/10.1016/S0308-8146(01)00301-6.

Altieri, Fabio, Francesco Cairone, Flavia Giamogante, Simone Carradori, Marcello Locatelli, Silvia Chichiarelli and Stefania Cesa. 2019. "Influence of Ellagitannins Extracted by Pomegranate Fruit on Disulfide Isomerase PDIA3 Activity." *Nutrients* 11(1), 186.

Al-Zoreky, N. S. 2009. "Antimicrobial activity of pomegranate (*Punica granatum* L.) fruit." *International of Journal of Food Microbiology* 134:244–248.

Altemimi, Ammar, Naoufal Lakhssassi, Azam Baharlouei, Dennis G. Watson and David A. Lightfoot. (2017). "Phytochemicals: Extraction, isolation, and identification of bioactive compounds from plant extracts." *Plants* 6(4), 42.

Attanayake, Rekha, Rasu Eeswaran, Ranil Rajapaksha, Palitha Weerakkody, Pradeepa C. G. Bandaranayake. 2018. "Biochemical composition and expression of anthocyanin biosynthetic genes of a yellow peeled and pinkish ariled pomegranate (*Punica granatum* L.) cultivar are differentially regulated in response to agro-climatic conditions." *Journal of Agricultural and Food Chemistry 66(33):8761-8771.* https://doi.org/ 10.1021/acs.jafc.8b02909.

Avello, Marcia, and Mario Suwalsky. 2006. "Radicales libres, antioxidantes naturales y mecanismos de protección." *Atenea (Concepción) 494*, 161-172. https://doi.org/ 10.4067/S0718-04622006000200010.

Bakkiyaraj, Dhamodharan, Janarthanam Rathna Nandhini, Balakumar Malathy and Shunmugiah Karutha Pandian. 2013. "The antibiofilm potential of pomegranate (*Punica granatum* L.) extract against human bacterial and fungal pathogens" *Biofouling* 29:929-937.

Bar-Ya'akov, Irit, Li Tian, Rachel Amir, and Doron Holland. 2019. "Primary Metabolites, Anthocyanins, and Hydrolyzable Tannins in the Pomegranate Fruit." *Frontiers in Plant Science* 10(620). Accessed February 20, 2020. https://doi.org/10.3389/fpls.2019.00620.

Bassiri-Jahromi, Shahindokht. 2018. "*Punica granatum* (pomegranate) activity in health promotion and cancer prevention." *Oncology reviews* 12(345):1-7.

Betanzos-Cabrera, Gabriel, Perla Y. Montes-Rubio, Héctor E. Fabela-Illescas, Helen Belefant-Miller and Juan C. Cancino-Diaz. 2015. "Antibacterial activity of fresh pomegranate juice against clinical strains of *Staphylococcus epidermidis*." *Food & nutrition research* 59(1):27620.

Çam, Mustafa, and Yasar Hışıl Y. 2010. "Pressurised water extraction of polyphenols from pomegranate peels." *Food Chemistry* 123(3):878-885.

Cao, Ke, Jie Xu, Wenjun Pu, Zhizhong Dong, Lei Sun, Weijin Zang, Feng Gao, Yong Zhang, Zhihui Feng and Jiankang Liu. 2015. "Punicalagin, an active component in pomegranate, ameliorates cardiac mitochondrial impairment in obese rats via AMPK activation." *Scientific Reports* 5(14014):1-12. https://doi.org/10.1038/srep14014.

Costantini, Susan; Rusolo, Fabiola; De Vito, Valentina; Moccia, Stefania; Picariello, Gianluca; Capone, Francesca; Guerriero, Eliana; Castello, Giuseppe; Volpe, Maria G. 2014. "Potential Anti-Inflammatory Effects of the Hydrophilic Fraction of Pomegranate (*Punica granatum* L.) Seed Oil on Breast Cancer Cell Lines." *Molecules* 19, no. 6: 8644-660. https://doi.org/10.3390/molecules19068644.

Cowan, Marjorie Murphy. 1999. "Plant products as antimicrobial agents." *Clinical microbiology reviews 12*(4):564-582.

Cristani, Mariateresa, Manuela D'arrigo, Giuseppina Mandalari, Francesco Castelli, Maria Grazia Sarpietro, Dorotea Micieli, Vincenza Venuti, Giuseppe Bisignano, Antonella Saija and Domenico Trombetta. 2007. "Interaction of four monoterpenes contained in essential oils with model membranes: implications for their antibacterial activity." *Journal of Agricultural and Food Chemistry* 55:6300-6308.

Dai, Zhaoli, Vidhya Nair, Marun Khan, and Henry P. Ciolino. 2010. "Pomegranate extract inhibits the proliferation and viability of MMTV-Wnt-1 mouse mammary cancer stem cells *in vitro*." *Oncology Reports* 24(4):1087-1091. https://doi.org/10.3892/or_00000959.

Dell'Agli, Mario, Germana V. Galli, Yolanda Corbett, Donatella Taramelli, Leonardo Lucantoni, Annette Habluetzel, Omar Maschi, Donatella Caruso, Flavio Giavarini, Sergio Romeo, Deepak Bhattacharya, Enrica Bosisio. 2009. "Antiplasmodial activity of *Punica granatum* L. fruit rind." *Journal of Ethnopharmacology* 125:279-285.

Deng, Yuanle, Yali Li, Fangfang Yang, Anqui Zeng, Shubuping Yang, Yi Luo, Yiwen Zhang, Yongmei Xie, Tinghong Ye, Yong Xia, and Wenya Yin. 2017. "The extract from *Punica granatum* (pomegranate) peel induces apoptosis and impairs metastasis in prostate cancer cells." *Biomedicine & pharmacotherapy* 93:976-984. https://doi.org/10.1016/j.biopha.2017.07.008.

Derakhshan, Zahra, Margherita Ferrante, Marzieh Tadi, Farnoosh Ansari, ali Heydari, Motahreh Sadat Hosseini, Gea Oliveri Conti and Elham khalili Sadrabad. 2018. "Antioxidant activity and total phenolic content of ethanolic extract of pomegranate peels, juice and seeds." *Food and Chemical Toxicology* 114:108-111.

Díaz-Mula, Huertas M., Francisco A. Tomás-Barberán, and Rocío García-Villalba. 2019. "Pomegranate fruit and juice (cv. Mollar), rich in ellagitannins and anthocyanins, also provide a significant content of a wide range of proanthocyanidins." *Journal of Agricultural and Food Chemistry* 67(33)9160-167.

Dikmen, Miris, Nilgun Ozturk and Yusuf Ozturk. 2011. "The antioxidant potency of *Punica granatum* L. fruit peel reduces cell proliferation and induces apoptosis on breast cancer." *Journal of Medicinal Food* 14(12):1638-1646.

Dițescu, Cristina, Narcisa Băbeanu, Laura Olariu, Brindusa Dumitriu, Raluca Papacocea, Ovidiu Popa, and Bruno Stefan Velescu. 2019. "Antioxidant effects on vascular endothelial cells induced by active components of *Punica granatum* and amaranthus sp. Extracts." *Farmacia* 67(4):722-728. doi:10.31925/farmacia.2019.4.22.

Durgadevi, K., J. Antonia Auxi Marueen, R. Gowri, and V. Ramamurthy. 2018. "A study of phytochemical screening and antibacterial activity of leaves extract of *Punica granatum." The Pharma Innovation Journal* 7(11): 173-177.

Elfalleh, Walid, Hédia Hannachi, Nizar Tlili, Yassine Yahia, Nizar Nasri and Ali Ferchichi. 2012. "Total phenolic contents and antioxidant activities of pomegranate peel, seed, leaf and flower." *Journal of Medicinal Plants Research 6*(32):4724-4730. https://doi.org/10.5897/JMPR11.995.

Elfalleh, Walid, Nizar Tlili, Nizar Nasri, Yassine Yahia, Hédia Hannachi, Nizar Chaira, Ma Ying, and Ali Ferchichi. 2011. "Antioxidant capacities of phenolic compounds and tocopherols from tunisian pomegranate (*Punica granatum*) fruits." *Journal of Food Science 76*(5):C1-C7. https://doi.org/10.1111/j.1750-3841.2011.02179.x.

El-Hadary, Abdalla E. and Mohamed Fawzy Ramadan. 2019. "Phenolic profiles, antihyperglycemic, antihyperlipidemic, and antioxidant properties of pomegranate (*Punica granatum*) peel extract." *Journal of Food Biochemistry 43*(4), e12803. https://doi.org/10.1111/jfbc.12803.

Erkan Mustafa, and Adem Dogan. (2020). Subtropical fruits: Pomegranates. In "*Controlled and Modified Atmospheres for Fresh and Fresh-Cut Produce."* (Academic press, pp: 477-486. Elsevier Inc. https://doi.org/10.1016/B978-0-12-804599-2.00036-3.

Fischer, Urike A., Reinhold Carle, and Dietmar R. Kammere. 2011. "Identification and quantification of phenolic compounds from pomegranate (*Punica granatum* L.) peel, mesocarp, aril and differently produced juices by HPLC-DAD-ESI/MSn." *Food Chemistry* 127(2):807-821. https://doi.org/10.1016/j.foodchem.2010.12.156.

Ganeshkumar, Moorthy, Muniram Sathishkumar, Thangavel Ponrasu, Murugan Girija Dinesh, and Lonchin Suguna. 2013. "Spontaneous ultra-fast synthesis of gold nanoparticles using *Punica granatum* for cancer targeted drug delivery." *Colloids and Surfaces B: Biointerfaces* 106: 208-216.

Gil, Maria I., Francisco A. Toma´s-Barberán, Betty Hess-Pierce, Deirdre M. Holcroft, and Adel A. Kader. 2000. "Antioxidant activity of pomegranate juice and its relationship with phenolic composition and processing." *Journal of Agricultural and Food chemistry* 48(10):4581-4589.

Goel, Gunjan, A. K. Puniya, C. N. Aguilar, and K. Singh 2005. "Interaction of gut microflora with tannins in feeds." *Naturwissenschaften* 92:497-503.

Gómez, Castillo José francisco. 2018. *Análisis comparativo del perfil polifenólico en cáscara de Punica granatum cultivada en Guanajuato y Oaxaca*. Tesis para obtener el grado de: Maestro en ciencias productos naturales y alimentos. Universidad Tecnológica de la

Mixteca. 60p. http://192.100.170.40:8080/bitstream/123456789/94/1/2018-MCPNA-JFGC.pdf. [*Comparative analysis of the polyphenolic profile in Punica granatum husk cultivated in Guanajuato and Oaxaca*. Thesis to obtain the degree of: Master of Science in Natural Products and Food. Technological University of the Mixteca.]

Gullon, Beatriz, Manuela E. Pintado, José A. Pérez-Álvarez, and Manuel Viuda-Martos. 2016. "Assessment of polyphenolic profile and antibacterial activity of pomegranate peel (*Punica granatum*) flour obtained from co-product of juice extraction." *Food Control* 59:94-98.

Haghighian, Mahdiyeh Khadem, Maryam Rafraf, Abdolvahab Moghaddam, Salar Hemmati, Mohammad Asghari Jafarabadi, Bahram Pourgha. 2016. "Pomegranate (*Punica granatum* L.) peel hydro alcoholic extract ameliorates cardiovascular risk factors in obese women with dyslipidemia: A double blind, randomized, placebo-controlled pilot study." *European Journal of Integrative Medicine* 8(5):676-682. https://doi.org/10.1016/j.eujim.2016.06.010.

Hashemi, Jamilah M. 2017. "Efficacy of pomergranate (*Punica granatum*) extract on experimentally metabolic syndrome rat with high fructose diet." *International Journal of Pharmaceutical Research & Allied Sciences* 6(2):227-237.

Hasnaoui, Nejib, Messaoud Mars, Sana Ghaffari, Mokhtar Trific, Pablo Melgarejo, Francisca Hernandez. 2011. "Seed and juice characterization of pomegranate fruits grown in Tunisia: Comparison between sour and sweet cultivars revealed interesting properties for prospective industrial applications." *Industrial Crops and Products* 33(2):374-381. https://doi.org/10.1016/j.indcrop.2010.11.006.

Hasnaoui, Nejib, Bernand Wathelet, and Ana Jiménez-Araujo. 2014. "Valorization of pomegranate peel from 12 cultivars: Dietary fibre composition, antioxidant capacity and functional properties." *Food Chemistry* 160:196-203.

Hossin, Fatma Labib Ahmed. 2009. "Effect of pomegranate (*Punica granatum*) Peel and it´s extracto n obede hypercholestoremic rats." *Pakistan Journal of Nutrition* 8(8):1251-1257.

Jha, Babita, Mugdha Rao, A. Chattopadhyay, A. Bandyopadhyay and K. Prasad, and Anal K. Jha. 2018. "*Punica granatum* fabricated paltinum nanoparticles: A therapeutic pil for breas cancer." In: *AIP Conference Proceedings* 1953(1):030087. https//doi.org/10.1063/1.5032422.

Johanningsmeier, D. Suzanne and G. Keith Harris. 2011. Pomegranate as a functional food and nutraceutical source. *Annual review of food science and technology* 2:181-201.

Kaderides, kyriakos, Ioannis Mourtzinos, and Athanasia M Goula. 2020. "Stability of pomegranate peel polyphenols encapsulated in orange juice industry by-product and their incorporation in cookies." *Food Chemistry* 310(125849):1-8. https://doi.org/10.1016/j.foodchem.2019.125849.

Karodi, S. Revan, C. B. Mahendrakumar, and Kiran Bhise. 2013. "Evaluation of anti-acne activity of hydroalcoholic extract of *Punica granatum* Linn." *Journal of pharmacognosy and phytotherapy* 5(9):160-163. Accessed: march 18, 2020. DOI:10.5897/JPP12.044.

Khabeer, Trichur Shamsiya, Akila Prashant, Manonmani Haravey Krishnan. 2019. "Dietary fatty acids from pomegranate seeds (*Punica granatum*) inhibit adipogenesis and impact the expression of the obesity-associated mRNA transcripts in human adipose-derived mesenchymal stem cells." *J Food Biochem* 43(3): e12739. Accessed: March 17, 2020. https://doi.org/10.1111/jfbc.12739.

Khadivi, Ali, Farhad Mirheidari, Younes Moradi, and Simin Paryan. 2020. "Morphological variability of wild pomegranate (*Punica granatum* L.) accessions from natural habitats in the Northern parts of Iran." *Scientia Horticulturae 264*(109165):1-11. https://doi.org/10.1016/j.scienta.2019.109165.

Kim, Nam Deuk, Rajendra Mehta, Weiping Yu, Ishak Neeman, Talia Livney, Akiva Amichay, Donald Poirier, Paul Nicholls, Andrew Kirby, Wenguo Jiang, Robert Mansel, Cheppail Ramachandran, Thangaiyan Rabi, Boris Kaplan, and Ephraim Lansky. 2002. "Chemopreventive and adjuvant therapeutic potential of pomegranate (*Punica granatum*) for human breast cancer." *Breast cancer research and treatment* 71:203-217. https://doi.org/10.1023/A:1014405730585.

Koh, Khee Hoon, and Fonge-Yee tham. 2011. "Screening of traditional Chinese medicinal plants for quorum-sensing inhibitors activity." *Journal of Microbiology, Immunology and Infection* 44(2):144-148.

Kyriacou, Marios C., Sotiroula Ioannidou, Nikolaos Nikoloudakis, Nicos Seraphides, Lambros C. Papayiannis, Angelos C. Kyratzis. 2020. "Physicochemical characterization and trait stability in a genetically diverse ex situ collection of pomegranates (*Punica granatum* L.) germplasm from Cyprus." *Scientia Horticulturae*, *263*(109116):1-14. https://doi.org/10.1016/j.scienta.2019.109116.

Lansky, Ephrain P., and Robert A. Newman. 2006. "*Punica granatum* (pomegranate) and its potential for prevention and treatment of inflammation and cancer." *Journal of Ethnopharmacology* 109: 177-206.

Lantzouraki, Damitra Z., Vassilia J. Sinanoglou, Panagiotis G. Zoumpoulakis, Jasmina Glamoclija, Ana Ciri, Marina Sokovic, George Heropoulos and Charalampos Proestos. 2015. "Antiradical–antimicrobial activity and phenolic profile of pomegranate (*Punica granatum* L.) juices from different cultivars: a comparative study." *Rsc Advances* *5*(4):2602-2614.

Lei, F., X. N. Zhang, W. Wang, D. M. Xing, W.D. Xie, H. Su, and L. J. Du. 2007. Evidence of anti-obesity effects of the pomegranate leaf extract in high-fat diet induced obese mice." *International Journal of Obesity* 31(6): 1023-1029. https://doi.org/10.1038/sj.ijo.0803502.

Li, Yingcan, Pan Gu, Liwei Wang, Shaoyang Wang, Hangyu Yang, Bolin Zhang, Baoqing Zhu, and chao Ma. 2017. "Comparison of amino acid profile in the juice of six pomegranate cultivars from two cultivation regions in China." *Journal of Food Processing and Preservation* 41(5):1-10, e13197. https://doi.org/10.1111/jfpp.13197

Li, Yunfeng, Changjiang Guo, Jijun Yang, Jingyu Wei, Jing Xu, Shuang Cheng. 2006. "Evaluation of antioxidant properties of pomegranate peel extract in comparison with pomegranate pulp extract." *Food Chemistry* 96:254-260.

Li, Yali, Fangfang Yang, Weidong Zheng, Mingxing Hu, Juanxiu Wang, Sisi Ma, Yuanle Deng, Yi Luo, Tinghong Ye, Wenya Yin. 2016. "*Punica granatum* (pomegranate) leaves extract induces apoptosis through mitochondrial intrinsic pathway and inhibits migration and invasion in non-small cell lung cancer *in vitro*." *Biomedicine & pharmacotherapy* 80:227-235. Accessed: January 20, 2020. https://doi.org/10.1016/j.biopha.2016.03.023.

Londoño, Londoño Julia. 2012. Antioxidantes: importancia biológica y métodos para medir su actividad. In *Desarrollo y Transversalidad serie Lasallista Investigación y Ciencia*. 129-162. Corporación Universitaria Lasallista. [Antioxidants: biological importance and methods to measure their activity. In *Development and Transversality series Lasallista Research and Science*.]

Martín, Gordo Dario. Alonso. 2018. "Los compuestos fenólicos, un acercamiento a su biosíntesis, síntesis y actividad biológica." *Revista de Investigación Agraria y Ambiental* 9(1):82-104. ["Phenolic compounds, an approach to their biosynthesis, synthesis and biological activity." *Journal of Agricultural and Environmental Research*]

Meléndez, P. A., and V. A. Capriles 2006. "Antibacterial properties of tropical plants from Puerto Rico." *Phytomedicine* 13:272-276.

Mira, Santiago. 2010. "La Granada: economía y comercialización." In: *el Granado*, editado por Pablo Melgarejo Moreno, Francisca Hernández García, y Pilar Legua Murcia. I Jornadas Nacionales sobre el granado, Universidad Miguel Hernández de Elche. 271p ["La Granada: economy and commercialization."]

Morzelle, Maressa Caldeira. 2016. *Efeito neuroprotetor da casca de romã (Punica granatum)*. PhD diss., Universidade de São Paulo. https://teses.usp.br/teses/disponiveis/11/11141/tde-30092016-182359/publico/Maressa_Caldeira_Morzelle_versao_revisada.pdf. [*Neuroprotective effect of pomegranate peel (Punica granatum)*.]

Mphahlele, Rebogile R., Marietjie A. Stander, Olaniyi A. Fawole, and Umezuruike Linus Opara. 2014. "Effect of fruit maturity and growing location on the postharvest contents of flavonoids, phenolic acids, vitamin C and antioxidant activity of pomegranate juice (cv. Wonderful)." *Scientia Horticulturae 179*, 36-45. https://doi.org/10.1016/j.scienta.2014.09.007

Naz, S., R. Siddiqi, S. Ahmad, S. A. Rasool, and S. A. Sayeed. 2007. "Antibacterial activity directed isolation of compounds from *Punica granatum*." *Journal of Food Science* 72: 341-5.

Nikdel, Kosar, Esmaiel Seifi, Hamed Babaie, Mehdi Sharifani, and Khodayar Hemmati. 2016. "Physicochemical properties and antioxidant activities of five Iranian pomegranate cultivars (*Punica granatum* L.) in maturation stage." *Acta Agriculturae Slovenica* 107(2), 277-286. https://doi.org/10.14720/aas.2016.107.2.02.

Olivas-Aguirre, Francisco Javier, Abraham Wall-Medrano, Gustavo A. González-Aguilar, Jose Alberto López-Díaz, Emilio Alvarez-Parilla, Laura A. De la Rosa, and Arnulfo Ramos-Jimenez. 2015. "Taninos hidrolizables; bioquímica, aspectos nutricionales y analíticos y efectos en la salud." *Nutrición hospitalaria 31*(1):55-66. https://doi.org/10.3305/nh.2015.31.1.7699. ["Hydrolyzable tannins; biochemistry, nutritional and analytical aspects and effects on health." *Hospital nutrition*]

Ono, Nadia N., Xiaoqiong Qin, Alexander E. Wilson, Gang Li, Li Tian. 2016. "Two UGT84 family glycosyltransferases catalyze a critical reaction of hydrolyzable tannin biosynthesis in pomegranate (*Punica granatum*)." *PLoS ONE 11*(5):1-25 e0156319. https://doi.org/10.1371/journal.pone.0156319.

Pareek, Sunil, Daniel Valero and María Serrano. 2015. "Postharvest biology and technology of pomegranate." *Journal of the Science of Food and Agriculture 95*(12):2360-2379. DOI 10.1002/jsfa.7069.

Pervaiz, Tariq, Jiu Songtao, Faezeh Faghihi, Muhammad Salman Haider and Jinggui Fang. 2017. "Naturally occurring anthocyanin, structure, functions and biosynthetic pathway in fruit plants." *Journal of Plant Biochemistry & Physiology* 5(2):1000187: 1-9 https://doi.org/10.4172/2329-9029.1000187.

Pinnamaneni, Rajasekhar. 2020. "Cell viability studies and anti-cancerous activity evaluation of pomegranate (*Punica granatum* L) extract." *Research J. Pharm. and Tech* 13(1):303-307. doi:10.5958/0974-360X.2020.00061. X.

Porras-Loaiza, A. P., and A López-Malo. 2009. "Importancia de los grupos fenólicos en los alimentos." *Temas Selectos de Ingenieria de Alimentos 3*(1), 121-134. ["Importance of phenolic groups in food." *Selected Food Engineering Topics*]

Puneeth, H. R., and Sharath Chandra, S. P. 2020. "A review on potential therapeutic properties of pomegranate (*Punica granatum* L.)." *Plant Science Today 7*(1):9-16. https://doi.org/10.14719/pst.2020.7.1.619.

Qin, Gaihua, Chunyan Xu, Ray Ming, Haibao Tang, Romain Guyot, Elena M. Kramer, Yudong Hu, Xingkai Yi, Yongjie Qi, Xiangyang Xu, Zhenghui Gao, Haifa Pan, Jianbo Jian, Yinping Tian, Zhen Yue, and Yiliu Xu. 2017. "The pomegranate (*Punica granatum* L.) genome and the genomics of punicalagin biosynthesis." *The Plant Journal 91*(6):1108-1128. https://doi.org/10.1111/tpj.13625.

Quiñones, M., M. Miguel and A. Aleixander. 2012. "Los polifenoles, compuestos de origen natural con efectos saludables sobre el sistema cardiovascular." *Nutricion Hospitalaria* 27(1):76-89. https://doi.org/10.3305/nh.2012.27.1.5418. ["Polyphenols, compounds of natural origin with healthy effects on the cardiovascular system." *Hospital Nutrition*]

Rao, Sreedevi Prabhakar and Vijayalakshmi Krishnamurthy. 2019. "Ethanol extract of *Punica granatum* L. leaf in combination with gallic acid ameliorates liver dysfunction in high-fat diet-induced obese rats." *Indian Journal of Pharmaceutical Sciences*, 81(4): 673-680.

Russo, Marina, Chiara Fanali, Giusy Tripodo, Paola Dugo, Rosario Muleo, Laura Dugo, Laura De Gara and Luigi Mondello. 2018. "Analysis of phenolic compounds in different parts of pomegranate (*Punica granatum*) fruit by HPLC-PDA-ESI/MS and evaluation of their antioxidant activity: application to different Italian varieties." *Analytical and bioanalytical chemistry* 410(15):3507-3520.

Salinas-Flores, Astrid, Alexandro Guevara-Aguilar, Ever Adrián Natividad-Torres, Ramiro Baeza-Jiménez, José Juan Buenrostro-Figueroa. 2019. "Efecto de las condiciones de extracción sobre la capacidad antioxidante de compuestos fenólicos de cáscara de granada." *Mexican journal of biotechnology 4*(2):33-46. https://doi.org/10.29267/mxjb.2019.4.2.33. ["Effect of extraction conditions on the antioxidant capacity of phenolic compounds from pomegranate peel."]

Seeram, N., R. Lee, M. Hardy, and D. Heber. 2005a. "Rapid large-scale purification of ellagitannins from pomegranate husk, a by-product of the commercial juice industry." *Separation and Purification Technology 41*(1):49-55. https://doi.org/10.1016/j.seppur.2004.04.003.

Seeram, Navindra P., Lynn S. Adams, Susanne M. Henning, Yantao Niu, Yanjun Zhang, Muraleedharan G. Nair and David Heber. 2005b. "*In vitro* antiproliferative, apoptotic and antioxidant activities of punicalagin, ellagic acid and a total pomegranate tannin extract are enhanced in combination with other polyphenols as found in pomegranate juice." *The Journal of Nutritional Biochemistry 16*(6):360-367. https://doi.org/10.1016/j.jnutbio.2005.01.006.

Seeram, Navindra P., Michael Aviram, Yanjun Zhang, Susanne M. Henning, Lydia Feng, Mark Dreher, and David Heber. 2008. "Comparison of antioxidant potency of commonly consumed polyphenol-rich beverages in the United States." *Journal of agricultural and food chemistry*, 56(4):1415-1422.

Singh, Balwinder, Jatinder Pal Singh, Amritpal Kaur, and Narpinder Singh. 2018. "Phenolic compounds as beneficial phytochemicals in pomegranate (*Punica granatum* L.) peel A review." *Elsevier Enhanced Reader 261:*75-86. https://doi.org/10.1016/j.foodchem.2018.04.039.

Sreedevi, P., K. Vijayalakshmi, and R. Venkateswari. 2017. "Phytochemical evaluation of *Punica granatum* L. leaf extract." *International Journal of Current Pharmaceutical Research 9*(4):14-18. https://doi.org/10.22159/ijcpr.2017v9i4.1159.

Stefano, Vita Di, Rosa Pitonzo, Maria Eugenia Novara, David Bongiorno, Serena Indelicato, Carla Gentile, Giuseppe Avellone, Rosaria Bognanni, Salvatore Scandurra and Maria Grazia Melilli. 2019. "Antioxidant activity and phenolic composition in pomegranate (*Punica granatum* L.) genotypes from south Italy by UHPLC–Orbitrap-MS approach." *Journal of the Science of Food and Agriculture 99*(3):1038-1045. https://doi.org/10.1002/jsfa.9270.

Tang, Dan, Liu Liu, Dildar Ajiakber, Jianping Ye, Jianjun Xu, Xuelei Xin and Haji Akber Aisa1. 2018. "Anti-diabetic effect of *Punica granatum* flower polyphenols extract in type 2 diabetic rats: activation of Akt/GSK-3β and inhibition of IRE1α-XBP1 pathways." *Frontiers in endocrinology* 9, 586.

Tapia-Campos, Ernesto, Ma Claudia Castañeda-Saucedo, Jessica del Pilar Ramírez-Anaya, Karina Alarcón-Dominguez, Elia Herminia Valdés-Miramontes, and Octavio Núñez-Maciel. 2016. "Physical-chemical characterization of fourteen pomegranate genotypes of southern Jalisco, Mexico." *Scientia Horticulturae 199:*163-169. https://doi.org/10.1016/j.scienta.2015.12.022.

Tavallali, Hossein, Atefeh Bahmanzadegan, Vahid Tavallali and Vahid Rowshan. 2020. "Phytochemical investigation of *Punica granatum* pomace: As source of bioactive and medicinal natural products." *Adv Med Plant Res 8*(1):1-13.

Toda, Keisuke, Mai Ueyama, Shomu Tanaka, Izumi Tsukayama, Takuto Mega, Yuka Konoike, Asako Tamenobu, Februadi Bastian, Iria Akai, Hideyuki Ito, Yuki Kawakami, Yoshitaka Takahashi and Toshiko Suzuki-Yamamoto 2020. "Ellagitannins from *Punica granatum* leaves suppress microsomal prostaglandin E synthase-1 expression and induce lung cancer cells to undergo apoptosis." *Bioscience, Biotechnology and Biochemistry* Accessed: March 17, 2020 84(4):757-763. doi:10.1080/09168451.2019.1706442.

Ustun-Argon, Zeliha. 2019. "Phenolic compounds, fatty acid compositions and antioxidant activity of commercial cold-pressed pomegranate (*Punica granatum*) seed oils from Turkey." *International Journal of Scientific & Engineering Research 10*(4):1347-1352.

Vázquez-Flores, Alma A., Emilio Álvarez-Parrilla, José Alberto López-Díaz, Abraham Wall-Medrano, and Laura A. De la Rosa. 2012. "Taninos hidrolizables y condensados: naturaleza química, ventajas y desventajas de su consumo." *Tecnociencia Chihuahua* 6(2):84-93. ["Hydrolyzable and condensed tannins: chemical nature, advantages and disadvantages of their consumption." *Tecnoscience Chihuahua*]

Viswanath, M., P Sridevi, BVK Bhagavan, K Ravindra Kumar and P Subbaramamma. 2019. "Toxicological, Pharmacological and Cellular properties of Pomegranate (*Punica granatum* L.): A Review." *Journal of Pharmacognosy and Phytochemistry 8*(2):172-176.

Viuda-Martos, M., J. Fernández-López, and J. A. Pérez-Álvarez. 2010. "Pomegranate and its many functional components as related to human health: a review." *Comprehensive Reviews in Food Science and Food Safety* 9(6):635-654.

Waly, Mostafa. I., Amanat Ali, Nejib Guizani, Amani S. Al-Rawahi, Sadar S. Farooq, and Mohammad S.Rahman. 2012. "Pomegranate (*Punica granatum*) peel extract efficacy as a dietary antioxidant against azoxymethane-induced colon cancer in rat." *Asian Pacific Journal of Cancer Prevention* 13(8):4051-4055. https://doi.org/10.7314/apjcp. 2012.13.8.4051.

Wu, Sheng and Li Tian. 2017. "Diverse phytochemicals and bioactivities in the ancient fruit and modern functional food pomegranate (*Punica granatum*)." *Molecules 22*(10):1-17. https://doi.org/10.3390/molecules22101606.

Yaranga, Farfán Gaudencio, Edson Morales Vásquez and Fany Villanueva Huamán 2019. Cuantificación y determinación de la actividad antioxidante de los polifenoles del extracto hidroalcohólico de la cáscara, semilla y el zumo del fruto de *Punica granatum* var. wonderfull. B diss. Universidad Maria auxiliadora.

Yusefi, Mostafa, Kamyar Shameli, Roshafima Rasit Ali, Siew-Wai Pang, Sin-Yeang Teow. 2020. "Evaluating anticancer activity of plant-mediated synthesized iron oxide nanoparticles using *Punica granatum* fruit peel extract." *Journal of Molecular Structure*. Accessed: March 20, 2020. 1204 doi: 10.1016/j.molstruc.2019.127539.

Zahin, Maryam, Smeena Hasan, Farrukh Aqil, Mohd Sajjad A. Khan, Fohad Mabood Husain and Iqbal Ahmad. 2010. "Screening of certain medicinal plants from India for their anti-quorum sensing activity." *Indian Journal of Experimental Biology* 48(12):1219-1224.

Zeynalova, Aydan M., Eldar N. Novruzov, and Biancaelena Maserti. 2019. "Studies on the physico-chemical characteristics, antioxidant activity and juice organic compound composition in Azerbaijan wild pomegranate fruits." *Plant & Fungal Research 2*(1):40-46.

Zhao, X., Z. Yuan, Y, Fang, Y Yin, and L Feng. 2014. "Flavonols and Flavones Changes in Pomegranate (*Punica granatum* L.) Fruit Peel during Fruit Development." *J. Agr. Sci. Tech. 16:*1649-1659.

In: Super and Nutraceutical Foods
Editors: Sandra Teresita Martín-del-Campo et al.
ISBN: 978-1-53619-082-3

Chapter 5

Raspberry's (Rubus idaeus L) Production, Nutritional Composition, Bioactive Compounds and Their Effect on Health

Elia Herminia Valdés Miramontes*, Zyanya Reyes-Castillo, Norma Piña Contreras, Berenice Sánchez Caballero, Carmen Alejandrina Virgen – Carrillo and Patricia Martín del Campo–Rayas
Behavioral Feeding and Nutrition Research Institute,
Southern University Center of the University of Guadalajara, Jalisco, México

Abstract

Raspberry is a representative of the fruits called berries. According to their economic and commercial importance, the most produced berries in the world are strawberries, blueberries, and raspberries. Mexican berry exports are mainly as fresh fruit, where the first months of the year (January to April), most of the berry production in Mexico is for export purposes. Numerous studies have shown the beneficial effects on the health of some compounds present in raspberry fruits and other berries. These biological properties include its antioxidant, lipid-lowering, hypoglycemic, anti-inflammatory capacity, among others, this being a factor that has allowed the rapid growth of its production and global marketing. Other reasons that explain the interest in the production of these fruits are high profitability, fast return on investment, intensive use of labor, the versatility of the fruits for consumption, and their export potential.

Keywords: raspberry, nutritional composition, biological properties

* Corresponding Author's E-mail: eliav@cusur.udg.mx.

Introduction

Berries refer to the various fruits of different botanical groups, which are characterized by their reduced shelf-life, tend to be small, sweet-acidulated, juicy, and brightly colored, which makes them very attractive for consumption (González et al. 2019).

The raspberry plant was recognized in the wild, on Mount Ida on the Crete island, Greece, by Carl Linnaeus, who was the first to name the species with this name; however, archaeological evidence shows that the inhabitants of the Paleolithic caves already ate wild raspberries. During the Middle Ages, this plant spread throughout Europe, from Greece to Italy, to the Netherlands and England. In the 18th century, it entered the United States, and at the beginning of the 19th century, it was already cultivated in South America (García Rubio, García González, and Ciordia Ara 2014).

Raspberry is a reddish fruit that comes from a plant belonging to the Rosaceae family, whose scientific name is Rubus idaeus L. This fruit is consumed fresh and is used for the preparation of preserves, desserts, drinks, and other products (Cés 2016).

In Mexico, raspberry is a typical representative of berries. This species production success is largely due to their sensory appeal, health benefits, fashion, supply control, marketing, and economic profitability (González et al. 2019).

Raspberries cultivation has acquired great commercial relevance worldwide in recent years, derived from its high economic value, its unique organoleptic characteristics, and the presence of bioactive compounds. It contains anthocyanins and flavonoids, which have health benefits, such as eliminating free radicals, antioxidants, anti-inflammatory, antimicrobial and anticancer activity. Some studies have shown that raspberry fruit decreases oxidation markers, improves lipid metabolism, and reduces adipose tissue inflammation (Četojević-Simin et al. 2015). Recent studies have shown that the dietary intake of red raspberry is associated with a lower cardiometabolic risk in Zucker Fatty Rats and other possible diseases related to T2DM (Snyder 2016).

Another study in obese rats showed that the consumption of raspberry ketones with a hypocaloric diet was able to modulate the neurodegenerative changes induced by obesity, including Alzheimer's disease (Mohamed et al. 2018).

Raspberry Production

In Mexico, the main berries produced are, in order of importance, strawberry, blackberry, raspberry, and blueberry. The main varieties of raspberry grown worldwide are Heritage, Amity, Autumn bliss, Ruby, Santa Clara, Santa Catalina, Santa Teresa, Meeker, Chilliwack, Coho, Comox, Tulameen, Glen magna, Glen ample, and Skeena. At present, Mexico is the largest raspberries producer in America; in the Asian continent, raspberry production is largely concentrated in Russia, which is primarily destined for industrial processing. In Europe, the main producers are Serbia and Poland, and Chile in the southern hemisphere. Table 1 (Food and Agriculture Organization Corporate Statistical Database [FAOSTAT], 2017).

Table 1. Leading raspberry production countries

Country	Output (ton)	Land used (ha)	Production (ton/ha)
Russian Federation	146.377	20.185	7.25
Mexico	120.184	6.390	18.80
Serbia	109.742	21.861	5.01
United States	106.100	8.272	12.82
Poland	104.482	29.317	3.65
Spain	43.529	2.482	17.5
Ukraine	34.160	5.000	6.83
Bosnia Herzegovina	22.671	3.296	6.87
Chile	19.237	4.784	4.02
Portugal	17.880	1.108	16.13

In 2017 Mexico increased raspberry production by 6.7%, with Jalisco being the largest producer in the country with 91,984 tons. Table 2 describes the production of raspberries from the main federal entities in Mexico. Figures 1, 2 show raspberry crops from Jalisco. Table 3 shows the raspberry production between 2009 and 2018 in Jalisco, according to FAOSTAT.

The main destination of Mexican berry exports is the United States since this market represents more than 90% of our country's strawberry, raspberry, blueberry, blackberry, and blackberry exports. In this sense, US imports of berries exceeded an average of 336 thousand tons during the 2011-2015 period, growing at a rate of 13.17%; 73.16% was marketed as fresh fruit and 26.84% as frozen fruit. This trade exceeded one billion dollars during that period, showing a growing trend over the past five years, both in volume and value. It is important to note that the exported volume during the period from January to April 2016 was equivalent to 73% of fresh berries exported throughout 2015 (González et al. 2019).

Table 2. Mexico's raspberry production by state

State	Output (ton)	Land used (ha)	Production by ha	Production's value in millions $
Jalisco	91.984	5.154	17.8	2.043
Michoacán	23.384	1.265	18.5	1.132
Baja California	14.149	649	21.8	1.738
Puebla	572	72	8	17
México	45	9	5.3	7
Colima	14	1	14	< 1

Conservation Methods and Raspberry Products

The raspberry has high respiration rates, which, together with its high sugar content, makes it one of the most perishable fruits, with a post-harvest life of only 3-5 days at room temperature (Jin et al. 2012). However, one of the most important obstacles to increasing its production and marketing are the high post-harvest losses that are generated due to microbial deterioration caused by the pathogen Botrytis cinerea and the physiological damage of the fruit that results in softening and bleeding from it (Forney et al. 2015).

Figure 1. Raspberry crop in the municipality of Zapotlán el Grande, Jalisco, México. Own authorship.

Figure 2. Raspberry plant with fruit in the municipality of Zapotlán el Grande, Jalisco, México. Own authorship.

Table 3. Mexico's raspberry production by year

Year	Output(ton)
2009	13,559
2010	14,344
2011	21,468
2012	17,009
2013	30,411
2014	35,627
2015	65,388
2016	112,661
2017	120,184
2018	130,187

Conservation Methods

The undesirable effects linked to the maturation and senescence processes can be delayed through the application of different technologies, one of which is refrigerated storage. In this sense, Visioni Tezotto-Uliana et al. (2018) determined that the post-harvest quality of Autumn bliss variety raspberries at 0°C was maintained for 7 days. Refrigerated storage is also a fundamental factor in preserving the stability of functional compounds. The temperature of 0°C is recommended for storing raspberries of the 'Autumn Bliss' variety since it allows preserving the parameters related to the physical-chemical quality and bioactive compounds for 12 days (Visioni Tezotto-Uliana et al. 2018).

On the other hand, dehydration is one of the food preservation methods that allow increasing availability of these products in the market. In a study carried out by Goncalves and collaborators, the dehydration process of raspberry was evaluated by lyophilization and drying (in a conventional oven at 40°C) in which some fruits were previously blanched, and others were not. The physical parameters evaluated were water activity, color, and moisture content. Conventional drying reduced water content to 20.7-35.8%, while dehydration by lyophilization reduced the water content to 11.6-14%. The lyophilization process presented higher color difference indices than conventional dehydration (Gonçalves et al. 2018).

Sun or microwave drying is another method used in raspberry preservation. In a study carried out by Anabel Rodriguez, the efficiency of drying in the sun and through microwaves was evaluated in terms of weight reduction as a function of time and some quality properties such as color, texture, antioxidant capacity, and the total content of polyphenols. The results showed that solar drying and microwave drying allowed good preservation of the surface color in the final product. Only sun-drying achieved a desirable texture of dried raspberries than microwave drying, and neither allowed high retention of antioxidant capacity. However, both processes allow stable final fruits (Rodriguez et al. 2019).

Products

Raspberries are used to make different products, such as jams, preserves, juices, syrups, ice cream, flavored liqueurs, pastries, etc. Some raspberry extracts are used in food supplements for their functional properties, both for the large industry and for the artisan type, obtaining products with quality nutritional and sensory properties. 90% of the world's raspberry production is destined to elaborate on these products (García Rubio et al. 2014).

Canned sweetened products obtained by cooking fruits, vegetables or tubers (whole or fractionated), their juices and / or pulps, with sugars (sugar, dextrose, invert sugar, glucose syrup, or mixtures) with or without the addition of other sweeteners, additives or ingredients are called "confiture." In jams, sweets, and jellies, the soluble solids content is regulated, which must contain between 64 to 66° Brix. Raspberry jams are made from the pulp of the fruit added with sugars either in crystallized form or sugars in the form of syrup, generally with added pectin or modified starches to promote gel consistency; once the mixture is prepared, it is subjected to a thermal process and some physical, chemical and nutritional parameters related to the quality of the product (pH and Brix degrees) are determined, as well as some organoleptic properties such as the gelation state and color, and then it is packaged at a temperature that allows it to generate a vacuum in the container (Figure 3 and 4) (Dervisi and Lamb 2001).

RASPBERRY NUTRITIONAL CONTENT

The caloric intake of raspberry is quite low since its carbohydrate content is moderate, and that of fat is minimal. Raspberry is a highly recommended fruit due to its vitamin content, including vitamin B9 or folic acid that is necessary for the process of cell mitosis; it also contains vitamin C that contributes to the formation of collagen for the proper functioning of bone tissue, cartilage, skin elasticity. It also contains dietary fibers, tocopherol, tocotrienol, calcium, potassium, magnesium, carotenoids, and linoleic acid (Chang-Ho, Choi, and Heo 2008). Tables 4 and 5 show the nutritional content of the raspberry fruit.

Figure 3. Jam made with raspberries grown in the municipality of Zapotlán el Grande, Jalisco, Mexico. Self-authorship.

Table 4. Nutritional content of raspberry per 100 grams.

Components	Raspberry fruit	Raspberry pulp
Energy (kcal)	52	6
Lipids (g)	0.65	0.44
Carbohydrates calculated by difference (g)	11.49	0.20
Fiber (g)	6.50	2.20
Proteins (g)	1.20	0.26
Total ash	1.70	0.4
Water (g)	84.96	96.5

Table 5. Mineral content of the raspberry fruit in every 100 g

Component	Quantity
Calcium	25 mg/100 g
Iron	0.69 mg/100 g
Magnesium	22 mg/100 g
Phosphorus	29 mg/100 g
Potassium	151 mg/100 g
Sodium	1 mg/100 g
Zinc	0.42 mg/100 g
Copper	0.090 mg/100 g
Manganese	0.670 mg/100 g
Selenium	0.2 µg/100 g
Vitamin C	26.2 mg/100 g
Thiamin	0.032 mg/100 g
Riboflavin	0.038 mg/100 g
Niacin	0.598 mg/100 g
Pantothenic acid	0.329 mg/100 g
Vitamin B6	0.055 mg/100 g
Total folate	21 µg REA/100 g
Total choline	12.3 mg/100 g
Vitamin A	2 mg/100 g
α-Tocopherol	0.87 mg/100 g
Vitamin K	7.8 µg/100 g

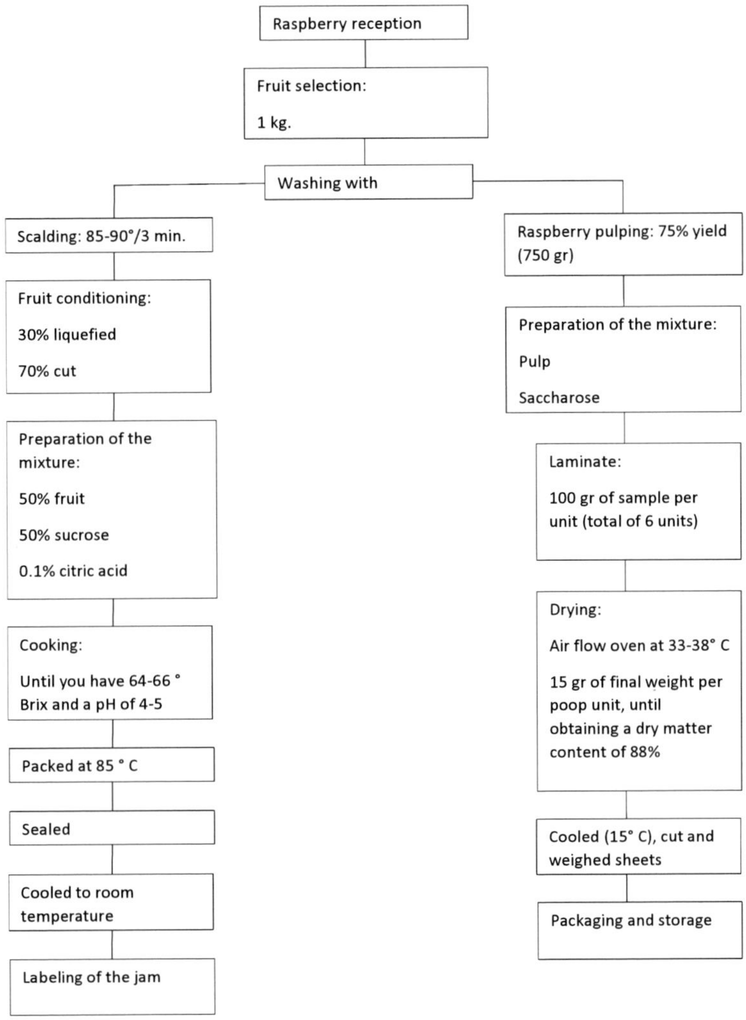

Figure 4. Diagram of the process for making raspberry jam and candy. Own elaboration.

HEALTH EFFECT OF RASPBERRY FRUITS AND LEAVES

Since ancient times, plants have been primary sources of compounds used in the treatment and prevention of diseases. So far, natural products in the form of herbal remedies or medicinal plants, functional foods, and their bioactive compounds have been effective in the treatment of various diseases (Kyoung-Tae et al. 2018). Interest in the use of bioactive plant compounds for the treatment of diseases has increased in recent years. This was probably due to the adverse side effects associated with conventional medications, which led to the investigation of natural products that may be a viable alternative to synthetic drugs.

The berries are famous for their high concentrations of bioactive compounds such as anthocyanins, flavonols, catechins, and hydroxybenzoic acids (Figure 5 and Table 6) with proven antioxidant, antimicrobial, anti-inflammatory, vasodilatory, antiproliferative, and anticancer activities. Raspberry is known as a rich source of dietary antioxidants, such as phenolic acids (ellagic acid and its conjugates, ellagitannin- lambertianin C and sanguine H-6), flavonoids (flavan-3-ols and their oligomers, quercetin) and anthocyanins (cyanidin-3-sophoroside, cyanidin-3-(2-glucosylrutinoside), cyanidin-3-glucoside, pelargonidin-3-sophoroside, cyanidin-3-rutinoside, pelargonidin-3-(2-glucosylrutinoside), pelargonidin-3-glucoside, pelargonidin-3-rutinoside) (Četojević-Simin et al. 2015).

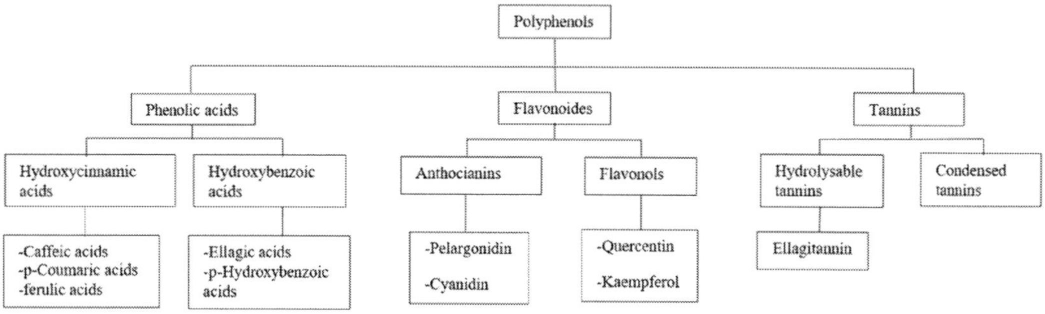

Figure 5. Diagram of the polyphenols present in the raspberry fruit.

Table 6. Concentration of some phenolic compounds from raspberry (dried wight)

Standard	Phenolic compounds (mg/100g)
Total	5.79
Quercetin	1.52
Rutin	0.00
Gallic acid	1.23
Syringic acid	2.76
Caffeic acid	0.00
Cinnamic acid	0.00
(+)-Catechin	0.27
Resveratrol	0.00
Chlorogenic acid	0.00

Numerous studies have evaluated the role of red fruits and their bioactive compounds in the prevention and treatment of chronic, degenerative, and metabolic diseases, such as type 2 diabetes mellitus, cardiovascular diseases, atherosclerosis, dyslipidemias, and systemic inflammatory processes. Some studies have shown that raspberry supplementation improves basal motor function, cognitive function, and has neuroprotective effects (Schell et al. 2019).

There is evidence of the important role that polyphenols in berries such as raspberry have in reducing cardiovascular risk, dyslipidemia, oxidative stress, blood glucose, and inflammation. These findings are promising, since the incorporation of berries as part of the habitual diet of individuals can improve their health; besides, this fruit is highly popular in several countries (Schell et al. 2019).

Several epidemiological studies have shown that quercetin has a neuroprotective effect against Parkinson's disease and Alzheimer's disease (Kyoung-Tae et al. 2018). On the other hand, chlorogenic acids are polyphenols that have gained interest in recent years. Several studies relate their properties with an improvement in glucose and insulin metabolism, besides, they have antioxidant and anti-inflammatory properties, but their mechanism it is not well understood yet. Other studies have indicated that they affect blood pressure; this effect is attributed to various metabolites of these acids with different mechanisms of action. Chlorogenic acid in the diet hydrolyzes in the small intestine to form trans-cinnamic acids (such as caffeic and ferulic acids). Most metabolites are absorbed from the large intestine after interacting with the microflora. Caffeic acid can be converted to ferulic acid by a transferase or reduced to dihydrocaffeic acid. Research suggests that the most potent hypotensive effect is produced by ferulic acid with a decrease of approximately 9 times better than caffeic acid and a reduction of 17 times better than quinic acid (Lee, McGeer, and McGeer, 2016).

Some scientific evidence has shown that catechins can reduce cholesterol absorption, prevent LDL-C oxidation, and improve nitric oxide production, which is essential to reduce the risk of cardiovascular disease (Snyder et al. 2016).

Raspberry leaves and roots have traditionally been used for their medicinal properties. Leaf teas are used to relieve muscle spasms and nausea. The root of the plant is traditionally used to treat wounds and sore throats. Raspberry juice has shown inhibition of the growth of bacteria such as *Staphylococcus aureus*, *Escherichia coli*, *Mycobacterium phlei*, *Clostridium*, *Enterococcus faecalis*, *Shigella soneri*, as well as some *Salmonella serovars*. Raspberry fruit extracts cause the death of *Helicobacter pylori*, inhibition of *Bacillus cereus*, *S. aureus*, and *Staphylococcus epidermidis*; this antimicrobial activity is mainly attributed to ellagitannins (Četojević-Simin et al. 2015). Table 7 shows some bioactive compounds of raspberry and their health effects.

Table 7. Bioactive compounds of raspberry and their health effects

Bioactive compound	Effects on health	Authors
Gallic acid	Hypoglycemic, cardioprotective, neuroprotective, antitumor.	Sun et al. 2014
Catechin	Cardiovascular diseases, hypoglycemic, protection against free radicals.	Murray et al. 2015
Syringic acid	Antibacterial, antistatic, anti-inflammatory, and neuroprotective.	Shi et al. 2015; Tokmark et al. 2017.
Cinnamic acid	Antitumor properties.	Mielecki and Lesyng, 2016
Chlorogenic acid	Improves vascular function, antihypertensive effect. Protection of vascular cells against oxidative, neuroprotective damage.	Rodriguez-Mateos et al, 2016; Taram, Winter and Liseman 2016.
Ferulic acid	It prevents Alzheimer's disease, antitumor, antimicrobial and anti-inflammatory properties.	Park and Lee, 2016
Quercetin	Protective effect on retinal degeneration, antitumor, hypoglycemic effect, cardioprotective effects.	Park and Lee 2016; Snyder et al, 2016.
Rutin	Hypoglycemic, anti-inflammatory, antitumor, neuroprotective.	Gautam et al. 2016.
Anthocyanin	Hypoglycemic, lipid lowering, anti-inflammatory, cardioprotective.	Li et al. 2015; Chen et al. 2016.

Red Raspberries Oxidative Stress and Inflammation

Cell cultures, chemical assays, and enzymatic activity studies, animal and human models have been used to study the effect of the bioactive compounds of red raspberries. Evidence of the potential benefits of red raspberries in oxidative stress and inflammation, especially using in vivo animal models, is a subject of great interest (Zhang, Luna-Vital, and Gonzalez de Mejia 2019).

A study by Kowalska et al. (2019) evaluated the ability of raspberry fruit extract to reduce adipose tissue dysfunction using adipocytes with 3T3-L1 hypertrophy. The results showed that the raspberry fruit juice extract decreased the generation of intracellular reactive oxygen species (ROS) in hypertrophied adipocytes, and the raspberry fruit extract significantly inhibited the expression of lipoprotein lipase (LPL) mRNA, apoprotein aP2, FAS, PLIN, likewise presented a high anti-inflammatory potential through the negative regulation of the expression of inflammation mediators (IL-6, TNF- α, IL-1 β, MCP-1, and leptin) (Kirakosyan et al. 2018).

A study in rats tested the effect of an extract of raspberry and lyophilized fruit added to the diet; results showed a decrease in oxidative stress-mediated by decreased damage to lipids and proteins and a decrease in inflation after one-month exposure (Burton-Freeman, Sandhu, and Edirisinghe 2016).

In an inflammation model of rats with arthritis, the red raspberry extract significantly reduced the development of clinical signs of arthritis, soft tissue inflammation, and osteophyte formation, thus preventing joint destruction in animals that received the raspberry extract (Kshatriya et al. 2019).

In another study, the effect of the intake of lyophilized raspberry added to the diet of obese-diabetic mice was investigated, the antioxidant effect was evaluated using biomarkers of risk of obesity and diabetes. The results showed that the intake of raspberry improved the antioxidant status and decreased IL-6 in the plasma of obese diabetic rats with respect to the control (Noratto, Chew, and Atienza 2017).

Table 8. Summary of research studies that tested polyphenolic components of red raspberry in models of oxidative stress and systemic inflammation

Model	Study details	Treatment	Mechanism	References
In vivo (mice)	The antioxidant, antidiabetic and anti-obesogenic effect of lyophilized raspberry was evaluated in an animal model	27.4 ± 1.8 g and 31.7 ± 1.9 g of freeze-dried raspberry	Blood biomarkers of oxidative stress and inflammation: ↓MDA; ↓IL-6 ↓TNF-α↓MCP1 ↑CAT activity Plasma biomarkers associated with metabolic syndrome and obesity: ↓Glucose Insulin ↓TGL; ↑TC; ↑HDL; ↑Leptin	Noratto et al. 2017.
In vitro	Raspberry extract as potential antioxidant	Raspberry extract of 2 mg/ml concentration in 24 h	Caco-2 cell line: Mitochondrial membrane potential ↓GSH ↓ROS	Chen et al, 2016
In vitro	Anthocyanin raspberry extract modulate proteins that involved in the antioxidant effects in cultured cell	Cyanidin-3-0-sophoroside 50- 200 μg/ml, 24 h Cyanidin-3-0(2G-0-glucosyl) 50-200 μg/ml, 24 h	HepG2 cell line: ↑SOD; ↓ROS; ↑NrF2; ↑P13K; ↓p38; ↓JNK ↓ERK	Mazewski, Kim, and Gonzalez de Mejia 2019
In vivo (mice)	Raspberry powder is inversely related to cardiac risk factors, as inflammation	2% of red raspberry freeze-dried powder in 20 g of diet per day	Plasma inflammatory markers: ↓IL-6 ↓CRP ↓TNF-α ↓8-Isoprostane	Kirakosyan et al, 2018

Few studies in the literature have been identified in humans evaluating the effect of raspberries on markers of oxidative stress and inflammation. A study showed that the consumption of the juice made from different berries by trained cyclists, using an exercise model to induce oxidative stress, the results showed reduced DNA and protein damage, after consumption of the juice, compared with the control, which is shown in Table 8.

Red Raspberries and Dyslipidemias

Dietary intake of raspberry ketones (RK) has been reported to exert anti-obese actions and alter lipid metabolism in *in vivo* and human studies. To elucidate a possible mechanism for the anti-obese actions of RK, Park and Lee, (2016), investigated the effects of RK on adipogenic and lipogenic gene expression in 3T3-L1 adipocytes. For this, mature 3T3-L1 preadipocytes and mature adipocytes were treated from day 2 to day 8 of differentiation for 24 h, on day 12 with 1, 10, 20, and 50 μM of RK. Triacylglycerols were evaluated by spectrophotometry and gene expression by a quantitative polymerase chain reaction in real-time (qRT-PCR). The results showed that RK treatment of adipocytes suppressed adipocyte differentiation and fat accumulation depending on concentration. RKs suppressed the expression of the main genes involved in the adipogenesis pathway, including peroxisome proliferator-activated receptor γ (PPARγ) and CCAAT enhancer-binding protein α (C/EBPα), leading to a decreased regulation of the protein binding to adipocyte fatty acids -2 (aP2). Furthermore, treatment with 10 μM RK also reduced the levels of mRNA from lipogenic genes such as acetyl-CoA carboxylase-1 (ACC1), fatty acid synthase (FASN), and stearoyl-CoA desaturase-1 (SCD1). In mature adipocytes, RK increased the transcriptional activities of genes involved in lipolysis and oxidative pathways, including triglyceride adipose lipase (ATGL), hormone-sensitive lipase (HSL), and carnitine palmitoyl transferase-1B (CPT1B). These results suggest that RKs hold great promise with biological activities that alter lipid metabolism in 3T3-L1 adipocytes. See Table 9.

Raspberry and Diabetes

In a case study by Cheang et al. in 2016, the effect of consuming raspberry leaves to treat hyperglycemia in a woman with gestational diabetes was analyzed, in which the results showed its hypoglycemic action as it reduced the patient's insulin requirements, and after delivery, the mother did not present hyperglycemia hypoglycemia in the son.

Another investigation by Schell et al. (2016) determined the effect of raspberry consumption on glucose concentrations and chronic inflammation in adults with T2DM. Results showed a significant decrease in both blood glucose and inflammation markers such as IL-6 and TNF-α 4 hours after the fruit was ingested, which suggests it can contribute to the control of both conditions.

In an investigation conducted in Wistar rats induced to myocardial infarction risk by the drug isoproterenol and treated with raspberry, the results showed that cardiac inflammation was diminished after the administration of raspberry because it neutralized the free radicals that caused damage in cardiac cells, as well as promoted the production of antioxidant enzymes and consequently, reduced the risk of myocardial infarction in the rats of the experimental group (Khan et al. 2017).

Table 9. Summary of research studies that tested polyphenolic components of red raspberry in models of obesity and dyslipidemia

Model	Study details	Treatment	Mechanism	Authors
In vitro	Elucidates a possible mechanism for anti-obese actions of raspberry ketones in 3T3-L1 adipocytes treated from days 2 to 8 of differentiation and mature adipocytes for 24 h on day 12 with different doses of raspberry ketones	1, 10, 20 or 50 μM raspberry ketones	Dose-dependent adipocyte differentiation Lipid accumulation Dose- and stage-specific effects: ↓PPARγ, ↓C/EBPα, ↓cP2, ↓FASN, ↓CAA1,↓SCD1 Mature cells: ↑ATGL, ↑HSL, ↑CPT1B.	Park, 2015.
In vivo (Mice)	Effect of dietary treatment of obese rats with natural raspberry ketone and its relationship with neurodegeneration.	44 mg / kg body weight / day raspberry ketones	Serum lipids: ↓TGL ↓C-LDL, ↓C-VLDL ↑C-HDL ↓TC Lipid peroxidation: ↓MDA	Mohamed et al, 2018.
In vitro	Use the ability of raspberry fruit extract (RBE) to mitigate adipose tissue dysfunction using hypertrophied 3T3-L1 adipocytes.	Extract concentration (anthocyanins conjugates, flavanols, catechin, ellagic acid, quercetin) of 5, 10 and 20 mg/ml	3T3-L1 preadipocytes: ↑SOD ↑GPX ↑Catalase, ↑Adiponectin ↑IL-10 Hypertrophied 3T3-L1 adipocytes: ↓ROS ↓NOX4 ↓Lipids ↓LPL ↓FAS ↓aP2 ↓Leptin ↓IL-6 ↓Resistin ↓IL-1β ↓TNF-α ↓MCP-1	Kowalska, 2019

The consumption of two doses of ketones obtained from raspberry fruits in obese and diabetes-induced Wistar rats caused a reduction in serum TG, LDL-C, and LDL-C levels and an increase in HDL-C in the experimental group. In contrast, in the control group, the changes were less significant; the authors attribute the changes in blood lipid concentrations to the consumption of raspberry ketones (Mehanna et al. 2018).

In another study carried out by Kowalska, Olejnik, Zielińska-Wasielica, and Olkowicz (2019), where they used polyphenols of raspberry fruit extract that was administered in an in vitro culture of hypertrophied adipocytes, it was found that the fat cells reduced its size significantly. Likewise, it reduced the intracellular production of free radicals, increased the expression of antioxidant enzymes, and decreased cellular inflammation by inhibiting some pro-inflammatory factors such as IL-1 and TNF-α. The authors concluded that the consumption of raspberry fruits in a regular diet might be a potential protector against metabolic disorders. See Table 10.

Table 10. Summary of research studies that tested polyphenolic components of red raspberry in models of diabetes

Model	Study details	Treatment	Mechanism	Source
In vivo (human)	This was a randomized crossover study with 2 different phases with the objective of examine the effects of daily dietary raspberries on postprandial and 4-week fasting glucose, lipids and biomarkers of inflammation in obese adults with type 2 diabetes.	250 g of frozen raspberries	Biochemical and vascular profiles: Glucose ↓TGL ↓CRP ↓IL-6 ↓IL-1β ↓TNF-α ↓PAI-1 ↓ALT ↓AST	Schell et al, 2019
In vivo (ICR normal mice)	A new α-glucosidase inhibitor (pelargonidin-3-O-rutinoside, a natural anthocyanin) was isolated from the berries. The structural determinants for the inhibition of α-glucosidase by Pg3R and the SAR of natural anthocyanins were revealed in detail by enzymatic kinetics and molecular coupling analysis.	50 mg / kg and 150 mg / kg of pelargonidine-3-O-rutinoside before the glucose tolerance test.	Øα-gluco-sidase ↓postprandial glucose	Xu et al., 2019
In vivo (human)	Consumption of raspberry leaf may lead to reduced insulin requirements in gestational diabetes mellitus (GDM). Women with GDM should be cautioned about its use, and their glucose levels more closely monitored.	Raspberry leaf tea consumed was two servings per day (2 cups)	Mother: Glucose ↓ → Baby: Glucose ↓ →	Cheang et al. 2016

ACKNOWLEDGMENTS

We thank Ana Cristina Espinoza Gallardo for the English translation of this chapter.

REFERENCES

Burton-Freeman, Britt M., Amandeep K. Sandhu, and Indika Edirisinghe. 2016. "Red Raspberries and Their Bioactive Polyphenols: Cardiometabolic and Neuronal Health Links." *Advances in Nutrition* (Bethesda, Md.) 7(1):44-65.

Četojević-Simin, Dragana D., Aleksandra S. Velićanski, Dragoljub D. Cvetković, Siniša L. Markov, Gordana S. Ćetković, Vesna T. Tumbas Šaponjac, Jelena J. Vulić, Jasna M. Čanadanović-Brunet, and Sonja M. Djilas. 2015. "Bioactivity of Meeker and Willamette Raspberry (*Rubus idaeus* L.) Pomace Extracts." *Food Chemistry* 166:407-13.

Cheang, Kai I, Thanh T Nguyen, Nicole W Karjane, and Kelsey E S Salley. 2016. "Raspberry Leaf and Hypoglycemia in Gestational Diabetes Mellitus." *Obstetrics and Gynecology* 128: 1421-1424. doi:10.1097/AOG.0000000000001757.

Chen, Wei, Yang Xu, Lingxia Zhang, Ya Li, and Xiaodong Zheng. 2016. "Wild Raspberry Subjected to Simulated Gastrointestinal Digestion Improves the Protective Capacity against Ethyl Carbamate-Induced Oxidative Damage in Caco-2 Cells." *Oxidative Medicine and Cellular Longevity* 2016: 3297363. doi: 10.1155/2016/3297363.

Chen, Yu Feng, Marthandam Asokan Shibu, Ming-Jen Fan, Ming-Cheng Chen, Vijaya Padma Viswanadha, Yi-Lin Lin, Chao-Hung Lai, Kuan-Ho Lin, Tsung-Jung Ho, Wei-Wen Kuo, Chih-Yang Huang. 2016. "Purple rice anthocyanin extract protects cardiac function in STZ-induced diabetes rat hearts by inhibiting cardiac hypertrophy and fibrosis." *The Journal of Nutritional Biochemistry*;31:98-105. doi: 10.1016/j.jnutbio.2015.12.020.

Dervisi, Panagiota and Jack Lamb. (2001). "High pressure processing in jam manufacture: Effects on textural and colour properties." *Food Chemistry*. 73: 85-91. doi: 10.1016/S0308-8146(00)00289-2.

Food and Agriculture Organization Corporate Statistical Database (FAOSTAT). 2017. *Raspberry production*. Accesed August 2017. http://www.fao.org/faostat/en/#data/QC (Google 2017) (FAOSTAT 2017).

Forney, Charles F., Andrew R. Jamieson, Kathleen D. Munro Pennell, Michael A.Jordan, and Sherry A.E.Fillmore. 2015. "Relationships between fruit composition and storage life in air or controlled atmosphere of red raspberry." *Postharvest Biology and Technology* 110, 121–130.

García Rubio, Juan, Guillermo García Gonzalez de Lena, and Marta Ciordia Ara. (2014). El cultivo del frambueso. Spain: SERIDA, *Consejería de Agroganadería y Recursos Autóctonos del Principado de Asturias*. [*The cultivation of raspberry. Spain: SERIDA, Ministry of Livestock Farming and Native Resources of the Principality of Asturias*]

Gautam, Raju, Manjari Singh, Swetlana Gautam, Jitendra Kumar Rawat, Shubhini A Saraf, and Gaurav Kaithwas. 2016. *Rutin attenuates intestinal toxicity induced by Methotrexate linked with anti-oxidative and anti-inflammatory effects*. 2016. 10;16:99. doi: 10.1186/s12906-016-1069-1.

Gonçalves, Otília, Marco Alves, Joana Grácio and Valentim Nunes. 2018. "A comparative study of raspberry dehydration by lyophilisation or conventional drying." *International Advanced Researches and Engineering Journal* 02(03): 267-272. https://dergipark.org.tr/en/download/article-file/593466.

González, Felipe, Samuel Rebollar, Juvencio Hernández, José L. Morales and Osohe Ramírez. 2019. "Situación actual y perspectivas de la producción de berries en México." *Revista Mexicana de Agronegocios* 44: 260-72 https://www.redalyc.org/jatsRepo/141/14161295012/html/index.html. ["Current situation and prospects for berry production in Mexico." *Mexican Agribusiness Magazine*]

Jeong, Chang-Ho, Sung-Gil Choi, and Ho-Jin Heo. 2018. "Analysis of Nutritional Compositions and Antioxidative Activities of Korean Commercial Blueberry and Raspberry." *Journal of the Korean Society of Food Science and Nutrition* 37:1375–81. doi:10.3746/jkfn.2008.37.11.1375.

Jin, Peng, Shiow Y. Wang, Haiyan Gao, Hangjun Chen, Yonghua Zheng, and Chien Y. Wang. 2012. "Effect of cultural system and essential oil treatment on antioxidant capacity in raspberries." *Food Chemistry* 132(1), 399–405. doi:10.1016/j.foodchem.2011.11.011.

Khan, Vasim, Sumit Sharma, Uma Bhandari, Mansoor Syed, and Syed Haque. 2017. "Raspberry ketone protects against isoproterenol-induced myocardial infarction in rats." *Life Sciences*. 194:205-212. doi: 10.1016/j.lfs.2017.12.013.

Kirakosyan, Ara, E. Mitchell Seymour, Nicholas Kondoleon, Enrique Gutierrez, Janet Wolforth, and Steven Bolling. 2018. "The Intake of Red Raspberry Fruit Is Inversely

Related to Cardiac Risk Factors Associated with Metabolic Syndrome." *Journal of Functional Foods* 41:83-89.

Kowalska, Katarzyna, Anna Olejnik, Joanna Zielińska-Wasielica, and Mariola Olkowicz. 2019. "Raspberry (*Rubus idaeus* L.) Fruit Extract Decreases Oxidation Markers, Improves Lipid Metabolism and Reduces Adipose Tissue Inflammation in Hypertrophied 3T3-L1 Adipocytes." *Journal of Functional Foods* 62:103568. doi: 10.1016/j.jff. 2019.103568.

Kshatriya, Dushyant, Xinyi Li, Gina M. Giunta, Bo Yuan, Danyue Zhao, James E. Simon, Qingli Wu, and Nicholas T. Bello. 2019. "Phenolic-Enriched Raspberry Fruit Extract (*Rubus Idaeus*) Resulted in Lower Weight Gain, Increased Ambulatory Activity, and Elevated Hepatic Lipoprotein Lipase and Heme Oxygenase-1 Expression in Male Mice Fed a High-Fat Diet." *Nutrition Research* 68:19-33. doi: 10.1016/j.nutres.2019.05.005.

Kyoung Sik Park. 2010. "Raspberry ketone increases both lipolysis and fatty acid oxidation in 3T3-L1 adipocytes." *Planta Med* 76(15):1654-1658. doi:10.1055/s-0030-1249860.

Kyoung-Tae, Kim, Nam Taek-Kyun, Park Yong-Sook, Kim Young-Baeg, and Park Seung-Won. 2011. "Neuroprotective Effect of Anthocyanin on Experimental Traumatic Spinal Cord Injury." *Journal of Korean Neurosurgical Society* 2011; 49(4): 205-211. doi: 10.3340/jkns.2011.49.4.205.

Lee, Moonhee, Edith G. McGeer, and Patrick L. McGeer. 2016. "Quercetin, Not Caffeine, Is a Major Neuroprotective Component in Coffee." *Neurobiology of Aging* 46:113-23. doi: 10.1016/j.neurobiolaging.2016.06.015.

Li, Dan, Yuhua Zhang, Yan Liu, Ruifang Sun, and Min Xia. 2015. "Purified anthocyanin supplementation reduces dyslipidemia, enhances antioxidant capacity, and prevents insulin resistance in diabetic patients." *The Journal of Nutrition* 145(4):742-8. doi: 10.3945/jn.114.205674.

Mazewski, Candice, Morgan Sanha Kim, and Elvira Gonzalez de Mejia. 2019. "Anthocyanins, delphinidin-3-O-glucoside and cyanidin-3-O-glucoside, inhibit immune checkpoints in human colorectal cancer cells in vitro and in silico." *Scientific Reports* 9: 11560. doi: 10.1038/s41598-019-47903-0.

Mehanna Raja, Hubert H. Fernandez, Aparna Wagle Shukla, and Jawad A. Bajwa. 2018. "Deep Brain Stimulation in Parkinson's Disease." *Parkinsons Dis.* 2018:9625291. doi:10.1155/2018/9625291.

Mielecki, Marcin, and Bogdan Lesyng. 2016. "Cinnamic Acid Derivatives as Inhibitors of Oncogenic Protein Kinases--Structure, Mechanisms and Biomedical Effects." *Current Medicinal Chemistry* 23(10):954-82. doi: 10.2174/0929867323666160316123609.

Mohamed Hoda E, Dina M Abo-ELmatty, Noha M Mesbah, Samy M Saleh, Abdel-Moniem A Ali, and Amr T Sakr. 2018. "Raspberry ketone preserved cholinergic activity and antioxidant defense in obesity induced Alzheimer disease in rats." *Biomedicine & pharmacotherapy = Biomedecine & pharmacotherapie* 107:1166-1174. doi:10.1016/j. biopha.2018.08.034.

Murray, Morgan, Chelsey Walchuk, Miyoung Suh, and Peter J.Jones. 2015. "Green tea catechins and cardiovascular disease risk factors: Should a health claim be made by the United States Food and Drug Administration?." *Trends in Food Science & Technology* 41(2):188-197. doi: 10.1016/j.tifs.2014.10.004.

Noratto, Giuliana D., Boon P. Chew, and Liezl M. Atienza. 2017. "Red Raspberry (*Rubus idaeus* L.) Intake Decreases Oxidative Stress in Obese Diabetic (Db/Db) Mice." *Food Chemistry* 227:305-14. doi: 10.1016/j.foodchem.2017.01.097.

Park, Kyoung S. 2015. "Raspberry ketone, a naturally occurring phenolic compound, inhibits adipogenic and lipogenic gene expression in 3T3-L1 adipocytes." *Pharmaceutical Biology* 56(6):870-875. doi: 10.3109/13880209.2014.946059.

Park, Eunmi, and Hyuna Lee. 2016. "Ferulic acid as a natural bioactive compound enhances PARP inhibitor sensitivity in breast cancer cells." *Cancer Research* 76(14 Supplement): 1338-133. doi: 10.1158/1538-7445.AM2016-1338.

Rodriguez, Anabel, Estela Bruno, Carlos Paola, Laura Campañone, and Rodolfo H Mascheroni. 2019. "Experimental study of dehydration processes of raspberries (*Rubus idaeus*) with microwave and solar drying." *Food Science and Technology* 39(2): 336-343.

Rodriguez-Mateos, Ana, Rodrigo P Feliciano, Albert Boeres, Timon Weber, Claudia Nunes Dos Santos, M Rita Ventura, and Christian Heiss. 2016. "*Cranberry (poly)phenol metabolites correlate with improvements in vascular function: A double-blind, randomized, controlled, dose-response, crossover study*." 60(10):2130-2140. doi: 10.1002/mnfr.201600250.

Schell, Jace, Nancy M. Betts, Timothy J. Lyons, and Arpita Basu. 2019. "Raspberries Improve Postprandial Glucose and Acute and Chronic Inflammation in Adults with Type 2 Diabetes." *Annals of Nutrition and Metabolism* 74(2):165-74. doi: 10.1159/000497226.

Shi, Chao, Yi Sun, Zhiwei Zheng, Xiaorong Zhang, Kaikuo Song, Zhenyu Jia, Yifei Chen, Miaochun Yang, Xin Liu, Rui Dong, and Xiaodong Xia. 2016. "Antimicrobial activity of syringic acid against *Cronobacter sakazakii* and its effect on cell membrane." *Food Chemistry* 197(Pt A):100-6. doi: 10.1016/j.foodchem.2015.10.100.

Snyder, Sarah M., Bingxin Zhao, Ting Luo, Clive Kaiser, George Cavender, Jill Hamilton-Reeves, Debra K. Sullivan, and Neil F. Shay. 2016. "Consumption of Quercetin and Quercetin-Containing Apple and Cherry Extracts Affects Blood Glucose Concentration, Hepatic Metabolism, and Gene Expression Patterns in Obese C57BL/6J High Fat–Fed Mice." *The Journal of Nutrition* 146(5):1001-7. doi: 10.3945/jn.115.228817.

Sun, Jing, Yun-Zi Li, Yin-Hui Ding, Jin Wang, Ji Geng, Huan Yang, Jie Ren, Jin-Yan Tang, and Jing Gao. 2014. "Neuroprotective effects of gallic acid against hypoxia/reoxygenation-induced mitochondrial dysfunctions *in vitro* and cerebral ischemia/reperfusion injury in vivo." *Brain Research* 17;1589:126-39. doi: 10.1016/j. brainres.2014.09.039.

Taram, Faten, Aimee N Winter, and Daniel A Linseman. 2016. "Neuroprotection comparison of chlorogenic acid and its metabolites against mechanistically distinct cell death-inducing agents in cultured cerebellar granule neurons." *Brain Research* 1648(Pt A):69-80. doi: 10.1016/j.brainres.2016.07.028.

Tokmak, Mehmet, Muserref Hilal Sehitoglu, Yasemin Yuksel, Mustafa Guven, Tarik Akman, Adem Bozkurt Aras, Umut Yaka, Cengiz Gomleksiz, Serdar Baki Albayrak, and Murat Cosar. 2017. "The Axon Protective Effects of Syringic Acid on Ischemia/Reperfusion Injury in a Rat Sciatic Nerve Model." *Turkish Neurosurgery* 27(1):124-132. doi: 10.5137/1019-5149.JTN.14656-15.5.

Visioni Tezotto-Uliana, Jaqueline, Natalia Dallocca Berno, Ana Cecilia Silveira Gómez and Ricardo Alfredo Kluge. 2018. "Evaluación del comportamiento poscosecha de Frambuesas en diferentes condiciones de almacenamiento refrigerado." *Revista*

Iberoamericana de Tecnología Postcosecha (19)1: 63-73. http://www.redalyc.org/articulo.oa?id=81355612005. ["Evaluation of the postharvest behavior of Raspberries in different conditions of refrigerated storage." *Ibero-American Journal of Postharvest Technology*]

Xu, Yang, Lianghua Xie, Jiahong Xie, Yu Liub and Wei Chen. 2020. "Pelargonidin-3-O-rutinoside as a novel α-glucosidase inhibitor for improving postprandial hyperglycemia." *Chemical Communications* 55:39-42. doi: 10.1039/C8CC07985D.

Zhang, Qiaozhi, Diego Luna-Vital, and Elvira Gonzalez de Mejia. 2019. "Anthocyanins from Colored Maize Ameliorated the Inflammatory Paracrine Interplay between Macrophages and Adipocytes through Regulation of NF-KB and JNK-Dependent MAPK Pathways." *Journal of Functional Foods* 54:175-86.

SECTION 2 CEREALS, PSEUDO CEREALS, AND LEGUMES

In: Super and Nutraceutical Foods
Editors: Sandra Teresita Martín-del-Campo et al.
ISBN: 978-1-53619-082-3

Chapter 6

AMARANTH: COMPOSITION AND PROCESSING TECHNOLOGIES

J. E. Borges-Martínez, D. del R. Moguel-Concha, M. S. Cid-Gallegos, D. I. Tellez-Medina and Cristian Jiménez-Martínez*
Escuela Nacional de Ciencias Biológicas, Instituto Politécnico Nacional, Ciudad de México, México

ABSTRACT

Amaranth is an old crop that originated on the American continent. It is one of the few plants from which grain, leaves, and stem are used in human nutrition. The leaves show high content of betacyanins and betalains, the stem is used as forage. The grain is the most used product of the plant, but it is necessary to apply pre-treatments (bursting, germination) to eliminate antinutrients and potentialize its bioactive components. Phenolic compounds (such as ruthin), squalene, and proteins (globulin 11s) are some compounds whose bioactivity has been described. Among the most important bioactivities are the antioxidant, antihypertensive, antidiabetic and anticarcinogenic. Another useful component is the amaranth starch, on which studies have carried out to be able to apply it as an alternative material for microencapsulation.

Keywords: amaranth, nutritional quality, functional properties, peptides, antioxidant capacity

INTRODUCTION: HISTORY, DISTRIBUTION AND HEALTH BENEFITS

Amaranth is a pseudocereal, which has a history of 6000 years; pre-Hispanic cultures used the grain as a staple of their diet as well as a religious connotation. Amaranth cultivation grew in the early 1980s; this was due to its agronomic, nutritional and nutraceutical characteristics (Barba de la Rosa et al., 2009; Corke, Cai, and Wu 2015; Santra and

* Corresponding Author's E-mail: crisjm_99@yahoo.com.

Schoenlechner 2017). This grain is native of the American continent, and it has spread to various countries in Asia and Africa (Escudero et al., 2004). China is currently the main producer, followed by India, Peru, Mexico and the United States (Akinola & Eresama, 2010).

According to taxonomic classification, amaranth belongs to the class of dicotyledons subclass Caryophyllidae, order Caryophyllales, Amaranthaceae family, and genus Amaranthus, consisting of about 800 species, with single characteristics that depend on the environment in which they grow, and this complicates taxonomic identification (Arent and Dal Bello 2008; Perales-Sanchez, Reyes et al., 2015). Amaranth is a highly efficient crop that thrives in drought conditions, saline, alkaline, and acid soils, and resist high temperatures. Besides, its cultivar is resistant to diseases and pests and has higher efficiency in CO_2 fixation and can produce the same amount of biomass with less amount of water than most of the cereals (Wrigley et al., 2004). Another essential feature is that it can be produced in a short time (150-180 days) (Huerta-Ocampo and Barba de la Rosa 2011).

This pseudocereal is useful in human food, especially in the preparation of flour, snacks, pasta, and biscuits. Their vegetative parts, such as leaves, can be used in human food (vegetables) or animal feed (fodder), but also as a crop fertilizer (Martirosyan et al., 2007). The characteristic red color of its vegetative tissue presents high betacyanin levels, which can be utilized as natural colorant in food (Corke et al., 2015). At present, amaranth has attracted worldwide attention as a crop with high nutritional and biological potential, with multiple uses, due to the diversity of compounds it possesses; as well, it can be consumed as nutraceutical food to prevent chronic-degenerative diseases (Klimczak et al., 2002; Plate and Aréas 2002; Paśko et al., 2009).

General Features

Amaranth is one of the few plants that are known as 'multipurpose plants' i.e, one of the few crops in which the leaves and grains consumed (Coelho et al., 2018). The genus Amaranthus includes the species: *A. caudatus, A. hypochondriacus, and A. cruentus*, from which the most considerable amount of grain for human consumption is obtained, hence its economic importance in Mexico (Ayala Garay et al., 2014) and other countries.

The size of the amaranth plant varies according to its branching, reaching up to 3.5 m in height. The leaves usually are elliptical or oval oblong with acute or accumulated apex and the base cradled or acute. The inflorescence is of considerable size, with lateral spikes and panicles; it is very dense, erect, and thorny.

The size of the bracts is equal to that of the branches; they are pointed and give the inflorescence its spiny appearance and feel. The amaranth flower has five sepals, slightly curved, and longer than those of related species (Peiretti 2018; D'Amico and Schoenlechner 2017).

Amaranth seed (Figure 1) is different from any typical cereal. It has been considered a pseudo-cereal because it shares similar chemical characteristics respect to those of cereals. The color of the grain varies accordingly to the species and can be black, brown, yellow, or white. The seeds have campylotropic embryo and present a circular form since the ends tend to join, almost enclosing the perisperm. The seeds are tiny, a single plant can produce more than 500 g of grain; however, this yield varies according to the variety used and soil fertility conditions (Corke et al., 2015; Wrigley et al., 2004).

The chemical composition of amaranth is similar to corn, rice or wheat (see Table 1) (Black 2001), its main component is starch (50-65%), composed of a low proportion of amylose and little or no retrograded (López et al., 1994). It also has a significant content of protein (14-17%) and fiber (4-10%), of which about 60% is insoluble fiber and the rest is soluble fiber (Escudero et al., 2004).

Amaranth leaves contain between 27.8-48.6% of crude protein, this percentage is higher than the protein content of spinach leaves. The lipid content varies from 2.9 to 7.1%, linoleic acid contained in higher proportion (40-55%). Besides, it has been found that the red color present in the leaves of certain species has high levels of betalains (amaranthine and isoamaranthine) as natural pigments (Pavokovi and Krsnik-Rasol 2011; Gengatharan et al., 2015; Strack et al., 2003). Amaranth leaves and stems are a good source of fiber (11.1-22.9%) and are used for human foodstuff in Countries such as China and India (Corke, Cai, and Wu 2015).

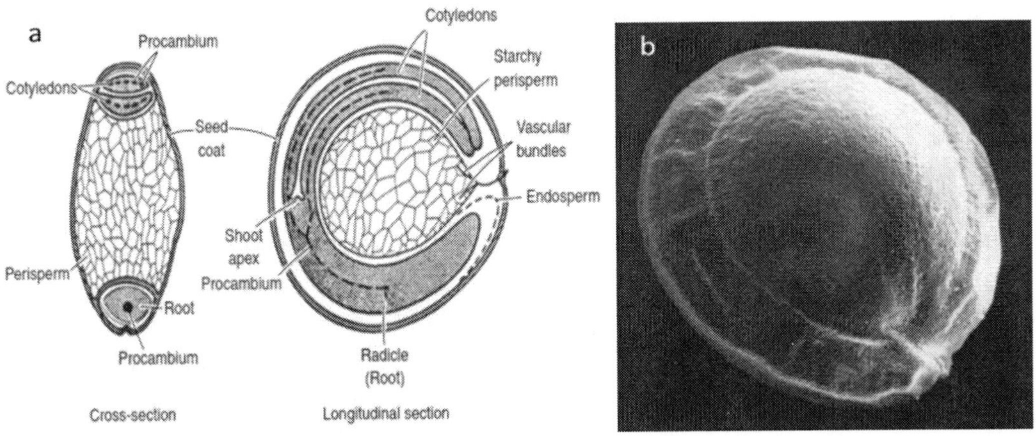

Figure 1. a) Cross and longitudinal sections of amaranth seed; b) Scanning electron micrograph of amaranth seed (Corke et al., 2015).

Table 1. Proximal composition of amaranth seed and some cereals

	Composition (% d.w.)			
	Amaranth	Wheat	Corn	Rice
Carbohydrates	59.2	66.9	67.7	75.4
Crude Protein	16.6	14	10.3	8.5
Lipids	7.2	2.1	4.5	2.1
Crude fiber	4.1	2.6	2.3	0.9
Ash	3.3	1.9	1.4	1.4
Moisture	9.6	12.5	13.6	11.7

(D'Amico and Schoenlechner 2017; Corke et al., 2015; Barba de la Rosa et al. 2009; Wrigley et al., 2004).

Chemical Composition

Amaranth species have been adapted and diversified according to cultivation conditions. Each species (Table 2) has its own varieties; for example, *A. hypochondriacus* represents the varieties Aztec, Mixtec, Nepalese, and Pico's race, and cultivated in Mexico. A. cruentus refers to South American, Mexican, Guatemalan, and African varieties, and also is planted in Central America and Mexico. *A. caudatus* is planted in the Andes region (Solano et al., 2010).

Table 2. Chemical composition of various species of amaranth grains

Composition (% d.w.)	A. caudatus	A. cruentus	A. hybridus	A. hypochondriacus
Carbohydrate	59.6-62.8	60.7-62.6	58.6	57.0
Crude Protein	17.6-18.4	13.2-18.2	14.0	17.9
Fat	6.9-8.1	6.3-8.1	6.7	7.7
Crude fiber	3.2-5.8	3.6-4.4	6.6	2.2
Ash	3.1-4-4	2.8-3.9	3.6	4.1
Moisture	9.5-11.6	6.2-8-8	10.5	11.1

(Peiretti 2018; Paredes-López 1994; Wrigley et al., 2004).

Starch

Starch is the main component of the amaranth grain, and it represents approximately 49-68% of starch on a dry basis. Starch is a molecule that is composed of two glucose polymers, amylose, and amylopectin. The ratio of these two molecules is essential for the final use of starch, this ratio is influenced by genetic diversity within amaranth species (Perez-Rea and Antezana-Gomez 2018). In addition, amaranth starch granules have polygonal, round, lenticular and elliptical shapes, and are very small in size (0.8-2.5 mm) compared to potato or corn starch granules. The small size and distribution of the granules influence the functional properties of starch (Santra and Schoenlechner 2017). Since 1970, several studies have been reported on the physicochemical and functional attributes of amaranth starch, for instance, its wide range of viscosity, its shear strength, and its gelatinization temperatures. The uses of amaranth starch include its application as food thickener, cosmetic and laundry powder, and even for obtaining biodegradable products (Singh and Singh 2011; Perez-Rea and Antezana-Gomez 2018; D'Amico and Schoenlechner 2017).

Proteins

The protein content of the amaranth seed depends on the variety and the environmental conditions of the crop. The amaranth protein fractions are divided, in accordance to their solubility, into albumins (40%), globulins (20%), glutelins (25-30%), and prolamins (2-3%). The 11s globulins are the most abundant reserve protein in amaranth, followed by the 7s ones. The molecular mass of 11s globulin is approximately 56 KDa, while 7s globulin has a total of 200 KDa (Alvarez-Jubete et al., 2010; Bhat et al., 2015). Albumins are rich in sulfur and amino acids such as valine and lysine. Globulins have a high presence of valine, leucine, prolamine, threonine, phenylalanine, leucine, glycine, and histidine. The protein fraction 11s is the most present due to the high content of essential amino acids (tryptophan, methionine, lysine, histidine, phenylalanine, valine, and isoleucine) (Rodríguez et al., 2020; Venskutonis and Kraujalis 2013). In amaranth, the gliadin content is less than 0.01%, which would make it easy to assimilate in food products for people with coeliac disease. Another important aspect is that the leucine content is higher in the amaranth grain than in rice and corn. The presence of lysine (5%) in amaranth, as well as the amount of sulphur amino acids, means that the amaranth grain has a high nutritional potential (Coelho et al., 2018; Tosi et al., 2002; Suárez et al., 2020; Valcárcel-Yamani and Lannes 2012). Amaranth seed also has peptides with various biological activities (Table 3).

Table 3. Biological activity of peptides in amaranth proteins

Peptides	Activity	Reference
SFNLPILR; FNLPILR; QAFEDGFEWVSFK; AFEDGFEWVSFK	Antirenin Antihypertensive	Nardo et al., 2020; Suárez et al., 2020; Suárez & Añón, 2019
SSEDIKE	Antiinflammatory	Moronta et al., 201
IQAEAGLTEV, TEVWDSNEQE, AFEDGFE	Antithrombotic	Sabbione et al., 201
AWEEREQGSR; YLAGKPQQEH; IYIEQGNGITGM; TEVWDSNEQ	Antioxidant	Orsini-Delgado et al., 2016

Lipids

Amaranth seed has a much higher lipid content compared to other cereals. The most abundant fatty acid in amaranth is linoleic (33%-55.9%), followed by oleic (18.7%-38.9%) palmitic (14.04%-26%) and stearic (3.11%-4.47%) acids and linoleic acid (0.20%-1.97%) (Kaur et al., 2010; Gimplinger et al., 2007). In this lipid fraction, the amount of squalene is remarkable (2.26-5.94%) (Table 4).

Squalene (Figure 2) is an unsaponifiable lipid that acts as a biosynthetic precursor to all plant and animal steroids (Tang et al., 2016; Palombini et al., 2013).

Table 4. Comparison of squalene content between amaranth and other seeds*

Species	Squalene (%)
cruentus	4.52-5.44
hybridus	5.23
hypochondriacus	3.62-5.01
Maize	0.03
Rice	0.3
Olive	0.4
Peanut	0.03

*Dry weight basis (Corke et al., 2015).

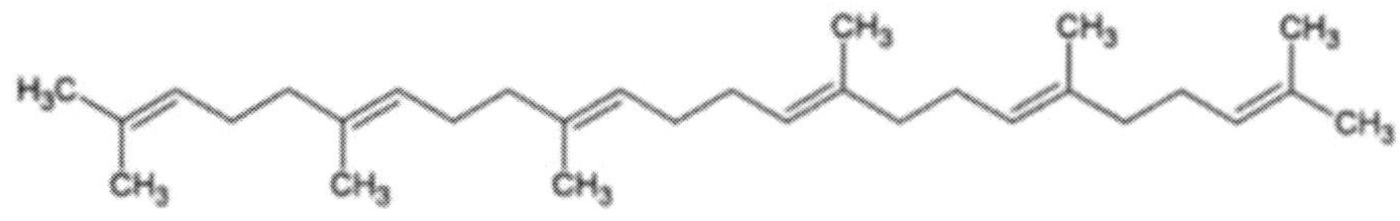

Figure 2. Squalene structure (Spanova and Daum 2011).

Berger et al., 2003 suggest that amaranth has hypocholesterolemic, antioxidant, and anticarcinogenic effects due to its unsaponifiable profile and unsaturated fatty acid content.

In amaranth, tocopherols and tocotrienols have been reported, which have a regulatory role in cholesterol metabolism, as well as high antioxidant capacity (Yelisyeyeva et al., 2014).

Specific studies that have been carried out to determine the biological activity of squalene. It has been demonstrated that squalene has an anti-tumorigenic effect on the colon, lung, and skin; when this compound is administered together with anti-tumor drugs, a growth decrease and inhibition of the proliferation of cancerous cells occur (Murakoshi et al., 1992; Smith et al., 1998). Another study focused on the effect of squalene on breast cancer cells with different degrees of involvement reported that, although squalene does not affect the rate of cell formation, it can protect against H_2O_2-induced oxidative damage in DNA (Warleta et al., 2010).

The potential contribution of squalene on drug effects has also been observed, for example, in the treatment of cholesterol and triacylglycerides in the blood (Smith et al., 1998). Chan et al., 1996, observed that consumption of squalene and pravastatin (a hypocholesterolemic drug) decreased low-density lipoprotein (LDL) levels and increased high-density lipoprotein (HDL) levels, compared to the control group. Richter & Schäfer, 1982, observed that the application of squalene produces an effect like that of substances used for cholesterol reduction. Squalene has also shown to protect eye tissues such as the retina, especially photoreceptors (Fliesler and Keller 1997). Alcohol consumption also causes lipid peroxidation and damage to the retina of the fetus during pregnancy, so observed that by administering squalene to chicken embryos, this damage decreased (Aguilera et al., 2005). It could also use in vaccines because of the high surface tension it has, which causes the formation of stable emulsions. Furthermore, due to its biocompatibility, it improves solubility and prolonged release of liposoluble substances facilitating their bioavailability for the cells (Fox, 2009).

Vitamins and Minerals

Amaranth seed is an important source of vitamins and minerals. Murakami et al., 2014 identified B-type vitamins in amaranth seed, riboflavin (B2) (147 ± 9.3 μg/100g), niacin (B3) (340 ± 16.45 μg/100g), pyridoxine (B6) (24.8 ± 3.7 μg/100g) and pantothenic acid (1150 ± 60 μg/100g). Valcárcel-Yamani & Lannes, 2012, quantified riboflavin (0.19-0.23, mg/100g), niacin (1.17-1.45 mg/100g) and thiamin (B1) (0.07-0.10 mg/100g). Pseudocereals such as amaranth or quinoa, have a higher mineral content than common cereals such as wheat or corn. Calcium is the mineral with the highest concentration in amaranth, which has five times more than quinoa. Iron, another mineral with a high content in amaranth (4.6-10.7 mg/100 g), presents 2 to 3 times more than wheat. Other minerals such as phosphorus, potassium, and sodium are present in lesser amounts, while zinc, copper, and manganese in moderate concentrations (Murakami et al., 2014; Palombini et al., 2013; Bolaños et al., 2016; Alvarez-Jubete et al., 2010).

PHYTOCHEMICALS

Among the most important secondary metabolites given their pharmacological effect, the phenolic compounds are those found in the highest concentration in amaranth seed. Phenolic compounds are any compound containing a benzene ring with one or more hydroxyl groups, including phenolic acids, flavonoids, tannins, and coumarins. These compounds affect the appearance, odor, taste, and oxidative stability (Dorf and Miller 2007). There are several studies in which phenolic acids (Table 5), flavonoids, and glycosylated compounds have been identified and quantified in amaranth seed in which shown that amaranth is rich in polyphenols, mostly flavonoid type, like ruthin. Other phenolic compounds that have been identified are quercetin, p-hydroxybenzoic acid, ferulic acid, and caffeic acid. The presence of these compounds in amaranth may be important for the prevention of chronic degenerative diseases (Repo-Carrasco-Valencia et al., 2010; Ogrodowska et al. 2014; Kalinova and Dadakova 2009; Klimczak et al., 2002; Barba de la Rosa et al., 2009). Nonetheless, pre-treatment to the amaranth seed is recommended because it contains some compounds that can act as antinutritional agents, specifically the tannins and phytates. Non-nutritional compounds might form insoluble complexes with other molecules and hinder the absorption of proteins and minerals, in particular the iron, zinc, and calcium (Kunyanga et al., 2012; Ogrodowska et al., 2014).

Table 5. Phenolic compounds identified in amaranth species

Species	Phenolic compounds	Reference
A. paniculatus *A. caudatus*	Gallic acid, protocatechuic acid., caffeic acid., p-coumaric acid, ferulic acid, sinapic acid	Klimczak et al., 2002
A. hypochondriacus	Ruthin, isoquercitine, nicotiflorine, 4-hydroxybenzoic acid	Barba de la Rosa et al., 2009
A. caudatus, A. hybrid, *A. hypochondriacus*	Ruthin, quercetin	Kalinova & Dadakova, 2009
A. caudatus	Caffeic acid, ferulic acid, p-coumaric acid, sinapic acid, protocatechuic acid, benzoic acid	Repo-Carrasco-Valencia et al., 2010
A. cruentus	Vanillin acid, p-hydroxybenzoic acid, ferulic acid, caffeic acid, sinapic acid, cinnamic acid	Ogrodowska et al., 2014

AMARANTH PROCESSING

Food processing induces significant changes in the chemical composition and improvements in the bioactivity of amaranth seed. There are several methods for amaranth seed processing. Traditionally, amaranth seed undergoes a bursting process (Figure 3), which is performed on metal comal hot mud, while the branches of the same plant are used to turn the grains up and down. This process is not controllable, and sometimes grains are left unburnt or burnt (Murakami et al., 2014).

Burst capacity of amaranth grain is determined by the presence of perisperm in the seed, which is formed by starch molecules. The general purpose of bursting the amaranth grain is to obtain a pleasant taste, color, and aroma. Factors such as humidity, grain size, bursting temperature, and exposure time are involved in this process (Barrales &Torres, 1998).

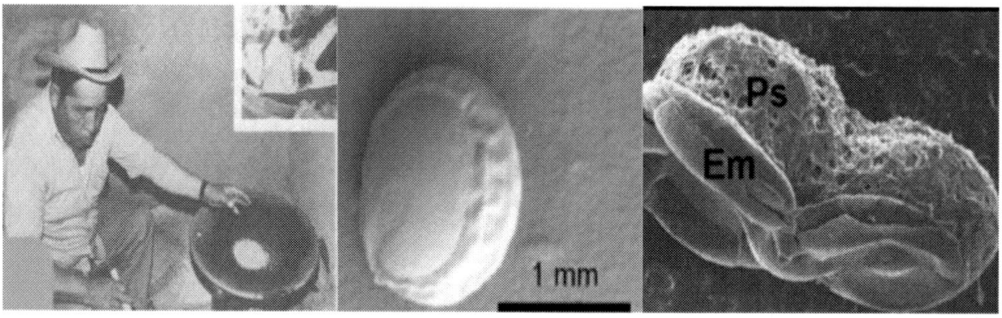

Figure 3. Amaranth seed bursting process (Murakami et al., 2014).

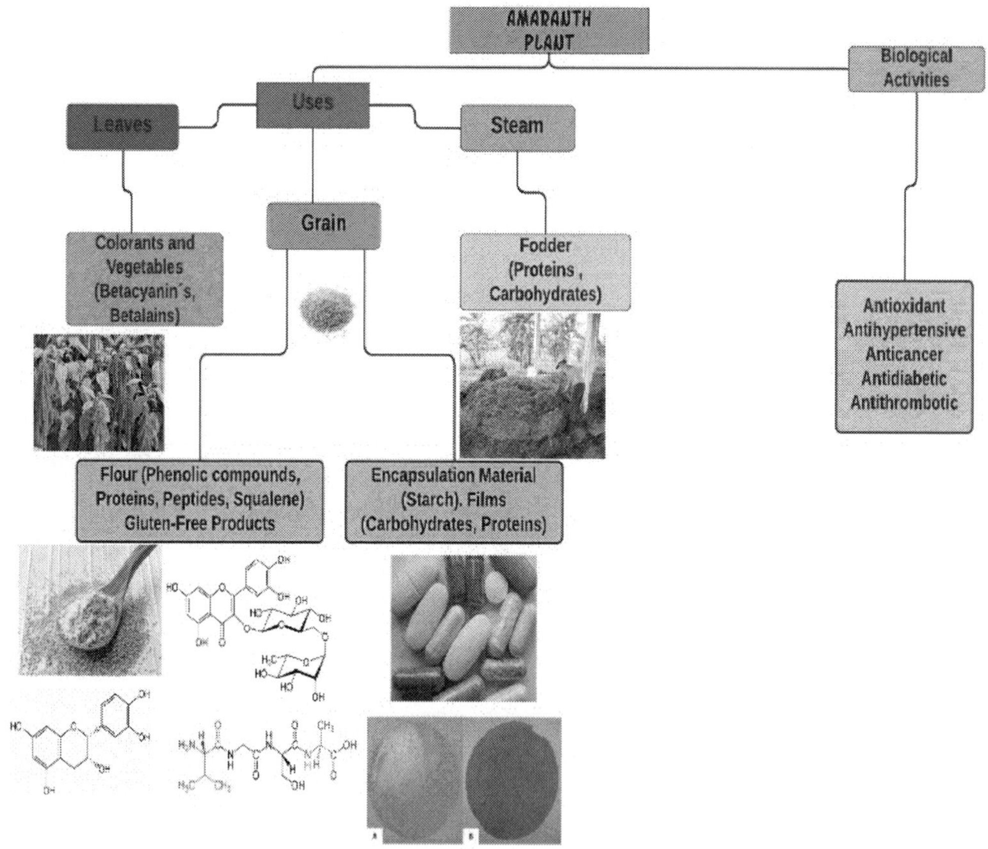

Figure 4. Biological activities and properties of the amaranth plant.

Shevkani & Singh, 2014, showed the influence of lipids and proteins on amaranth flour, demonstrating a change from lipid removal to degreasing, improving the stability of flour and starch during heating. Muyonga et al., 2014, analyzed various amaranth protein concentrates and described that *A. cruentus* species has a higher protein content (which is related to the antioxidant activity of the concentrate) and, observed that both protein content and antioxidant activity are affected by exposure of the grain to high temperatures (60°C). González et al., (2007) showed that during the amaranth seed extrusion process, the oxidative activity of the grain is not reduced compared to the raw grain, which would increase the antioxidant activities of the grain. Perales-Sánchez et al., (2014) indicated that a process of amaranth seed germination at 30°C for 78 h increases the content of phenolic acids and flavonoids; therefore, this would lead to an increase in antioxidant activity. Montoya-Rodríguez et al., (2015), demonstrated through an in silico analysis that peptides (TSAPELIY,AMPAGVSHW) derived from amaranth globulins 11s could be used as functional food ingredients to assist in the prevention and reduction of chronic diseases. Sabbione et al., (2015), analyzed protein fractions from amaranth seeds, showing that antithrombotic activity was due to the glutelin fraction. Corke et al., (2015) performed an extraction of betacyanin's from amaranth leaves for use in gelatin, ice cream. A model beverage containing pigments ffrom amaranth was evaluated by comparisson with other commercial dyes, showing that the amaranth leaf dye had better characteristics and potent antioxidant activity (Miguel 2018). This could lead to the possibility of adding amaranth pigments in the diet to be used as potential natural antioxidants. Amaranth seed has various uses that could potentially use in the food industry (Figure 4).

Ferreira et al., (2016) and Oliveira et al., (2017) have focused on the development of films and coatings from amaranth proteins and polysaccharides. In such studies, they have also focused on the optimization of these films to improve their mechanical characteristics such as resistance to breakage and adhesion (Otoni et al., 2017). Another feature of amaranth is that it has been used as a food thickener and as a fat substitute (Ott et al., 2015). Tari et al., (2003) used amaranth starch as an encapsulation material; their results showed that *A. paniculatus* starch in combination with a linking agent (gum arabic) had better characteristics to encapsulate vanillin than other compounds, such as carboxymethyl cellulose or carrageenan.

Conclusion and Perspectives

Amaranth has importance for human nutrition; as well as its bioactive compounds and its encapsulating properties offer technological benefits and several potential applications. Its grains and leaves have an attractive chemical composition and a higher nutritional and functional value than cereals. These characteristics confer important biological activities to amaranth (antioxidant, antithrombotic, antihypertensive, antidiabetic, anticancer) so that its leaves and seeds can be used for alternative treatments in chronic-degenerative diseases. It is certain that amaranth has attracted the world's attention and is a new crop with high potential in various areas of research.

References

Aguilera, Yolanda, Manuel E. Dorado, Francisco A. Prada, Juan J. Martínez, Adela Quesada, and Valentina Ruiz-Gutiérrez. 2005. The Protective Role of Squalene in Alcohol Damage in the *Chick Embryo Retina*. *Experimental Eye Research* 80 (4): 535–43. https://doi.org/10.1016/j.exer.2004.11.003.

Akinola, A. A., and P. C. Eresama. 2010. Economics of *Amaranthus* Production under Tropical Conditions. *International Journal of Vegetable Science* 16 (1): 32–43. https://doi.org/10.1080/19315260903178655.

Alvarez-Jubete, L., E. K. Arendt, and E. Gallagher. 2010. Nutritive Value of Pseudocereals and Their Increasing Use as Functional Gluten-Free Ingredients. *Trends in Food Science and Technology* 21 (2): 106–13. https://doi.org/10.1016/j.tifs.2009.10.014.

Alvarez-Jubete, L., H. Wijngaard, E. K. Arendt, and E. Gallagher. 2010. Polyphenol Composition and in Vitro Antioxidant Activity of Amaranth, Quinoa Buckwheat and Wheat as Affected by Sprouting and Baking. *Food Chemistry* 119 (2): 770–78. https://doi.org/10.1016/j.foodchem.2009.07.032.

Arent, Elke, and Fabio Dal Bello. 2008. *Gluten-Free Cereal Products and Beverages. Academic Press*. 1st ed. Vol. 1. Burlington, USA: Elsevier. https://doi.org/10.1108/nfs.2012.01742eaa.010.

Ayala Garay, Alma, Patricia Rivas Valencia, Lorena Cortes-Espinoza, Micaela de la Olán, Diana Escobedo-López, and Eduardo Espitia Rangel. 2014. The Profitability of the Amaranth (*Amaranthus spp.*) Cultivation in the Central Region of Mexico. *CIENCIA Ergo-Sum* 21 (1): 47–54.

Barba de la Rosa, A. P., Inge S. Fomsgaard, Bente Laursen, Anne G. Mortensen, L. Olvera-Martínez, C. Silva-Sánchez, A. Mendoza-Herrera, J. González-Castañeda, and A. De León-Rodríguez. 2009. Amaranth (*Amaranthus hypochondriacus*) as an Alternative Crop for Sustainable Food Production: Phenolic Acids and Flavonoids with Potential Impact on Its Nutraceutical Quality. *Journal of Cereal Science* 49 (1): 117–21. https://doi.org/10.1016/j.jcs.2008.07.012.

Barrales-Domínguez, J. S., and L. Torres-Hernández. 1998. Amaranth Granule *(Amaranthus hypochondriacus L.*) Trapping Capacity Produced in Two Temporary Environments. *Revista Chapingo Serie Horticultura* IV (01): 63–66. https://doi.org/10.5154/r.rchsh.1997.03.027.

Berger, Alvin, Gérard Gremaud, Marcel Baumgartner, Dietrich Rein, Irina Monnard, Elena Kratky, Wolfgang Geiger, et al. 2003. Cholesterol-Lowering Properties of Amaranth Grain and Oil in Hamsters. *International Journal for Vitamin and Nutrition Research* 73 (1): 39–47. https://doi.org/10.1024/0300-9831.73.1.39.

Bhat, Anshula, Gouri Satpathy, and Rajinder K Gupta. 2015. Evaluation of Nutraceutical Properties of *Amaranthus Hypochondriacus L* . Grains and Formulation of Value Added Cookies. *Journal of Pharmacognosy and Phytochemistry* 3 (5): 51–54.

Black, J L. 2001. Variation in Nutritional Value of Cereal Grains across Livestock Species. *Proceedings of the Australian Poultry Science Symposium* 13: 22–29. https://pdfs.semanticscholar.org/4baf/7e405b1adae59dfaa301be1ac79b6fa90f9c.pdf.

Bolaños, Daniela, Eduardo J. Marchevsky, and José M. Camiña. 2016. Elemental Analysis of Amaranth, Chia, Sesame, Linen, and Quinoa Seeds by ICP-OES: Assessment of

Classification by Chemometrics. *Food Analytical Methods* 9 (2): 477–84. https://doi.org/10.1007/s12161-015-0217-4.

Chan, Paul, Brian Tomlinson, Ching Bing Lee, and Ying Shiung Lee. 1996. Effectiveness and Safety of Low-Dose Pravastatin and Squalene, Alone and in Combination, in Elderly Patients with Hypercholesterolemia. *Journal of Clinical Pharmacology* 36 (5): 422–27. https://doi.org/10.1002/j.1552-4604.1996.tb05029.x.

Coelho, Laylla Marques, Pedro Miguel Silva, Joana T. Martins, Ana C. Pinheiro, and António A. Vicente. 2018. Emerging Opportunities in Exploring the Nutritional/Functional Value of Amaranth. *Food and Function* 9 (11): 5499–5512. https://doi.org/10.1039/c8fo01422a.

Corke, Harold, Y. Z. Cai, and H. X. Wu. 2015. Amaranth: Overview. *Encyclopedia of Food Grains: Second Edition* 1–4: 287–96. https://doi.org/10.1016/B978-0-12-394437-5.00032-2.

D'Amico, Stefano, and Regine Schoenlechner. 2017. Amaranth: Its Unique Nutritional and Health-Promoting Attributes. Gluten-Free Ancient Grains: Cereals, Pseudocereals, and Legumes: Sustainable, Nutritious, and Health-Promoting Foods for the 21st Century. *Elsevier Ltd.* https://doi.org/10.1016/B978-0-08-100866-9.00006-6.

Dorf, ERC, and SL Miller. 2007. Phenolic Compounds in Cereal Grains and Their Health Benefits. *Cereal Foods World* 52 (3): 105–11. https://doi.org/10.1094/CFW-52-3-0105

Escudero, N. L., M. L. De Arellano, J. M. Luco, M. S. Giménez, and S. I. Mucciarelli. 2004. "Comparison of the Chemical Composition and Nutritional Value of Amaranthus Cruentus Flour and Its Protein Concentrate." *Plant Foods for Human Nutrition* 59 (1): 15–21. https://doi.org/10.1007/s11130-004-0033-3.

Ferreira, Ana R. V., Vítor D. Alves, and Isabel M. Coelhoso. 2016. Polysaccharide-Based Membranes in Food Packaging Applications. *Membranes* 6 (2): 1–17. https://doi.org/10.3390/membranes6020022.

Fliesler, Steven J., and R. Kennedy Keller. 1997. Isoprenoid Metabolism in the Vertebrate Retina. *International Journal of Biochemistry and Cell Biology* 29 (6): 877–94. https://doi.org/10.1016/S1357-2725(97)00018-6.

Fox, Christopher B. 2009. Squalene Emulsions for Parenteral Vaccine and Drug Delivery. *Molecules* 14 (9): 3286–3312. https://doi.org/10.3390/molecules14093286.

Gengatharan, A., G. A. Dykes, and W. S. Choo. 2015. Betalains: Natural Plant Pigments with Potential Application in Functional Foods. *Lwt* 64 (2): 645–49. https://doi.org/10.1016/j.lwt.2015.06.052.

Gimplinger, D. M., G. Dobos, R. Schönlechner, and H. P. Kaul. 2007. Yield and Quality of Grain Amaranth (*Amaranthus sp.*) in Eastern Austria. *Plant, Soil and Environment* 53 (3): 105–12. https://doi.org/10.17221/2224-pse.

González, R., C. Carrara, E. Tosi, M. C. Añón, and A. Pilosof. 2007. Amaranth Starch-Rich Fraction Properties Modified by Extrusion and Fluidized Bed Heating. *LWT - Food Science and Technology* 40 (1): 136–43. https://doi.org/10.1016/j.lwt.2005.09.003.

Huerta-Ocampo, Jose Angel, and Ana Paulina Barba de la Rosa. 2011. Amaranth: A Pseudo-Cereal with Nutraceutical Properties. *Current Nutrition & Food Science* 7 (1): 1–9. https://www.ingentaconnect.com/content/ben/cnf/2011/00000007/ 00000001/art00001.

Kalinova, Jana, and Eva Dadakova. 2009. Rutin and Total Quercetin Content in Amaranth (*Amaranthus spp.*). *Plant Foods for Human Nutrition* 64 (1): 68–74. https://doi.org/10.1007/s11130-008-0104-x.

Kaur, Shubhpreet, Narpinder Singh, and Jai Chand Rana. 2010. *Amaranthus hypochondriacus* and *Amaranthus caudatus* Germplasm: Characteristics of Plants, Grain and Flours. *Food Chemistry* 123 (4): 1227–34. https://doi.org/10.1016/j.foodchem.2010.05.091.

Klimczak, I., M. Małecka, and B. Pachołek. 2002. Antioxidant Activity of Ethanolic Extracts of Amaranth Seeds. *Nahrung - Food* 46 (3): 184–86. https://doi.org/10.1002/1521-3803(20020501)46:3<184::AID-OOD184>3.0. CO;2-H.

Kunyanga, Catherine N., Jasper K. Imungi, Michael W. Okoth, Hans K. Biesalski, and Vellingiri Vadivel. 2012. Total Phenolic Content, Antioxidant and Antidiabetic Properties of Methanolic Extract of Raw and Traditionally Processed Kenyan Indigenous Food Ingredients. *LWT - Food Science and Technology* 45 (2): 269–76. https://doi.org/10.1016/j.lwt.2011.08.006.

Martirosyan, Danik M., Lidia A. Miroshnichenko, Svetlana N. Kulakova, Ala V. Pogojeva, and Vladimir I. Zoloedov. 2007. "Amaranth Oil Application for Coronary Heart Disease and Hypertension." *Lipids in Health and Disease* 6 (February 2007). https://doi.org/10.1186/1476-511X-6-1.

Miguel, Maria Graça. 2018. Betalains in Some Species of the *Amaranthaceae* Family: A Review. *Antioxidants* 7 (4): 1–33. https://doi.org/10.3390/antiox7040053.

Montoya-Rodríguez, Alvaro, Mario A. Gómez-Favela, Cuauhtémoc Reyes-Moreno, Jorge Milán-Carrillo, and Elvira González de Mejía. 2015. Identification of Bioactive Peptide Sequences from Amaranth (*A. hypochondriacus*) Seed Proteins and Their Potential Role in the Prevention of Chronic Diseases. *Comprehensive Reviews in Food Science and Food Safety* 14 (2): 139–58. https://doi.org/10.1111/1541-4337.12125.

Moronta, Julian, Paola L. Smaldini, Guillermo H. Docena, and María Cristina Añón. 2016. Peptides of Amaranth Were Targeted as Containing Sequences with Potential Anti-Inflammatory Properties. *Journal of Functional Foods* 21: 463–73. https://doi.org/10.1016/j.jff.2015.12.022.

Murakami, Taro, Aiko Yutani, Tetsuo Yamano, Hiroyuki Iyota, and Yotaro Konishi. 2014. Effects of Popping on Nutrient Contents of Amaranth Seed. *Plant Foods for Human Nutrition* 69 (1): 25–29. https://doi.org/10.1007/s11130-013-0392-7.

Murakoshi, Michiaki, Hoyoku Nishino, Harukuni Tokuda, Akio Iwashima, Junichi Okuzumi, Hirokazu Kitano, and Ryozo Iwasaki. 1992. Inhibition by Squalene of the Tumor-promoting Activity of 12-O-Tetradecanoylphorbol-13-acetate in Mouse-skin Carcinogenesis. *International Journal of Cancer* 52 (6): 950–52. https://doi.org/10.1002/ijc.2910520620.

Muyonga, John H., Brian Andabati, and Geoffrey Ssepuuya. 2014. "Effect of Heat Processing on Selected Grain Amaranth Physicochemical Properties." *Food Science & Nutrition* 2 (1): 9–16. https://doi.org/10.1002/fsn3.75.

Nardo, Agustina E., María C. Añón, and Alejandra V. Quiroga. 2020. "Identification of Renin Inhibitors Peptides from Amaranth Proteins by Docking Protocols." *Journal of Functional Foods* 64 (July): 103683. https://doi.org/10.1016/j.jff.2019.103683.

Ogrodowska, Dorota, Ryszard Zadernowski, Sylwester Czaplicki, Dorota Derewiaka, and Beata Wronowska. 2014. "Amaranth Seeds and Products - The Source of Bioactive Compounds." *Polish Journal of Food and Nutrition Sciences* 64 (3): 165–70. https://doi.org/10.2478/v10222-012-0095-z.

Oliveira, Sandra Prestes Lessa Fernandes, Larissa Canhadas Bertan, Christiane Maciel Vasconcellos Barros De Rensis, Ana Paula Bilck, and Priscila Cristina Bizam Vianna.

2017. "Whey Protein-Based Films Incorporated with Oregano Essential Oil." *Polimeros* 27 (2): 158–64. https://doi.org/10.1590/0104-1428.02016.

Orsini Delgado, María C., Agustina Nardo, Marija Pavlovic, Hélène Rogniaux, María C. Añón, and Valeria A. Tironi. 2016. "Identification and Characterization of Antioxidant Peptides Obtained by Gastrointestinal Digestion of Amaranth Proteins. *Food Chemistry* 197: 1160–67. https://doi.org/10.1016/j.foodchem.2015.11.092.

Otoni, Caio G., Roberto J. Avena-Bustillos, Henriette M.C. Azeredo, Marcos V. Lorevice, Márcia R. Moura, Luiz H. C. Mattoso, and Tara H. McHugh. 2017. Recent Advances on Edible Films Based on Fruits and Vegetables—A Review. *Comprehensive Reviews in Food Science and Food Safety* 16 (5): 1151–69. https://doi.org/10.1111/1541-4337.12281.

Ott, Cristina, Ioana Lacatusu, Gabriela Badea, Iulia Adriana Grafu, Daniela Istrati, Narcisa Babeanu, Raluca Stan, Nicoleta Badea, and Aurelia Meghea. 2015. Exploitation of Amaranth Oil Fractions Enriched in Squalene for Dual Delivery of Hydrophilic and Lipophilic Actives. *Industrial Crops and Products* 77: 342–52. https://doi.org/10.1016/j.indcrop.2015.08.057.

Palombini, Sylvio Vicentin, Thiago Claus, Swami Arêa Maruyama, Aline Kirie Gohara, Aloisio Henrique Pereira Souza, Nilson Evelázio de Souza, Jesuí Vergílio Visentainer, Sandra Terezinha Marques Gomes, and Makoto Matsushita. 2013. Evaluation of Nutritional Compounds in New Amaranth and Quinoa Cultivars. *Food Science and Technology* 33 (2): 339–44. https://doi.org/10.1590/S0101-20612013005000051.

Paredes-López, Octavio. 1994. *Amaranth Biology, Chemistry and Technology*. Edited by Octavio Paredes-López. *Chemical Ruber Company*. 1st ed. Irapuato, México: Taylor & Francis Group.

Paśko, Paweł, Henryk Bartoń, Paweł Zagrodzki, Shela Gorinstein, Maria Fołta, and Zofia Zachwieja. 2009. Anthocyanins, Total Polyphenols and Antioxidant Activity in Amaranth and Quinoa Seeds and Sprouts during Their Growth. *Food Chemistry* 115 (3): 994–98. https://doi.org/10.1016/j.foodchem.2009.01.037.

Pavokovi, Dubravko, and Marijana Krsnik-Rasol. 2011. Complex Biochemistry and Biotechnological Production of Betalains. *Food Technology and Biotechnology* 49 (2): 145–55.

Peiretti, P. G. 2018. Amaranth in Animal Nutrition: A Review. *Livestock Research for Rural Development* 30 (5).

Perales-Sánchez, Janitzio X. K., Cuauhtémoc Reyes-Moreno, Mario A. Gómez-Favela, Jorge Milán-Carrillo, Edith O. Cuevas-Rodríguez, Angel Valdez-Ortiz, and Roberto Gutiérrez-Dorado. 2014. Increasing the Antioxidant Activity, Total Phenolic and Flavonoid Contents by Optimizing the Germination Conditions of Amaranth Seeds. *Plant Foods for Human Nutrition* (Dordrecht, Netherlands) 69 (3): 196–202. https://doi.org/10.1007/s11130-014-0430-0.

Perez-Rea, Daysi, and Raquel Antezana-Gomez. 2018. *The Functionality of Pseudocereal Starches. Starch in Food: Structure, Function and Applications: Second Edition*. Vol. 570. Elsevier Ltd. https://doi.org/10.1016/B978-0-08-100868-3.00012-3.

Plate, Andrea Y. A., and Josê A. G. Aréas. 2002. Cholesterol-Lowering Effect of Extruded Amaranth *(Amaranthus caudatus L.)* in Hypercholesterolemic Rabbits. *Food Chemistry* 76 (1): 1–6. https://doi.org/10.1016/S0308-8146(01)00238-2.

Repo-Carrasco-Valencia, Ritva, Jarkko K. Hellström, Juha Matti Pihlava, and Pirjo H. Mattila. 2010. Flavonoids and Other Phenolic Compounds in Andean Indigenous Grains: Quinoa (*Chenopodium quinoa*), Kañiwa (*Chenopodium pallidicaule*) and Kiwicha (*Amaranthus caudatus*). *Food Chemistry* 120 (1): 128–33. https://doi.org/10.1016/j.foodchem.2009.09.087.

Richter, E., and S. G. Schäfer. 1982. The Effect of Squalane on the Absorption of Dietary Cholesterol by the Rat. *Research in Experimental Medicine* 180 (2): 189–91. https://doi.org/10.1007/BF01851058.

Rodríguez, Juan Pablo, Hifzur Rahman, Sumitha Thushar, and Rakesh K. Singh. 2020. Healthy and Resilient Cereals and Pseudo-Cereals for Marginal Agriculture: Molecular Advances for Improving Nutrient Bioavailability. *Frontiers in Genetics* 11 (February): 1–29. https://doi.org/10.3389/fgene.2020.00049.

Sabbione, Ana Clara, Agustina Estefanía Nardo, María Cristina Añón, and Adriana Scilingo. 2016. Amaranth Peptides with Antithrombotic Activity Released by Simulated Gastrointestinal Digestion. *Journal of Functional Foods* 20: 204–14. https://doi.org/10.1016/j.jff.2015.10.015.

Sabbione, Ana Clara, Adriana Scilingo, and María Cristina Añón. 2015. Potential Antithrombotic Activity Detected in Amaranth Proteins and Its Hydrolysates. *LWT - Food Science and Technology* 60 (1): 171–77. https://doi.org/10.1016/j.lwt.2014.07.015.

Santra, D. K., and R. Schoenlechner. 2017. *Amaranth Part 2-Sustainability, Processing, and Applications of Amaranth. Sustainable Protein Sources*. Elsevier Inc. https://doi.org/10.1016/B978-0-12-802778-3.00016-0.

Shevkani, Khetan, and Narpinder Singh. 2014. Influence of Kidney Bean, Field Pea and Amaranth Protein Isolates on the Characteristics of Starch-Based Gluten-Free Muffins. *International Journal of Food Science and Technology* 49 (10): 2237–44. https://doi.org/10.1111/ijfs.12537.

Singh, Narpinder, and Prabhjeet Singh. 2011. Amaranth: Potential Source for Flour Enrichment. *Flour and Breads and Their Fortification in Health and Disease Prevention*, 101–11. https://doi.org/10.1016/B978-0-12-380886-8.10010-8.

Smith, Theresa J., Guang Yu Yang, Darren N. Seril, Jie Liao, and Sungbin Kim. 1998. Inhibition of 4-(Methylnitrosamino)-1-(3-Pyridyl)-1-Butanone-Induced Lung Tumorigenesis by Dietary Olive Oil and Squalene. *Carcinogenesis* 19 (4): 703–6. https://doi.org/10.1093/carcin/19.4.703.

Solano, J Porfirio Legaria. 2010. Genetic Diversity in Some Species of Amaranth (*Amaranthus spp* .) *Revista Fitotec. Mex.* 33 (2): 89–95.

Spanova, Miroslava, and Günther Daum. 2011. Squalene - Biochemistry, Molecular Biology, Process Biotechnology, and Applications. *European Journal of Lipid Science and Technology* 113 (11): 1299–1320. https://doi.org/10.1002/ejlt.201100203.

Strack, Dieter, Thomas Vogt, and Willibald Schliemann. 2003. Recent Advances in Betalain Research. *Phytochemistry* 62 (3): 247–69. https://doi.org/10.1016/S0031-9422(02)00564-2.

Suárez, Santiago, and María Cristina Añón. 2019. Amaranth Proteins Emulsions as Delivery System of Angiotensin-I Converting Enzyme Inhibitory Peptides. *Food Hydrocolloids* 90 (November 2018): 154–61. https://doi.org/10.1016/j.foodhyd.2018.11.046.

Suárez, Santiago, Paula Aphalo, Gustavo Rinaldi, Maria Cristina Añón, and Alejandra Quiroga. 2020. Effect of Amaranth Proteins on the RAS System. *In Vitro*, *In Vivo* and *Ex*

Vivo Assays. *Food Chemistry* 308: 125601. https://doi.org/10.1016/j.foodchem.2019.125601.

Tang, Yao, Xihong Li, Peter X. Chen, Bing Zhang, Ronghua Liu, Marta Hernandez, Jamie Draves, Massimo F. Marcone, and Rong Tsao. 2016. Assessing the Fatty Acid, Carotenoid, and Tocopherol Compositions of Amaranth and Quinoa Seeds Grown in Ontario and Their Overall Contribution to Nutritional Quality. *Journal of Agricultural and Food Chemistry* 64 (5): 1103–10. https://doi.org/10.1021/acs.jafc.5b05414.

Tari, Tejashree A., Uday S. Annapure, Rekha S. Singhal, and Pushpa R. Kulkarni. 2003. Starch-Based Spherical Aggregates: Screening of Small Granule Sized Starches for Entrapment of a Model Flavouring Compound, Vanillin. *Carbohydrate Polymers* 53 (1): 45–51. https://doi.org/10.1016/S0144-8617(02)00293-X.

Tosi, E. A., E. D. Ré, R. Masciarelli, H. Sánchez, C. Osella, and M. A. De La Torre. 2002. Whole and Defatted Hyperproteic Amaranth Flours Tested as Wheat Flour Supplementation in Mold Breads. *LWT - Food Science and Technology* 35 (5): 472–75. https://doi.org/10.1006/fstl.2002.0892.

Valcárcel-Yamani, Beatriz, and Suzana Caetano Da Silva Lannes. 2012. Applications of Quinoa (*Chenopodium quinoa willd* .) and Amaranth (*Amaranthus spp* .) and Their Influence in the Nutritional Value of Cereal Based Foods. *Food and Public Health* 2 (6): 265–75. https://doi.org/10.5923/j.fph.20120206.12.

Venskutonis, Petras R., and Paulius Kraujalis. 2013. Nutritional Components of Amaranth Seeds and Vegetables: A Review on Composition, Properties, and Uses. *Comprehensive Reviews in Food Science and Food Safety* 12 (4): 381–412. https://doi.org/10.1111/1541-4337.12021.

Warleta, Fernando, María Campos, Yosra Allouche, Cristina Sánchez-Quesada, Jesús Ruiz-Mora, Gabriel Beltrán, and José J. Gaforio. 2010. Squalene Protects against Oxidative DNA Damage in MCF10A Human Mammary Epithelial Cells but Not in MCF7 and MDA-MB-231 Human Breast Cancer Cells. *Food and Chemical Toxicology* 48 (4): 1092–1100. https://doi.org/10.1016/j.fct.2010.01.031.

Wrigley, Colin, Harold Corke, and Charles Walker. 2004. Amaranth. In *Grain Science*, 1:1–11. Elsevier. https://doi.org/10.2307/j.ctt7zw89n.41.

Yelisyeyeva, O. P., K. O. Semen, G. V. Ostrovska, D. V. Kaminskyy, T. V. Sirota, N. Zarkovic, D. Mazur, O. D. Lutsyk, K. Rybalchenko, and A. Bast. 2014. The Effect of Amaranth Oil on Monolayers of Artificial Lipids and Hepatocyte Plasma Membranes with Adrenalin-Induced Stress. *Food Chemistry* 147: 152–59. https://doi.org/10.1016/j.foodchem.2013.09.119.

In: Super and Nutraceutical Foods
Editors: Sandra Teresita Martín-del-Campo et al.
ISBN: 978-1-53619-082-3

Chapter 7

CHICKPEA, COMPOSITION AND PROCESSING TECHNOLOGIES

D. del R. Moguel-Concha, J. E. Borges-Martínez, M. S. Cid-Gallegos, D. I. Tellez-Medina and C. Jiménez-Martínez*
Instituto Politécnico Nacional, Escuela Nacional de Ciencias Biológicas,
Unidad Profesional Adolfo López Mateos, Zacatenco,
Delegación Gustavo A. Madero, Ciudad de México, México

ABSTRACT

Chickpea is one of the most important grain legumes consumed globally due to its protein and carbohydrate nutritional elements, low lipid content, and a good source of vitamins and minerals. Like other pulses, chickpeas have some components denominated non-nutritional factors, which can be reduced or eliminated by different processes. However, technologies for the treatment of legumes also improve the concentration of some compounds with beneficial effects for human health; for instance, processing enhances the content of protein, bioactive components or secondary metabolites, and the starch's digestibility.

Keywords: chickpea, bioactive compounds, technologies

INTRODUCTION: HISTORY, DISTRIBUTION, AND HEALTH BENEFITS

Chickpea belongs to the Fabaceae family; it is the third most consumed legume, just after beans and peas worldwide. Two varieties are currently produced from this seed: kabuli (which is characterized by being a large seed, light color, and characteristic shape) and desi (smaller size, irregular shape, and dark color) (Allen 2013). Traditionally, the cultivation system has been developed in India, Iran, Bangladesh, Pakistan, Mexico, and Ethiopia.

* Corresponding Author's E-mail: crisjm_99@yahoo.com.

Additionally, with the genetic improvements applied to chickpea cultivar groups and the progress of technologies, there has been an expansion of production in countries like Turkey, Canada, and Australia (Siddique and Krishnamurthy 2016).

On the other hand, chickpea contains many proteins, fiber, starch, phytochemical compounds, good calcium, vitamin B complex, and iron source. They can be consumed differently, for example: boiled, autoclaved, roasted, fried (Retes et al., 2018; Yniestra Marure et al., 2019). Furthermore, recent studies on chickpea have shown beneficial health effects, particularly in decreasing cardiovascular diseases, diabetes, and cancer. Specifically, research works in cancer have found distinct phytochemical compounds and peptides with the therapeutic effect, which can be used as chemopreventive, especially in colon cancer (Merga and Haji 2019; Sánchez-Chino et al., 2015). These compounds have shown antioxidant, antihypertensive, anti-inflammatory, anticarcinogenic, antifungal, hypoglycemic, anti-obesity, and hypocholesterolemic activities (Real-Hernández and Gonzalez de Mejia 2019).

General Aspects: Morphology and Production

Chickpea seed is constituted of a layer denominated testa (Figure 1); this is an important component because it is the only protective barrier between the embryo and the external environment. Moreover, its protective function is due to external and internal cuticle presence, which is often impregnated with lipids and waxy substances, plus one or more thick wall protective cell layers. These protective layers produce mechanical protection through the synthesis of secondary cell walls, impregnated with lipids or waxes or lignin, and the production of polyphenols, which protect the seed against insects (Bewley et al., 2012). On the other hand, the seed coat's texture and color are characteristic of each seed and can change due to the environmental and genetic influence during seed development (Bewley et al., 2012). Inside this protective layer or test, there are the seed's two cotyledons (Figure 1B). Each cotyledon has only a complete layer of epidermal cells composed mainly of proteins, and it also has thickened peripheral walls. However, the cells differ depending on how they are distributed in the seed to be located adjacent to the seed coat (testa) or to the cotyledon that has paired (Wood, Knights, and Chocty 2011). Variety kabuli shows a higher content of β-glucan type polysaccharides than variety desi. Additionally, chickpea seed has hypocotyl or embryonic axis (Figure 1B), a structure between the cotyledons which contributes to the development of the stem and ends in the radicle or embryonic root (Figure 1B) (Wood, Knights, and Chocty 2011). On the outside is the hilum (Figure 1C), which is like a scar of the seed; from there, the seed adheres to the pod's wall during its development. The hilum is next to the micropyle (Figure 1B), a tiny pore that controls the humidity of the seed, and a crown surrounds both structures. Finally, the raphe extends in a line from the bottom of the crown to the *spermotilium* (Figure 1A), which contains the chalaza, this is, the base of the ovule (Wood, Knights, and Chocty 2011).

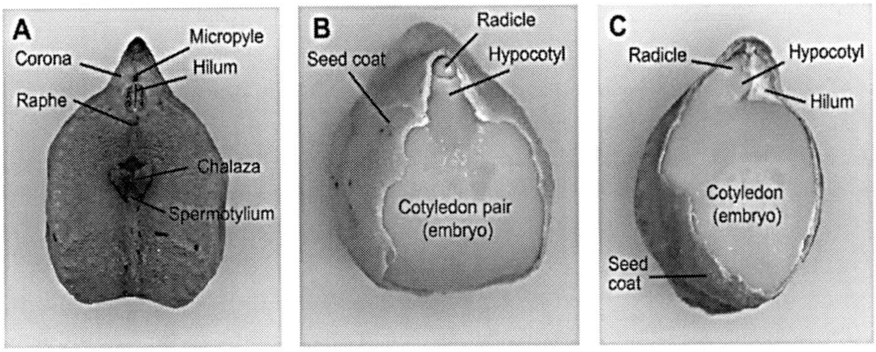

Figure 1. Chickpea seed morphology. A, view of the external characteristics of the seed; B and C main internal characteristics (Wood, Knights, and Chocty 2011).

Chemical Composition

The biochemical composition is directly related to the genotype of the seed and the environmental conditions; however, the amounts of nutritional compounds (Table 1) can also differ between the varieties of a given species or even between the seeds of the same plant (Allen 2013).

Table 1. Macronutrients of Chickpea *(C. arietinum)*

	de Almeida et al., 2006 *		Sarmento et al., 2014 ǂ		Rachwa-Rosiak et al., 2015
Nutriment	*Raw*	*Cooked* [1]	*Raw*	*Cooked*[2]	(%)
Carbohydrates	54.0	57.8	640.69⁼	704.35⁼	47.4-55.8
Proteins	18.5[a]	21.3[a]	283.27	198.68	24.4-25.4
Lipids	6.69	6.73	43.10	72.20	3.7-5.1

* g/100g.
ǂ g/kg dry weight.
⁼By difference.
[a] conversion factor: 5.40.
[1] Soaking 16 hours (1:2 a/v). Cooking in express pot 14.7 psi with water (1:3 a/v) for 40 minutes.
[2] Soaking 24 hours. Cooking in an express pot with distilled water for 15 minutes.

Proteins

Proteins are the second most abundant compound in chickpea; they range between 10 and 30% (Boye, Zare, and Pletch 2010; Hall, Hillen, and Robinson 2017; Dida Bulbula and Urga 2018). These are mainly divided into two groups: a) enzymatic, regulatory and structural proteins (Allen 2013), and b) storage proteins, which represent approximately 70% of the total nitrogen in the seed. Otherwise, the second group is classified according to their solubility properties in a) albumins, which are soluble in water at pH 6.6; b) globulins, that are soluble in saline solutions at pH 7.0; c) prolamines, soluble in 70% ethanol solutions; and d) glutelins, which are soluble in acid solutions at pH 2.0 and alkaline at pH 12 (Allen 2013; Badui Dergal 2006).

Besides, reserve proteins of legumes are relatively low in sulfur amino acids (methionine and cysteine); however, this deficiency can be compensated by the presence of essential amino acids such as lysine, which is found in higher amounts than in cereals (Duranti 2006).

Furthermore, globulins are the most abundant storage protein fractions in legumes since they constitute 53 to 60% of total protein (Hall, Hillen, and Robinson 2017). According to their sedimentation coefficient, these proteins can be classified as 7S (oligomeric proteins, usually represented as trimers) and 11S (also oligomeric proteins, but these are presented in the form of hexamers), which in turn are denominated vicilins and legumins, respectively. The proportion in which vicilins and legumins are present will depend on the cultivar group (Duranti 2006; Allen 2013). In dry chickpea seeds, three globulins of greater abundance have been identified: legumin, vicilin, and convicilin (Rachwa-Rosiak, Nebesny, and Budryn 2015; Chang et al., 2012).

Carbohydrates

The carbohydrate content is close to 60%; the starch represents between 37.5 and 50.8%, being the main component of the chickpea seed (Yniestra Marure et al., 2019). Monosaccharides (represent less than 1%) and α-galactosides (constitute 14%) represent approximately 15% in legumes. Moreover, monosaccharides and disaccharides represent between 30-35% of the total carbohydrates, being galactose, ribose, fructose, glucose, maltose, and sucrose, the most abundant ones. Stachyose, raffinose, and verbascose are the most abundant among α-galactosides (Hall, Hillen, and Robinson 2017; Jukanti et al., 2012). In turn, α-galactosides are not hydrolyzed during digestion, so they are fermented by bacteria found in the colon and produce methane, hydrogen, and CO_2 (Jukanti et al., 2012; Sánchez-Chino et al., 2015).

Finally, fiber represents about 4% of chickpea seed. The galacturonic acid content ranges 10-12%, and it has been reported that the content of this acid is compatible with pectin polysaccharides as the main component of soluble fiber (Hall, Hillen, and Robinson 2017). Similarly, hemicellulose sugar and arabinose/rhamnose, have been reported to be found in the insoluble fiber fractions of chickpea seeds, being xylose the main component of soluble fiber fractions in chickpea (Dalgetty and Baik 2003).

Lipids

Lipids are in a small amount that ranges from 2 to 7% (Jukanti et al., 2012; Dida Bulbula and Urga 2018; Hall, Hillen, and Robinson 2017). Among these are triacylglycerols, phospholipids, free fatty acids, sterols, and sterol esters. Moreover, triacylglycerols have been reported to represent 56-63% of the total content, while phospholipids range from17-20% (Hall, Hillen, and Robinson 2017). Compared to other legumes, chickpea stands out because of its linoleic, oleic, and palmitic acid contents being the former the one of the highest proportion in the seed (Jukanti et al., 2012). Also, the oil fraction of the chickpea seed has tocopherols, sterols, and tocotrienols, which are considered nutritionally important. The presence of α-tocopherol, together with δ-tocopherol, enhance chickpea oil oxidative stability and improve its shelf life. Sitosterol is the dominant sterol, followed by campesterol (Jukanti et al., 2012).

Vitamins and Minerals

Legumes are recognized for being a good source of the vitamin B complex, making them an ideal ingredient for other food products. Tetrahydrofolate, 5-methyltetrahydrofolate (5-MTHF), and 5-formyltetrahydrofolate (5-FTHF) have been identified in most legumes. In chickpea seeds, it has been determined that the content of 5-methyltetrahydrofolate ranges 42-537 μg/100 g; they are also a good source of thiamine (453 μg/ 100 g) and riboflavin (173 μg/ 100 g) (Hall, Hillen, and Robinson 2017). As well, the mineral content in chickpea has been reported for potassium (9.94 to 12.64 mg/g), phosphorus (3.94 to 4.52 mg/g), and calcium (0.82 to 2.72 mg/g) (Jukanti et al., 2012; Hall, Hillen, and Robinson 2017).

Other Nutraceutical Compounds

Pulses are excellent sources of bioactive compounds that can be used as ingredients in functional foods and other products; these components play metabolic roles in humans that improve health. Bioactive compounds can be from a protein (protease inhibitors, lectins, and peptides), or non-protein (tannins, phytic acid, saponins and phenolic) origin (Muzquis & Wood, 2007; Jukanti et al., 2012). Furthermore, these compounds can be eliminated, inactivated, or decreased in concentration by different processes such as soaking, germination, or fermentation. These bioactive components can positively impact human health depending on their concentration (Muzquiz et al., 2012; Eiwegger et al., 2004; Cardador-Martínez et al., 2002; Friedman and Brandon 2001). Table 2 shows some of the effects attributed to those compounds, which can be classified into two groups: the first corresponds to protein origin compounds, and the second group includes non-protein compounds (Sánchez-Chino et al., 2015).

Since ancient times, man has processed legumes for consumption, intending to generate attractive and nutritious products. During processing, digestibility and bioavailability of nutrients from legumes is increased (Avola, Patanè, and Barbagallo 2012), besides the inactivation of thermolabile phytochemicals (such as protease inhibitors and phytic acid) and the decrease of soluble compounds (such as α-galactosides) (Frias et al., 2000; Vidal-Valverde et al., 1994).

Due to their composition, legumes have been widely used either as whole or as a fraction (carbohydrates, fiber, proteins, phenolic compounds, etc.) in studies focused on the prevention and control of different chronic degenerative diseases, such as obesity and cardiovascular ones (Duranti and Gius 1997; Garrido G, De La Maza C, and Valladares B 2003; Macarulla et al., 2001; Zhu, Jiang, and Thompson 2012), diabetes (Abete, Parra, and Martinez 2009; Hutchins, Winham, and Thompson 2012), inflammation (Hermsdorff et al., 2011; Esmaillzadeh and Azadbakht 2012) and cancer (Bean et al., 2008; Bobe et al., 2008; Faris et al., 2009; Hangen & Bennink, 2002; Haydé et al., 2012; Murillo et al., 2004).

Table 2. Nutritional and non-nutritional compounds with beneficial health effect from chickpea (*C. arietinum*)

Compounds	Beneficial effect	Reference
Protease inhibitors	Anticarcinogenic, anti-inflammatory, antihypertensive and neurodegenerative diseases	Roy, Boye, and Simpson 2010
Lectins	Anticarcinogenic, obesity prevention	Gautam et al., 2020; Roy et al., 2010
Peptides	Inhibition of ACE-I, antioxidant activity, iron chelating activity.	Barbana & Boye, 2010; Ghribi et al., 2015; Kou et al., 2013; Torres-Fuentes et al., 2012, 2015; Yust et al., 2003
Tannins	Iron chelating activity	Muzquiz & Wood, 2007
Phytic acid	Decrease in glycemic index	Muzquiz & Wood, 2007
Phenolic compounds	Antioxidant activity (ferulic acid, caffeic acid, ρ-coumaric acid, chlorogenic acid, vanillic acid, protocatechuic acid, synergistic acid, synaptic acid).	Vaz Patto et al., 2015
Saponins	Promote the reduction of cholesterol, lipids, and glucose in the blood, reduction of cardiovascular diseases, and obesity.	Barakat et al., 2015
Isoflavones	Anti-inflammatory activity	Milán-Noris et al., 2018
α-galactosides	Increase the population of bifidobacteria, and reduce constipation, stimulate the immune system.	Xing et al., 2020; Muzquiz & Wood, 2007

Furthermore, phenolics have been highlighted for their different biological activities: antioxidant, antihypertensive, antimicrobial, and anticarcinogenic (Aguilera et al., 2011). Phenolics identified in chickpea include catechin, prodelfinidin, trans-ferulic acid, syringic acid, ρ-coumaric, caffeic acid, chlorogenic acid, gallic acid, hydroxybenzoic acid, protocatechuic acid, protocatechuic aldehyde, sinapic acid, and vanillic acid.

Carotenoids such as β-carotene, lutein, zeaxanthin, β-cryptoxanthin, lycopene, and carotene have also been identified (Hall et al., 2017; Jukanti et al., 2012). The importance of carotene is that it is converted to vitamin A more efficiently than other carotenoids. Besides, anthocyanins, including cyanidin, petunidin, and delphinidin, have been reported to be present in higher proportions in the testa of chickpea seed (Hallet al., 2017; Jukanti et al., 2012; Sreerama et al., 2010). The level of flavonoids contained in chickpeas, mainly isoflavones, is as significant as that found in soybean; actually, biochanin A and formononetin have been detected only in chickpeas (Jukanti et al., 2012; De Camargo et al., 2019), while the concentration of daidzein and genistein is similar in both seeds, making the chickpea an alternative source of these isoflavones (De Camargo et al., 2019).

Processing Methods: Thermal and Non-Thermal

Generally, before the consumption of legumes, a process is applied, such as soaking, cooking, fermentation, drying, or germination (Chandra-Hioe et al., 2016; Dida Bulbula and Urga 2018). These methods can be classified into thermal and non-thermal.

Thermal

Cooking is one of the most commonly used techniques before consuming legumes, as it results in improved digestibility of proteins (86.0-93.3%) and starch (84.0-90.4%) (Rehman and Shah 2005); however, it has the disadvantage that it decreases the content of polyphenolic compounds. To compensate for this decrease, legumes might be allowed to germinate before applying the thermal process (Haileslassie, Henry, and Tyler 2019).

The residual cooking water of chickpea is known as aquafaba. This product can be used as a good emulsifier in various bakery products (Alsalman et al., 2020; Xu et al., 2020; He et al., 2019) and desserts (Meurer et al., 2020) like meringues. Processing aquafaba with ultrasound at a frequency of 20 kHz generates positive effects on the expansion, stability, texture, and color of the foam, reduces bubble size, and improves foam homogeneity (Meurer, de Souza, and Ferreira Marczak 2020). Ultrasound is a technique that can induce physical and chemical modifications in the proteins of the chickpea seed, causing partial denaturation and improving the capacity of water retention because the sections of hydrophilic amino acids become open in the aqueous phase of the treated solution. Similarly, Mustafa et al., (2018) reported that aquafaba could replace egg white in baking recipes by imparting similar technological properties to the resulting product as a foaming agent and emulsifier.

Extrusion is another thermal method used in food, which has been applied in chickpea flour resulting in a decrease of α-galactosides, specifically, its application has reduced 42% stachyose content (Rachwa-Rosiak, Nebesny, and Budryn 2015). Another method that has been used to treat chickpea seeds is microwave cooking. Alajaji & El-Adawy (2006) compared this method with the boiling and autoclaving process and reported that microwaving improved nutritional quality by reducing trypsin inhibitors (80.50%), however autoclaving (83.87%) and boiling (82.27%) showed better inhibition. With respect to α-galactosides content, they reported total elimination of verbascose and reduction of raffinose and stachyose content, reaching concentrations similar to the other two methods. Also, these methods positively influenced the digestibility of proteins; however, the retention of vitamin B complex and minerals contents were higher than with microwaving.

Baik & Han, (2012) and Gupta et al., (2019), elaborated a high protein snack based on chickpea seeds showing good digestibility and deficiency of non-nutritional compounds applied from soaking, pressure cooking, baking, and toasting.

Non-Thermal

Fermentation is a simple, low-cost processing technique used to improve nutritional quality and eliminate undesired compounds. In chickpeas, fermentation allows the reduction of α-galactosides and improves the digestibility of proteins. An example of these applications is the elaboration of miso-based on cooked chickpeas, which being a product fermented with lactic acid bacteria, promotes the proliferation of intestinal microbiota. Fermentation with lactic acid bacteria improves some of the techno-functional properties of proteins, for example, the capacity and stability of foam, which allows them to be incorporated into various food applications (Chandra-Hioe et al., 2016). Xing et al., (2020), reported on chickpea fermentation and observed that, after 72 h of fermentation with lactic acid bacteria, raffinose (88.3%), stachyose (99.1%) and phytic acid (17%) contents decreased, verbascose was not detected and phenolic compounds content (119%) increased; this could be beneficial because of the biological activities that the product might possess. In addition, fermented chickpea flour has shown an improvement in its composition of essential amino acids such as methionine, cysteine, phenylalanine, tyrosine, and threonine (Rachwa-Rosiak et al., 2015).

The chickpea seed starch has been independently evaluated also as a yogurt stabilizer to improve viscosity and decrease syneresis, i.e., the separation of the whey during storage (Saleh et al., 2020).

Another popular non-thermal treatment is germination, which has been demonstrated to reduce raffinose, stachyose, and verbascose contents and improve the antioxidant potential of chickpeas seeds (Jukanti et al., 2012; Bruno et al., 2019). An increase in isoflavones such as biochanin A (19 μg/g on the fresh basis) and formononetin (1.0 μg/g on a fresh basis) in the ungerminated seed has also been reported. After 4 days of germination, the content increased to 702 and 1484 μg/g fresh basis, respectively (Wu et al., 2012). Besides, the incorporation of flour and protein isolates from germinated chickpea seeds has been tested and recommended to improve the rheological quality of various food products such as rice-based noodles, plus showing nutritional improvements such as in vitro protein digestibility (78.68-87.59%) and a decrease in the glycemic index of rice-based noodles from 70.83 to 57.49 (Sofi et al., 2020). Sprout legume seeds are foods that are considered highly nutritious because of their improved bioavailability of iron-related to reduce phytates (46.4%) and increased ascorbic acid (5.46%) (Bains et al., 2014). However, germination improves the nutritional aspects of the seed and helps in functional features such as solubility, water absorption, foaming, and emulsifying properties (Xu et al., 2020).

A characteristic shared by fermentation and germination is that both are suitable for reducing phytates' contents and condensed tannins in the chickpea seed (Dida Bulbula and Urga 2018).

Conclusion and Perspectives

Chickpea is an important seed for human consumption, due to its nutritional value (protein, carbohydrates, lipids, minerals, and vitamins). As well, it has non-nutritional compounds that can be reduced or eliminated with some treatment. Thermal (boiling, ultrasound, microwaving) and non-thermal processing (germination, fermentation) allow improving its biological value and to enhance the effect of bioactive compounds, particularly, the case of phenolic compounds whose beneficial effect is reflected in the prevention of chronic degenerative diseases. However, its necessary more research in vitro and in vivo, in order to continue to understand all the advantages generated by the consumption of chickpea seed on its own or added to other foods, thus improving both the functional and techno-functional quality of these.

REFERENCES

Abete, Itziar, Dolores Parra, and J. Alfredo Martinez. 2009. Legume-, Fish-, or High-Protein-Based Hypocaloric Diets: Effects on Weight Loss and Mitochondrial Oxidation in Obese Men. *Journal of Medicinal Food* 12 (1): 100–108. https://doi.org/10.1089/jmf.2007.0700.

Aguilera, Yolanda, Montserrat Dueñas, Isabel Estrella, Teresa Hernández, Vanesa Benitez, Rosa María Esteban, and María A. Martín-Cabrejas. 2011. Phenolic Profile and Antioxidant Capacity of Chickpeas (*Cicer Arietinum L.*) as Affected by a Dehydration Process. *Plant Foods for Human Nutrition* 66 (2): 187–95. https://doi.org/10.1007/s11130-011-0230-8.

Alajaji, Saleh A., and Tarek A. El-Adawy. 2006. Nutritional Composition of Chickpea (*Cicer Arietinum L.*) as Affected by Microwave Cooking and Other Traditional Cooking Methods. *Journal of Food Composition and Analysis* 19 (8): 806–12. https://doi.org/10.1016/j.jfca.2006.03.015.

Allen, LH. 2013. *Legume Types*. Vol. 2. https://doi.org/10.1016/B978-0-12-375083-9.00170-7.

Alsalman, Fatemah B, Mehmet Tulbek, and Michael Nickerson. 2020. Evaluation and Optimization of Functional and Antinutritional Properties of Aquafaba. *Legume Science*, no. January: 1–15. https://doi.org/10.1002/leg3.30.

Avola, Giovanni, Cristina Patanè, and Riccardo N. Barbagallo. 2012. Effect of Water Cooking on Proximate Composition of Grain in Three Sicilian Chickpeas (*Cicer Arietinum L.*). *LWT - Food Science and Technology* 49 (2): 217–20. https://doi.org/10.1016/j.lwt.2012.07.004.

Badui Dergal, S. 2006. *Química de Los Alimentos*. 4a ed. Pearson.

Bains, Kiran, Veny Uppal, and Harpreet Kaur. 2014. Optimization of Germination Time and Heat Treatments for Enhanced Availability of Minerals from Leguminous Sprouts. *Journal of Food Science and Technology* 51 (5): 1016–20. https://doi.org/10.1007/s13197-011-0582-y.

Barakat, Hassan, Valeria Reim, and Sascha Rohn. 2015. Stability of Saponins from Chickpea, Soy and Faba Beans in Vegetarian, Broccoli-Based Bars Subjected to Different Cooking Techniques. *Food Research International* 76 (P1): 142–49. https://doi.org/10.1016/j.foodres.2015.03.043.

Barbana, Chockry, and Joyce Irene Boye. 2010. Angiotensin I-Converting Enzyme Inhibitory Activity of Chickpea and Pea Protein Hydrolysates. *Food Research International* 43 (6): 1642–49. https://doi.org/10.1016/j.foodres.2010.05.003.

Bean, Dry, Matthew D Thompson, Henry J Thompson, Mark a Brick, John N Mcginley, Weiqin Jiang, Zongjian Zhu, and Pamela Wolfe. 2008. Mechanisms Associated with Dose-Dependent Inhibition of Rat Mammary Carcinogenesis by Dry Bean (*Phaseolus Vulgaris, L.*). *The Journal of Nutrition* 138 (August): 2091–97. https://doi.org/10.3945/jn.108.094557.such.

Bewley, J Derek, Kent J Bradford, Henk W. M. Hilhorst, and Hiro Nonogaki. 2012. *Seeds: physiology of development, germination and dormacy, 2-7.* Springer New York.

Bobe, Gerd, Kathleen G. Barrett, Roycelynn A. Mentor-Marcel, Umberto Saffiotti, Matthew R. Young, Nancy H. Colburn, Paul S. Albert, Maurice R. Bennink, and Elaine Lanza. 2008. Dietary Cooked Navy Beans and Their Fractions Attenuate Colon Carcinogenesis in Azoxymethane-Induced Ob/Ob Mice. *Nutrition and Cancer* 60 (3): 373–81. https://doi.org/10.1080/01635580701775142.

Boye, Joyce, Fatemeh Zare, and Alison Pletch. 2010. Pulse Proteins: Processing, Characterization, Functional Properties and Applications in Food and Feed. *Food Research International*. https://doi.org/10.1016/j.foodres.2009.09.003.

Bruno, Joseph A., David W. Konas, Evan L. Matthews, Charles H. Feldman, Kate M. Pinsley, and Adrian L. Kerrihard. 2019. Sprouted and Non-Sprouted Chickpea Flours: Effects on Sensory Traits in Pasta and Antioxidant Capacity. *Polish Journal of Food and Nutrition Sciences* 69 (2): 203–9. https://doi.org/10.31883/pjfns/109280.

Camargo, Adriano Costa De, Bruno Trevenzoli Favero, Maressa Caldeira Morzelle, Marcelo Franchin, Emilio Alvarez-Parrilla, Laura A. De La Rosa, Marina Vilar Geraldi, Mário Roberto Maróstica, Fereidoon Shahidi, and Andrés R. Schwember. 2019. Is Chickpea a Potential Substitute for Soybean? Phenolic Bioactives and Potential Health Benefits. *International Journal of Molecular Sciences*. Vol. 20. https://doi.org/10.3390/ijms 20112644.

Cardador-Martínez, Anaberta, Guadalupe Loarca-Piña, and B. Dave Oomah. 2002. Antioxidant Activity in Common Beans (*Phaseolus Vulgaris L.*). *Journal of Agricultural and Food Chemistry* 50 (24): 6975–80. https://doi.org/10.1021/jf020296n.

Chandra-Hioe, Maria V., Christina H. M. Wong, and Jayashree Arcot. 2016. The Potential Use of Fermented Chickpea and Faba Bean Flour as Food Ingredients. *Plant Foods for Human Nutrition* 71 (1): 90–95. https://doi.org/10.1007/s11130-016-0532-y.

Chang, Yu Wei, Inteaz Alli, Aline T. Molina, Yasuo Konishi, and Joyce I. Boye. 2012. Isolation and Characterization of Chickpea (*Cicer Arietinum L.*) Seed Protein Fractions. *Food and Bioprocess Technology* 5 (2): 618–25. https://doi.org/10.1007/s11947-009-0303-y.

Dalgetty, David D., and Byung Kee Baik. 2003. Isolation and Characterization of Cotyledon Fibers from Peas, Lentils, and Chickpeas. *Cereal Chemistry* 80 (3): 310–15. https://doi.org/10.1094/CCHEM.2003.80.3.310.

Dida Bulbula, Dejene, and Kelbessa Urga. 2018. Study on the Effect of Traditional Processing Methods on Nutritional Composition and Anti Nutritional Factors in Chickpea (*Cicer Arietinum*). *Cogent Food & Agriculture* 4 (1). https://doi.org/10.1080/23311932.2017.1422370.

Duranti, Marcello. 2006. Grain Legume Proteins and Nutraceutical Properties. *Fitoterapia* 77 (2): 67–82. https://doi.org/10.1016/j.fitote.2005.11.008.

Duranti, Marcello, and Cristina Gius. 1997. Legume Seeds: Protein Content and Nutritional Value. *Field Crops Research* 53 (1–3): 31–45. https://doi.org/10.1016/S0378-4290(97)00021-X.

Eiwegger, Thomas, Bernd Stahl, Joachim Schmitt, Günther Boehm, Marianne Gerstmayr, Josefa Pichler, Eleonora Dehlink, Christine Loibichler, Radvan Urbanek, and Zsolt Szépfalusi. 2004. Human Milk-Derived Oligosaccharides and Plant-Derived Oligosaccharides Stimulate Cytokine Production of Cord Blood T-Cells in Vitro. *Pediatric Research* 56 (4): 536–40. https://doi.org/10.1203/01.PDR.0000139411.35619.B4.

Esmaillzadeh, Ahmad, and Leila Azadbakht. 2012. Legume Consumption Is Inversely Associated with Serum Concentrations of Adhesion Molecules and Inflammatory

Biomarkers among Iranian Women. *The Journal of Nutrition* 142 (2): 334–39. https://doi.org/10.3945/jn.111.146167.

Faris, Mo'ez Al Islam E., Hamed R. Takruri, Maha S. Shomaf, and Yasser K. Bustanji. 2009. Chemopreventive Effect of Raw and Cooked Lentils (*Lens Culinaris L.*) and Soybeans (*Glycine Max*) against Azoxymethane-Induced Aberrant Crypt Foci. *Nutrition Research* 29 (5): 355–62. https://doi.org/10.1016/j.nutres.2009.05.005.

Frias, Juana, Concepción Vidal-Valverde, Cristina Sotomayor, Concepción Diaz-Pollan, and Gloria Urbano. 2000. Influence of Processing on Available Carbohydrate Content and Antinutritional Factors of Chickpeas. *European Food Research and Technology* 210 (5): 340–45. https://doi.org/10.1007/s002170050560.

Friedman, Mendel, and David L. Brandon. 2001. Nutritional and Health Benefits of Soy Proteins. *Journal of Agricultural and Food Chemistry* 49 (3): 1069–86. https://doi.org/10.1021/jf0009246.

Garrido G, Argelia, María Pía De La Maza C, and Luis Valladares B. 2003. Fitoestrógenos Dietarios y Sus Potenciales Beneficios En La Salud Del Adulto Humano. *Revista Medica de Chile* 131 (11): 1321–28. https://doi.org/10.4067/s0034-98872003001100015. [Dietary Phytoestrogens and Their Potential Health Benefits of the Human Adult. *Medical Journal of Chile*]

Gautam, Ajay Kumar, Divakar Sharma, Juhi Sharma, and Khem Chand Saini. 2020. Legume Lectins: Potential Use as a Diagnostics and Therapeutics against the Cancer. *International Journal of Biological Macromolecules* 142: 474–83. https://doi.org/10.1016/j.ijbiomac.2019.09.119.

Ghribi, Abir Mokni, Assaâd Sila, Rémi Przybylski, Naima Nedjar-Arroume, Ines Makhlouf, Christophe Blecker, Hamadi Attia, Pascal Dhulster, Ali Bougatef, and Souhail Besbes. 2015. Purification and Identification of Novel Antioxidant Peptides from Enzymatic Hydrolysate of Chickpea (*Cicer Arietinum L.*) Protein Concentrate. *Journal of Functional Foods* 12: 516–25. https://doi.org/10.1016/j.jff.2014.12.011.

Haileslassie, Hiwot Abebe, Carol J. Henry, and Robert T. Tyler. 2019. Impact of Pre-Treatment (Soaking or Germination) on Nutrient and Anti-Nutrient Contents, Cooking Time and Acceptability of Cooked Red Dry Bean (*Phaseolus Vulgaris L.*) and Chickpea (*Cicer Arietinum L.*) Grown in Ethiopia. *International Journal of Food Science and Technology* 54 (8): 2540–52. https://doi.org/10.1111/ijfs.14165.

Hall, Clifford, Cassandra Hillen, and Julie Garden Robinson. 2017. Composition, Nutritional Value, and Health Benefits of Pulses. *Cereal Chemistry* 94 (1): 11–31. https://doi.org/10.1094/CCHEM-03-16-0069-FI.

Hangen, L., and M. R. Bennink. 2002. Consumption of Black Beans and Navy Beans (*Phaseolus Vulgaris*) Reduced Azoxymethane-Induced Colon Cancer in Rats. *Nutrition and Cancer* 44 (1): 60–65. https://doi.org/10.1207/s15327914nc441_8.

Haydé, Vergara Castãeda, Guevara González Ramón, Guevara Olvera Lorenzo, Oomah B. Dave, Reynoso Camacho Rosalía, Wiersma Paul, and Loarca Pĩa Guadalupe. 2012. Non-Digestible Fraction of Beans (*Phaseolus Vulgaris L.*) Modulates Signalling Pathway Genes at an Early Stage of Colon Cancer in Sprague-Dawley Rats. *British Journal of Nutrition* 108 (SUPPL. 1). https://doi.org/10.1017/S0007114512000785.

He, Yue, Youn Young Shim, Rana Mustafa, Venkatesh Meda, and Martin J. T. Reaney. 2019. Chickpea Cultivar Selection to Produce Aquafaba with Superior Emulsion Properties. *Foods* 8 (12): 1–16. https://doi.org/10.3390/foods8120685.

Hermsdorff, Helen Hermana M., M. Ángeles Zulet, Itziar Abete, and J. Alfredo Martínez. 2011. A Legume-Based Hypocaloric Diet Reduces Proinflammatory Status and Improves Metabolic Features in Overweight/Obese Subjects. *European Journal of Nutrition* 50 (1): 61–69. https://doi.org/10.1007/s00394-010-0115-x.

Hutchins, Andrea M., Donna M. Winham, and Sharon V. Thompson. 2012. Phaseolus Beans: Impact on Glycaemic Response and Chronic Disease Risk in Human Subjects. *British Journal of Nutrition* 108 (SUPPL. 1). https://doi.org/10.1017/S0007114512000761.

Jukanti, A. K., P. M. Gaur, C. L. L. Gowda, and R. N. Chibbar. 2012. Nutritional Quality and Health Benefits of Chickpea (*Cicer Arietinum L.*): A Review. *British Journal of Nutrition* 108 (SUPPL. 1). https://doi.org/10.1017/S0007114512000797.

Kou, Xiaohong, Jie Gao, Zhijun Zhang, Hua Wang, and Xu Wang. 2013. Purification and Identification of Antioxidant Peptides from Chickpea (*Cicer Arietinum L.*) Albumin Hydrolysates. *LWT - Food Science and Technology* 50 (2): 591–98. https://doi.org/10.1016/j.lwt.2012.08.002.

Macarulla, M. Teresa, César Medina, M. Aránzazu De Diego, M. Chávarri, M. Ángeles Zulet, J. Alfredo Martínez, Catherine Nöel-Suberville, Paul Higueret, and María P. Portillo. 2001. Effects of the Whole Seed and a Protein Isolate of Faba Bean (*Vicia Faba*) on the Cholesterol Metabolism of Hypercholesterolaemic Rats . *British Journal of Nutrition* 85 (5): 607–14. https://doi.org/10.1079/bjn2000330.

Merga, Bulti, and Jema Haji. 2019. Economic Importance of Chickpea: Production, Value, and World Trade. *Cogent Food & Agriculture* 5 (1). https://doi.org/10.1080/23311932.2019.1615718.

Meurer, Mariana Cassel, Daiana de Souza, and Lígia Damasceno Ferreira Marczak. 2020. Effects of Ultrasound on Technological Properties of Chickpea Cooking Water (Aquafaba). *Journal of Food Engineering*. https://doi.org/10.1016/j.jfoodeng.2019.109688.

Milán-Noris, Ada Keila, Janet Alejandra Gutiérrez-Uribe, Arlette Santacruz, Sergio O. Serna-Saldívar, and Cristina Martínez-Villaluenga. 2018. Peptides and Isoflavones in Gastrointestinal Digests Contribute to the Anti-Inflammatory Potential of Cooked or Germinated Desi and Kabuli Chickpea (*Cicer Arietinum L.*). *Food Chemistry* 268 (March): 66–76. https://doi.org/10.1016/j.foodchem.2018.06.068.

Mustafa, Rana, Yue He, Youn Young Shim, and Martin J. T. Reaney. 2018. Aquafaba, Wastewater from Chickpea Canning, Functions as an Egg Replacer in Sponge Cake. *International Journal of Food Science and Technology* 53 (10): 2247–55. https://doi.org/10.1111/ijfs.13813.

Muzquiz, Mercedes, Alejandro Varela, Carmen Burbano, Carmen Cuadrado, Eva Guillamón, and Mercedes M. Pedrosa. 2012. Bioactive Compounds in Legumes: Pronutritive and Antinutritive Actions. Implications for Nutrition and Health. *Phytochemistry Reviews* 11 (2–3): 227–44. https://doi.org/10.1007/s11101-012-9233-9.

Rachwa-Rosiak, Danuta, Ewa Nebesny, and Grażyna Budryn. 2015. Chickpeas—Composition, Nutritional Value, Health Benefits, Application to Bread and Snacks: A Review. *Critical Reviews in Food Science and Nutrition* 55 (8): 1135–43. https://doi.org/10.1080/10408398.2012.687418.

Real Hernandez, Luis M., and Elvira Gonzalez de Mejia. 2019. Enzymatic Production, Bioactivity, and Bitterness of Chickpea (*Cicer Arietinum*) Peptides. *Comprehensive*

Reviews in Food Science and Food Safety 18 (6): 1913–46. https://doi.org/10.1111/1541-4337.12504.

Rehman, Zia Ur, and W. H. Shah. 2005. Thermal Heat Processing Effects on Antinutrients, Protein and Starch Digestibility of Food Legumes. *Food Chemistry* 91 (2): 327–31. https://doi.org/10.1016/j.foodchem.2004.06.019.

Retes, Rafael, Salomón Moreno, Fernando Arturo Ibarra, Martha Martín, and Fernando Agustín Ibarra. 2018. Analisis De Rentabilidad Del Garbanzo En Sonora, Ciclos 2012-2013 Al 2016-2017. *Revista Mexicana de Agronegocios* 43 (2018): 89–98.

Roy, F., J. I. Boye, and B. K. Simpson. 2010. Bioactive Proteins and Peptides in Pulse Crops: Pea, Chickpea and Lentil. *Food Research International* 43 (2): 432–42. https://doi.org/10.1016/j.foodres.2009.09.002.

Saleh, Ali, Abdellatif A. Mohamed, Mohammed S. Alamri, Shahzad Hussain, Akram A. Qasem, and Mohamed A. Ibraheem. 2020. Effect of Different Starches on the Rheological, Sensory and Storage Attributes of Non-Fat Set Yogurt. *Foods* 9 (1): 61. https://doi.org/10.3390/foods9010061.

Sánchez-Chino, Xariss, Cristian Jiménez-Martínez, Gloria Dávila-Ortiz, Isela Álvarez-González, and Eduardo Madrigal-Bujaidar. 2015. Nutrient and Nonnutrient Components of Legumes, and Its Chemopreventive Activity: A Review. *Nutrition and Cancer* 67 (3): 401–10. https://doi.org/10.1080/01635581.2015.1004729.

Siddique, K. H. M, and L. Krishnamurthy. 2016. Chickpea, Agronomy. *Reference Module in Food Science*. 2nd ed. Elsevier Ltd. https://doi.org/10.1016/b978-0-08-100596-5.00192-x.

Sofi, Sajad Ahmad, Jagmohan Singh, Navnidhi Chhikara, and Anil Panghal. 2020. Effect of Incorporation of Germinated Flour and Protein Isolate from Chickpea on Different Quality Characteristics of Rice-Based Noodle. *Cereal Chemistry* 97 (1): 85–94. https://doi.org/10.1002/cche.10192.

Sreerama, Yadahally N., Vadakkoot B. Sashikala, and Vishwas M. Pratape. 2010. Variability in the Distribution of Phenolic Compounds in Milled Fractions of Chickpea and Horse Gram: Evaluation of Their Antioxidant Properties. *Journal of Agricultural and Food Chemistry* 58 (14): 8322–30. https://doi.org/10.1021/jf101335r.

Torres-Fuentes, Cristina, Manuel Alaiz, and Javier Vioque. 2012. Iron-Chelating Activity of Chickpea Protein Hydrolysate Peptides. *Food Chemistry* 134 (3): 1585–88. https://doi.org/10.1016/j.foodchem.2012.03.112.

Torres-Fuentes, Cristina, María Del Mar Contreras, Isidra Recio, Manuel Alaiz, and Javier Vioque. 2015. Identification and Characterization of Antioxidant Peptides from Chickpea Protein Hydrolysates. *Food Chemistry* 180: 194–202. https://doi.org/10.1016/j.foodchem.2015.02.046.

Vaz Patto, Maria C., Ryszard Amarowicz, Alberta N. A. Aryee, Joyce I. Boye, Hyun Jung Chung, Maria A. Martín-Cabrejas, and Claire Domoney. 2015. Achievements and Challenges in Improving the Nutritional Quality of Food Legumes. *Critical Reviews in Plant Sciences* 34 (November 2014): 105–43. https://doi.org/10.1080/07352689.2014.897907.

Vidal-Valverde, Concepcion, Juana Frias, Isabel Estrella, Maria J. Gorospe, Raquel Ruiz, and Jim Bacon. 1994. Effect of Processing on Some Antinutritional Factors of Lentils. *Journal of Agricultural and Food Chemistry* 42 (10): 2291–95. https://doi.org/10.1021/jf00046a039.

Wood, J. A., E. J. Knights, and M. Chocty. 2011. Morphology of Chickpea Seeds (*Cicer Arietinum L.*): Comparison of Desi and Kabuli Types. *International Journal of Plant Sciences* 172 (5): 632–43. https://doi.org/10.1086/659456.

Wu, Ziyun, Lixia Song, Shengbao Feng, Yuancai Liu, Guangyuan He, Yoecelyn Yioe, Shao Quan Liu, and Dejian Huang. 2012. Germination Dramatically Increases Isoflavonoid Content and Diversity in Chickpea (*Cicer Arietinum L.*) Seeds. *Journal of Agricultural and Food Chemistry* 60 (35): 8606–15. https://doi.org/10.1021/jf3021514.

Xing, Qinhui, Susanne Dekker, Konstantina Kyriakopoulou, Remko M. Boom, Eddy J. Smid, and Maarten A. I. Schutyser. 2020. Enhanced Nutritional Value of Chickpea Protein Concentrate by Dry Separation and Solid State Fermentation. *Innovative Food Science and Emerging Technologies*. https://doi.org/10.1016/j.ifset.2019.102269.

Xu, Minwei, Zhao Jin, Zixuan Gu, Jiajia Rao, and Bingcan Chen. 2020. Changes in Odor Characteristics of Pulse Protein Isolates from Germinated Chickpea, Lentil, and Yellow Pea: Role of Lipoxygenase and Free Radicals. *Food Chemistry*. https://doi.org/10.1016/j.foodchem.2020.126184.

Yniestra Marure, Lucero M., M. Carmen Núñez-Santiago, Edith Agama-Acevedo, and Luis A. Bello-Perez. 2019. Starch Characterization of Improved Chickpea Varieties Grown in Mexico. *Starch/Staerke* 71 (3–4): 1–8. https://doi.org/10.1002/star.201800139.

Yust, María M., Justo Pedroche, Julio Girón-Calle, Manuel Alaiz, Francisco Millán, and Javier Vioque. 2003. Production of Ace Inhibitory Peptides by Digestion of Chickpea Legumin with Alcalase. *Food Chemistry* 81 (3): 363–69. https://doi.org/10.1016/S0308-8146(02)00431-4.

Zhu, Zongjian, Weiqin Jiang, and Henry J. Thompson. 2012. Edible Dry Bean Consumption (*Phaseolus Vulgaris L.*) Modulates Cardiovascular Risk Factors and Diet-Induced Obesity in Rats and Mice. *British Journal of Nutrition* 108 (SUPPL. 1): 66–73. https://doi.org/10.1017/S0007114512000839.

In: Super and Nutraceutical Foods
Editors: Sandra Teresita Martín-del-Campo et al.
ISBN: 978-1-53619-082-3

Chapter 8

LENTILS AS SUPERFOOD

Paola Cáceres-Challapa[1], Kimberly Molina-Bean[1], Sheccid Acevedo-Juárez[1], Anaberta Cardador-Martínez[1], Carmen Téllez-Pérez[1,2] and Maritza Alonzo-Macías[1,*]
[1]Tecnologico de Monterrey, School of Engineering and Science, Queretaro, Mexico
[2]University of La Rochelle, Intensification of Transfer Phenomena on Industrial Eco-Processes, Laboratory of Engineering Science for Environment LaSIE-UMR-CNRS 7356, La Rochelle, France

ABSTRACT

Lentils (*Lens culinaris*) are an ancient crop that represents an important part of various civilizations' diets throughout history. Although for lost time, it is now becoming increasingly important around the world for its great benefits and properties. This chapter aims to describe lentils as a superfood due to its components and the beneficial activities it has, such as antidiabetic, anticancer, antioxidant, antimicrobial, among others. Moreover, economic importance is developed, as well as the different technologies used to treat them.

Keywords: lentils, DIC, *Lens culinaris*, phenolics compounds

INTRODUCTION

Lentil (*Lens culinaris*) is an annual legume herb crop whose seeds are used as a valuable human food rich in protein, carbohydrates, iron, zinc, and other minerals and vitamins. The genus has seven taxa. This crop is one of the oldest crops in history (Wong et al. 2015). The first evidence of their existence comes from the Neolithic. It was found in different regions of Asia, Europe, and Africa, in countries like Syria, Israel, Iran, and Cyprus (Sandhu and Singh 2007). However, nowadays, this crop is distributed all over the world, and it is increasingly

* Corresponding Author's E-mail: malonzoma@tec.mx.

relevant. From 1996 to 2018, there was an annual growth of 4.6% in lentil production. And now, the most prominent producers are North America (57%), South Asia (23%), and Australia (3%). The increase in its production is mainly due to the fact that it is now seen as a highly potential food in the diet of vegans, vegetarians, and the rest of the people (FAO 2019).

Lentils are recognized for their high nutritional level. They are mainly made up of carbohydrates, representing approximately 60% of its total composition, followed by the protein that represents between 17 to 40% of its composition, and lentils are mainly rich in lysine (Hefnawy 2011, Wang et al. 2014). Talking about the number of fatty acids is relatively little; the total percentage varies between 1.5 to 8% of the total (Zhang et al. 2014b). Among the micronutrients present in lentils are potassium, phosphorus, magnesium, and calcium. The composition of the lentil positions it as a superfood, among its benefits is its antidiabetic, anticancer, antioxidant, and antimicrobial. However, duckweed also has antinutritional compounds that cause some problems, such as reducing protein bioavailability (Hefnawy 2011).

Nowadays, it is imperative to take care of food processing technologies and kept the food quality until it reaches people's tables. Among the most promising methods of treating lentil are drying, canning, boiling, freezing, fermentation, extrusion, instantaneous controlled pressure drop, and lyophilization, which will be seen throughout this chapter.

The objective of this chapter is to take a tour of the most important aspects of one of the oldest crops in history: lentil. Throughout the chapter, its history, botanical description, and current importance will be seen. In addition to its components, its functional properties, and finally, some of the most widely used food processing methods and their possible application for lentil will be evaluated.

Botanical Description

Lentil (*Lens culinaris*) is an annual legume herb crop whose seeds are used as a valuable human food rich in protein, carbohydrates, iron, zinc, and other minerals and vitamins (Wong et al. 2015). The plant height value varies from 20 to 45 cm tall. It has both male and female organs, which confers the characteristic of being naturally self-pollinated due to its cleistogamous flowers (Fratini, Ruiz, and De La Vega 2004, Joshi, Timilsena, and Adhikari 2017). In June or early July, two to four pale blue flowers bloom in the leaf's axils, which subsequently give rise to small pods containing one to two seeds in each one. Lentil seeds have a double convex lens shape, and their size is about 4 to 6 mm, the color may vary among yellow, red, orange, brown, green or black (Britannica 2020).

According to the United States Department of Agriculture (USDA 2020), the taxonomic description of *Lens culinaris* is shown in Table 1.

The genus Lens has seven taxa arranged into four species, namely *L. culinaris* ssp. culinaris, *L. culinaris* ssp. orientalis, *L. culinaris* ssp. tomentosus, *L. culinaris* ssp. odemensis, *L. ervoides*, *L. lamottei*, and *L. nigricans* (Wong et al. 2015). The cultivated species of *L. culinaris* has two groups, the microsperma and the macrosperma. The microsperma group is also known as the green lentils group; they are often present in Southern Europe, North Africa, and America. In this group, plants are characterized by being higher and for having longer pods with larger and flattened seeds. The plants belonging to the macrosperma group are shorter, and they have smaller and rounder seeds. Lentils in this category are red with cotyledons in which color may vary from yellow to black. This type of lentils is found in Asia, Ethiopia, and Afghanistan (Zucconi 2011).

Table 1. Taxonomic information on the species *Lens culinaris* (USDA 2020)

Taxonomy	
Domain	Eukaryota
Kingdom	Plantae
Subkingdom	Tracheobionta
Superdivision	Spermatophyta
Division	Magnoliophyta
Class	Magnoliopsida
Subclass	Rosidae
Order	Fabales
Family	Fabaceae / Leguminosae
Genus	*Lens*
Species	*culinaris*

History of Lentils

Lentils are one of the oldest crops in history. According to several researchers, history dates from the year 13000 B.C. and some charred seeds belonging to the Neolithic Era have been found, specifically in the year 11000 B.C. in some caves in Greece (Sandhu and Singh 2007), for which it is known as one of the first crops domesticated by man. It is worth mentioning that the evidence is also ancient, between the years 8500-2500 B.C. It was found in different regions of Asia, Europe, and Africa, in countries like Syria, Israel, Iran, and Cyprus.

Although now its cultivation is widely distributed around the world, it is mainly present in countries such as India, Bangladesh, Pakistan, Egypt, Greece, Italy and several countries in the Mediterranean region and it is precisely because of its wide distribution that it has been difficult to identify where exactly was tamed. However, it is accepted that lentils were domesticated in an area called "the cradle of agriculture," located in the Near East (Sonnante, Hammer, and Pignone 2009). There are several theories about the domestication of this crop. However, the most prevalent hypothesis is the "weedy/dump heap hypothesis," which says that humans brought wild lentil seeds to their villages and, without intention, dispersed the seed, and thanks to the soil fertility, the plant became stronger. However, there is still discussion about this topic, mainly because this hypothesis may seem to contrast with the intolerance that wild lentils show nowadays to the altered habitats (Sonnante, Hammer, and Pignone 2009).

Lentils are historical of great importance, as they are said to have been one of the main nutritional contributors to the population in the Neolithic Era. However, in the next years, its importance and consumption have been different in the continents. In the classical period in Europe, they would become food for poor people, since it was mixed with barley to make bread in times of scarcity. Even in the 18th century, lentils were fed almost exclusively on horses. On the other hand, lentils constituted an important part of the culture of countries of Asia and Africa, since they were used for the preparation of different traditional foods. Even now, lentils are used as a salad, first plate, side dish or starter. It could be that one of the reasons why lentils were so popular in Asia and Africa is because they provide nutritional security to low-income consumers. Nowadays, lentils are still an important part of the culture of several Asian and African countries. In fact, the countries with higher lentil production are Turkey, Syria, Iran, Morocco, and Ethiopia. These countries mainly produce red and green lentils. The preference of productions of red lentils or green lentils depends completely on the consumption habits of each country (Sarker and Kumar 2011).

It is interesting the evolution of lentils in different regions. Today, lentils represent a very promising food for the food industry due to their beneficial properties for health, such as anti-cancer properties, antioxidant activity, and among others, and their high nutritional value (Faris, Takruri, and Issa 2013).

Agricultural Production and Economic Value of the Crop

One of the ways to combat malnutrition and hunger in the world is a balanced diet, and it is estimated that pulses are proposed due to their high nutritional value special proteins. Lentils contain 20-25% of protein than cereals, which have around 6-10 protein percent; also, lentils are an excellent dietary fiber source, vitamin and mineral-rich, and low-fat content. Because lentils contain lower levels of sulfur amino acids, cereals' combination made a rich and balanced diet (Akibode and Maredia 2012). In general, the developed world regions' diets have lower caloric consumption of legumes (~2.51-3%) compared to cereals (50%). It is worth highlighting developing countries such as India, Ethiopia, and Cuba's legumes cover more than 10% of their total caloric consumption (Akibode and Maredia 2012).

Culture and tendency may impact the production and consumption of pulses, especially in developing countries. For example, India has the most lentils consumers (FAO 2020). It has one of the highest percentages of vegan and vegetarians worldwide with 30%, compared with the number of vegetarians in Americas 5%, Canada 8%, or Germany 4.3% (Paslakis et al. 2020). However, in the recent past years, the vegan and vegetarian tendency has been increasing, especially in developed countries, which causes increases in legumes consumption because of protein content, to reduce meat consumption (FAO 2020). Simultaneously, these world zones and legumes production have increased due to different factors such as vegan-vegetarian tendency.

In general, legumes production in 2014 was 77 million tons, in which dry bean produced 24, chickpea 13, dry pea 11, cowpea 7, lentil 5, pigeon pea 4, faba bean 4, and other legumes 7 million tons (FAO 2019). However, legumes production was a small portion compared to 2,712 million tons of cereal (FAO 2019). In fact, in the same year, North America became one of the major pulses producers, especially for lentils. The major three legumes producer regions were Sub-Sahara-Africa (22.8%), North America (10.7%), and Latin America and the Caribbean (8.9%). 43.2% of the total lentil production was produced in North America, while South Asia produces 30.2% (FAO 2019). Also, soybean has demonstrated considerable increases in tendency production from 1996 to 2018 (FAO 2020). It obtained an annual growth rate of 7.17% (Akibode and Maredia 2012). It was the legume most cultivated in

2018, followed by chickpea, green pea, and lentils. The grater legumes producers in 2018 were soybean by the USA and Brazil, followed by chickpea for India and Australia, green peas, and lentils for China, India, and Canada (FAO 2019).

Lentil tendency production has been increased. From 1996 to 2018, an annual growth of 4.6% in lentil world production was reported. In 2016, around 6.5 million tons of lentils were produced, leaving a net production value of 2 billion Int.$ (international dollar) in the world, in which North America produces 57%, South Asia 23%, and Australia 3% of the total production value (FAO 2020). From North America, Canada and the USA produced around 1.3 billion Int.$, and 0.1 billion Int.$, respectively, which means lentil as an important crop, especially to Canada, where production value reaches more than half of worldwide net production value. Probably due to Canada obtained the lower producer price (307 USD/ton) in 2018, it means that for Canada, the costs to produce lentils are cheaper than in other countries (FAO 2020). World lentil area harvested showed an intermediate value compared with other pulses such as chickpea and faba bean, with 18 and 2 million ha in 2018, respectively (FAO 2019).

A wide range of legumes (70%) is cultivated over rainfed conditions globally, compared with cereals, in which almost all the crops are under irrigation system and only 30% by rainfed (Akibode and Maredia 2012). This means that the major lentil crop depends on the rain cycles and not on the irrigation systems. It is probably the principal reason why lentil yield suffered fluctuation over time. Other factors that may affect lentil yield are agriculture practices, such as fertilization, disease control, seed variety, crop protection, and tilled soil. For example, countries with more protected lentil crops are Canada and Australia, compared with India and Bangladesh; Australia also used more fertilizers than Canada or India (FAO 2019).

It is important to know that Canada in 2018 did not obtain the best yield value (~1400kg/ha) but producer price, with 307 USD/ton (FAO 2020). The countries where is cheaper lentil production is Canada, followed by the USA, Turkey, and Mexico, with a range of 360-550 USD/ton (FAO 2020).

Culinary culture, socioeconomic status, and cultural trends impact the legume's consumption. The developing countries consumed more lentils and pulses in general. In the Indian diet, around 13% of the protein consumed belongs to pulses' contributions; also, legume's protein consumed in Central and South America represents approximately 10-13% (Akibode and Maredia 2012). In general, lentils are also widely consumed in Asia and Europe (especially for Mediterranean cuisine) countries (FAO 2019). Besides, the socioeconomic level has demonstrated a negative correlation with pulses consumption. For example, in America, low-income households consumed up to 20% more legumes than middle and high-income families. On the other hand, culture and country economic status may alter this pattern. For instance, research demonstrated Indian high-income households as major bean consumers, while in Tanzania, the middle-income families were the principal pulses consumers (FAO 2019).

Abiotic and biotic factors usually have an impact on agricultural yield, and consequently, economic value. It has become important to consider these aspects to improve the quality and yield of crops. Each plant reacts in its way to external factors stimulus. Hence, lentils are cultivated worldwide, especially in Canada, Australia, the USA, Mexico, India, and China. Each region differs in climate conditions, new varieties of lentil impact in the phenotypic and genotypic characteristics. For example, some varieties are more resistant to salinity stress, increasing biomass than non-salt-resistant varieties (Ouji et al. 2015). Hence, lentil crop characteristics are affected by environmental conditions, nutritional requirements, and diseases.

Temperature

The optimal growth temperature of lentils varies from 18°C to 25°C (Nleya, Vandenberg, and Walley 2004). The temperature range that does not significantly affect seed production is from 16°C to 23°C (Hojjat and Galstayan 2012). High temperatures can affect yield lentil production, especially in the reproductive stage. In Southeast, Australia lentils were exposed to a high-temperature wave of 36°C for more than 6 days in the reproductive phase. They resulted in a negative effect of 70% on lentil quality and yield production (Delahunty et al. 2015). Thus, it is vital to take into consideration high-temperature genotypic characteristics.

Water

The most important stages in which water is a key factor to improve yield and quality, besides germination, is the reproductive stage and seed filling. Drought stress limited the development and production of lentils, affecting cell membranes, photosynthetic traits, and water relations, impacting the biochemical process, and seed filling (Sehgal et al. 2017). Besides, drought affects mineral seed composition, especially for Fe and Zn (Sehgal et al. 2019). However, lentil crop is considered relatively resistant to water deficiency. At these drought stages, increases in antioxidant and peroxidase activity, ascorbate peroxidase, and proline betide are a lentil mechanism to resist water deficiency (Ahmadi et al. 2019). This water resistance capacity makes lentils able to use as rainfed crop, which around 70% of the world cultivated lentils depends on seasonal rains (FAO 2019).

Salinity

Soil salinity is the abundance of salts such as sodium chloride, calcium and magnesium sulfates and/or bicarbonates in soil and water and are generally more pronounced in arid and semi-arid regions (Gupta et al. 2008). High salt concentration affects crop yield. Different salt concentrations of NaCl (1%, 2%, 3% and 4%) demonstrated a negative effect in lentils on the percentajes of germination yield and yield contributing characters in the rhizosphere soil (Rahman et al. 2017). In order to resist stress, plants produce metabolites to decrease the adverse effects caused by high saline soils, for example, in research, salinity induced decreases of volatile organic acids, accumulation of sugars and polyols in leaves and accumulation of other key-metabolites, such as L-asparagine, D-trehalose, allantoin and urea in the roots (Skliros et al. 2018). In addition, mycorrhizal inoculation alleviated stress caused by a high concentration of salt and recovered the target yield (Rahman et al. 2017).

Soil

Soil characteristic has demonstrated to impact the viability of nutrients (Lauchli and Grattan 2008), rhizosphere population (Ibekwe et al. 2010), and agrochemical application (McMurray et al. 2019). Consequently, causes physiological and morphological changes in lentils. For example, research showed pH, salinity, and boron content affected plant biomass and rhizosphere population, each and in conjunction (Ibekwe et al. 2010). In addition, it has become important to maintain pH range since mineral availability spectrum, and toxicity depends on the soil pH. For lentils, pH and lentil type have demonstrated to have an impact on seed production and quality, being the nutritional flow and absorption the main characteristic. Therefore, it is recommended a neutral to alkaline soil for an optimal growth of lentils. However, it can be growth in wide range of pH soil (5.5-9.0 pH), and well-drained clay to sandy loam soils (Nleya, Vandenberg, and Walley 2004, Ford et al. 2007).

Fertilization

Nitrogen is the most important compound in lentil crops, soil composition from around 70-90% of N is needed for biological fixation. Other important macronutrients for optimal growth in lentils are P, K and S (Nleya, Vandenberg, and Walley 2004). The soil can be enriched with these compounds by compost or chemical applications. In chemical applications, it is recommended N_2 (10-15 kg/ha), P (20 kg/ha) and K (2.7 kg (K_2O)/100 kg) of seed produced, but it is essential to do not overapply recommended doses, because could cause adverse effects. Micronutrients also play a key role in getting target yield, Zn, Mo, and B have demonstrated to be wide necessary for lentil crops (Islam et al. 2018). This micronutrient supports biochemical reactions required for N fixation; in fact, micronutrient deficiency is usually confounded with N deficiency. The well micronutrient application can lead to an increase of 52% of yield production in lentils (Islam et al. 2018).

Pathogens

Plant pathogens contribute to important economic losses in the agriculture sector; around US$220 billion of losses by plant disease around the world were estimated (FAO 2019). Crop disease can cause yield losses from 20 to 40%, and in severe cases, up to 70% (Gossen and Morrall 1984, Bwye, Jones, and Proudlove 1994). Lentils are affected by a wide range of diseases; fungi disease contributes to the most important yield losses in lentils by affecting leaves, steam, pods, and roots and by discoloring the seeds (Taylor et al. 2007).

Ascochyta blight is a disease in lentils caused by *Ascochyta lentils Vassilievsky,* an ascomycete fungus. *Ascochyta* spp. is able to infect a wide range of crops belonging to *Lens* genera, such as chickpea, faba bean, field pea, hairy vetch, and other legumes; also can infect a wide range of lentil types including, both, wild and domestic types such as *L. culinaris* subsp. *orientalis, L. culinaris* subsp. *odemensis, L. ervoides, L. lamottei, L. nigricans, and L. tomentosa* (Bayaa, Erskine, and Hamdi 1994).

Another well-known legume disease is Fusarium wilt that causes loses from 23–50% in lentils, depending on the region and the climate conditions (Dubey and Singh 2018, Singh et al. 2017). Disease such as Botrytis grey mold, lentil rust, Stemphylium blight, and anthracnose, are also caused by fungi and infects lentil leaves, flowers and pods affecting physiological function on the plant development, and causing important lentil losses, however, some techniques are known to reduce the infection incidence. Wild-type lentils are known to possess higher resistance against pathogens.

Viruses also contribute to substantial crop yield losses in legumes ranging from 40-80% (Bwye, Jones, and Proudlove 1994, Jones et al. 2008). Some viruses persist in lentil seed and spread via insects and nematodes as vectors and may persevere in weeds or volunteers. Typical lentil family viruses correspond to *Bromoviridae, Luteoviridae, Potyviridae, Comoviridae, Geminiviridae,* and *Nanoviridae*, corresponding to diseases such as Faba bean necrotic yellows virus, Bean mosaic yellow virus, Chickpea chlorotic drown virus, Bean leaf roll virus, Alfalfa mosaic virus among others (Porta-Puglia and Aragona 1997).

Hence, well practices impact the micro and macro cosmos of agricultural soil, which, principally, reflect in a favorable economic impact in lentil production (Makkouk et al. 2014).

PHYTOCHEMICAL COMPOSITION

Lentil seed is divided into three main components: cotyledon (90%), seed coat (8%), and embryo (2%). Each part is responsible for different functions, for seed protection and development. In cotyledons, proteins and carbohydrates are stored for plant growth supply. In addition, secondary metabolites, such as phenolic compounds, are present in both cotyledons, such as non-flavonoid phenolic compounds (Dueñas, Hernández, and Estrella 2002), and the seed coat, such as flavonoids and other phenolic compounds. Lentils are known for their high content of minerals (Wang et al. 2009, Kaur, Sandhu, and Lim 2010), and vitamins (USDA 2018b, Porres et al. 2003). Lentils also contain several antinutritional compounds (Wang et al. 2009, Hefnawy 2011) such as tannins, phytic acid, and trypsin inhibitors. However, cooking methods are used to reduce antinutritional compounds. In almost all the cultures, lentils preparation involves a boiling method for 15-20 minutes. In this section, phytochemical composition of raw material (RM) and cooked lentils are analyzed.

Carbohydrates

Carbohydrates represent the major component of lentil seeds (Wang et al. 2014), representing ~60% of the total seed weight of lentils. Dietary fiber is only present in vegetables, cereals, fruits, and nuts. It is known for its potent behavior against cardiovascular diseases and cancer (Wang et al. 2009). Dietary fiber has resistance to human enzyme hydrolysis and can not be digested in the small intestine but has complete or partial digestion on the large intestine (Dhingra et al. 2012). There are various classifications of dietary fiber. The most accepted is the one described by Dhingra et al. (2012), in which dietary properties are defined by solubility and fermentation capacity. Soluble and more fermentable fiber is represented by pectin (present in fruits, vegetables, legumes), gums (present in leguminous seeds, seaweed extracts), and mucilage (present in plant extracts). While insoluble and less fermented fiber is divided on cellulose, hemicellulose, and lignin, that is present in major concentrations on plants (principal component of plant cell wall) cereal grains and woody plants, respectively (Dhingra et al. 2012). Insoluble dietary fiber (IDF) represents 83-93% of total dietary fiber of lentil seeds (Wang et al. 2009, Kaur, Sandhu, and Lim 2010, Dhingra et al. 2012, Jarpa-Parra 2018) with 3.0, 2.2, 3.1 g/100 g dry weigh (d.w.) of cellulose, hemicellulose, and lignin, respectively (Srivastava and Vasishtha 2013). Physico-chemical properties of dietary fiber are part of the digestibility capacity, some process such as grinding, drying, and heating cause changes in the composition of dietary fiber, affecting hydration characteristics, particle size, and porosity. In addition, the environmental condition also affects the dietary composition, and hydration characteristics (Dhingra et al. 2012). Conventional lentil cooking (boiling for 15-20 min) can cause losses in soluble dietary fiber (SDF) and increases on IDF (Table 2). Lentils have a low concentration of soluble fiber compared to withe beans, flaxseed, fava bean, and chickpea (Wang et al. 2009, Kaur, Sandhu, and Lim 2010, Dhingra et al. 2012, Jarpa-Parra 2018), while insoluble fiber is relatively high compared to beans, fava bean and chickpea (Table 2). Lentils showed to have 3, 2.2, 3.1, and 4.9 g/100 g of cellulose, hemicellulose, lignin, and pectin, respectively (Srivastava and Vasishtha 2013). Cooking methods caused an increase in cellulose, lignin, and pectin content on lentils, while a reduction content on hemicellulose (Srivastava and Vasishtha 2013).

Table 2. Dietary fiber content (g/100 g d.w.) in different legumes

Legume	Soluble fiber (SDF)	Reference	Insoluble fiber (IDF)	Reference
Lentil	1-1.9	(Dhingra et al. 2012, Wang et al. 2009)	10-12.8	(Dhingra et al. 2012)
White beans	13.4	(Dhingra et al. 2012)	4.3	
Flaxseed	12.18		10.15	

Lentils show to have a wide range of phytochemical compounds. Thus, crop cultivation conditions, such as soild nutrient content, lentil type, and environmental factors, can change the nutritional composition (Muscolo et al. 2020, Johnson et al. 2013). In addition, lentils contain low digestible carbohydrates, including sugar alcohols (SA), galactooligosaccharides (GAO), fructooligosaccharides (FO), and resistant starch (RS), which are used by plants against abiotic stress such as cold (Liu et al. 2007, Nishizawa, Yabuta, and Shigeoka 2008), drought and salinity (Tang, Peng, and Newton 2005, Zhifang and Loescher 2003). Further, SA and GAO act as antioxidants (Liu et al. 2007, Nishizawa, Yabuta, and Shigeoka 2008) and osmolytes (Bartels and Sunkar 2005) that maintain cell structure preventing cell damage. And, they confer human intestinal health modulating beneficial gut microbes (Johnson et al. 2013, Siva, Thavarajah, and Thavarajah 2018).

Oligosaccharides are molecules containing a small number of monosaccharides residues connected by glycosidic linkages. They are known for their microflora nutrition because the small intestine enzymes can not digest them. However, these oligosaccharides, when they are gut microbes fermented, cause human gastrointestinal discomfort and flatulence (Siva, Thavarajah, and Thavarajah 2018). Galactooligosaccharides, are known for their probiotic capacity. Thus, they can increase intestinal gut bacteria such as lactic acid and bifidobacteria (Yadav and Kayastha 2020). Lentils contain significant quantities of raffinose, stachyose, and verbascose molecules (Table 3), which are part of the GAO family. GAO in lentils represent 53% of total sugars (Hefnawy 2011). Its concentration was higher than beans and chickpea

(4.1, 3.0, 2.1 g/100 g d.w., respectively) (Siva et al. 2019). Different types of lentils, previously soaked and cooked, were analyzed by Wang et al. (2009). Cooking caused a significant ($p<0.05$) reduction in sucrose, raffinose, and stachyose levels, but verbascose increased. Increases and decreases in some compounds after cooking methods are due to the enzymatic activation and/or chain-compound breakdown into monomers or other compounds. For example, Njoumi et al. (2019) soaked and cooked lentils, fava beans, fenugreek, and chickpea. They analyzed carbohydrate compounds and found a reduction of 25-35% in raffinose, stachyose, and verbascose, but galactose increased by 54%. Srivastava and Vasishtha (2013) obtained similar results; in soaked lentils, cellulose increased by 10%, while the cooked one increased by 16.67%.

Table 3. Content of galactooligossaccharides family (g/100 g d.w.) in lentils, fava bean, and chickpea

Legume	Raffinose		Stachyose		Verbascose	
	RM	Cooked	RM	Cooked	RM	Cooked
Lentil Range [a]	0.3-0.6	0.2-0.3	1.8-3.0	1.0-2.5	0.5-1.0	0.3-0.95
Fava bean [b]	2.561		0.516		1.824	
Chickpea [b]	2.533		0.29		0.96	

[a] (Hefnawy 2011, Njoumi et al. 2019, Wang et al. 2009); [b] (Njoumi et al. 2019); RM= raw material.

Fructooligosaccharides are composed of linear β-1,2 fructosidic bonds. Some of the recent interests of this oligosaccharides are due to the low-sweetness intensity and calorie-free. In addition, studies found that FO have carcinogenic effects, an increase mineral absorption, and decrease cholesterol levels on serum (Sabater-Molina et al. 2009). FO founded in lentils include kestose and nytose. Siva, Thavarajah, and Thavarajah (2018) compared different types of lentils, divided into red and green under different treatments (cooking, cooling, and reheating procedures), finding that lentils contain 1.845 g/100 g of kestose and ~0.058 g/100 g of nytose. The study demonstrated a higher concentration of kestose in green lentils compared to red one. FO concentration in lentils is low compared to white onion (2.24 g/100 g w.w.), quite similar compared to garlic (1.2 g/100 g w.w.), but high compared to other fruits and vegetables (range from 0.1-1 g/100 g w.w.). However, FO in lentils were higher than other legumes as bean and chickpea (Siva et al. 2019).

Sugar alcohols are sugars with and additional hydroxyl group, are photosynthetic products, and store, normally in the leaves (Chibbar and Båga 2003). SA are important by their metabolic behavior (used in diet control and dental healt) (Rapaille, Goosens, and Heume 2003). However, SA may cause gastrointestinal discomfort as laxation and flatulence (especially at high intakes). The health profile of the individual limits SA consumption. SA are not recommended to those with a gastrointestinal disorder, while they are recommended to those individuals suffering from constipation (Jovanovic-Malinovska, Kuzmanova, and Winkelhausen 2014). SA found in lentil seeds were sorbitol and mannitol. SA in cooked lentil ranged from 0.353 to 0.813 g/100 g d.w. (Siva, Thavarajah, and Thavarajah 2018). Different types of commercial lentils were tested, and red lentils demonstrated that contain higher quantities of SA compared to green lentils, with 0.75 and 0.572 g/100 g d.w., respectively. Also, sorbitol showed to be present in higher concentrations than mannitol, 87% compared to 12.5% of total SA in lentils (Siva, Thavarajah, and Thavarajah 2018). Lentils contain a low concentration of SA compared to those in fruits (blackberry, pear, and nectarine) tested by Jovanovic-Malinovska, Kuzmanova, and Winkelhausen (2014), and SA amounts greater than 3 g/100 g w.w.

Resistant starch. It is known as the starch, including its degradation products, that escapes of the degradation process. Resistant starch is divided into 3: readily digestive starch (RDS) (digested in the small intestine), slowly digestive starch (SDS) (digested in the small intestine at a lower rate), and resistant starch (not digested in the small intestine). RS family has similar physiological behavior as dietary fiber, not digested at all in the small intestine, but wholly or partially fermented by colon microorganisms (Okumus et al. 2018). Concentration ranged from 1.964 to 3.026 g/100 g of total resistant starch in cooked lentils were obtained by Siva, Thavarajah, and Thavarajah (2018), while 1.65 g/100 g were obtained in raw material by Dhingra et al. (2012). The total amount of starch and uncooked lentils contain is shown in Table 4. RDS, SDS, and RS were analyzed by their digestibility capacity, showing cooked red lentil higher resistance starch content than green lentils. Green type lentils as Asterix and Greenland were unique by their high SDS activity; this was possible for their high content of phenolic and α-glucoside inhibitory of SDS (Lu et al. 2018). RDS fast capacity of digestion is correlated to elevations in postprandial plasma glucose (Kaur, Sandhu, and Lim 2010). Thus, concerning dietary starches, SDS is desirable to be higher than RDS, especially to those people diagnosticated with diabetes. In general, SDS has been shown higher concentrations than RDS on lentils seed; this beneficial property helps diabetic people due to the low glycemic index of this type of starch (Kaur, Sandhu, and Lim 2010).

Starch digestion properties have been attributed to the interaction of many factors, such as starch source, amylose and amylopectin retrogradation, and degree of crystallinity (Kaur, Sandhu, and Lim 2010). Amylose is a linear chain of α-1,4 glycosidic bonds; amylopectin is a highly branched polymer with glucose linked bonds of α-1,4 and α-1,6. Both are water-insoluble; however, amylopectin exhibits higher solubility than amylose due to the branched structure (Sun et al. 2017). Glycemic index, expressed about the digestibility, is affected by the composition of the starch and play an essential role in the human alimentation, not only for those people with diabetes but healthy individuals (Kaur, Sandhu, and Lim 2010). Lentils showed a range of 20–31 g/100 g of amylose (Joshi et al. 2013, Kaur, Sandhu, and Lim 2010). However, Siva et al. (2019) found 17 and 25 g/100 g of amylose and amylopectin, respectively. They were directly correlated with the total amount of RS. Moreover, cooked lentils with vegetable oil can form the amylose-lipid complex, resulting in lower postprandial plasma glucose and insulin responses in human subjects (Okumus et al. 2018).

Table 4. Range content of total starch in lentils (%) determined by Kaur, Sandhu, and Lim (2010), and Lu et al. (2018)

Resistant starch (RS)	Readily Digestive Starch (RDS)	Slowly Digestive Starch (SDS)
40-60	6-15.6	20-30

Proteins

Legumes are an important source of protein for the human diet, ranged from 17-40% (Wang et al. 2014). Legumes have demonstrated to be rich in lysine but lower sulfur-containing amino acids, contrasting results in cereal amino acids. Thus, the combinations of both make a nutritional source of protein content. Lentils contain around 23 to 28 g/100 g of nitrogen-protein on a dry basis and 2.4 g/100 g of non-nitrogen protein (Table 5) (Hefnawy 2011). Quantities of lentil protein demonstrated to be higher than other legumes such as chickpea and common bean (Abd EI-Azim et al. 2018). Most of the proteins are in the seeds' cotyledons and have a low percentage of sulfur-containing amino acids; also the proteins are consumed by the germ during the germination (Hefnawy 2011).

Table 5. Aminoacid profile and total protein content of lentils chickpea and eggs

Amino acid	Lentils-Range (g/16g N d.w.) [a]	Decreased (-) or increased (+) content of boiling lentils (%) [a]	Chickpea (g/16gN d.w.) [b]	Egg (g/16g N protein) [c]
Alanine	4.2-4.3	2.80	4.69	9.2316
Aspartic acid	11-11.5	-0.86	11.53	16.69224
Glutamic acid	15.5-22	7.10	16.17	21.01288
Glycine	3.5 -4	9.75	3.55	5.42592
Histidine	2.5-3	-6.66	2.88	3.88104
Isoleucine	3.8-4.8	-2.13	4.04	8.4152
Leucine	7-7.8	8.11	7.26	13.5648
Lysine	7-7.3	-2.85	7.04	11.45472
Methionine	1-1.5	-14.28	2.07	4.7728
Cystine	1-1.2	-8.33	2.4	3.41632
Phenylamine	4.8-6.2	-4.92	5.9	8.5408
Tyrosine	2-3.5	34.78	3.22	6.1544
Proline	4-4.2	-7.50	3.77	6.43072
Serine	4.5-5	-8.69	4.15	12.19576
Threonine	3.6-3.8	21.05	3.55	6.98336
Tryptophan	0.5-1	-28.57		2.09752
Valine	5-5.2	0	5.73	10.77648
Total protein (g/100 g d.w.) [d]	23-28	2.35	18.3	12.56
Non-N protein [e]	2.4	12.50		

[a] (Abd EI-Azim et al. 2018, Haq et al. 2011, Hefnawy 2011); [b] (Abd EI-Azim et al. 2018); [c] (USDA 2018a); [d](Wang et al. 2009, Kaur, Sandhu, and Lim 2010, Srivastava and Vasishtha 2013, Hefnawy 2011); [e] (Hefnawy 2011).

Lentils accumulate globulins, albumins, and glutenins principally. Globulins are water-insoluble and salt-soluble. Due to these physicochemical properties, globulins are the most abundant protein in lentils and represent 70% of the total protein content (Boye, Zare, and Pletch 2010). In human health, globulins are known as immune proteins. Albumins are water-soluble and account for 16% of total protein content in lentils (Boye, Zare, and Pletch 2010), and take part enzymatic and structural function in humans (Grela et al. 2017). Gluteins are soluble in diluted acid and base solutions and accounts for 3% of total protein in lentils (Boye, Zare, and Pletch 2010).

Amino acids uptake become important to human and animal feed and health. Humans can synthesize some amino acids, while others must be obtained from other sources. These non-synthetized amino acids are known as essential amino acids. Animal products (e.g., eggs and milk) contain complete amino acid profiles, although some people cannot access to animal protein quality for different reasons, cultures, traditions, trends, or economic factors. It is important to consume essential diet nutrition requirements (FAO 2011). They recommend an uptake range from 15 to 90 mg/g of each amino acid, depending on the age of the individual and or type of amino acid. Lentils demonstrated to be a good source of proteins and amino acids. They are low in cysteine and methionine (sulfur-containing amino acids), a contrast to results obtained on cereals (e.g., rice), which are abundant on these amino acids. Lentils, chickpea, and eggs amino acid content are shown in Table 5 (Nosworthy et al. 2018).

Several studies on cooked lentils demonstrated slight changes in the protein content. Some amino acids were demonstrated to decrease, while others showed to increase after cooking treatments (Table 5). Similar results are obtained in other pulses. The increase in the amino acid content may be due to the carbohydrate losses and/or enzymatic activity during the boiling process. Nowadays, new protein quality measures replace Protein Digestibility-Corrected Amino Acid Score (PDCAAS) to Digestible Indispensable Amino Acid Score (DIAAS). DIAAS range reflects the amounts of amino acids absorbed so, determinate not only by quantity but the quality (FAO 2011). Nosworthy et al. (2018) compared DIAAS, between different lentils types (red and green) under cooking conditions and casein protein. The average amount of DIAAS in lentils was 51%, while casein was 92%. However, these results also conclude that boiled lentils have more quality protein, which obtains more DIAAS value than the other process. In conclusion, lentils are an excellent protein source, and it is known to maintain a good protein dietary alimentation, even to gain muscle mass. Traditional cuisines (boiling for 15 to 20 min) also demonstrated to increase amino acid digestibility in lentil seeds.

Fatty Acids

Cell membranes are composed of phospholipid fatty acids in organisms. Unsaturated fatty acids are known for containing one or more C-C double bones, while monounsaturated fatty acid are those which include one C-C double bond in the aliphatic chain. A diet containing these types of fatty acids lowers the risk of cancer and heart disease. Dietary sources of unsaturated fatty acids include olive oil, avocado, nuts, peanuts, and seeds (Moghadasian and Shahidi 2017). Lentils demonstrated to contain a low percentage in the total oil content ranging from 1.5 to 8% (Kaur, Sandhu, and Lim 2010, Padhi et al. 2017, Pal et al. 2017, Zhang et al. 2014b). Though, the quality structure of fatty acids was found, such as omega3 to omega6 ratio (Pal et al. 2017, Zhang et al. 2014b). Fatty acid profile including unsaturated fatty acids (UFA, 80-85%), monounsaturated fatty acids (MUFA, 17-31%), polyunsaturated fatty acids (PUFA, 52-64%), and saturated fatty acids (SFA, low %) are shown in Table 6.

However, the range of oil concentrations varies between lentil types. Palmitic acid, stearic acid, oleic acid, and linolenic acid, demonstrated to be the most abundant fatty acids in all lentil types (Pal et al. 2017, Zhang et al. 2014a, Zhang et al. 2014b). Nevertheless, the thermal process showed to reduce lipid content by 2.5-3.8% (Pal et al. 2017, Zhang et al. 2014b), while dehulling showed to increase by 3.2%, caused for the removal of the hull (Pal et al. 2017).

Table 6. Fatty acid profile in raw lentils and after cooking process

Fatty acid	Concentration range (%) in dry basis [a]	Cooking [b]
14:0	0.28-48	Boiling, dehulling and germination caused an increase, decrease or not affected fatty acid profile
15:0	0.17-0.20	
16:0	11.5-17.15	
16:1	0.1-0.41	
17:0	0.09-0.15	
17:1	0.06-0.17	
18:0	0.96-4.9	
18:1	15.36-31	
19:0	0.02-0.05	
18:2n-6	32.19-49.3	
20:0	0.51-0.63	
20:1	0.88-1.02	
18:3n-3	10.3-14	
21:0	0.22-0.32	
20:2n-6	0.08-0.14	
22:0	0.3-0.6	
20:3n-6	0.01-0.03	
22:1	0.14-0.24	
23:0	0.13-0.23	
24:0	0.21-0.23	
22:5n-6	0.04-0.22	
26:0	0.03-0.04	
22:5n-3	0.01-0.08	
24:ln-9	0.03-0.04	
22:3n-6	0.02-0.1	
26:0	0.25-0.46	
22:5n-3	0.02-0.08	
Σ SFA	16-21	(+) 7.1-15%
Σ MUFA	17-31.4	(+) 2-7%
Σ PUFA	50-64	(-) 3.2-7.5%
Σ UFA	79.3-84.4	(-) 1.5-1.9%

[a] (Pal et al. 2017, Zhang et al. 2014a); [b] (Zhang et al. 2014a).

Micronutrients: Vitamins and Minerals

Lentils contain a high amount of ashes with 2.8–5 g/100 g. Potassium demonstrated to be the most abundant mineral in lentils, followed by phosphorus, magnesium, and calcium (Table 7). Due to the high K levels and low sodium ratios, lentils are the ideal diet for patients with hypertension (Hefnawy 2011).

Lentils have characterized as a good source of minerals and are gaining particular attention for its high content on Se, Fe, and Zn; these minerals are malnutrition major global health problems (Thavarajah et al. 2011). However, seed minerals are dependent on the agricultural practices, and crop location found higher Se concentration in Canadian lentils than in other five world regions (Thavarajah et al. 2011). Also, lentils contain major minerals comparing to pea, chickpea, and beans.

The thermal process is known to decrease mineral content (Table 7). However, antinutritional compounds, especially those that are chelating agents, decrease mineral absorption. A recent study showed the Ca bioavailability increased with the thermal process (Porres et al. 2003).

Water-soluble vitamins found in green lentils include B complex: B1, B2, B5, B6, and B9 (Table 8) (Margier et al. 2018). Margier et al. (2018) showed B5 vitamin from green lentil as the major content among kidney beans, white beans, chickpeas, brown lentils, and flageolets. Thiamin (B1), Riboflavin (B2), and Niacin (B3) showed a significant reduction after treatments, soaking, and cooking (alone and in conjunction). The most significant losses were those treatments in which soaking and cooking took place, nearly to 18, 4, and 46% of losses, respectively (Prodanov, Sierra, and Vidal-Valverde 2004).

Table 7. Mineral concentration in lentil

Mineral	Content-Range (mg/100 g) [a]	Decrease of minerals content in cooked lentils (%) [a]	Notes
Ca	56-82	Ns	K is the most abundant. Cooking affects all minerals, but only Fe, K, Mg, and P had significant losses.
K	844-930	~40	
Mg	97-120	~12	
P	330-456	~13.4	
Na	64-156	Ns	
Fe	6-8.5	~14	
Cu	0.8-1.1	ns	
Zn	2.5-3.8	ns	
Mn	1-1.9	ns	
Se	0.05-0.1	No data	
Total Ashes	0.2-2.7 g/100 g d.w.	0.2-0.3	

[a] (Hefnawy 2011, Kaur, Sandhu, and Lim 2010, Thavarajah et al. 2011, Wang et al. 2009, USDA 2018b); ns = non significant.

Table 8. Recommended micronutrient intake compared to the micronutrient content in lentils

Micronutrient	Lentil range content (mg/100 g) [a]	FAO recommendation adult's (mg/day) [b]
B1	0.072	1.2
B2	0.032	1.3
B3	4.127	16
B5	0.318	5
B6	0.136	1.3
Folate	0.174-0.361	0.4
Vitamin E	3.0-6.5	10
Vitamin C	4.5	45
Vitamin K	0.005	0.065

[a] (Margier et al. 2018, Zhang et al. 2019, Zhang et al. 2014b, Padhi et al. 2017); [b] (FAO 2001).

Folates have also been characterized in lentils. Domestic genotype lentils (*Lens culinaris*) contain 174–361 μg/100 g of total folate when wildtype contains 195–497 μg/100 g (Zhang et al. 2019). USDA (2018b) also reported vitamin C, A, and K (Table 8). Moriyama and Oba

(2008) reported a range of 17-27 mg/100 g of vitamin C in different legumes, including yard bean, kidney bean, green pea, soybean, and peanut. However, the thermal process dismisses total vitamin C content to a range of 0.26 to 9.42 mg/100 g. Table 8 shows the recommended micronutrients intake compared to the micronutrients content in lentils.

Antinutritional Compounds

Legumes are classified as a high nutritional value food; however, antinutritional compounds are present in almost all the pulses. Enzymes, enzyme inhibitors, and lectins are secondary metabolites of plants that have a role in the defensive system. In lentils, these compounds showed to reduce the digestibility of proteins by 79-92% (Hefnawy 2011).

Bowman-Brik family (BBF) compounds, are known for their trypsin inhibition and are capable of binding and, consequently, deactivating the digestive enzyme (Hefnawy 2011). Lectins and protease inhibitors are part of the BBF. Lentils raw material contains around 2 to 2.7 mg/g d.w. of trypsin inhibitory activity (TIA) (Hefnawy 2011, Wang et al. 2009). Nevertheless, the cooking process has positive impact over TIA. Several studies demonstrated that levels of TIA decrease by ~78% applying thermal methods. Cooked lentils have no more than 0.8 mg/g of TIA, but concentrations may depend on the specie and other factors (Wang et al. 2009). And, lectin wholly removed after 72 hrs of fermentation (Jarpa-Parra 2018). In addition, trypsin inhibitors in lentils were characterized for their thermal and ionic denaturation resistance due to their tertiary structure (Hefnawy 2011). This characteristic is ideal in medical applications, such as cancer inhibition and therapies in acquired immunodeficiency syndrome; those capacities are due to their proteolytic effect against reverse transcriptase enzyme involved in viral replication (Hefnawy 2011).

Tannins are polyphenolic compounds that bind proteins by hydrogen bonds and hydrophobic interactions, ending in the lowering of protein content (Hefnawy 2011). The binding occurs in the amino groups, causing insoluble tannin complex and, consequently, in the resistance to digestive enzymes. Besides, tannins have shown to act as chelating agents, reducing iron bootability (Fratianni et al. 2014). Tannins concentrate in the plant testa and play an essential antioxidative role in seeds (Hefnawy 2011). Studies reported from 1.3 to 6 mg/g d.w. of tannin content in lentils (Table 9). Thermal treatment showed to have around ~43% in tannin losses (Table 9), but not in all the types, Lentil Eston contained 3.4, and 3.1 mg/g d.w., of raw and cooked lentils, respectively, meaning an 8.8% of tannin decreases after the thermal process.

Table 9. Antinutritional compunds in lentil seeds and the effect of thermal processing [a]

	Trypsin inhibitor activity (TIA) [a]	Tannins [a]	Phytic acid [a]	Notes
Content range (mg/g d.w.)	2-2.7	2.8-5.7	5.7-8.8	Cooking showed to have a positive effect in lentils seeds. Decreasing TIA, Tannins and Phytic Acid.
Decreased in cooked lentil (%)	~78.8%	~43.53%	~10.7%	

[a] (Hefnawy 2011, Wang et al. 2009).

Phytic acid, also known as phytate and inositol hexaphosphate, is the hexaphosphoric ester of cyclohexane. Tissue plants used these phytates to store phosphorus and are present in certain dietary fibers, especially in cereal, legumes, and oily seeds (Bansal et al. 2018). Ruminant animals can digest phytic acid; however, other non-ruminant animals cannot digest phytates (Higuchi 2014). Consequently, they act as an antinutritional compound, decreasing mineral bioavailability, and acting as a chelating agent, especially for iron, zinc, and calcium ions (Shitan and Yazaki 2013b). Lentils contain around 7 g/100 g of d.w. on phytic acid (Hefnawy 2011, Wang et al. 2009). However, it depends on the variety; lentils could reach up to 8.8 or 4.11 g/100 g (Wang et al. 2009) (Hefnawy 2011). The thermal process demonstrated to have a positive effect in the reduction of phytic acid in some legumes, around 36% (Hefnawy 2011), while a decrease from 6% to 15% in lentils boiled for 15-20 min (Wang et al. 2009). However different methods are described to be more efficient than cooking, dehulling was the most effective method to decrease tannin compound (~90%) in lentils (Pal et al. 2017).

Several studies have found the potential of phytic acid, acting as a chemopreventive, and anticancer agent via upregulating metabolic pathways. Phytic acid is demonstrated to regulate arachidonic acid-dependent pathways. Therefore, the prevention of colon carcinoma via altered the COX (Cyclooxygenase) and LOX (Lipoxygenase) genes in humans (Bansal et al. 2018). Besides, phytates demonstrate to reduce the incidence of tumor cells in a model with rats. The study suggests the inhibition and antiproliferative effects via crosstalk between P13K /Akt and Wnt pathways (Yu et al. 2017).

Bioactive Compounds in Lentils

Phenols

Phenolic compounds are known to have antioxidant capacity like ascorbic acid, but they are not classified as vitamins. There are about 8000 forms presented in foods; most polyphenols cause orange, red, and purple pigments and can reduce the damage of different radiations, including UV light (Dai and Mumper 2010). The antioxidant capacity of polyphenols can potentially prevent neurodegenerative disease (Vauzour, Kerr, and Czank 2014). Flavonoids are digested to phenolic acids by gut microbes splitting the heterocyclic ring of flavonoids, which can be absorbed and, consequently, anti-inflammatory and antiallergenic reactions, and chemopreventive role, via modulating the immune system (Fratianni et al. 2014). Also, polyphenols have demonstrated to induce apoptosis, decrease platelet aggregation, increase blood vessel dilation, modulate intercellular signaling, and modulate enzyme activities associated with carcinogen activation (Gollücke et al. 2018). Fruits and vegetables contain free phenolic acid, whole grains and seeds are often in the bound form. Lentils have a high concentration of phenols, compared to the other 14 types of legumes (Kalogeropoulos et al. 2010), with ranges of 1.1 to 7.45 mg/g d.w. (Fratianni et al. 2014, Padhi et al. 2017). More than twenty single fraction compounds have been characterized in lentils (Fratianni et al. 2014, Pal et al. 2017, Zhang et al. 2014b). The majority are flavonols fractions (kaempferol glycosides and catechin/epicatechin glucosides) in contrast to Dueñas, Hernández, and Estrella (2002), that found procyanidin dimer as the major fraction, followed by catechins, flavones, flavonols, procyanidin trimers, cinnamic acids, and benzoic acid fractions (Figure 1). Five bounded compounds were found in lentils, including gallic acid, catechin, epicatechin, protocatechuic, and 3-hydroxycinnamic acid (Zhang et al. 2014b). Thought, Fratianni et al. (2014) found gallic acid, chlorogenic acid, epicatechin, caffeic acid, coumaric acid, rutin, quercetin, luteolin, and naringenin in lentils; even though the phenolic profile and concentrations between lentil types vary.

The difference in phenolic compounds among lentil types could be because of the pigments on lentils, considering the different types (yellow, brown, green, red, and black), and environmental and agricultural conditions (Xu and Chang 2010). Phenolic flavonoids are present in the cotyledon, while flavanols' glycosides and flavones are present in the lentil's seed coat (Dueñas, Hernández, and Estrella 2002). Padhi et al. (2017) found 1.22 mg/g of total phenolic compounds in red lentils, while green lentils demonstrated to contain up to 7.30 mg/g d.w. Zhang et al. (2014b) found flavonols, mainly kaempferol glycosides, as the most abundant in lentils. Fratianni et al. (2014) analyzed total phenolic content in raw lentils, chickpea, and grass pea, with concentration from 1098 to 1594, 147 to 183, and 206 to 213 μg/g d.w., respectively. Besides, lentils have a higher antioxidant activity (2.2 mg Ec50), proportional to the polyphenolic concentration (Fratianni et al. 2014, Zhang et al. 2014b). Gallic acid concentration suffered a reduction of 0.05-29.1% in five different lentil cultivars after the cooking process (Zhang et al. 2014b); yet, it demonstrated fewer losses than germination dehulling (25.11-50.25%, and 13.57-40.41%, respectively). This result suggests that gallic acid storages mainly in the seed coat (Pal et al. 2017, Xu and Chang 2010). Cooking demonstrated a reduction of 3.9-48.32 and 22.4-45.04% of quercetin and catechins, respectively, while dehulling and germination originated 4 to 68% of losses (Pal et al. 2017). Antioxidant activity decreases significantly after cooking and dehulling; however, several studies have demonstrated that germination enhances antioxidant activity in different legumes, including bean, lentil, and chickpea (Padhi et al. 2017, Pal et al. 2017).

Carotenoids and Tocopherols

Carotenoids are a group of polyisoprenoid hydrocarbons synthesized in all the photosynthetic organisms such as alga, bacteria, and plants, acting as photoprotective and antioxidant agents in the organisms. Carotenoids are known for their visual pigments, responsible for red, orange, or yellow colors. More than 600 carotenoids have been characterized in nature and can act as a vitamin A precursor (Abdalla 2003). However, carotenoids found in lentils are not converted to vitamin A but act as antioxidants in humans preventing diseases including cardiovascular, cancer, and other chronic diseases (Astorg et al. 1997). Tocopherols are fatty-soluble antioxidants present in cell membranes; tocopherol, also known as vitamin E, with the isoforms β, γ, and δ-tocopherols, act by breaking free-radical chain reactions through scavenge peroxyl and alkoxyl radicals (Engelking 2015). Carotenoids found in lentils include 15-*cis*-lutein, 13-cis-lutein, 13'-*cis*-lutein, all-*trans*-lutein, all-*trans*-zeaxanthin, 9-*cis*-lutein, 9'-cis-lutein, 9-*cis*-zeaxanthin with a range of total carotenoid content from 7.24 to 25.9 μg/g (Padhi et al. 2017, Zhang et al. 2014a, Zhang et al. 2014b). The range of carotenoids content depends on the lentil variety. The color index is a determinant factor in predicting lutenin content. In a colorimetric study, Zhang et al. (2014a) found that the yellow index value in lentils determined the lutein content.

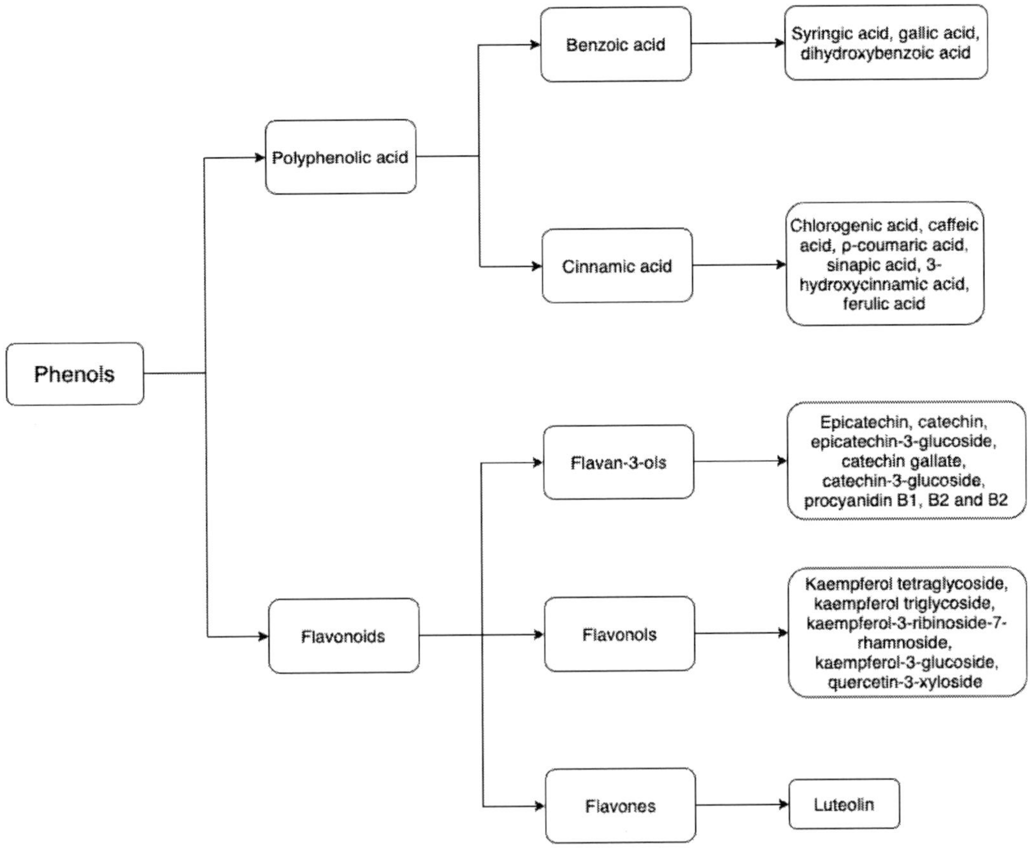

Figure 1. Classification of phenols in lentils (Fratianni et al. 2014) (Zhang et al. 2014b).

Also, lentils demonstrated the major carotenoid concentrations against chickpea, beans, and fava beans (Qudah 2014). Tocopherols found in lentils were α, γ, and δ-tocopherol (Padhi et al. 2017, Zhang et al. 2014a, Zhang et al. 2014b). Lentils contain about 30–65 μg/g of total tocopherols (Padhi et al. 2017, Zhang et al. 2014b). The most abundant tocopherol in all types of lentils was γ-tocopherol, accounting for more than 90% of total tocopherol content (Zhang et al. 2014b). Moreover, antioxidant activity was found to correlate to the carotenoid plus tocopherols content in the lipophilic extracts by DPPH assay (Zhang et al. 2014a). Zhang Zhang et al. (2014b), reported increases in carotenoids and tocopherols, after thermal treatment, probably due to the disruption of the cell membrane and, consequently, the release of lipophilic compounds, increasing bioavailability of carotenoids and tocopherols. However, more assays in the thermal effects on antioxidant components in lentils are required.

Phytosterols

Plant sterols are structurally related to mammalian cholesterol. They are present in vegetable oils, cereals, nuts, and vegetables. In humans, they can modulate low-density cholesterol in the plasma acting against atherosclerosis (Murphy, Marques-Lopes, and Sánchez-Tainta 2018). Phytosterols found in lentils include β-sitosterol, campesterol, stigmasterol, Δ^5-avenasterol. β-sitosterol is the most abundant phytosterol in legumes, lentils contain from 150 to 200 μg/g d.w. of β-sitosterol, while chickpea and black-eye beans 385 and 67 μg/g d.w., respectively (Kalogeropoulos et al. 2010).

Terpenoids

Squalene is a terpenoid polyunsaturated hydrocarbon naturally occurring in foods like fish liver oils and vegetable oils. Squalene is well-known for its use in cosmetic dermatology; it increases skin hydration. It is known for its emollient, antioxidant, and antitumor properties (Huang, Lin, and Fang 2009). Lentils contained around 1.4–1.6 μg/g d.w. of squalene and were shown to be poor squalene sources compared to other legumes (Kalogeropoulos et al. 2010). However, Mostafa, Rahma, and Rady (1987) found that squalene concentration increased on germinated soybean seeds.

Triterpenoids are isoprenoids derived from squalene, found in nature as cyclic triterpenes. These compounds are then converted into sterols, steroids, and saponins (Kushiro and Ebizuka 2010). Triterpenoids in legumes are oleanolic acid, ursolic acid, and maslinic acid. In lentils contain 3628.1 to 4814.6 μg/100 g (wet weigth) w.w. of total triterpenic acid, wich makes it a good source compared to other legumes (Kalogeropoulos et al. 2010).

Saponins naturally occurr in plants, especially in legumes, which are active sterol or triterpene glycosides. Dietary saponins research has demonstrated effective prevention of heart disease and lower plasma cholesterol levels (Oakenfull 1981). Lentils demonstrated to be a good source of saponins (~10g/100 g d.w.) compared to other food like quinoa, soybean, and lupin, containing less than 5.5 g/100 g d.w., less by fenugreek which contains 12.9 g/100 g d.w. However, lentils obtained a major percentage of bioavailability in water (83%) (Navarro del Hierro et al. 2018).

Functional Properties of Lentils

Antidiabetic Activity

Diabetes mellitus is a chronic metabolic disorder and one of the most important endocrine diseases around the world. High blood glucose levels characterize this disease due to inadequate insulin action or failure in its secretion, which may lead to severe damage in different organs such as the heart, blood vessels, eyes, nerves, and kidney (WHO 2020). Insulin is a hormone produced by the pancreas' beta cells that allow the body to use glucose as a source of energy and metabolize it, helping to maintain the levels of blood sugar within normal limits (Zakaria et al. 2015). There are two types of diabetes, namely diabetes type 1 and diabetes type 2. The most common is type 2 diabetes, in which the body becomes insulin resistant or does not produce the right amount of insulin. This type of diabetes is mainly suffered by adults and can be caused by several factors such as genetic information and an unhealthy lifestyle.

On the other hand, type 1 diabetes is characterized by the fact that the pancreas produces insufficient or no insulin at all by itself. Type 1 mainly affects the young population.

According to the International Diabetes Federation (2019), about 463 million people worldwide have diabetes, and another 1.1 million children and adolescents younger than 20 years have diabetes type 1. Each year 1.6 million deaths are directly attributed to diabetes, and as time goes on, the numbers in statistics seem to be increasing. The two main treatment drugs for type 2 diabetes are insulin and metformin, but some studies prove the possible development of severe side effects like hypoglycemia. Due to the unwanted after-effects of drugs, using dietary therapies and having a healthy lifestyle are good ways to improve the living condition and the quality of life of those who have the condition in the early stage. Legumes, such as lentils, are an excellent consumption choice due to their multiple positive health effects derived from their high content of non-starch polysaccharides and their low glycemic index (Bai et al. 2020).

Several scientists recommend the consumption of low glycemic index foods for the prevention and control of diabetes. Currently, legumes and particularly lentils have become a possible alternative that could be used as a therapeutic food ingredient to treat diabetic patients (Shams et al. 2008).

Around the world, legumes are processed in different forms to be consumed. However, nutraceutical and nutritional quality improve by sprouting. Lentils are not the exception as the sprouts contain several components that, once ingested, help reduce the risk of cardiovascular diseases, obesity, diabetes, inflammation, and cancer. Swieca, Baraniak, and Gawlik-Dziki (2013) observed the changes in starch content and expected glycemic index caused by different germination techniques and conditions and determined the potential in vitro antidiabetic effect. Numerous characteristics were evaluated during the experiment. These included the sensory aspect, total starch content, resistant and potentially bioavailable starch content, in vitro starch digestibility, and activity of a-amylase inhibitors. The results showed a decrease in total starch content, a-amylase inhibitors activity, and expected glycemic index values on the germinated plants. Swieca, Baraniak, and Gawlik-Dziki (2013) concluded that the regular consumption of lentil sprouts is beneficial and efficient for the prevention and management of diabetes.

Aslani et al. (2015) observed the response of lentil sprouts on lipid profiles in overweight and obese patients with type 2 diabetes. Forty-eight patients with the desired characteristics were subjected to the clinical trial and randomly divided into two groups; the lentil sprout group and the control. Patients belonging to the first group were given a dose of 60 g of lentil sprout that was consumed alone or with salads for eight weeks along with routine medication. Otherwise, patients in the control group only received regular drug therapy. By the end of the trial, only 39 patients completed the entire protocol and submitted it to the data analysis to evaluate the results. The outcome of the tests showed that after the specified period, serum levels of high-density lipoprotein cholesterol were higher. The oxidized low-density lipoprotein was lower for the lentil sprouts group compared to the control. Therefore, it is possible to conclude that the ingestion of lentil sprouts as a complementary dietary treatment combined with a regular medical treatment could positively affect lipid metabolism and in management and reduction of metabolic disorders in patients with diabetes type 2.

Tefera et al. (2020) developed a different experiment, in which an 80% aqueous methanol extract of *Lens culinaris* germinated seeds was used in Streptozotocin-induced diabetic mice to evaluate its antidiabetic activity. The doses used were 100, 200, and 400 mg/kg of the extract for three weeks, and during this period, blood glucose levels and body-weight changes were measured weekly. The results showed that the extract increased body weight and decreased blood sugar levels. It was possible to conclude that the extract of germinated *L. culinaris* seeds presented antidiabetic activity in mice with diabetes because it effectively reduced blood glucose levels.

Hypocholesterolemic Activity

Cholesterol is a fatty compound that is naturally present in the body and is necessary to develop some essential functions such as making the membrane cell and as a precursor to several hormones. Only a small amount of fat needs to be ingested by the regular diet to make enough cholesterol and keep it at the right levels (Harvard Medical School 2019b, National Heart Lung and Blood Institute 2018).

The fat eaten through the diet is absorbed in the intestine and transported to the liver, where it is transformed into cholesterol and released into the bloodstream. Mainly, there are two different types of cholesterol; low-density lipoprotein (LDL) cholesterol, also known as the bad cholesterol, and the high-density lipoprotein (HDL) cholesterol or "good" cholesterol. In general, when a person has high HDL cholesterol levels, he or she is less likely to develop cardiovascular disease. Nevertheless, if cholesterol levels are high for both types, HDL and LDL, treatment needs to be set to decrease the LDL level (Harvard Medical School 2019b).

The expression used to refer to a high blood cholesterol level is hypercholesterolemia. It is defined as an LDL cholesterol level higher than 190 mg/dL for a person without any significant risk factor. There are different reasons why hypercholesterolemia can develop, but the two most important are genetic causes or a high intake of fat. The genetic disorder that causes a high blood cholesterol level is called familial hypercholesterolemia. People who suffer from this condition present a problem with the LDL receptor, or it is absent due to mutations. Without the receptor, the liver can not use and metabolize cholesterol (Ibrahim and Jialal 2020). According to the Harvard Medical School (2019b) about 1 out of 500 people have this inherited disorder. And, around 73.5 million adults have high levels of LDL cholesterol.

Most people who have hypercholesterolemia do not develop any symptoms until the damages are enormous. Therefore, the only efficient way to measure the cholesterol levels in the blood is by performing a lipidic profile. Those whose results indicate having hypercholesterolemia are highly recommended to modify their diet by avoiding foods rich in saturated fat and instead choosing healthier meals. Also, it is critical to have an active lifestyle and to exercise frequently. If, after following these recommendations, the blood cholesterol level remains high, the patient may be prescribed proper medication (Mandal 2019).

On the other hand, due to this common problem affecting people all around the world, there has been an important effort to study different foods that stand out for their hypocholesterolemic activity that helps to reduce levels of cholesterol in the blood. Studies have shown that lentils have compounds that help keep blood cholesterol levels within the suitable status. Such is the case of the work done by Micioni Di Bonaventura et al. (2017). They produced a 70% aqueous ethanol lentil extract and tested it for the hypocholesterolemic effect in diet-induced hypercholesterolemic rats. Animals were randomly divided into two groups. During the trial, rats in the first group received water and rats for the second group received the lentil extract. In the end, the plasma cholesterol level was examined. The results show a decrease of 16.8% of the cholesterol level in rats treated with the lentil extract compared with the control. Therefore, it is possible to conclude that the extract has hypocholesterolemic properties, and further investigation and experiments need to be done to use it in some areas of interest in the nutraceutical field.

Several researches have demonstrated during the same experiment that lentils have both antidiabetic and hypocholesterolemic activities. For instance, the one mentioned before developed by Aslani et al. (2015), in which it is concluded that the ingestion of lentil sprouts may have a positive reaction on lipid metabolism but also in the reduction of metabolic disorders in patients with diabetes type 2. In this case, Al-Tibi, Takruri, and Ahmad (2010), diabetic rats were used to analyze *L. culinaris* on serum glucose and lipid levels. There were five groups, each one with a different experimental diet, a casein diet (control), whole raw lentil, cooked whole lentil, raw dehulled lentil and cooked dehulled lentil. Each group contained eight rats that were assigned randomly. After six weeks of the experimental trial, blood samples were taken to evaluate the established parameters. The obtained results suggested that the intake of cooked lentils is more useful to decrease the blood glucose level and to upgrade the HDL cholesterol in diabetic rats.

Anticancer Activity

There are several types of cancer, more than 100, and the number seems to increase as time goes on. Cancer is defined as a group of diseases in which cells change atypically and proliferate out of control. Almost all types of cancer are characterized by their ability to form tumors, a mass or agglomeration of abnormal cells. There are two types of tumors; benign and malignant. Benign or not-malignant tumors are not life-threatening and do not spread from one part of the body to another. On the other hand, malignant tumors may experiment with a process called metastasis, which is characterized by the release of some cancerous cells that spread out to other organs or body parts (Harvard Medical School 2019a). According to the International Agency for Research on Cancer (2018), cancer is the second cause of death worldwide, with 9.6 million deaths, which means that 1 in 6 deaths is due to cancer. The most frequent types of cancer are lung, female breast, and colorectum cancer. Each of these three types represents 11.6% of the total number of cases and an estimated 2.1 million diagnoses in 2018. Together, they are accountable for one-third of the cancer incidence and mortality globally (International Agency for Research on Cancer 2018, World Health Organization 2018). The number of cancer incidence for 2040 is expected to be around 29.5 million cases (Ferlay et al. 2018).

Currently, lentils are considered as one of the most important legumes for human nutrition due to their composition and nutritional content, but also for the positive impact in diseases such as different types of cancer including breast cancer, colorectal cancer, among others. *Lens culinaris* offers a bunch of bioactive compounds that confer this food to anticancer activity such as antioxidants, polyphenols, dietary fiber, resistant starches, phytosterols, and oligosaccharides. These compounds are strongly related to cancer mortality prevention.

A study by Adebamowo et al. (2005) conducted an epidemiologic trial with 90,630 women, demonstrating the inverse correlation between the ingestion of dietary flavonols through lentils and the risk of breast cancer. Flavonols are a classification of phenolic compounds, and one of the main sources are lentils. They could act as antioxidants, antiestrogenic, cell signaling pathways, or mitochondrial toxicity to reduce the risk of cancer. Likewise, through these bioactive compounds, legumes have many relevant functions, and one of the most distinguished is the chemoprotective ability to inhibit several tumors. Chemoprevention is a promising approach to the leading problem of cancer worldwide, and that affects all kinds of people. Spanou et al. (2007) reported the elaboration of methanolic and aqueous extracts from 11 plants of the Leguminosae family, including *L. culinaris*. The extracts were evaluated using in vitro assays to test the chemopreventive activity. Results showed that lentil extract had an essential protective activity against DNA damage.

The study carried out by Faris et al. (2009) used sixty male rats randomly divided into six different groups (10 animals per group). According to the diet assigned, control, raw whole lentils, raw split lentils, cooked whole lentils, cooked split lentils, and raw soybeans, that were used for comparison. After a period, rats received a carcinogen injection. At the end of the experiment, results were evaluated through an aberrant crypt foci assessment showing that lentils intake may be protective against colon carcinogenesis and that the cooking process improves and enhance the chemopreventive potential of lentil seeds.

Antioxidant Activity

In recent years, the popularity of antioxidants has been increasing. They are substances whose main objective is to protect cells from the damage caused by the free radicals, preventing premature lipid oxidation that, when consumed in the diet, could have anti-aging effects on the human body. The antioxidant work is slowing or preventing the oxidation of other molecules. They remove free radicals to inhibit other oxidation reactions. Nowadays, the study and usage of antioxidants in different industries is worldwide distributed; some examples are the pharmacology industry (Berger et al. 2012, Hamid et al. 2010).

Antioxidants are present in fruits, vegetables, legumes, nuts, grains, and fish. The most common antioxidants found in foods are lutein (presented in green vegetables), beta-carotene (presented in orange vegetables), and lycopene (presented in tomatoes). However, these are just some examples. In the case of the lentils, they are known to have mainly phenolic compounds, which give these antioxidants properties to this legume.

The antioxidant properties of lentils are mainly based on the content of phenolic compounds that they possess. These phenolic compounds have at least one aromatic ring and come from the secondary metabolism of plants. These are important and recognized for their high antioxidant activity since they can act as protectors against different diseases caused by oxidative stress (Ganesan and Xu 2017).

On the other hand, it is necessary to evaluate what happens after the lentils' domestic cooking with the phenolic compounds. Zhang et al. (2014b) investigated the antioxidant activity in lipophilic and hydrophilic fractions of lentils before and after cooking. They concluded that cooking reduces the flavanols concentration, which can affect the antioxidant activity of lentils.

As was shown previously, varieties of *Lens culinaris* can be visibly differentiated by the background color of the seed coat. Following this topic, a study used Liquid Chromatography-Mass Spectrometry (LC-MS) to compare the phenolic compounds present in the genotypes of different types of lentils. Four seed coat background colors (green, gray, tan, and brown) were analyzed in this study, as well as two cotyledon colors (yellow and red). The study concluded that the values of the phenolic compounds present in the seeds vary depending on the color. Finally, it was determined that the seed coats of lentils with gray and green seed coats (genotypes with the homozygous recessive tgc allele) had higher antioxidant activity (Mirali, Purves, and Vandenberg 2017). Another study conducted by Xu and Chang (2008) used 11 profiles of lentils and showed that among all the phenolic compounds present, synaptic acid was the predominant phenolic acid. This study also agreed with the one previously presented in that there are considerable variations in the phenolic compounds present in different types of duckweed. Therefore, it could be beneficial to study the different types of lentils to choose the one with the highest amount of antioxidants and are most beneficial for human health.

Currently, there are already some products made from lentils that seek to take advantage of their antioxidant properties. Given this, the need arises to evaluate this antioxidant capacity in ending products. Bioactive compounds and antioxidant capacity of extruded snack-type products developed from novel formulations of lentil, and nutritional yeast flours were determined (Ciudad-Mulero et al. 2018). There was a decrease in the total phenolic compounds, which was directly attributed to the extrusion temperature. It showed that the care of the treatment process is essential to avoid the decrease of phenolic compounds that are important for the desired antioxidant activity.

Antimicrobial Activity

Antimicrobials are essential, especially in modern medicine. Since its development, the decline in infectious disease mortality has decreased markedly. It is surprising that in the past, one of the leading causes of death was postoperative infections. Today it is impossible for us to think that something like this could happen without considering it medical negligence. Thus, with a simple example, we can understand the great importance of antimicrobials. Given this, the need to find new sources of antimicrobials arises. In this regard, food represents something promising. Plants due to the tremendous genetic and structural diversity of their components, making them an optimal source for obtaining antimicrobial compounds for possible applications in industry. Among the most commonly studied sources are oregano, thyme, and basil (Dixon and Duncan 2014) (Sakkas and Papadopoulou 2017).

Lentil is known for its various applications. Its antimicrobial activity is something that is still under study. The antimicrobial peptides are the key to this. They usually have between 15 and 45 amino acid residues and epithelial cells, neutrophils and macrophages typically produce them. One of the most studied antimicrobial peptides is the defensins. *Lens culinaris* has the defensin protein, which is one of the most studied antimicrobial peptides. Its primary role is the direct lysis of microorganisms. However, its properties are still under constant study (Rivas-Santiago et al. 2006). A new type of defensin (Lc-def) was isolated from the sprouted seeds of lentil. Its antimicrobial activity was studied, which showed that Lc-def is active for fungi; this activity could be mediated by the electrostatic interaction with components anionic lipids of fungal membranes (Shenkarev et al. 2014). On the other hand, it is believed that Lc-def has little or no activity in the growth of Gram (+) and Gram (-) bacteria. Khan et al. (2013) evaluated the microbial activity of different species against various Bacterial species using the diffusion method. Results showed that *Lens culinaris* had not an activity with *Pseudomonas aeruginosa*.

The antimicrobial activity of lectins has recently been studied in the following Egyptian cultivars: beans, lentils, and peas. The lectins showed antimicrobial activity, indicating that the purified lentil lectin was the one with a more significant zone of inhibition (35 mm) against *Staphylococcus aureus* ATCC 6538. This study is critical because it is the first to demonstrate antifungal activity against *Candida albicans*, demonstrating its potential as a microbial agent against human pathogens. It could also show that the lectins of Egyptian legumes are different from the others, so it would be worthwhile to study this also in lentils present elsewhere (El-Araby et al. 2020).

Finally, the *Lens culinaris* did not have an antimicrobial effect. However, more research about activity and scope is still required before it can be brought into an industrial application.

Anti-Inflammatory Activity

Inflammation is a normal response of the immune system to infections. It is mediated through the regulation of various inflammatory cytokines. However, it also can be dangerous when it continues. It can lead to autoimmune or autoinflammatory disorders, neurodegenerative disease, or even cancer (Dinarello 2010). Therefore, its study is of great importance, and researchers are looking for the best alternatives for new anti-inflammatory therapeutics.

Therefore, the anti-inflammatory activity of lentil attracts attention. This activity in legumes is attributed to lectins and peptides (Zhu, Du, and Xu 2018). There are some studies focused on this. In one of them, crude lentil methanolic extracts were obtained using a Prostaglandin E 2 inhibitor assay (PGE2, a vital lipid mediator in inflammatory responses) that produces COX-2 (the enzyme responsible for inflation and pain). A 74.47% inhibition of PGE2 was obtained in lentil at 20 m/ml (Zia-Ul-Haq et al. 2013) (Agard, Asakrah, and Morici 2013). This first study could show the COX-2 inhibitory potential of lentil. It is also interesting to mention that anti-inflammatory activity is found in certain parts of the duckweed. Boudjou et al. (2013) carried out a study where lentil seeds were separated into cotyledons and hulls to evaluate their anti-inflammatory and antioxidant activity. In this study, the lentil hulls demonstrated to have high inhibitory activity on the COX pathway.

López-Barrios, Antunes-Ricardo, and Gutiérrez-Uribe (2016) analyzed the effect of germinating black bean seeds on the production of cotyledon protein hydrolysates with the antioxidant and anti-inflammatory activities. The results showed that oxygen radical absorbance capacity (ORAC) of cotyledon protein hydrolysates obtained was lower than observed in raw cotyledon. The results also showed that the peptides derived from the cell wall could be responsible for anti-inflammatory activity.

These previously presented studies show the anti-inflammatory capacity of lentils. However, there is still a lot of work to be done in this regard to finally determine whether it is possible to use lentil with its anti-inflammatory property.

Food Processing Technologies Applied to Lentils

Various processes can be applied to lentils, depending on what is required, formulations, food supplements, etc. These technologies could be drying, freeze-drying, extrusion, boiling, freezing, and instant controlled pressure drop (DIC). They are used to increase shelf-life, reduce humidity, or more specific tasks such as removing unwanted compounds or further applying protein in food.

Drying

The oldest method described to preserve food, and one of the most common processes to which many foods are subjected for different purposes is drying. Drying can be defined as a process in which there is a simultaneous exchange of heat and mass between the ambient air and the solid. In other words, two processes are co-occurring during drying. One is a transfer of energy in the form of heat from the surroundings to evaporate the moisture from the surface of the solid. The second one is a transfer of the internal moisture to the surface of the solid. Several factors affect the drying process in different ways, which must always be taken into account and adjusted to the optimal requirements depending on what is desired and the material you are treated. The conditions are divided into external and internal. Some examples of external conditions are the temperature of the surroundings, air-flux, pressure, and air humidity. On the other hand, the moisture content and the physicochemical conditions of the material or food to be dried are examples of internal conditions (Sánchez 2006, Iordache et al. 2017).

The study of the physical properties is essential for the designing of the equipment required for drying. The design should be based on the seed dimensions' measurements and the bulk density, kernel density, and porosity of the seed (Ghosh et al. 2007). When drying pulses, an effective way to reduce moisture is natural air in conjunction with suitable parameters for the temperature and relative humidity, always taking care of not exceeding 45°C for the air temperature to keep seeds viable to preserve germination. After drying with heat, it is necessary to cool down the pulses to achieve safe storage (Barker 2018).

Drying by convection is used mainly to increase the shelf life of many products and foods with considerable moisture content, giving them great stability and making them easy to store, which makes it remarkably easy to market (Alvarado 2016). Usually, after harvesting, lentils need to be transported and stored until they arrived at the final consumer. The time that the product remains stored varies from a few weeks to even more than a year, and drying is one of the most used processes to preserve the seeds for extended periods (Ghosh et al. 2007).

In addition to conventional drying, different methods have been evaluated, as in the study developed by González and Pérez (2002). They showed that microwave irradiation and extrusion cooking of lentils caused a decrease in crude protein and fiber content, moisture, solubility, water absorption, starch retrogradation, and an increase of ash content.

Canning

Another traditional technique used to store food is canning, which is defined as the process of packaging food in a container and applies heat once it is hermetically closed to destroy any microorganisms, especially those that may cause food spoils and pathogens (Berry and Pflug 2003, Iordache et al. 2017).

Nicolas Appert created the canning process in 1810. His method has been improved and developed into advanced processes resulting in safer food to consume and with a long shelf life, which represents a lower economic loss (Berry and Pflug 2003).

During an optimal canning technique, the food is heated at a specific temperature for an exact period, which kills microorganisms and does not significantly affect the physical and nutritional characteristics. That is why the establishment of the thermal process results indispensable (Berry and Pflug 2003).

There are two major methods for canning. According to Boyer (2019), the first one is the water bath canning, mainly used to treat high acid foods in which the filled cans are submerged into a boiler with boiling water, and the cans are heated, reaching an internal temperature of 100°C for a specific time. The second method is called pressure canning and is used to process vegetables and low acid foods (NCHFP 2009). This method uses a steamer where the jars are introduced to reach an internal temperature of 115°C under specific pressure.

The study realized by Donoso and Alvarez (2016) highlighted a prevalent problem with canned foods, the large amount of sodium they contain, and which results are harmful to health. In their research, the changes in the physical, chemical, and organoleptic properties of a canned lentil stew reduced in sodium were evaluated. For this purpose, instead of using sodium chloride, they used other salts such as potassium chloride and magnesium chloride to prepare some canned food samples and compare it with the control that. In this case, was the canned lentil stew with a regular amount of salt. After evaluating the physicochemical aspects and viewing the results of sensory tests for the different samples, it was concluded that it is possible to manufacture a canned lentil stew containing less sodium than the control and with a positive acceptance by consumers.

Boiling

Lentils are a high nutritive leguminous. However, they need to be processed mainly because of the anti-nutritional factors, lower protein digestibility, and flatulence that they can cause. Therefore, boiling is one of the preferred processes for lentils due to its low cost and facility (Joshi, Timilsena, and Adhikari 2017). Nevertheless, it is essential to evaluate the capacity of this technology to preserve bioactive compounds in lentils and because boiling as food processing technology can produce changes in structure-forming components.

Xu and Chang (2008) evaluated the effects of soaking, boiling, and steaming on phenolic components and antioxidant activity of different legumes: lentils. All the steps caused a significant decrease in the total phenolic content (TPC) and DPPH free radical scavenging activity (DPPH) in lentils. However, when treating the samples with single or combined steps, the obtained result was different. They demonstrated that all soaking and atmospheric treatment also caused a significant decrease in oxygen radical absorbing capacity (ORAC).

However, the steaming treatments showed better retention of ORAC, DPPH, and TPC values in comparison with boiling treatments. This study showed that steaming treatment is better than boiling when preserving bioactive compounds (specifically phenolic content) of lentils and when referring to the retention of appearance and texture of lentils. However, when comparing traditional boiling cooking with microwave cooking of lentils, it is shown that boiling treatment reduces the total concentration of amino acids (lysine, tryptophan, and aromatic sulfur amino acids) in lentils. This information was presented by Hefnawy (2011). However, it was previously mentioned by several authors. Another critical point is that boiling has shown excellent results when reducing antinutritional factors such as trypsin, showing a reduction of 80.27%.

Lentils are of great importance in different areas and aspects. They are also recognized for having the characteristics of reduction of post-prandial blood glucose response (PPBG) (post-prandial hyperglycemia is an abnormality related to diabetes type 2). Following this, a study evaluated the boiling processing effects in the PPBG response to human essays. This study demonstrated that people's glycemic index was significant minor when lentils were boiling processed than when spray-dried processed (Ramdath et al. 2018).

The studies presented are showing that boiling is still an optimal technology for lentils processing. However, it is necessary to evaluate what characteristics or properties of the lentils we want to preserve.

Freezing

Freezing is one of the most ancient technologies for preserving food. Frost stops spoilage activity caused by microorganisms and helps to keep food for more extended periods. This technology kills microorganisms by physical and chemical effects. The studies suggest that many pathogenic microorganisms may be sublethally injured by freezing (Archer 2004).

The process of freezing includes three key steps. The first one is the pre-cooling phase, in which sensible heat is removed from the product, cooling continues, but the product continues in a supercooled state. Eventually, ice nucleation occurs. Then the transition phase occurs; in this step, the temperature remains almost constant, while the latent heat of crystallization is removed, and ice crystals grow. Once most of the water has become ice, the tempering phase arrives. In this phase, sensible heat is removed (Sutariya and Sunkesula 2020). Even though freezing is sufficient most of the time, the traditional methods have the disadvantage of low freezing efficiency and generation of large ice crystals, which can affect the final quality of the product.

The study carried out by Slupski (2011), the content of vitamin C in immature seeds of five bean cultivars was evaluated. Beans were treated with three different methods. The first was a traditional freezing treatment (blanching-freezing-frozen storage-cooking). The second was a modified freezing method (cooking–freezing–frozen-storage–thawing and heating in a microwave oven). The third was the product ready for consumption at room temperature, and the fourth was canned products obtained by sterilization. The results showed that the modified freeze has better vitamin C retention and that canned seeds conserved less vitamin C than freezing seeds.

Another study worked with seeds of the grass pea, one of the preferred legume seeds of two varieties: Derek and Krab, to determine the effect of freezing and canning in the content of vitamins C, B1, and B2 and of carotenoids, beta-carotene, and chlorophylls. The study was made after six months of storage and after cooking (in frozen products) and after six months of storage (in canned products). The results revealed that both treatments produced a significant reduction of vitamins C, B1, B2, carotenoids, beta-carotene, and chlorophylls (Korus, Lisiewska, and Kmiecik 2002). Therefore, freezing seems not to be the best option for legumes preservation.

The studies presented above where freezing is evaluated in different legumes show that freezing affects the content of vitamins and other essential nutrients. Therefore, it could be explained why no studies are assessing the effect of freezing on lentils specifically.

Extrusion

Extrusion is a process in which the food mixture is subjected to continuous cooking, mixing, texturing, and forming the ingredients (Guy 2001). Initially, cold extrusion was used for pasta products (Harper 1989). Nowadays, extrusion is a high-temperature short-time process, is a versatile, high efficient and low energy technology in food processing, with a wide range of products and food ingredients, such as breakfast, cereals, baby food, and ready to eat food (Akdogan 1999).

In extrusion, the food pass through a rotational screw from the feed hopper to the shaping die. The extrusion process is a multi-step operation that occurs before passing to the shaping die, including conveying, mixing, kneading, cooking, compression shearing, and stream alignment (Bouvier and Campanella 2014). During extrusion, raw material suffers molecular transformation and chemical reaction, such as starch gelatinization, protein denaturation, complex formation between amylose and lipids, and degradation reactions of vitamins, pigments, etc. (Kamau, Nkhata, and Ayua 2020). Some of the final product's physicochemical properties are determined by the independent variables of the process, such as barrel temperature of the extruder, rotational screw speed, feed rate, and features, die design, etc. (Osen 2017).

High nutritional and healthier industrial food is becoming the trend of choice of costumers. In general, lentils and pulses have gained attention, not only due to its high protein content but also for dietary fiber, mineral, and vitamin content. Nevertheless, lentils contain anti-nutritional factors that can affect the bioavailability of some proteins, vitamins, and minerals. Extruded legumes have been evaluated to see the configurational changes in the physicochemical properties, such as proteins, minerals, fat, phenolic compounds, and carbohydrates.

Antinutritional Factors

In foods, extrusion has shown an increase in protein availability due to the inactivation of some non-nutritional compounds such as trypsin inhibitors, tannins, and phytates (Nikmaram et al. 2017). Some of the denaturation processes are due to the high temperatures from 110-160°C carried in the process. For example, trypsin inhibitors activity (TIA) lower by either elevated temperatures or increasing residence time at lower temperatures (Ti et al. 2015). Similar results obtained Frias et al. (2011), where levels of TIA in raw pea (*Pisum sativum* L.

var. Laguna) were 1.84 U/mg d.w., while extruded pea at 129°C showed a reduction of 90% and no activity of TIA at 135°C and 142°C. Thermal processing caused destruction or conformational changes in covalent bonds, including disulfide bones triggering TIA inactivation (Adams 1991, Osen 2017).

Extruded beans also showed decreased phytic acid from 4.1 mg/g in raw material to 1.5-2.1 mg/g in the processed material. Phytic acid is believed to suffer dephosphorylation through mechanical or thermal processes. For example, a different operational process carried by Sumargo (2016) in which experiment with different moisture percentage (17%-20.1%) at a maximum temperature of 110°C, obtained a significant reduction in the final phytic acid content of extruded beans. Frias et al. (2011) tested different temperature barrel (129-142°C) and obtained slightly or not a reduction in the phytic acid extruded peas content. These results showed operational parameters might affect the final composition of the phytic acid content (Rathod and Annapure 2016). For example, some authors associate the phytic acid hydrolysis partly for the interplay of thermal/chemical factors (Nwabueze 2007).

Extrusion operational variables, especially temperature, have demonstrated positive behavior on antinutritional compounds, such as tannins (Sumargo 2016, Rathod and Annapure 2016). In an extrusion process with temperatures from 140-180°C, and moisture from 14%-22% showed an effective reduction (98%) on tannins content in lentils (Rathod and Annapure 2016). In comparison, temperatures from 80-110°C and 17-20% of moisture accounted for 86% on decreases on tannins content in dry beans (Sumargo 2016).

Micronutrients

In general, in extruded legumes, minerals have not shown significant reduction (Alonso et al. 2001), but an increase in bioavailability assays resulting from the inactivation of anti-nutritional compounds in the extrusion process. Phytic acid and tannins have shown to act as chelating agents and decrease the biosorption of certain minerals. However, extrusion has demonstrated to reduce this anti-nutritional compound and subsequently, the affinity to the minerals. Sumargo et al. (2016) showed that Fe bioaccessibility had a significant negative correlation with phytic acid concentration, while Mg bioaccessibility was negatively correlated with tannin concentration. Besides, many authors have reported a slight increase in minerals such as Fe, Ca, Co, Mg, K, and P.

The vitamins more sensitive to the extrusion process are A, E, C, B1, and folic acid. In contrast, other vitamins tend to be more resistant against degradation to extrusion, such as B2, B6, B12 (Riaz, Asif, and Ali 2009). However, there is little information on the effects of extrusion on legumes vitamins. Frias et al. (2011) reported a negative impact on vitamins such as B1 and B2 in peas. Riboflavin remains constant, and thiamin decreases up to 50% because thiamin has low heat-labiality. These results are consistent with the information mentioned above, which referee to the extrusion process of other food, such as cereals.

Proteins

In legumes, the amino acid composition did not show a significant change under thermal extrusion conditions below 140°C and high moisture conditions (Sobota, Sykut-Domańska, and Rzedzicki 2010, Frias et al. 2011, Osen 2017). However, conformational changes in protein have described due to thermal denaturation. For example, legumin, which is present in most legumes, after extrusion, the band of this protein disappears in SDS-PAGE analysis. Also, legumin is known for its macromolecular network cross-linked by disulfide bonds, that may be affected by thermal processes. Interesting, vicilin and convicilin proteins were not affected, and both lack cysteine residues (Osen 2017). In lupin seeds, covalent and non-covalent bonds were destroyed, and new conformation intramolecular bones took place after the extrusion process. This changed the physiological properties of the final product emulsifying and hydrophobicity characteristics.

The biological assay showed no effect in net protein utilization, protein digestibility, biological value, net protein ratio, and relative net protein ratio in extruded peas compared to raw pea (Frias et al. 2011). However, in some cereals, a bioassay is demonstrated to improve protein availability (especially sulfur amino acid), due to the inactivation of trypsin inhibitor after extrusion. For example, lima bean (*Phaseolus lunatus*) and corn (*Zea mays*) mix obtained 82% of protein digestibility compared to 77% on raw flours. Besides thermal anti-nutritional inactivation, the shear forces applied in extrusion can modify protein structure and improve digestibility (Kamau, Nkhata, and Ayua 2020).

Carbohydrates

The extrusion process demonstrated to cause changes in legumes carbohydrates compositions. Ribose, fructose, and glucose monosaccharides showed a significant increase in extruded lentils. This effect may be for the hydrolysis of complex carbohydrates due to the extrusion process (Berrios 2010). Nevertheless, sucrose in lentils showed a slight decrease, while maltose increases considerably after the extrusion process (Berrios 2010). In oligosaccharides, the effect of the extrusion process reminds unclear. Frias et al. (2011) reported a decrease in the a-oligosaccharides content, especially at the highest temperatures (134 and 142°C). After extrusion of a pea, the principal affected was raffinose, followed by stachyose and verbascose, 46, 26, and 21%, respectively. In contrast, other studies demonstrated an increase in stachyose a decrease in raffinose in extruded legumes. Bounds breakage and Millard reaction could cause a reduction in oligosaccharides in the extrusion process (Frias et al. 2011).

Total starch products tend to increase in extruded products, flour banana, and lentil mix, which demonstrated an increase from 34% to 64% in total starch after extrusion (Hernández-Nava et al. 2011). Nevertheless, the extruded food is used to increase amylose peaks, while decreases on amylopectin peaks (Vanier et al. 2016), so the increases are for the high digestive starch, causing higher glycemic index foods when creating reducing sugars from complex carbohydrates (Arribas, Cabellos B Fau - Sánchez, et al. 2017). These results suggest that digestive enzymes have more affinity to the substrate after extrusion due to the partial gelatinization, starch polymer fragmentation, and depolymerization of starch (Kamau, Nkhata, and Ayua 2020).

Besides, insoluble dietary fiber decreases while soluble dietary fiber increases after extrusion. Sobota, Sykut-Domańska, and Rzedzicki (2010) reported an average increase of 15.7 mg/g for soluble dietary fiber and 46 mg/g for insoluble dietary fiber in extruded beans. These increases are the result of shearing forces, and high temperature triggered the substantial degradation of the lignin fraction of insoluble dietary fiber (Sobota, Sykut-Domańska, and Rzedzicki 2010, Cheftel 1986).

However, after the extrusion process, there is more bioavailability of nutrients, not only for antinutritional compound inactivation but, for the reduction on insoluble dietary fiber, which tends to form the large complex and inhibits the bioavailable nutrients (Kamau, Nkhata, and Ayua 2020, Sumargo et al. 2016, Sumargo 2016) including free phenolics (Zhang et al. 2018). Furthermore, the increases of soluble dietary fiber are negatively correlated to the accessibility of heavy metals such as Cd and Pb, leading to a desirable effect of extrusion o element bioaccessibility (Sumargo 2016). Starch retrogradation, which occurs after extrusion, led to the formation of resistant starch type 3 after gelatinization in extruded cereals (Arribas, Cabellos B Fau - Sánchez, et al. 2017) allowing the trap of some phenolics (Lu and Luthria 2016).

Phenolic Composition

The phenolic composition may vary among extrusion processes and or a food sample. For example, antioxidant capacity demonstrated to increase after adding extruded broccoli flour to snack food (Bisharat et al. 2015). Similar results were obtained by Zhang et al. (2018), which showed an increase in free phenolic a decrease in bounded phenolics and increased up to 7.3% of the total phenolic content of extruded rice bran. Soybean and dry bean also demonstrated to increase the total phenolic compounds after the extrusion process (Zilić et al. 2014, Korus, Gumul, and Czechowska 2007). However, polished rice, brown rice showed different behavior when free, bounded, and total phenolic content decreased after extrusion These changes may occur due to the food physicochemical characteristics after extrusion, such as gelatinization ability. For example, during the gelatinization of starch, phytochemicals are trapped in the matrix complex, increasing the bioaccessibility of these phytochemicals (Ti et al. 2015). Thus, selecting appropriate process conditions and food type to improve the quality of antioxidant capacity.

In conclusion, the extrusion process is suitable for the elaboration of base legumes products. However, the different operational conditions, such as barrel temperature, pressure (shear forces), moisture, and others, affect the nutrients composition and bioavailability, thus selecting an appropriate process to dismiss antinutritional compounds without changing vitamin, mineral and protein composition. On the other hand, extruded legumes have not been used due to the low expansion and physical characteristics, being unattractive to the consumer and not feasible the industry. Nevertheless, due to the rich protein legumes content, the mix of legumes and cereals (or others) in extruded products have been gaining industrial interest.

Freeze-Drying

Freeze drying is an emergent technology that is being used for food processing since the 1950s. Freeze drying is currently a preservation method used primarily for foods and pharmaceuticals. This technology is growing 7.4% per year, and it is used mainly for fruits, representing 32% of the total. However, its applications are still being studied (McHugh 2018).

The process consists of the crystallization of the solvent at low temperatures and its remotion later. There are three main stages in this process: freezing, sublimation drying, and desorption drying. The freezing stage is essential to take care of the operation's velocity to avoid the formation of ice crystals that could affect the quality of the final product. The final product is expected to have from 1% to 4% of moisture.

In a study by Joshi et al. (2011)there was a comparison of the technologies: spray drying, freeze-drying, and vacuum drying by observing the differences in particle size distribution, protein subunit composition, and color and surface morphology of lentil protein isolate extract. The obtained results showed that the different drying methods no influence the chemical composition; instead, they affect the physical characteristics. The freeze-drying method showed that the mean diameter of the particles was 45.4 ± 0.70 μm, being the medium size obtained compared with the other methods. It also showed better gelation ability and higher gel strength and solubility of 78%, which was better than that of vacuum-dried powders (50%).

Therefore, it is possible to say that any drying method is accepted to maintain the stability of proteins when methods are conducted in their gentle states.

Even though freeze-drying is gaining popularity in the food industry by great results when it refers to food conservation and food quality, it is necessary to evaluate its effect on lentils. Pieniazek and Messina (2016) studied the impact of freeze-drying on the microstructure and texture of lentils, rice, potato, and carrots using scanning electron microscopy and image analysis. For processing lentils, 100 g of them were cooked in water by 38 minutes, after cooling them, they were freeze-dried with a temperature of -50 °C at 0.346 Pa for 11 hours. The results showed that boiling caused a deterioration of the external appearance and softened texture. In contrast, lentils also showed homogeneity, adhesiveness, gumminess, and chewiness.

Some other researchers have been working with lentil flour to evaluate them as supplement salad dressing emulsions. In the study carried out by Ma et al. (2013), raw lentil flour, thermally treated lentil flour, and seeds were subjected to different thermal processing treatments, including freeze-drying. The objective was to evaluate the change in color, physical stability, rheological, microstructural, and sensory properties. The obtained results showed that salad dressings supplemented with freeze/spray-dried lentil powder had higher consistency and viscosity, resulting as the optimal for dressing emulsions.

Currently, there are not many articles reporting the use of freeze-drying for lentil preservation. However, based on its use for other products, it is possible to say that it would be worth trying it in lentil.

Fermentation

The fermentation is one of the older methods for the conservation and production of food, such as alcoholic beverages. Fermentation also helps in the reduction of the raw material, increase of nutritional value, increase in the food shelf life, and produce sensory food changes (Kaur, Ghoshal, and Banerjee 2019). In the industry, fermentation is a classic process frequently used for nutraceutical biomolecules extraction, cost reduction, and desirable food composition changes.

Antinutritional Compounds

Legumes contained a diverse type of nutritional compounds such as proteins, minerals, and vitamins. However, the disponibility of the nutrients is reduced by antinutritional compounds. Food fermentation is known that it reduces proteases inhibitors, tannins, and phytic acid. Bautista-Expósito et al. (2018) carried out the fermentation of lentils by *Lactobacillus plantarum* and *savinase* at different pH ranges. The protein and peptides were measured at different times. The results showed that both pH and time decrease the concentration of soluble protein in fermented lentils and increases in peptide concentration, due to proteolysis. The results also showed a positive correlation between pH and peptide concentration suggesting a proteolytic efficiency at alkaline pH. Similar results were obtained by Rizzello et al. (2015), they analyzed six different lentils types among other legumes and demonstrated an increase in the concentration of peptides and free amino acid in the fermented lentils compared to the controls. The increment of the peptide in lentils is desirable in human nutrition. In addition, the proteolytic effect in lentils after fermentation may explain the inactivation of the antinutritional protease inhibitor compounds. This inactivation has been demonstrated in many other fermented legumes (Patterson, Curran, and Der 2017).

Other antinutritional lentil compounds such as raffinose, part of the α-galactosidase family, and phytic acid have reported decreasing after fermentation. Curiel et al. (2015) analyzed six types of lentils and other legumes in sourdough fermentation, starting with *Lactobacillus plantarum* C48 and *Lactobacillus brevis* AM7. The results showed a significant decrease in raffinose (up to 64%) in all the analyzed legumes. Lactobacilli can produce raffinose degrading enzymes hydrolysis (i.e., levansucrase), which explain the decrease of raffinose after fermentation with mentioned strains (Teixeira, McNeill, and Gänzle 2012). Phytase activity, known for the phytic acid hydrolysis, was also analyzed confirming a major activity after fermentation than in control. The authors suggest that the activation of these enzymes occurs during lactic acidification, where lactic acid bacteria can extend the phytase activity (Curiel et al. 2015).

Antioxidant Compounds

Fermentation is known to conserve and/or enhance antioxidant activity. Phenolic compounds are the most abundant antioxidants present in legumes, such as lentils. Fermentation proceeded with different strains of soybean flours produced an increase in total phenolic compounds content and did not affect or reduce glutathione content (Fernandez-Orozco et al. 2007). Similar results were obtained by Magro et al. (2019), antioxidant activity increased by 107 and 81% in *Aspergillus oryzae* and *Aspergillus niger*, respectively. Compared to the non-fermented lentils, which correspond to the content of the total phenolic compounds. This result might be for the release and solubilization of phenolic compounds because of the degrading cell wall caused by hydrolytic enzymes produced by microorganisms (Magro et al. 2019).

The phenolic composition changed, a reduction in gallic acid decreased dramatically in chickpea. However, some compounds tended to increase during fermentation, such as quercetin, which, in natural fermentation, was about 11, while acidic lactic fermentation was 22 μg/g (Dueñas et al. 2005).

Saharan et al. (2020) demonstrated an increase of antioxidant activity and content on phenolics and flavonoids in lentils after *Aspergillus awamori* fermentation. Similar results were obtained by Bautista-Expósito et al. (2018), the soluble fraction of lentils showed an increase of phenolic compounds. It was observed after fermentation by *L. plantarum*. The increase in the pH variable correlates to a higher phenolic measurement where the breakage of the cell-wall and release of the phenolic compounds is favorable at alkaline conditions (Rizzello et al. 2015). They did not find any effect on the time and the antioxidant activity among the time but pH. It reached a maximum value of antioxidant activity at pH 8.5 compared to pH 6.8, 7.5 and 8.2.

However, microorganisms and fermentation methods showed different profiles of extracts' biological activities from the fermented samples. For example, water-soluble extractions of liquid state fermentation contained higher free amino acid content, c-aminobutyric acid, antioxidant, and angiotensin I-converting enzyme inhibitory activities than solid-state fermentation by *Lactobacillus plantarum*. Thus, in this case, lentil liquid state fermentation is particularly promising as functional ingredients in preventing hypertension (Torino et al. 2013). While Magro et al. (2019) demonstrated an improvement in antioxidant and antidiabetic properties in a solid-state fermentation carried by *Aspergillus oryzae* and *Aspergillus niger*. Both showed different behavior, being *Aspergillus niger* lentil extracts, which exhibit higher (2,2′-azinobis-3-ethyl-benzothiazoline-6-sulfonic acid) ABTS activity, confirming the importance in the selection of adequate microorganism for desirable results.

Instant Controlled Pressure Drop, DIC

Legumes are a good source of nutrients, such as proteins (Wang et al. 2014). However, non and anti-nutritional compounds are present in lentils seeds and decrease the bioavailability of protein and minerals (Hefnawy 2011). Further research is required to comprehend how Instant Controlled Pressure-Drop (DIC) affects lentil and legumes bio compounds. Hence, the information described in this section shows how DIC could be a novel industrial process for lentil products.

Antinutritional Compounds

Phytic acid (IP6), presented in lentils, is known to be a chelating agent specially to be bounded to Fe, Ca and Zn (Shitan and Yazaki 2013a). Therefore, some industrial processes are used to decrease this antinutritional compounds in food. Temperature and pressure may play an essential role in the inactivation of IP6 in legumes. However, some research found in DIC treatment a significant correlation in the reduction of the IP6 and steam pressure and time. For example, after one minute of DIC treatment, a reduction of 16 and 19% in tow lupin species (*Lupin albunis* and *Lupin mutabilis*) while after 7 min both achieved up to 50% of IP6 reduction (Haddad, Greiner, and Allaf 2007). Similar results were obtained in a sank containing wheat and chickpea flour with a total decrease in IP6 of 40% (Yağcı and Evci 2015b). Pedrosa et al. (2012) demonstrated a reduction of 37% of IP6 in lentils after DIC treatment. A maximum reduction was reported of 92% total phytates in common vetch (*Vicia sativa* spp) treated with DIC (0.41 MPa and 78-312 s), which corroborate correlation in the pressure and the time on the reduction of phytic acid. This reduction is probably due to the dephosphorylation caused by the high-pressure used. For example, an increase of IP4 and IP5 due to the IP6 dephosphorylation after DIC treatment in lupin, soybean, and peanut was observed (Haddad, Greiner, and Allaf 2007, Pedrosa et al. 2012).

Trypsin inhibitors are also antinutritional compounds present in legumes that decrease protein availability and are present in legumes (Hefnawy 2011). After DIC treatment, a reduction from 95-99.5% in legumes such as soybean, black bean lentil, and chickpea was observed (Haddad and Allaf 2007, Cardador-Martínez et al. 2020, Pedrosa et al. 2012). Pressure, processing time, and temperature may play a role in conjunction or separate inactivation of trypsin inhibitory activity. For example, a decrease of 94% of trypsin inhibitory activity was achieved after one-minute of treatment. After the six-minutes of DIC treatment, a reduction of 99% was reported (Haddad and Allaf 2007).

Also, other anti-nutritional compounds in legumes decreased with DIC, such as flavonoids (65%), tannins (23-73%) and lectins (98.8%) in vetch, black bean, and lentils, respectively (Hernandez-Aguirre et al. 2019, Cardador-Martínez et al. 2020, Pedrosa et al. 2012). High temperatures and steam pressure made DIC a promising technology in treating legumes for their reduction in the anti-nutritional compound.

Proteins and Minerals

It is well-known that high temperatures by prolonged times can denaturalize protein and affect mineral in foods (Hefnawy 2011, Kaur, Sandhu, and Lim 2010). Even no evidence was found; it is suspected that the DIC process may not significantly affect protein and minerals in lentils, probably due to short thermal exposures (less than 3 minutes) and the apparent inactivation of antinutritional compounds which may increase protein and mineral digestibility.

Phenolic Compounds

The phenolic compound, such as phenolic acid, exhibit antioxidant activity in vivo that are beneficial for human health (Ratnavathi 2019). Legumes contain a wide variability of these compounds. In some cases, thermal processes decrease phenolic content in legumes. After DIC (0.13-0.3 MPa) treatment, black beans decreased 4-14% in phenolics and were related to temperature and pressure. However, chickpeas DIC treatment (0.3-0.5 MPa) had a positive correlation between phenolic content with time and pressure, where chickpeas showed an increase of two-fold in total phenolic content (Yağcı and Evci 2015a). This results may be due to the cell wall disruption, and consequently, the liberation of phenolic acid (Anton, Gary Fulcher, and Arntfield 2009).

Carbohydrates

Carbohydrates, in general, presented noticeable changes after DIC treatment. DIC increased starch digestibility from 75 to 205 mg maltose/g d.w. in chickpea compared to unprocessed raw (Yağcı and Evci 2015a). This result was observed in legumes treated with a thermal process such as boiling and extrusion (Johnson et al. 2015, Arribas, Cabellos, et al. 2017). Sucrose also tends to increase in some legumes like lentils, chickpeas, and soybean by 4-10%, while in lupin, a decrease of 23% was reported after DIC treatment (Pedrosa et al. 2012). Besides, DIC increased legumes oligosaccharides, which may play human gut health for its probiotic behavior. An improvement of 150 and 173% in the Leguminosae *Tephrosia purpurea* of stachyose and ciceritol, respectively (Amor et al. 2008), while raffinose and stachyose increase their content in black beans by 50 and 80%, respectively (Cardador-Martínez et al. 2020). And verbascose by 13.9% in lentils after DIC treatment (Pedrosa et al. 2012). After DIC treatment, the increase in the oligosaccharides content is possibly due to the seed's structural modification, such as increased porosity and disruption of the cell wall, which improves contact between the inside seed and the extraction solvent (Pedrosa et al. 2012).

Conclusion

Lentils are highly consumed food worldwide, performing an important role in human nutrition due to their vast quantities of protein, carbohydrates, vitamins, minerals, dietary fiber, and other bioactive compounds. Due to the high demand for this food, production and consumption have become increasingly relevant. Other compounds of great importance present in lentils are antioxidants, polyphenols, dietary fiber, resistant starches, phytosterols, and oligosaccharides. Numerous properties have been attributed, such as antidiabetic, anticarcinogenic, hypocholesterolemic, antioxidant, antimicrobial, among others, which have been demonstrated in several studies. On the other hand, different processes can apply to lentils, depending on what is required. It can be from increasing shelf-life, reducing humidity, or more specific tasks such as removing unwanted compounds or for further application of its protein in food supplements or formulations.

References

Abd El-Azim, M. A., Nashwa A. l. Abo El-Azam, Afaf O. Serage, and M. M. F. Abdallah. 2018. Sprouting using saline water on chemical composition, antinutritional compounds and aminoacid profile of chikcpea and lentil seeds. *Arab Universities Journal of Agricultural Sciences* 26 (Special issue (2D)):2239-2251. doi: 10.21608/ajs.2018.35339.

Abdalla, D. S. P. 2003. Coronary hearth disease | Antioxidant Status. In *Encyclopedia of Food Sciences and Nutrition (Second Edition)*, edited by Benjamin Caballero, 1654-1663. Oxford: Academic Press.

Adams, J. B. 1991. Review: Enzyme inactivation during heat processing of food-stuffs. *International Journal of Food Science & Technology* 26 (1):1-20. doi: 10.1111/j.1365-2621.1991.tb01136.x.

Adebamowo, C. A., E. Cho, L. Sampson, M. B. Katan, D. Spiegelman, W. C. Willett, and M. D. Holmes. 2005. Dietary flavonols and flavonol-rich foods intake and the risk of breast cancer. *International Journal of Cancer* 114 (4):628-33. doi: 10.1002/ijc.20741.

Agard, Mallory, Saja Asakrah, and Lisa Morici. 2013. PGE2 suppression of innate immunity during mucosal bacterial infection. *Frontiers in Cellular and Infection Microbiology* 3 (45). doi: 10.3389/fcimb.2013.00045.

Ahmadi, A., M. Amini Dehaghi, M. H. Fotokian, M. Sedghi, and C. Mansourifar. 2019. The effect of drought stress on antioxidant enzyme activity and chlorophyll content of some advanced genotypes of lentil (*Lens culinaris* Medik).

Akdogan, Hülya. 1999. High moisture food extrusion. *International Journal of Food Science & Technology* 34 (3):195-207. doi: 10.1046/j.1365-2621.1999.00256.x.

Akibode, Comlanvi Sitou, and Mywish K Maredia. 2012. Global and regional trends in production, trade and consumption of food legume crops. *Department of Agricultural, Food, and Resource Economics*: 89.

Al-Tibi, A. M., Jr., H. R. Takruri, and M. N. Ahmad. 2010. Effect of dehulling and cooking of lentils (*Lens Culinaris*, L.) on serum glucose and lipoprotein levels in streptozotocin-induced diabetic rats. *Malaysian Journal of Nutrition* 16 (3):409-18.

Alonso, R., L. A. Rubio, M. Muzquiz, and F. Marzo. 2001. The effect of extrusion cooking on mineral bioavailability in pea and kidney bean seed meals. *Animal Feed Science and Technology* 94 (1-2):1-13.

Alvarado, M. B. 2016. Study of the strawberry drying process using microwave dryer. *Prospectiva* 15:29-34. doi: 10.15665/rp.v15i1.658.

Amor, Bouthaina Ben, Cécile Lamy, Patrice Andre, and Karim Allaf. 2008. Effect of instant controlled pressure drop treatments on the oligosaccharides extractability and microstructure of *Tephrosia purpurea* seeds. *Journal of Chromatography A* 1213 (2):118-124. doi: 10.1016/j.chroma.2008.10.065.

Anton, Alex A., R. Gary Fulcher, and Susan D. Arntfield. 2009. Physical and nutritional impact of fortification of corn starch-based extruded snacks with common bean (*Phaseolus vulgaris* L.) flour: Effects of bean addition and extrusion cooking. *Food Chemistry* 113 (4):989-996. doi: 10.1016/j.foodchem.2008.08.050.

Archer, D. L. 2004. Freezing: an underutilized food safety technology? *International Journal of Food Microbiology* 90 (2):127-38. doi: 10.1016/s0168-1605(03)00215-0.

Arribas, C., C. Cabellos B Fau - Sánchez, C. Sánchez C Fau - Cuadrado, E. Cuadrado C Fau - Guillamón, M. M. Guillamón E Fau - Pedrosa, and M. M. Pedrosa. 2017. *The impact of extrusion on the nutritional composition, dietary fiber and in vitro digestibility of gluten-free snacks based on rice, pea and carob flour blends*. (2042-650X (Electronic)).

Arribas, C., B. Cabellos, C. Sánchez, C. Cuadrado, E. Guillamón, and M. M. Pedrosa. 2017. The impact of extrusion on the nutritional composition, dietary fiber and in vitro digestibility of gluten-free snacks based on rice, pea and carob flour blends. *Food & Function* 8 (10):3654-3663. doi: 10.1039/c7fo00910k.

Aslani, Zahra, Parvin Mirmiran, Beitollah Alipur, Zahra Bahadoran, and Mahdie Abbassalizade Farhangi. 2015. Lentil sprouts effect on serum lipids of overweight and obese patients with type 2 diabetes. *Health promotion perspectives* 5 (3):215-224. doi: 10.15171/hpp.2015.026.

Astorg, Pierre, Sandra Gradelet, Raymond Bergès, and Marc Suschetet. 1997. Dietary lycopene decreases the initiation of liver preneoplastic foci by diethylnitrosamine in the rat. *Nutrition and Cancer* 29. doi: 10.1080/01635589709514603.

Bai, Zhouya, Jinxia Meng, Xiaojun Huang, Guangjie Wu, Sheng Zuo, and Shaoping Nie. 2020. Comparative study on antidiabetic function of six legume crude polysaccharides.

International Journal of Biological Macromolecules 154:25-30. doi: 10.1016/j.ijbiomac. 2020.03.072.

Bansal, Megha, Neha Singh, Saurabh Pal, Indra Dev, and Kausar Mahmood Ansari. 2018. Chemopreventive role of dietary phytochemicals in colorectal cancer. In *Advances in Molecular Toxicology*, edited by James C. Fishbein and Jacqueline M. Heilman, 69-121. Elsevier.

Barker, Bruce. 2018. Post-harvest storage of pulses. *Saskatchewan Pulse Growers.*

Bartels, Dorothea, and Ramanjulu Sunkar. 2005. Drought and salt tolerance in plants. *Critical Reviews in Plant Sciences* 24 (1):23-58. doi: 10.1080/07352680590910410.

Bautista-Expósito, Sara, Elena Peñas, José Manuel Silván, Juana Frias, and Cristina Martínez-Villaluenga. 2018. pH-controlled fermentation in mild alkaline conditions enhances bioactive compounds and functional features of lentil to ameliorate metabolic disturbances. *Food Chemistry* 248:262-271. doi: 10.1016/j.foodchem.2017.12.059.

Bayaa, B., W. Erskine, and A. Hamdi. 1994. Response of wild lentil to *Ascochyta fabae* f.sp. lentis from Syria. *Genetic Resources and Crop Evolution* 41 (2):61-65. doi: 10.1007/BF00053049.

Berger, R. G., S. Lunkenbein, A. Ströhle, and A. Hahn. 2012. Antioxidants in food: mere myth or magic medicine? *Critical Reviews in Food Science and Nutrition* 52 (2):162-71. doi: 10.1080/10408398.2010.499481.

Berrios, Jose. 2010. Extrusion cooking: Legume pulses. *Encyclopedia of Agricultural, Food, and Biological Engineering, 2nd ed.; CRC Press: Boca Raton, FL, USA*:453-464.

Berry, M. R., and I. J. Pflug. 2003. Canning | Principles. In *Encyclopedia of food sciences and nutrition (Second Edition)*, edited by Benjamin Caballero, 816-824. Oxford: Academic Press.

Bisharat, G. I., A. E. Lazou, N. M. Panagiotou, M. K. Krokida, and Z. B. Maroulis. 2015. *Antioxidant potential and quality characteristics of vegetable-enriched corn-based extruded snacks*. (0022-1155 (Print)).

Boudjou, Souhila, B. Dave Oomah, Farid Zaidi, and Farah Hosseinian. 2013. Phenolics content and antioxidant and anti-inflammatory activities of legume fractions. *Food Chemistry* 138 (2):1543-1550. doi: 10.1016/j.foodchem.2012.11.108.

Bouvier, J. M., and O. H. Campanella. 2014. Generic extrusion processes. *Extrusion Processing Technology*: 1-12. doi: 10.1002/9781118541685.ch1.

Boye, Joyce, Fatemeh Zare, and Alison Pletch. 2010. Pulse proteins: Processing, characterization, functional properties and applications in food and feed. *Food Research International* 43 (2):414-431. doi: 10.1016/j.foodres.2009.09.003.

Boyer, Renee 2019. Boiling water bath canning. In *Virginia Cooperative Extension.* Virginia Tech: The National Center for Home Food Preservation.

Britannica. 2020. *Lentil.* Encyclopædia Britannica, Inc., accessed May 29, 2020. https://www.britannica.com/plant/lentil-plant.

Bwye, A. M., R. A. C. Jones, and W. Proudlove. 1994. Effects of sowing seed with different levels of infection, plant density and the growth stage at which plants first develop symptoms on cucumber mosaic-virus infection of narrow-leafed lupins (*Lupinus angustifolius*). *Australian Journal of Agricultural Research* 45 (7):1395-1412. doi: 10.1071/AR9941395.

Cardador-Martínez, Anaberta, Yara Martínez-Tequitlalpan, Tzayhri Gallardo-Velazquez, Xariss M. Sánchez-Chino, Jorge Martínez-Herrera, Luis Jorge Corzo-Ríos, and Cristian

Jiménez-Martínez. 2020. Effect of instant controlled pressure-drop on the non-nutritional compounds of seeds and sprouts of common black bean (*Phaseolus vulgaris* L.). *Molecules* 25 (6):1464. doi: 10.3390/molecules25061464.

Cheftel, Jean Claude. 1986. Nutritional effects of extrusion-cooking. *Food Chemistry* 20 (4):263-283. doi: 10.1016/0308-8146(86)90096-8.

Chibbar, R. N., and M. Båga. 2003. Genetic modification of primary metabolism | Carbohydrates. In *Encyclopedia of Applied Plant Sciences*, edited by Brian Thomas, 449-459. Oxford: Elsevier.

Ciudad-Mulero, María, Lillian Barros, Ângela Fernandes, José De J. Berrios, Montaña Cámara, Patricia Morales, Virginia Fernández-Ruiz, and Isabel C. F. R. Ferreira. 2018. Bioactive compounds and antioxidant capacity of extruded snack-type products developed from novel formulations of lentil and nutritional yeast flours. *Food & Function* 9 (2):819-829. doi: 10.1039/C7FO01730H.

Curiel, José Antonio, Rossana Coda, Isabella Centomani, Carmine Summo, Marco Gobbetti, and Carlo Giuseppe Rizzello. 2015. Exploitation of the nutritional and functional characteristics of traditional Italian legumes: The potential of sourdough fermentation. *International Journal of Food Microbiology* 196:51-61. doi: 10.1016/j.ijfoodmicro.2014.11.032.

Dai, Jin, and Russell J. Mumper. 2010. Plant phenolics: extraction, analysis and their antioxidant and anticancer properties. *Molecules* 15 (10):7313-7352. doi: 10.3390/molecules15107313.

Delahunty, Audrey, James Nuttall, Marc Nicolas, and Jason Brand. 2015. Genotypic heat tolerance in lentil. *Paper presented at the 17th Australian Agronomy Conference Hobart, Tasmania.*

Dhingra, Devinder, Mona Michael, Hradesh Rajput, and R. T. Patil. 2012. Dietary fibre in foods: a review. *Journal of Food Science and Technology* 49 (3):255-266. doi: 10.1007/s13197-011-0365-5.

Dinarello, C. A. 2010. Anti-inflammatory agents: present and future. *Cell* 140 (6):935-50. doi: 10.1016/j.cell.2010.02.043.

Dixon, J., and C. J. Duncan. 2014. Importance of antimicrobial stewardship to the English National Health Service. *Infection and Drug Resistance* 7:145-52. doi: 10.2147/idr.S39185.

Donoso, Natalia, and Ruth Cecilia Alvarez. 2016. Sodium reduced canned lentil stew. *Revista de la Facultad de Ciencias Químicas*, 57-65.

Dubey, Khushboo, and Sushil K. Singh. 2018. Efficacy of different soil amendments on disease incidence wilt of lentil. *International Journal of Chemical Studies* 6 (5):72-74.

Dueñas, Montserrat, Dolores Fernández, Teresa Hernández, Isabel Estrella, and Rosario Muñoz. 2005. Bioactive phenolic compounds of cowpeas (*Vigna sinensis* L). Modifications by fermentation with natural microflora and with *Lactobacillus plantarum* ATCC 14917. *Journal of the Science of Food and Agriculture* 85 (2):297-304. doi: 10.1002/jsfa.1924.

Dueñas, Montserrat, Teresa Hernández, and Isabel Estrella. 2002. Phenolic composition of the cotyledon and the seed coat of lentils (*Lens culinaris* L.). *European Food Research and Technology* 215 (6):478-483. doi: 10.1007/s00217-002-0603-1.

El-Araby, Magda M., Einas H. El-Shatoury, Mervat M. Soliman, and Hanan F. Shaaban. 2020. Characterization and antimicrobial activity of lectins purified from three Egyptian leguminous seeds. *AMB Express* 10 (1):90. doi: 10.1186/s13568-020-01024-4.

Engelking, Larry R. 2015. Chapter 46 - Vitamin E. In *Textbook of Veterinary Physiological Chemistry (Third Edition)*, edited by Larry R. Engelking, 294-298. Boston: Academic Press.

FAO. 2001. FAO/WHO expert consultation on human vitamin and mineral requirements. In *Report of a joint FAO/WHO expert consultation Bangkok, Thailand*.

FAO. 2011. Dietary protein quality evaluation in human nutrition. *FAO Food and Nutrition Paper* 92:1-66.

FAO. 2019. *The global economy of pulses*.

FAO. 2020. Crops. http://www.fao.org/faostat/en/#data/QC: FAO.

Faris, Mo'ez Al-Islam E., Hamed R. Takruri, Maha S. Shomaf, and Yasser K. Bustanji. 2009. Chemopreventive effect of raw and cooked lentils (*Lens culinaris* L) and soybeans (*Glycine max*) against azoxymethane-induced aberrant crypt foci. *Nutrition Research* 29 (5):355-362. doi: 10.1016/j.nutres.2009.05.005.

Faris, Mo'ez Al-Islam Ezzat, Hamed Rabah Takruri, and Ala Yousef Issa. 2013. Role of lentils (*Lens culinaris* L.) in human health and nutrition: a review. *Mediterranean Journal of Nutrition and Metabolism* 6:3-16. doi: 10.3233/s12349-012-0109-8.

Ferlay, J., M. Ervik, F. Lam, M. Colombet, L. Mery, M. Piñeros, A. Znaor, I. *Soerjomataram, and F. Bray. 2018. Global cancer observatory: cancer tomorrow*. International Agency for Research on Cancer, accessed May 29, 2020. https://gco.iarc.fr/tomorrow/home.

Fernandez-Orozco, Rebeca, Juana Frias, Rosario Muñoz, Henryk Zielinski, Mariusz K. Piskula, Halina Kozlowska, and Concepción Vidal-Valverde. 2007. Fermentation as a bio-process to obtain functional soybean flours. *Journal of Agricultural and Food Chemistry* 55 (22):8972-8979. doi: 10.1021/jf071823b.

Ford, R., Rubeena, R. J. Redden, M. Materne, and P. W. J. Taylor. 2007. Lentil. In *Pulses, Sugar and Tuber Crops*, edited by Chittaranjan Kole, 91-108. Berlin, Heidelberg: Springer Berlin Heidelberg.

Fratianni, Florinda, Federica Cardinale, Autilia Cozzolino, Tiziana Granese, Donatella Albanese, Marisa Di Matteo, Massimo Zaccardelli, Raffaele Coppola, and Filomena Nazzaro. 2014. Polyphenol composition and antioxidant activity of different grass pea (*Lathyrus sativus*), lentils (*Lens culinaris*), and chickpea (*Cicer arietinum*) ecotypes of the Campania region (Southern Italy). *Journal of Functional Foods* 7:551-557. doi: 10.1016/j.jff.2013.12.030.

Fratini, Richard, Ma Ruiz, and Marcelino De La Vega. 2004. Intra-specific and inter-sub-specific crossing in lentil (*Lens culinaris* Medik.). *Canadian Journal of Plant Science* 84:981-986. doi: 10.4141/P03-201.

Frias, J., S. Giacomino, E. Peñas, N. Pellegrino, V. Ferreyra, N. Apro, M. Olivera Carrión, and C. Vidal-Valverde. 2011. Assessment of the nutritional quality of raw and extruded *Pisum sativum* L. var. *laguna* seeds. *LWT - Food Science and Technology* 44 (5):1303-1308. doi: 10.1016/j.lwt.2010.12.025.

Ganesan, K., and B. Xu. 2017. Polyphenol-rich lentils and their health promoting effects. *International Journal of Molecular Sciences* 18 (11). doi: 10.3390/ijms18112390.

Ghosh, P. K., D. S. Jayas, C. Srivastava, and A. N. Jha. 2007. Drying and storing lentils: Engineering and entomological aspects. In *Lentil: An Ancient Crop for Modern Times*, edited by Shyam S. Yadav, David L. McNeil and Philip C. Stevenson, 385-414. Dordrecht: Springer Netherlands.

Gollücke, Andréa Pittelli Boiago, Rogério Correa Peres, Daniel Araki Ribeiro, and Odair Aguiar. 2018. Chapter 2 - Polyphenols as supplements in foods and beverages: recent discoveries and health benefits, an update. In *Polyphenols: Mechanisms of action in human health and disease (Second Edition)*, edited by Ronald Ross Watson, Victor R. Preedy and Sherma Zibadi, 11-18. Academic Press.

González, Zurima, and Elevina Pérez. 2002. Evaluation of lentil starches modified by microwave irradiation and extrusion cooking. *Food Research International* 35 (5):415-420. doi: 10.1016/S0963-9969(01)00135-1.

Gossen, B. D., and R. A. A. Morrall. 1984. Seed quality loss at harvest due to ascochyta blight of lentil. *Canadian Journal of Plant Pathology* 6 (3):233-237. doi: 10.1080/07060668409501559.

Grela, Eugeniusz R., Bożena Kiczorowska, Wioleta Samolińska, Jan Matras, Piotr Kiczorowski, Wojciech Rybiński, and Ewa Hanczakowska. 2017. Chemical composition of leguminous seeds: part I—content of basic nutrients, amino acids, phytochemical compounds, and antioxidant activity. *European Food Research and Technology* 243 (8):1385-1395. doi: 10.1007/s00217-017-2849-7.

Gupta, R. K., I. P. Arbol, C. W. Finkl, M. B. Kirkham, M. C. Arbestian, F. Macías, W. Chesworth, J. Germida, R. H. Loeppert, M. G. Cook, G. O. Achwab, K. Konstankiewicz, J. Pytka, J. J. Oertli, A. Singer, W. J Edmonds, Y. Feng, S. B. Feldman, C. Shang, L. W. Zelazny, P. W. Ford, B. E. Clothier, and M. A. Arshad. 2008. Soil salinity and salinization. In *Encyclopedia of Soil Science*, 699-704. Dordrecht: Springer Netherlands.

Guy, R. C. E. 2001. *Extrusion cooking: technologies and applications.* Edited by Fla: CRC Press. Boca Raton.

Haddad, Joseph, and Karim Allaf. 2007. A study of the impact of instantaneous controlled pressure drop on the trypsin inhibitors of soybean. *Journal of Food Engineering* 79 (1):353-357. doi: 10.1016/j.jfoodeng.2006.01.066.

Haddad, Joseph, Ralf Greiner, and Karim Allaf. 2007. Effect of instantaneous controlled pressure drop on the phytate content of lupin. *LWT - Food Science and Technology* 40 (3):448-453. doi: 10.1016/j.lwt.2006.02.008.

Hamid, A. A., O. O. Aiyelaagbe, L. A. Usman, O. M. Ameen, and A. Lawal. 2010. Antioxidants: Its medicinal and pharmacological applications. *African Journal of Pure and Applied Chemistry* 4 142-151.

Haq, M. Z. U., S. Ahmad, M. A. Shad, and M. Qayum. 2011. Compositional studies of lentil (*lens culinaris* medik) cultivars commonly grown in Pakistan. *Pakistan Journal of Botany* 43 (3):1563-1567.

Harper, Judson M. 1989. Food extrusion. In *Food Properties and Computer-Aided Engineering of Food Processing Systems*, edited by R. Paul Singh and Augusto G. Medina, 271-297. Dordrecht: Springer Netherlands.

Harvard Medical School. 2019a. *Cancer overview*. Harvard Health Publishing accessed May 29, 2020. https://www.health.harvard.edu/a_to_z/cancer-overview-a-to-z.

Harvard Medical School. 2019b. *High cholesterol (hypercholesterolemia).* Harvard Health Publishing accessed 29 May 2020. https://www.health.harvard.edu/a_to_z/high-cholesterol-hypercholesterolemia-a-to-z.

Hefnawy, T. H. 2011. Effect of processing methods on nutritional composition and anti-nutritional factors in lentils (*Lens culinaris*). *Annals of Agricultural Sciences* 56 (2):57-61. doi: 10.1016/j.aoas.2011.07.001.

Hernandez-Aguirre, Angel I., Carmen Téllez-Pérez, Alejandra San Martín-Azócar, and Anaberta Cardador-Martínez. 2019. Effect of instant controlled pressure-drop (DIC), cooking and germination on non-nutritional factors of common vetch (*Vicia sativa* spp.). *Molecules* 25 (1):151. doi: 10.3390/molecules25010151.

Hernández-Nava, R. G., L. A. Bello-Pérez, E. San Martín-Martínez, H. Hernández-Sánchez, and R. Mora-Escobedo. 2011. Effect of extrusion cooking on the functional properties and starch components of lentil/banana blends: Response surface analysis. *Revista mexicana de ingeniería química* 10:409-419.

Higuchi, Masashi. 2014. Antioxidant properties of wheat bran against oxidative stress. In *Wheat and Rice in Disease Prevention and Health*, edited by Ronald Ross Watson, Victor R. Preedy and Sherma Zibadi, 181-199. San Diego: Academic Press.

Hojjat, Seyed Saeid, and Merudjan Galstayan. 2012. Effects of different temperatures and duration on germination of Lentil (*Lens culinaris* Medik.) seeds. *Russian Agricultural Sciences* 38 (2):101-105. doi: 10.3103/S106836741202019X.

Huang, Zih-Rou, Yin-Ku Lin, and Jia-You Fang. 2009. Biological and pharmacological activities of squalene and related compounds: potential uses in cosmetic dermatology. *Molecules* 14 (1):540-554. doi: 10.3390/molecules14010540.

Ibekwe, A. M., J. A. Poss, S. R. Grattan, C. M. Grieve, and D. Suarez. 2010. Bacterial diversity in cucumber (*Cucumis sativus*) rhizosphere in response to salinity, soil pH, and boron. *Soil Biology and Biochemistry* 42 (4):567-575.

Ibrahim, Michael A, and Ishwarlal Jialal. 2020. *Hypercholesterolemia.* StatPearls, Last Modified January 7, accessed May 29, 2020. https://www.ncbi.nlm.nih.gov/books/NBK459188/.

International Agency for Research on Cancer. 2018. *Latest global cancer data: Cancer burden rises to 18.1 million new cases and 9.6 million cancer deaths in 2018.* Edited by World Health Organization. Geneva, Switzerland.

International Diabetes Federation. 2019. *IDF Diabetes Atlas.* Edited by IDF. Brussels, Belgium.

Iordache, Florin, Irina Gheorghe, Veronica Lazar, Carmen Curutiu, Lia Mara Ditu, Alexandru Mihai Grumezescu, and Alina Maria Holban. 2017. Chapter 9 - Nanostructurated materials for prolonged and safe food preservation. In *Food Preservation*, edited by Alexandru Mihai Grumezescu, 305-335. Academic Press.

Islam, M. Md, R. Md Karim, M. Md Oliver, A. Tahmina Urmi, A. Md Hossain, and M. M. Haque. 2018. Impacts of trace element addition on Lentil (*Lens culinaris* L.) Agronomy. *Agronomy* 8 (7). doi: 10.3390/agronomy8070100.

Jarpa-Parra, Marcela. 2018. Lentil protein: a review of functional properties and food application. An overview of lentil protein functionality. *International Journal of Food Science & Technology $V 53* (4):892-903.

Johnson, Casey R., Dil Thavarajah, Gerald F. Combs, and Pushparajah Thavarajah. 2013. Lentil (*Lens culinaris* L.): A prebiotic-rich whole food legume. *Food Research International* 51 (1):107-113. doi: 10.1016/j.foodres.2012.11.025.

Johnson, Casey R., Dil Thavarajah, Pushparajah Thavarajah, Scott Payne, Jayma Moore, and Jae-Bom Ohm. 2015. Processing, cooking, and cooling affect prebiotic concentrations in lentil (*Lens culinaris* Medikus). *Journal of Food Composition and Analysis* 38:106-111. doi: 10.1016/j.jfca.2014.10.008.

Jones, R. A. C., B. A. Coutts, L. J. Latham, and S. J. McKirdy. 2008. Cucumber mosaic virus infection of chickpea stands: temporal and spatial patterns of spread and yield-limiting potential. *Plant Pathology* 57 (5):842-853. doi: 10.1111/j.1365-3059.2008.01838.x.

Joshi, M., B. Adhikari, P. Aldred, J. F. Panozzo, and S. Kasapis. 2011. Physicochemical and functional properties of lentil protein isolates prepared by different drying methods. *Food Chemistry* 129 (4):1513-1522. doi: 10.1016/j.foodchem.2011.05.131.

Joshi, M., P. Aldred, S. McKnight, J. F. Panozzo, S. Kasapis, R. Adhikari, and B. Adhikari. 2013. Physicochemical and functional characteristics of lentil starch. *Carbohydrate Polymers* 92 (2):1484-1496. doi: 10.1016/j.carbpol.2012.10.035.

Joshi, Matina, Yakindra Timilsena, and Benu Adhikari. 2017. Global production, processing and utilization of lentil: A review. *Journal of Integrative Agriculture* 16 (12):2898-2913. doi: 10.1016/S2095-3119(17)61793-3.

Jovanovic-Malinovska, Ruzica, Slobodanka Kuzmanova, and Eleonora Winkelhausen. 2014. Oligosaccharide profile in fruits and vegetables as sources of prebiotics and functional foods. *International Journal of Food Properties* 17 (5):949-965. doi: 10.1080/10942912.2012.680221.

Kalogeropoulos, Nick, Antonia Chiou, Maria Ioannou, Vaios T. Karathanos, Maria Hassapidou, and Nikolaos K. Andrikopoulos. 2010. Nutritional evaluation and bioactive microconstituents (phytosterols, tocopherols, polyphenols, triterpenic acids) in cooked dry legumes usually consumed in the Mediterranean countries. *Food Chemistry* 121 (3):682-690. doi: 10.1016/j.foodchem.2010.01.005.

Kamau, Elijah Heka, Smith G. Nkhata, and Emmanuel Owino Ayua. 2020. Extrusion and nixtamalization conditions influence the magnitude of change in the nutrients and bioactive components of cereals and legumes. *Food Science & Nutrition* 8 (4):1753-1765. doi: 10.1002/fsn3.1473.

Kaur, Maninder, Kawaljit Singh Sandhu, and Seung-Taik Lim. 2010. Microstructure, physicochemical properties and in vitro digestibility of starches from different Indian lentil (*Lens culinaris*) cultivars. *Carbohydrate Polymers* 79 (2):349-355. doi: 10.1016/j.carbpol.2009.08.017.

Kaur, Prabhjot, Gargi Ghoshal, and Uttam C. Banerjee. 2019. Chapter 3 - Traditional bio-preservation in beverages: fermented beverages. In *Preservatives and Preservation Approaches in Beverages*, edited by Alexandru Mihai Grumezescu and Alina Maria Holban, 69-113. Academic Press.

Khan, D. A., F. Hassan, H. Ullah, S. Karim, A. Baseer, M. A. Abid, M. Ubaidi, S. A. Khan, and G. Murtaza. 2013. Antibacterial activity of *Phyllantus emblica*, *Coriandrum sativum*, *Culinaris medic*, *Lawsonia alba* and *Cucumis sativus*. *Acta Poloniae Pharmaceutic* 70 (5):855-9.

Korus, A., Z. Lisiewska, and W. Kmiecik. 2002. Effect of freezing and canning on the content of selected vitamins and pigments in seeds of two grass pea (*Lathyrus sativus* L.) cultivars at the not fully mature stage. *Nahrung* 46 (4):233-7. doi: 10.1002/1521-3803(20020701)46:4<233::Aid-food233>3.0.Co;2-u.

Korus, Jaroslaw, Dorota Gumul, and Kamila Czechowska. 2007. Effect of extrusion on the phenolic composition and antioxidant activity of dry beans of *Phaseolus vulgaris* L. *Food Technology and Biotechnology* 45.

Kushiro, Tetsuo, and Yutaka Ebizuka. 2010. Triterpenes. In *Comprehensive Natural Products II*, edited by Hung-Wen Liu and Lew Mander, 673-708. Oxford: Elsevier.

Lauchli, Andre., and Stephen. R. Grattan. 2008. Soil pH extremes. In *Plant Stress Physiology* 194-2009. Massachusetts: CABI International.

Liu, Hui-li, Xiao-yan Dai, Yun-yuan Xu, and Kang Chong. 2007. Over-expression of OsUGE-1 altered raffinose level and tolerance to abiotic stress but not morphology in *Arabidopsis*. *Journal of Plant Physiology* 164 (10):1384-1390. doi: 10.1016/j.jplph.2007.03.005.

López-Barrios, L., M. Antunes-Ricardo, and J. A. Gutiérrez-Uribe. 2016. Changes in antioxidant and antiinflammatory activity of black bean (*Phaseolus vulgaris* L.) protein isolates due to germination and enzymatic digestion. *Food Chemistry* 203:417-424. doi: 10.1016/j.foodchem.2016.02.048.

Lu, Yingjian, and Devanand Luthria. 2016. Influence of gelatinization on the extraction of phenolic acids from wheat fractions. *Food Chemistry* 194:1138-1142. doi: 10.1016/j.foodchem.2015.08.074.

Lu, Zhan-Hui, Elizabeth Donner, Rong Tsao, D. Dan Ramdath, and Qiang Liu. 2018. Physicochemical and digestion characteristics of flour and starch from eight Canadian red and green lentils. *International Journal of Food Science & Technology* 53 (3):735-746. doi: 10.1111/ijfs.13649.

Ma, Zhen, Joyce I. Boye, Jacinthe Fortin, Benjamin K. Simpson, and Shiv O. Prasher. 2013. Rheological, physical stability, microstructural and sensory properties of salad dressings supplemented with raw and thermally treated lentil flours. *Journal of Food Engineering* 116 (4):862-872. doi: 10.1016/j.jfoodeng.2013.01.024.

Magro, Ana Elisa Alves, Laura Carvalho Silva, Gabriela Boscariol Rasera, and Ruann Janser Soares de Castro. 2019. Solid-state fermentation as an efficient strategy for the biotransformation of lentils: enhancing their antioxidant and antidiabetic potentials. *Bioresources and Bioprocessing* 6 (1):38. doi: 10.1186/s40643-019-0273-5.

Makkouk, Khaled M., Safaa G. Kumari, Joop A. G. van Leur, and Roger A. C. Jones. 2014. Control of plant virus diseases in cool-season grain legume crops. In *Advances in Virus Research*, edited by Gad Loebenstein and Nikolaos Katis, 207-253. Academic Press.

Mandal, Ananya. 2019. What is hypercholesterolemia?, Last Modified February 26, accessed May 29, 2020. https://www.news-medical.net/health/What-is-Hypercholesterolemia.aspx.

Margier, Marielle, Stéphane Georgé, Noureddine Hafnaoui, Didier Remond, Marion Nowicki, Laure Du Chaffaut, Marie-Josèphe Amiot, and Emmanuelle Reboul. 2018. Nutritional composition and bioactive content of legumes: Characterization of pulses frequently consumed in France and effect of the cooking method. *Nutrients* 10 (11):1668. doi: 10.3390/nu10111668.

McHugh, Tara. 2018. Freeze drying fundamentals. *Food Technology Magazine*, February 1, 2018.

McMurray, Larn S., C. Preston, A. Vandenberg, Dili Mao, and Jefrrey G. Paull. 2019. Soil and weather conditions associated with plant damage from post-emergent metribuzin in lentil (*Lens culinaris*) in southern Australia. *Crop and Pasture Science* 70 (11):958-968.

Micioni Di Bonaventura, M. V., C. Cecchini, P. Vila-Donat, G. Caprioli, C. Cifani, M. M. Coman, A. Cresci, D. Fiorini, M. Ricciutelli, S. Silvi, S. Vittori, and G. Sagratini. 2017. Evaluation of the hypocholesterolemic effect and prebiotic activity of a lentil (*Lens culinaris* Medik) extract. *Molecular Nutrition and Food Research* 61 (11). doi: 10.1002/mnfr.201700403.

Mirali, Mahla, Randy W. Purves, and Albert Vandenberg. 2017. Profiling the phenolic compounds of the four major seed coat types and their relation to color genes in lentil. *Journal of Natural Products* 80 (5):1310-1317. doi: 10.1021/acs.jnatprod.6b00872.

Moghadasian, Mohammed H., and Fereidoon Shahidi. 2017. Fatty acids. In *International Encyclopedia of Public Health (Second Edition)*, edited by Stella R. Quah, 114-122. Oxford: Academic Press.

Moriyama, Michie, and Kazuko Oba. 2008. Comparative study on the vitamin C contents of the food legume seeds. *Journal of Nutritional Science and Vitaminology* 54 (1):1-6. doi: 10.3177/jnsv.54.1.

Mostafa, M. M., E. H. Rahma, and A. H. Rady. 1987. Chemical and nutritional changes in soybean during germination. *Food Chemistry* 23 (4):257-275. doi: 10.1016/0308-8146(87)90113-0.

Murphy, Karen J., Iva Marques-Lopes, and Ana Sánchez-Tainta. 2018. Cereals and legumes. In *The Prevention of Cardiovascular Disease through the Mediterranean Diet*, edited by Almudena Sánchez-Villegas and Ana Sánchez-Tainta, 111-132. Academic Press.

Muscolo, Adele, Antonella Calderaro, Teresa Papalia, Giovanna Settineri, Carmelo Mallamaci, and Maria Rosaria Panuccio. 2020. Soil salinity improves nutritional and health promoting compounds in three varieties of lentil (*Lens culinaris* Med.). *Food Bioscience* 35:100571. doi: 10.1016/j.fbio.2020.100571.

National Heart Lung and Blood Institute. 2018. *High Blood Cholesterol*. National Institude of Health, accessed 29 May 2020. https://www.nhlbi.nih.gov/health-topics/high-blood-cholesterol.

Navarro del Hierro, Joaquín, Teresa Herrera, Mónica R. García-Risco, Tiziana Fornari, Guillermo Reglero, and Diana Martin. 2018. Ultrasound-assisted extraction and bioaccessibility of saponins from edible seeds: quinoa, lentil, fenugreek, soybean and lupin. *Food Research International* 109:440-447. doi: 10.1016/j.foodres.2018.04.058.

NCHFP. 2009. General Canning Information. National Center for Home Food Preservation.

Nikmaram, Nooshin, Sze Leong, Mohamed Koubaa, Zhenzhou Zhu, Francisco Barba, Ralf Greiner, Indrawati Oey, and Shahin Roohinejad. 2017. Effect of extrusion on the anti-nutritional factors of food products: An overview. *Food Control* 79. doi: 10.1016/j.foodcont.2017.03.027.

Nishizawa, A., Y. Yabuta, and S. Shigeoka. 2008. Galactinol and raffinose constitute a novel function to protect plants from oxidative damage. *Plant Physiology* 147 (3):1251-63. doi: 10.1104/pp.108.122465.

Njoumi, Sondos, Marie Josephe Amiot, Isabelle Rochette, Sihem Bellagha, and Claire Mouquet-Rivier. 2019. Soaking and cooking modify the alpha-galacto-oligosaccharide and dietary fibre content in five Mediterranean legumes. *International Journal of Food Sciences and Nutrition* 70 (5):551-561. doi: 10.1080/09637486.2018.1544229.

Nleya, T., A. Vandenberg, and F. L. Walley. 2004. Lentil | Agronomy. In *Encyclopedia of Grain Science*, 150-157. Oxford: Elsevier.

Nosworthy, Matthew G., Gerardo Medina, Adam J. Franczyk, Jason Neufeld, Paulyn Appah, Alphonsus Utioh, Peter Frohlich, and James D. House. 2018. Effect of processing on the in vitro and in vivo protein quality of red and green lentils (*Lens culinaris*). *Food Chemistry* 240:588-593. doi: 10.1016/j.foodchem.2017.07.129.

Nwabueze, Titus. 2007. Effect of process variables on trypsin inhibitor activity (TIA), phytic acid and tannin content of extruded African breadfruit–corn–soy mixtures: A response surface analysis. *LWT - Food Science and Technology* 40:21-29. doi: 10.1016/j.lwt. 2005.10.004.

Oakenfull, David. 1981. Saponins in food—A review. *Food Chemistry* 7 (1):19-40. doi: 10.1016/0308-8146(81)90019-4.

Okumus, Bahar Nur, Zeynep Tacer-Caba, Kevser Kahraman, and Dilara Nilufer-Erdil. 2018. Resistant starch type V formation in brown lentil (*Lens culinaris* Medikus) starch with different lipids/fatty acids. *Food Chemistry* 240:550-558. doi: 10.1016/j.foodchem. 2017.07.157.

Osen, Raffael. 2017. Texturization of pea protein isolates using high moisture extrusion cooking.

Ouji, Ali, Safia El Bok, Mohieddine Mouelhi, Ben Mongi, Younes, and Mohamed Kharrat. 2015. Effect of salinity on germination of five tunisian lentil (*Lens culinaris* L.) genotypes. *European Scientific Journal* 1111:1857-7881.

Padhi, Emily M. T., Ronghua Liu, Marta Hernandez, Rong Tsao, and D. Dan Ramdath. 2017. Total polyphenol content, carotenoid, tocopherol and fatty acid composition of commonly consumed Canadian pulses and their contribution to antioxidant activity. *Journal of Functional Foods* 38:602-611. doi: 10.1016/j.jff.2016.11.006.

Pal, R. S., Anuradha Bhartiya, Pradhuman Yadav, Lakshmi Kant, K. K. Mishra, J. P. Aditya, and A. Pattanayak. 2017. Effect of dehulling, germination and cooking on nutrients, anti-nutrients, fatty acid composition and antioxidant properties in lentil (*Lens culinaris*). *Journal of food science and technology* 54 (4):909-920. doi: 10.1007/s13197-016-2351-4.

Paslakis, Georgios, Candice Richardson, Mariel Nöhre, Elmar Brähler, Christina Holzapfel, Anja Hilbert, and Martina de Zwaan. 2020. Prevalence and psychopathology of vegetarians and vegans – Results from a representative survey in Germany. *Scientific Reports* 10 (1):6840.

Patterson, Carol Ann, Julianne Curran, and Tanya Der. 2017. Effect of processing on antinutrient compounds in pulses. *Cereal Chemistry* 94 (1):2-10. doi: 10.1094/cchem-05-16-0144-fi.

Pedrosa, Mercedes M., Carmen Cuadrado, Carmen Burbano, Karim Allaf, Joseph Haddad, Eva Gelencsér, Krisztina Takács, Eva Guillamón, and Mercedes Muzquiz. 2012. Effect of instant controlled pressure drop on the oligosaccharides, inositol phosphates, trypsin inhibitors and lectins contents of different legumes. *Food Chemistry* 131 (3):862-868. doi: 10.1016/j.foodchem.2011.09.061.

Pieniazek, Facundo, and Valeria Messina. 2016. Scanning electron microscopy combined with image processing technique: Microstructure and texture analysis of legumes and vegetables for instant meal. *Microscopy Research and Technique* 79 (4):267-275. doi: 10.1002/jemt.22626.

Porres, Jesús M., María López-Jurado, Pilar Aranda, and Gloria Urbano. 2003. Effect of heat treatment and mineral and vitamin supplementation on the nutritive use of protein and calcium from; entils (*Lens culinaris* M.) in growing rats. *Nutrition* 19 (5):451-456. doi: 10.1016/S0899-9007(02)01011-0.

Porta-Puglia, Angelo, and Maria Aragona. 1997. Improvement of grain legumes general part: diseases. *Field Crops Research* 53 (1):17-30. doi: 10.1016/S0378-4290(97)00020-8.

Prodanov, Marin, Isabel Sierra, and Concepción Vidal-Valverde. 2004. Influence of soaking and cooking on the thiamin, riboflavin and niacin contents of legumes. *Food Chemistry* 84 (2):271-277. doi: 10.1016/S0308-8146(03)00211-5.

Qudah, Jafar. 2014. Estimation of carotenoid contents of selected mediterranean legumes by HPLC. *World Journal of Medical Sciences* 10:89-93.

Rahman, M., M. Bhuiyan, F. Alam, and R. Begum. 2017. Effect of arbuscular mycorrhizal fungi on the tolerance to sodium chloride levels, and on growth and yield of lentil (*Lens culinaris*). *The Agriculturists* 15 (1):156-169.

Ramdath, D. D., T. M. S. Wolever, Y. C. Siow, D. Ryland, A. Hawke, C. Taylor, P. Zahradka, and M. Aliani. 2018. Effect of processing on postprandial glycemic response and consumer acceptability of lentil-containing food items. *Foods* 7 (5). doi: 10.3390/foods7050076.

Rapaille, A., J. Goosens, and M. Heume. 2003. Sugar alcohols. In *Encyclopedia of Food Sciences and Nutrition (Second Edition)*, edited by Benjamin Caballero, 5665-5671. Oxford: Academic Press.

Rathod, Rahul, and Uday Annapure. 2016. Effect of extrusion process on antinutritional factors and protein and starch digestibility of lentil splits. *LWT - Food Science and Technology* 66:114-123. doi: 10.1016/j.lwt.2015.10.028.

Ratnavathi, C. V. 2019. Chapter 12 - Grain structure, quality, and nutrition. In *Breeding Sorghum for Diverse End Uses*, edited by C. Aruna, K. B. R. S. Visarada, B. Venkatesh Bhat and Vilas A. Tonapi, 193-207. Woodhead Publishing.

Riaz, Mian N., Muhammad Asif, and Rashida Ali. 2009. Stability of vitamins during extrusion. *Critical Reviews in Food Science and Nutrition* 49 (4):361-368. doi: 10.1080/10408390802067290.

Rivas-Santiago, Bruno, Eduardo Sada, Rogelio Hernández-Pando, and Víctor Tsutsumi. 2006. Péptidos antimicrobianos en la inmunidad innata de enfermedades infecciosas. *Salud Pública de México* 48:62-71. [Antimicrobial peptides in the innate immunity of infectious diseases. *Public Health of Mexico*]

Rizzello, Carlo Giuseppe, Blanca Hernández-Ledesma, Samuel Fernández-Tomé, José Antonio Curiel, Daniela Pinto, Barbara Marzani, Rossana Coda, and Marco Gobbetti. 2015. Italian legumes: effect of sourdough fermentation on lunasin-like polypeptides. *Microbial Cell Factories* 14 (1):168. doi: 10.1186/s12934-015-0358-6.

Sabater-Molina, M., E. Larqué, F. Torrella, and S. Zamora. 2009. Dietary fructooligosaccharides and potential benefits on health. *Journal of Physiology and Biochemistry* 65 (3):315-28. doi: 10.1007/bf03180584.

Saharan, Pooja, Pardeep K. Sadh, Surekha Duhan, and Joginder S. Duhan. 2020. Bio-enrichment of phenolic, flavonoids content and antioxidant activity of commonly used pulses by solid-state fermentation. *Journal of Food Measurement and Characterization* 14 (3):1497-1510. doi: 10.1007/s11694-020-00399-z.

Sakkas, H., and C. Papadopoulou. 2017. Antimicrobial activity of basil, oregano, and thyme essential oils. *Journal of Microbioly and Biotechnology* 27 (3):429-438. doi: 10.4014/jmb.1608.08024.

Sánchez, A. 2006. *Drying coriander seeds in a fluidized vacuum bed, using superheated steam.* Bachelor of Science in Mechanical Engineering, Industrial and Mechanical Engineering Department, Universidad de las Américas Puebla.

Sandhu, J. S., and Sarvjeet Singh. 2007. History and origin. In *Lentil: An Ancient Crop for Modern Times*, edited by Shyam S. Yadav, David L. McNeil and Philip C. Stevenson, 1-9. Dordrecht: Springer Netherlands.

Sarker, Ashutosh, and Shiv Kumar. 2011. Lentils in production and food systems in West Asia and Africa. *Grain Legumes* 56:46-48.

Sehgal, Akanksha, K. Sita, J. Kumar, S. Kumar, S. Singh, K. Siddique, and H. Nayyar. 2017. Effects of drought, heat and their interaction on the growth, yield and photosynthetic function of Lentil (*Lens culinaris* Medikus) genotypes varying in heat and drought sensitivity. *Frontiers in Plant Science* 8:1776.

Sehgal, Akanksha, Kumari Sita, Kalpna Bhandari, Shiv Kumar, Jitendra Kumar, P. V. Vara Prasad, Kadambot H. M. Siddique, and Harsh Nayyar. 2019. Influence of drought and heat stress, applied independently or in combination during seed development, on qualitative and quantitative aspects of seeds of lentil (*Lens culinaris* Medikus) genotypes, differing in drought sensitivity. *Plant, Cell & Environment* 42 (1):198-211. doi: 10.1111/pce.13328.

Shams, Hamidreza, Farideh Tahbaz, Mohammad Hassan Entezari, and Alireza Abadi. 2008. Effects of cooked lentils on glycemic control and blood lipids of patients with type 2 diabetes. *ARYA Atherosclerosis Journal* 3.

Shenkarev, Zakhar O., Albina K. Gizatullina, Ekaterina I. Finkina, Ekaterina A. Alekseeva, Sergey V. Balandin, Konstantin S. Mineev, Alexander S. Arseniev, and Tatiana V. Ovchinnikova. 2014. Heterologous expression and solution structure of defensin from lentil Lens culinaris. *Biochemical and Biophysical Research Communications* 451 (2):252-257. doi: 10.1016/j.bbrc.2014.07.104.

Shitan, Nobukazu, and Kazufumi Yazaki. 2013a. Chapter 9 - New insights into the transport mechanisms in plant vacuoles. In *International Review of Cell and Molecular Biology*, edited by Kwang W. Jeon, 383-433. Academic Press.

Shitan, Nobukazu, and Kazufumi Yazaki. 2013b. New insights into the transport mechanisms in plant vacuoles. In *International Review of Cell and Molecular Biology*, edited by Kwang W. Jeon, 383-433. Academic Press.

Singh, Mohar, Jai C. Rana, Badal Singh, Sandeep Kumar, Deep R. Saxena, Ashok Saxena, Aqeel H. Rizvi, and Ashutosh Sarker. 2017. Comparative agronomic performance and reaction to *Fusarium* wilt of *Lens culinaris* × *L. orientalis* and *L. culinaris* × *L. ervoides* derivatives. *Frontiers in Plant Science* 8:1162-1162. doi: 10.3389/fpls.2017.01162.

Siva, Niroshan, Pushparajah Thavarajah, Shiv Kumar, and Dil Thavarajah. 2019. Variability in prebiotic carbohydrates in different market classes of chickpea, common bean, and lentil collected from the American local market. *Frontiers in Nutrition* 6:38-38. doi: 10.3389/fnut.2019.00038.

Siva, Niroshan, Pushparajah Thavarajah, and Dil Thavarajah. 2018. The impact of processing and cooking on prebiotic carbohydrates in lentil. *Journal of Food Composition and Analysis* 70:72-77. doi: 10.1016/j.jfca.2018.04.006.

Skliros, D., C. Kalloniati, G. Karalias, G. N. Skaracis, H. Rennenberg, and E. Flemetakis. 2018. Global metabolomics analysis reveals distinctive tolerance mechanisms in different plant organs of lentil (*Lens culinaris*) upon salinity stress. *Plant and Soil* 429:451-468.

Slupski, Jacek. 2011. Effect of freezing and canning on the content of vitamin C in immature seeds of five cultivars of common bean (*Phaseolus vulgaris* L.). *Acta Scientiarum Polonorum: Technologia Alimentaria* 10 (2):197-208.

Sobota, A., Emilia Sykut-Domańska, and Zbigniew Rzedzicki. 2010. Effect of extrusion-cooking process on the chemical composition of corn-wheat extrudates, with particular emphasis on dietary fibre fractions. *Polish Journal of Food and Nutrition Sciences* 60:251-259.

Sonnante, Gabriella, Karl Hammer, and Domenico Pignone. 2009. From the cradle of agriculture a handful of lentils: History of domestication. *Rendiconti Lincei* 20 (1):21-37. doi: 10.1007/s12210-009-0002-7.

Spanou, C., D. Stagos, L. Tousias, A. Angelis, N. Aligiannis, A. L. Skaltsounis, and D. Kouretas. 2007. Assessment of antioxidant activity of extracts from unique Greek varieties of *Leguminosae* plants using in vitro assays. *Anticancer Res* 27 (5a):3403-10.

Srivastava, Rajendra, and Hina Vasishtha. 2013. Dietary fiber, protein and lectin contents of lentils (*Lens culinaris*) on soaking and cooking. *Current Advances in Agricultural Sciences* 5:238-241.

Sumargo, F., P. Gulati, S. A. Weier, J. Clarke, and D. J. Rose. 2016. Effects of processing moisture on the physical properties and in vitro digestibility of starch and protein in extruded brown rice and pinto bean composite flours. *Food Chemistry* 211:726-33. doi: 10.1016/j.foodchem.2016.05.097.

Sumargo, Franklin. 2016. *Improving the utilization of dry edible beans in a ready-to-eat snack product by extrusion cooking*. Degree of Master of Science, Food Science and Technology Department, University of Nebraska-Lincoln (Spring 4-18-2016).

Sun, Y., G. Jiao, Z. Liu, X. Zhang, J. Li, X. Guo, W. Du, J. Du, F. Francis, Y. Zhao, and L. Xia. 2017. Generation of high-amylose rice through CRISPR/Cas9-mediated targeted mutagenesis of starch branching enzymes. *Frontiers in Plant Science* 8:298. doi: 10.3389/fpls.2017.00298.

Sutariya, Suresh G., and Venkateswarlu Sunkesula. 2020. Food Freezing: emerging techniques for improving quality and process efficiency a comprehensive review. In *Reference Module in Food Science*. Elsevier.

Swieca, Michal, Barbara Baraniak, and Urszula Gawlik-Dziki. 2013. *In vitro* digestibility and starch content, predicted glycemic index and potential *in vitro* antidiabetic effect of lentil sprouts obtained by different germination techniques. *Food Chemistry* 138 (2):1414-1420. doi: 10.1016/j.foodchem.2012.09.122.

Tang, W., X. Peng, and R. J. Newton. 2005. Enhanced tolerance to salt stress in transgenic loblolly pine simultaneously expressing two genes encoding mannitol-1-phosphate dehydrogenase and glucitol-6-phosphate dehydrogenase. *Plant Physioly Biochemistry* 43 (2):139-46. doi: 10.1016/j.plaphy.2005.01.009.

Taylor, Paul, Kurt Lindbeck, Weidong Chen, and Rebecca Ford. 2007. Lentil diseases. In *Lentil: An Ancient Crop for Modern Times*, edited by Shyam S. Yadav, David L. McNeil and Philip C. Stevenson, 291-313. Dordrecht: Springer Netherlands.

Tefera, Mulugeta Mihrete, Birhanetensay Masresha Altaye, Ebrahim M. Yimer, Derbew Fikadu Berhe, and Senait Tadesse Bekele. 2020. Antidiabetic effect of germinated *Lens culinaris* Medik seed extract in streptozotocin-induced diabetic mice. *Journal of Experimental Pharmacology* 12:39-45. doi: 10.2147/JEP.S228834.

Teixeira, Januana S., Victoria McNeill, and Michael G. Gänzle. 2012. Levansucrase and sucrose phoshorylase contribute to raffinose, stachyose, and verbascose metabolism by lactobacilli. *Food Microbiology* 31 (2):278-284. doi: 10.1016/j.fm.2012.03.003.

Thavarajah, Pushparajah, Asoka Wejesuriya, Michael Rutzke, Raymond P Glahn, Gerald F Combs, and Albert Vandenberg. 2011. The potential of lentil (*Lens culinaris* L.) as a whole food for increased selenium, iron, and zinc intake: preliminary results from a 3 year study. *Euphytica* 180 (1):123-128.

Ti, H., R. Zhang, M. Zhang, Z. Wei, J. Chi, Y. Deng, and Y. Zhang. 2015. Effect of extrusion on phytochemical profiles in milled fractions of black rice. *Food Chemistry* 178:186-94. doi: 10.1016/j.foodchem.2015.01.087.

Torino, Maria Inés, Rocío I. Limón, Cristina Martínez-Villaluenga, Sari Mäkinen, Anne Pihlanto, Concepción Vidal-Valverde, and Juana Frias. 2013. Antioxidant and antihypertensive properties of liquid and solid state fermented lentils. *Food Chemistry* 136 (2):1030-1037. doi: 10.1016/j.foodchem.2012.09.015.

USDA. 2018a. *Egg, whole, raw, fresh*. Agricultural Research Service Last Modified 2018, accessed 4/12. https://fdc.nal.usda.gov/fdc-app.html#/food-details/171287/nutrients.

USDA. 2018b. *Lentils, raw*. Agricultural Research Service Last Modified 2018, accessed 4/12. https://fdc.nal.usda.gov/fdc-app.html#/food-details/172420/nutrients.

USDA. 2020. Classification for kingdom plantae down to species *Lens culinaris* Medik. In *The Plants Database* Natural Resources Conservation Service.

Vanier, Nathan Levien, Varatharajan Vamadevan, Graziella Pinheiro Bruni, Cristiano Dietrich Ferreira, Vânia Zanella Pinto, Koushik Seetharaman, Elessandra da Rosa Zavareze, Moacir Cardoso Elias, and Jose De J. Berrios. 2016. Extrusion of rice, bean and corn starches: Extrudate structure and molecular changes in amylose and amylopectin. *Journal of Food Science* 81 (12):E2932-E2938. doi: 10.1111/1750-3841.13545.

Vauzour, David, Jason Kerr, and Charles Czank. 2014. Chapter 27 - Plant polyphenols as dietary modulators of brain functions. In *Polyphenols in Human Health and Disease*, edited by Ronald Ross Watson, Victor R. Preedy and Sherma Zibadi, 357-370. San Diego: Academic Press.

Wang, N., D. W. Hatcher, R. Toews, and E. J. Gawalko. 2009. Influence of cooking and dehulling on nutritional composition of several varieties of lentils (*Lens culinaris*). *LWT - Food Science and Technology* 42 (4):842-848. doi: 10.1016/j.lwt.2008.10.007.

Wang, Ning, Tom D. Warkentin, Bert Vandenberg, and Deng J. Bing. 2014. Physicochemical properties of starches from various pea and lentil varieties, and characteristics of their noodles prepared by high temperature extrusion. *Food Research International* 55:119-127. doi: 10.1016/j.foodres.2013.10.043.

WHO. 2020. *Diabetes.* Last Modified 15 May accessed 29 May 2020. https://www.who.int/news-room/fact-sheets/detail/diabetes.

Wong, Melissa M. L., Neha Gujaria-Verma, Larissa Ramsay, Hai Ying Yuan, Carolyn Caron, Marwan Diapari, Albert Vandenberg, and Kirstin E. Bett. 2015. Classification and characterization of species within the genus lens using genotyping-by-sequencing (GBS). *PloS One* 10 (3):e0122025-e0122025. doi: 10.1371/journal.pone.0122025.

World Health Organization. 2018. *Cancer*. accessed Mayo 29, 2020. https://www.who.int/news-room/fact-sheets/detail/cancer.

Xu, Baojun, and Sam K. C. Chang. 2008. Effect of soaking, boiling, and steaming on total phenolic contentand antioxidant activities of cool season food legumes. *Food Chemistry* 110 (1):1-13. doi: 10.1016/j.foodchem.2008.01.045.

Xu, Baojun, and Sam K. C. Chang. 2010. Phenolic substance characterization and chemical and cell-based antioxidant activities of 11 lentils grown in the Northern United States. *Journal of Agricultural and Food Chemistry* 58 (3):1509-1517. doi: 10.1021/jf903532y.

Yadav, Anjali, and Arvind M. Kayastha. 2020. *Lens culinaris* β-galactosidase (Lsbgal): Insights into its purification, biochemical characterization and trisaccharides synthesis. *Bioorganic Chemistry* 95:103543. doi: 10.1016/j.bioorg.2019.103543.

Yağcı, Sibel, and Tuba Evci. 2015a. Effect of instant controlled pressure drop process on some physicochemical and nutritional properties of snacks produced from chickpea and wheat. *International Journal of Food Science & Technology* 50 (8):1901-1910. doi: 10.1111/ijfs.12843.

Yağcı, Sibel, and Tuba Evci. 2015b. Effect of instant controlled pressure drop process on some physicochemical and nutritional properties of snacks produced from chickpea and wheat. *International Journal of Food Science & Technology* 50:1901-1910. doi: 10.1111/ijfs.12843.

Yu, Wenyang, Cuiping Liu, Xin Li, Fuguo Yang, Lixue Cheng, Chang Liu, and Yang Song. 2017. Inositol hexaphosphate suppresses colorectal cancer cell proliferation via the Akt/GSK-3β/β-catenin signaling cascade in a 1,2-dimethylhydrazine-induced rat model. *European Journal of Pharmacology* 805:67-74. doi: 10.1016/j.ejphar.2017.03.011.

Zakaria, Hesham, Nagwa Hassan, Hany Khalil, and Shawky Kerolles. 2015. *Antidiabetic effects of dietary formulas prepared from some grains and vegetables on type 2 diabetic rats*.

Zhang, Bing, Zeyuan Deng, Yao Tang, Peter Chen, Ronghua Liu, D. Dan Ramdath, Qiang Liu, Marta Hernandez, and Rong Tsao. 2014a. Fatty acid, carotenoid and tocopherol compositions of 20 Canadian lentil cultivars and synergistic contribution to antioxidant activities. *Food Chemistry* 161:296-304. doi: 10.1016/j.foodchem.2014.04.014.

Zhang, Bing, Zeyuan Deng, Yao Tang, Peter X. Chen, Ronghua Liu, D. Dan Ramdath, Qiang Liu, Marta Hernandez, and Rong Tsao. 2014b. Effect of domestic cooking on carotenoids, tocopherols, fatty acids, phenolics, and antioxidant activities of lentils (*Lens culinaris*). *Journal of Agricultural and Food Chemistry* 62 (52):12585-12594. doi: 10.1021/jf504181r.

Zhang, Haixia, Ambuj B. Jha, Devini De Silva, Randy W. Purves, Thomas D. Warkentin, and Albert Vandenberg. 2019. Improved folate monoglutamate extraction and application to folate quantification from wild lentil seeds by ultra-performance liquid chromatography-selective reaction monitoring mass spectrometry. *Journal of Chromatography B* 1121:39-47. doi: 10.1016/j.jchromb.2019.05.007.

Zhang, Ruifen, Sher Ali Khan, Jianwei Chi, Zhencheng Wei, Yan Zhang, Yuanyuan Deng, Lei Liu, and Mingwei Zhang. 2018. Different effects of extrusion on the phenolic profiles and antioxidant activity in milled fractions of brown rice. *LWT* 88:64-70. doi: 10.1016/j.lwt.2017.09.042.

Zhifang, G., and W. H. Loescher. 2003. Expression of a celery mannose 6-phosphate reductase in Arabidopsis thaliana enhances salt tolerance and induces biosynthesis of both mannitol and a glucosyl-mannitol dimer. *Plant, Cell & Environment* 26 (2):275-283. doi: 10.1046/j.1365-3040.2003.00958.x.

Zhu, F., B. Du, and B. Xu. 2018. Anti-inflammatory effects of phytochemicals from fruits, vegetables, and food legumes: A review. *Critical Reviews in Food Science and Nutrition* 58 (8):1260-1270. doi: 10.1080/10408398.2016.1251390.

Zia-Ul-Haq, M., P. Landa, Z. Kutil, M. Qayum, and S. Ahmad. 2013. Evaluation of anti-inflammatory activity of selected legumes from Pakistan: in vitro inhibition of cyclooxygenase-2. *Pakistan Journal of Pharmaceutical Sciences* 26 (1):185-7.

Zilić, S., B. A. Mogol, G. Akıllıoğlu, A. Serpen, N. Delić, and V. Gökmen. 2014. Effects of extrusion, infrared and microwave processing on Maillard reaction products and phenolic compounds in soybean. *Journal of the Science of Food and Agriculture* 94 (1):45-51. doi: 10.1002/jsfa.6210.

Zucconi, Yésica. 2011. *Breadmaking with lentil flour*. Degree in Nutrition Medical Science Department, FASTA University.

In: Super and Nutraceutical Foods
Editors: Sandra Teresita Martín-del-Campo et al.
ISBN: 978-1-53619-082-3

Chapter 9

MAIZE (*ZEA MAYS*) AS A SUPERFOOD

***Juan Leopoldo Pech-Almeida*[1], *Maritza Alonzo-Macías*[1], *Anaberta Cardador-Martínez*[1] *and Carmen Téllez-Pérez*[1,2,*]**
[1]Tecnologico de Monterrey, Escuela de Ingeniería y Ciencias, Querétaro México
[2]University of La Rochelle, Intensification of Transfer Phenomena on Industrial Eco-Processes, Laboratory of Engineering Science for Environment LaSIE - UMR-CNRS, La Rochelle, France

ABSTRACT

Maize (*Zea mays* L.) has been used in human nutrition as a staple food, especially in Latin America and African countries. According to FAO, maize is a primary source of calories in the diets of 230 million inhabitants of developing countries. Recent studies have also identified it as a good source of nutraceuticals compounds, such as phenolics, phospholipids, phytosterols, arabinoxylans, policosanols, carotenoids, vitamins, dietary fiber, among others, which have demonstrated positive health effects. Therefore, this chapter aims to present the potential of maize as a superfood, being highlighted its physicochemical composition, and its antioxidant, anticancer, antihypertensive, anti-diabetic, antiobesogenic, and anti-inflammatory activities. Additionally, it also focuses on the most common food processing methods to face maize mycotoxins, which is one of its main issues.

Keywords: maize (*Zea mays* L), superfood, nutraceutical compounds, mycotoxins

INTRODUCTION

Since around 20000 years B.C., maize (*Zea mays* L.) has been used in human nutrition as a staple food, especially in Latin America and African countries (Brown, Zuber, and Darrah 1985, García-Lara and Serna-Saldivar 2019). In fact, according to the Food and Agriculture Organization (FAO), maize is a primary source of calories in the diets of 230 million

* Corresponding Author's E-mail: ctellezperez@gmail.com.

inhabitants of developing countries (81 million in sub-Saharan Africa, 141 million in South Asia, and 8 million in Latin America) (FAO 2020b). Moreover, even if maize has been recognized primarily as caloric or starchy, recent studies have also identified it as a good source of nutraceuticals compounds like phenolics, phospholipids, phytosterols, arabinoxylans, policosanols, carotenoids, vitamins, dietary fiber, among others, which have demonstrated positive health effects (Ndolo and Beta 2014, Nile and Park 2014, Del Pozo-Insfran et al. 2006, FAO 1992). Furthermore, in recent years to better satisfy consumer's needs, new special corn varieties, as well as high-amylose, waxy and high-lysine specialty corn varieties, have been developed. Therefore, this chapter aims to present the most accurate picture of the potential of maize as a superfood. The physicochemical composition and the antioxidant (Salinas-Moreno et al. 2012), anticancer (Herrera-Sotero et al. 2017), antihypertensive (Li et al. 2019), anti-diabetic (Zhang et al. 2019), antiobesogenic (Luna-Vital et al. 2020) and anti-inflammatory activities of maize have been highlighted (Zhang, Luna-Vital, and Gonzalez de Mejia 2019). Additionally, it also focuses on the most common food processing methods to face maize mycotoxins (aflatoxins, ochratoxins, fumonisins, trichothecenes, and zearalenone).

History of Maize

Although the exact origin of maize (*Zea mays* L.) is unclear, the older paleoethnobotany evidence of its domestication has been found in Mexico, in the states of Tamaulipas, Puebla, Oaxaca, and Mexico. In Tamaulipas, corn findings of three primitive caves of the archeological site named "Nac Neish," recorded data between 6000- and 20000-years B.C. In Puebla, the oldest corn was found in the Valley of Tehuacan (7000 years B.C.). And in Oaxaca and Mexico, the oldest record dated 5000 years B.C. (Brown, Zuber, and Darrah 1985, García-Lara and Serna-Saldivar 2019). Mexico has the most extensive diversity of genetic resources in the world, with approximately 64 different races reported (59 native landraces and five previously reported in other regions) (Sanchez G, Goodman, and Stuber 2000, CONABIO 2011). Therefore, most experts recognize Mexico as the center of origin and domestication of maize. In this respect, numerous theories about the botanical origin of maize have been offered, being two which receive serious consideration: the first one and the most acceptable is that teosinte (*Zea mexicana*) is the wild progenitor of corn. The second one is that wild pod corn, now extinct, was the ancestor of domesticated corn (Brown, Zuber, and Darrah 1985). In fact, in both cases, the degree of genetic alteration has converted corn from a wild plant into a plant that is dependent upon humans for survival (Brown, Zuber, and Darrah 1985, Goodman and Galinat 1988). The earliest cobs (5000 B.C.) were tiny, with a relatively soft rachis and soft glumes. Beginning at about 2300 B.C., in several sites from the U.S to central Mexico, many maize cobs showed changes, especially in elongation and induration. Over about 3000 years, the maize ear became robust and much more productive. Perhaps, the most considerable degree of modification was produced by the indigenous people of Mexico (Goodman and Galinat 1988, Cabrera-Ponce, Valencia-Lozano, and Trejo-Saavedra 2019). Noteworthy, today, in Latin America, there are recognized 220 races of corn associated with a defined climatic and geographic region (García-Lara and Serna-Saldivar 2019).

BOTANICAL DESCRIPTION

Zea mays L. is the most economically important species among the other four wild species of the genus *Zea* found in Mexico and northern Central America (*Z. luxurians*, *Z. Mexicana*, *Z. perennis* and *Z. diploperennis*). *Zea mays* L. belongs to the grass family (*Poaceae*), which includes crops such as wheat (*Triticum aestivum*) and rice (*Orzya sativa*) (Arendt and Zannini 2013, Rouf Shah, Prasad, and Kumar 2016). The number of chromosomes in *Zea mays* is 2n = 20. The plant is classified as an *Angiosperm* that forms monocotyledon seeds borne in an ovary and to the tribe *Andropogonea*e. The tribe *Andropogoneae* comprises seven genera Coix (2n = 10/20), Chionachne (2n = 20), Sclerachne (2n = 20), Trilobachne (2n = 20), Polytoca (2n = 20), Zea and Tripsacum (Rouf Shah, Prasad, and Kumar 2016). Table 1 shows the taxonomy of the species *Zea mays*.

Table 1. Taxonomy of the species *Zea mays*

Domain	*Eukaryota*
Kingdom	*Plantae*
Phylum	*Spermatophyta*
Subphylum	*Angioespermae*
Class	*Monocotyledonae*
Order	*Cyperales*
Family	*Poaceae*
Subfamily	*Panicoideae*
Tribe	*Andropogoneae*
Genus	*Zea*
Species	*Zea mays*

Based on kernel characteristics, five domesticated varieties are commercially grown specifically for human consumption: *Zea mays var. indentata* Sturt (dent), *Zea mays var. indurate* Sturt (flint), *Zea mays var. amylacea* Sturt (floury), *Zea mays var. saccharata* Sturt (sweet) and *Zea mays var. everta* Sturt (popcorn) (Brown, Zuber, and Darrah 1985). Moreover, new special corn varieties as well as high-amylose, waxy and high-lysine specialty corn varieties have been developed to satisfy industrial needs better.

Maize reproduces via monoecious cross-pollination of male (tassel) and female (ear) organs. Ears are oblong and mature within protective green husks where numerous tiny flowers evolve into rows of kernels along the cob after reproduction. The corn seed is not merely a seed, but a single fruit called the kernel (Kiesselbach 1949). Botanically, the corn kernel is classified as a caryopsis, and it is composed of three main structures: pericarp, germ, and endosperm.

The pericarp (seed coat) is the transformed ovary wall. Its main function is to protect the true seed from abiotic and biotic stresses. It constitutes around 5-6% of the dry weight of the whole kernel, and it contributes to the maize kernel color. It is subdivided into three parts: epicarp, mesocarp, and endocarp, which are composed of dead cells that are cellulosic tubes, which play a crucial role in water distribution within the kernel (Arendt and Zannini 2013).

The germ is considered as the first reserve tissues necessary for germination and thus to develop the seedling and new plant. The germ consists of two major parts: embryo and scutellum. The embryo makes up 11% of the kernel dry weight, and it contains 83% of the total kernel lipids, 70% of the kernel sugars, 78% of kernel minerals, and 26% of the kernel proteins. The scutellum (cotyledon) is the first reserve tissue rich in protein lipids, and micronutrients and its walls cells facilitate the movement of material among the cells (Watson 2003, Arendt and Zannini 2013).

The mature endosperm makes up the more significant part of the kernel, with around 80%-85% of its weight. It represents a source of starch (86–89% dry bases), protein (7–10% dry bases), and lipids (Arendt and Zannini 2013). Thus, it is considered as the second reserve tissues necessary for germination. It is divided into the aleurone tissue and starchy endosperm. The aleurone tissue is a single layer of cells lying immediately under the pericarp. The starchy endosperm cells are constituted by a thin cell wall, starch granules embedded in a protein matrix, and spherical-shaped protein bodies (García-Lara, Chuck-Hernandez, and Serna-Saldivar 2019, Darrah, McMullen, and Zuber 2019). The starchy endosperm is subdivided into two zones: the vitreous endosperm characterized by a hard and compact structure and the floury endosperm characterized by a soft and loose structure (Yu et al. 2015).

Table 2. Main characteristics of the most commercialized *Zea mays* varieties

***Zea mays* variety**	**Characteristics**
Dent (*Zea mays var. indentata* Sturt)	Vitreous endosperm at the sides and back of the kernel. Soft and floury endosperm at central core extending to the crown. Grown mainly in US corn belt, Argentina, China, France, Italy, and Hungary (Darrah, McMullen, and Zuber 2019, Győri 2017).
Flint (*Zea mays var. indurate* Sturt)	Thick, hard, vitreous endosperm is surrounding a small granular center. Smooth and rounded kernels with no denting. Grown mainly in South America and northern Europe (Darrah, McMullen, and Zuber 2019, Győri 2017).
Floury (*Zea mays var. amylacea* Sturt)	Endosperm is composed entirely with soft starch with almost no vitreous material. Is one of the oldest types of corn grown in the world. Many indigenous populations in Latin America grow it as staple food (Darrah, McMullen, and Zuber 2019).
Sweet (*Zea mays var. saccharata* Sturt)	The rate of conversion of sugar into starch is reduced. Kernels are consumed in fresh, canned, and frozen form. Grown mainly in the United States, China, France, Hungary, and Ukraine (Darrah, McMullen, and Zuber 2019, Győri 2017).
Popcorn (*Zea mays var. everta* Sturt)	Consumed as snacks Kernels give a higher expansion volume. Vitreous endosperm expands while floury endosperm can remain intact. Pericarp plays an essential role in the popping phenomenon by acting as a pressure vessel. Grown mainly in United States and Argentina (Vázquez-Carrillo, Santiago-Ramos, and Figueroa-Cárdenas 2019).

According to environmental, cultural, genetic background, plant age, geographic location, among others parameters, maize kernels can vary in nutrient composition, color (e.g., white, yellow, orange, red, blue and black), quantity (300 to 1000 kernels per ear), weight (190 to 300 g per 1000 kernels) and spatiality (12 to 16 kernels per row) (Nuss and Tanumihardjo 2010). Therefore, depending on those parameters, the potential of maize as a superfood varied. Table 2 describes the main characteristics of the most commercialized *Zea mays* varieties.

Agricultural Production and Economic Value of the Crop

Maize has high value and economic importance worldwide as human food, as animal feed and as feedstock for a large number of industrial products and biofuels. Thanks to its ability to grow in a wide diversity of environments located from 58°N in Canada to 40°S in Argentina, it has been easily adopted around the world (Paliwal et al. 2000). According to FAO (2020b), in 2018 the total world production of corn was 1,147 million tons, being the American continent, which contributed with barely more than half of the whole production (50.4%), followed by the Asian (31.5%), the European (11.2%), the African (6.9%), and Oceania (0.1%). The top ten maize producers of the world were the United States of America, China, Brazil, Argentina, Ukraine, Indonesia, India, Mexico, Romania, and Canada (FAO 2020b). Figure 1 shows the production share of maize by world regions in 2018.

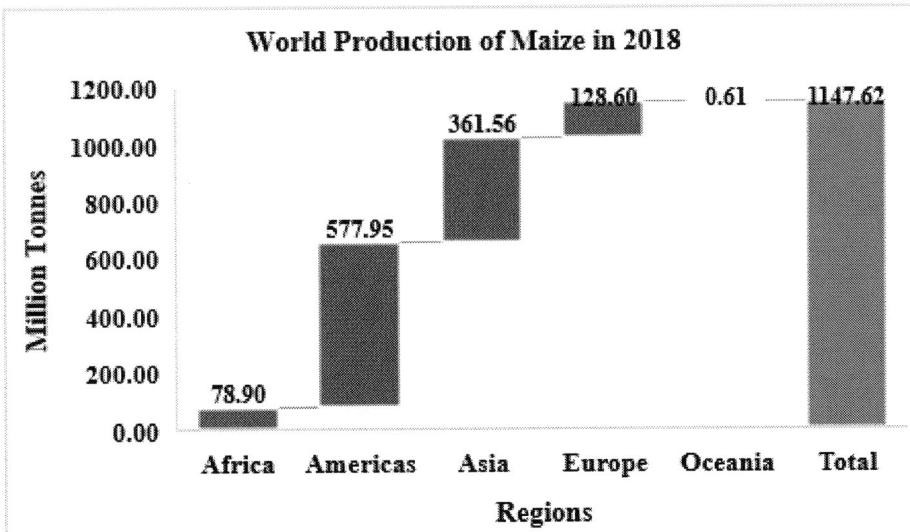

Figure 1. Production share of maize by world regions in 2018 (FAO 2020b).

Even if nowadays most of world corn production is used in developing countries as a feedstuff for animals and ethyl alcohol production, for 900 million poor consumers of many countries, especially in Latin America and Africa, its use in human nutrition is considered as a staple food (Listman and Ordóñez 2019). The FAO recognizes corn as a primary source of calories in the diets of 230 million inhabitants of developing countries (81 million in sub-Saharan Africa, 141 million in South Asia, and 8 million in Latin America) (FAO 2020a).

Maize-based food has been recognized primarily as caloric or starchy, and more recently, as a good source of nutraceuticals compounds as phenolics, phospholipids, phytosterols, arabinoxylans, policosanols, carotenoids, vitamins, dietary fiber, among others, which have demonstrated positive health effects.

PHYSICOCHEMICAL COMPOSITION OF MAIZE

Chemically, the kernel is constituted by macro and micronutrients distributed in the pericarp, germ, and endosperm. Among the macronutrients, starch, non-starch polysaccharides, proteins, and lipids are distributed heterogeneously in the maize kernel. Respect to micronutrients, recent studies have shown that maize kernel is rich in vitamins, minerals, carotenoids, phenolic compounds, phytosterols, phospholipids, arabinoxylans, among other phytochemicals. Moreover, during the last four decades, research works have emphasized into increase the nutritional potential of maize kernels by generating new types of corn as quality protein maize (QPM) and High-oil hybrids (HOC).

MACRONUTRIENTS OF MAIZE

On the basis that kernel composition varied within and between maize varieties, an average maize kernel contains about 72% of carbohydrates, 9% of proteins, 5% of lipids, and 2% of fiber (Camelo-Méndez et al. 2017, Nuss and Tanumihardjo 2010).

Carbohydrates

Carbohydrates in maize kernels could be found in the forms of starch, sugar, and non-starch polysaccharides (NSPs) (Győri 2017). Starch is the major carbohydrate and makes up around 72% of the total kernel weight and 85% of the kernel endosperm weight. The non-starch polysaccharides (contained in both the endosperm and pericarp layers) are principally cellulose, hemicellulose, and pectin (Hamaker, Tuncil, and Shen 2019).

Starch is found in the cells of the endosperm in the form of granules whose shape and size are characteristic of the grain species. Starch consists primarily of D-glucose linked together to form two kinds of polymers: amylose and amylopectin. Amylose is a smaller linear polymer of D-glucopyranose linked together with α-1,4-glycosidic bond and amylopectin. On the other is a larger branched polymer with linear α-1,4-glycosidic bond and branches connected by α-1,6-glycosidic bonds. Starch from the most commercialized corn (dent or flint) is composed of 27% amylose and 73% amylopectin. However, this proportion can be altered by genetic modifications. In fact, according to the genotype of maize, the amylose content could vary from 0% to 73%. For example, waxy corn (WCS) is composed of 1% amylose and 99% amylopectin, whereas high amylose corn (HACS) has 55%–70% amylose with only 30%–45% amylopectin (Nuss and Tanumihardjo 2010, Hamaker, Tuncil, and Shen 2019). In recent years, nutritional quality and health considerations around starch have centered on glycemic index and response. Corn starch differs in its glycemic profile depending on its ungelatinized or gelatinized state, whether it has been cooled and retrograded, or what kind of food matrix form is found (Hamaker, Tuncil, and Shen 2019). Therefore, in addition to the content and the kind of starch, the further processing of corn kernel impacts directly on its nutritional quality.

Maize kernels contain around 1% to 3% of sugars, being sucrose (glucose-fructose, α-1, 2 glycosidic bonds), maltose (glucose-glucose, α-1, 4 glycosidic bonds), glucose, fructose and raffinose (galactose, glucose, and fructose) the most representatives. These free sugars are almost exclusively located in the germ, with only 25% present in the endosperm. In the embryo, sugars occur to a greater extent as disaccharides (Győri 2017, Nuss and Tanumihardjo 2010).

Cellulose, hemicellulose, and pectic substances are major non-starch polysaccharides present in corn kernels, particularly in the pericarp. Cellulose is a linear polysaccharide of β-D-glucopyranose connected by β (1→4) glycosidic bonds. Cellulose content varies among phenotypes, e.g., yellow corn contains only 1.7% cellulose (Jaworski et al. 2015). By regarding the composition of corn bran, cellulose represents around 23% of total weight (Ai and Jane 2016). Hemicellulose consists of heterogeneous groups of pentose and hexose sugars, including xylose, mannose, glucose, and galactose in the backbone chain, and arabinose, galactose, glucose, and glucuronic acid in branches (Hamaker, Tuncil, and Shen 2019). Its content in corn kernel varied among 5.8–6.7%, and it represents around 70% of the total weight of corn bran (Eckhoff and Watson 2009). The main hemicelluloses in bran are arabinoxylans and glucuronoxylans. Both have a xylan backbone and a diverse group of derivative units. The xylan backbone is composed of β-D-xylopyranose linked by β (1→4) glycosidic bonds, and the derivative units of the backbone include arabinose, trisaccharides composed of arabinose, 4-O-methylglucuronic acid, xylose, and galactose (Ai and Jane 2016). Pectins are complex polymers of α-D-(1→4) galacturonic acid with occasional rhamnopyranosil residues bound in different points of the linear chain. In the corn kernel, pectin quantity represents around 0.2% (dry basis) (Hamaker, Tuncil, and Shen 2019, Bailoni et al. 2005).

Crude fiber is composed of pectic substances, hemicellulose, cellulose, and lignin, which constitute the cell wall and other non-structural materials of corn kernel (Győri 2017). Pericarp fiber and endosperm fiber are the two kinds of fiber in the corn kernel, mainly differentiated by the ratio of xylose to arabinose (Srinivasan et al. 2006). In common maize varieties, insoluble fiber is generally 12%, while soluble fiber is less than 2% (Nuss and Tanumihardjo 2010).

Lipids

Lipids are important nutritional components of the corn kernel, ranging from 3.5% to 6% of the total kernel weight. The germ is the most important reserve of lipids with around 76-83% of complete kernel content, followed by the endosperm with 14%–23%, the aleurone layer with 13%–15%, the pericarp with 1%–2%, and the tip cap with <1% (Wang and White 2019, Nuss and Tanumihardjo 2010). Lipid composition of corn kernel include triacylglycerols, phospholipids, glycolipids, phytosterols (sterols and stanols), free fatty acids, hydrocarbons, carotenoids, tocols (tocopherols and tocotrienols), sphingolipids and waxes (Wang and White 2019).

Triacylglycerols are the major storage form of kernel lipids. In corn oil, about 60% of fatty acids are polyunsaturated, followed by 25% and 15% of monounsaturated and saturated fatty acids, respectively. Among unsaturated fatty acids in maize oil, linoleic (34%-65%), oleic (20%-42%), and palmitic acid (8-16%) are the most predominant. Trace amounts (<1%) of lauric, myristic, margaric, margaroleic, arachidic, gadoleic, behenic, erucic, and lignoceric acids have been found in maize oil (Nuss and Tanumihardjo 2010, Barrera-Arellano, Badan-

Ribeiro, and Serna-Saldivar 2019). Oils with a high proportion of oleic or linoleic acid and reduced amounts of palmitic and stearic acids are desired and often recommended to reduced heart blockages. Table 3 shows the average fatty acid composition of maize oil.

Table 3. Typical fatty acid composition of maize oil[a]

Acid name (abbreviation)	% Total fatty acids
Palmitic (C16:0)	8.6–16.5
Palmitoleic (C16:1)	0.0–0.5
Stearic (C18:0)	0.0–3.3
Oleic (C18:1)	20.0–42.2
Linoleic (C18:2)	34.0–65.6
Linolenic (C18:3)	0.0–2.0
Eicosanoic (C20:0)	0.0–0.1

[a] Adapted from Barrera-Arellano, Badan-Ribeiro, and Serna-Saldivar (2019).

Phospholipids are functional constituents of cell membranes; they are composed of glycerol, fatty acids, phosphate, and an organic base. Maize oil contains around 0.7% to 2.0% of phospholipids, being phosphatidylcholine, phosphatidylinositol, and phosphatidylethanolamine the most predominant. The minor phospholipids in maize grain are phosphatidylglycerol, diphosphatidylglycerol, phosphatidylserine, phosphatidic acid, and N-acylphosphatidylethanolamine (Wang and White 2019, Akoh and Min 2002).

The glycolipids are formed when a 1,2-diacyl-*sn*-3-glycerol is linked via the *sn*-3 position to a carbohydrate molecule. The major glycolipids of maize grain are monogalactosyldiacylglycerols (MGDG), digalactosyldiacylglycerols (DGDG), and sulfoquinovosyldiacylglycerols (SQDG) (Wang and White 2019, Akoh and Min 2002).

Corn oil contains 1.3%–2.3% of unsaponifiable lipids, including main hydrocarbons, phytosterols, phytostanols, carotenoids, and tocols (tocopherols, tocotrienols) (Moreau 2011, Harrabi et al. 2008). Squalene is the major hydrocarbon presented mainly in aleurone and fiber layers of corn (Moreau et al. 2000). Phytosterols and phytostanols are mainly located in pericarp and endosperm maize fractions. Corn oil contains approximately 1% of phytosterols, which could be found in free (about 58%) and esterified with fatty acids (42%) forms. The main phytosterols are β-sitosterol (55%–67%), campesterol (19%– 24%), stigmasterol (4%–8%), and Δ-5-avenasterol (4%–8%) (Moreau 2005, Winkler-Moser and Breyer 2011). Phytostanols are localized mainly in the endosperm fraction, being the most predominant in maize oil, sitostanol (77%– 87%), and campestanol (13%–23%) and stigmasterol (4%–10%) (Wang and White 2019, Harrabi et al. 2008).

Vitamin E consists of tocopherols and tocotrienols, collectively known as tocols. Both classes of compounds are subdivided into four lipophilic isomers (α-, β-, γ-, and δ-) that are naturally found in several vegetable oils. Refined corn oil contains around 0.08% to 0.12% of tocopherols, being the most predominant γ-tocopherol, followed by α-tocopherol and δ-tocopherol (Gunstone 2005). The tocotrienols differ from the tocopherols in having three double bonds in the isoprenoid side chain. Usually, maize contains α-, δ- and γ- tocotrienols, being the last one the most abundant in oil extracted from the whole kernel. The germ contains the major proportion of tocopherols, and endosperm contains the most content of tocotrienols (Wang and White 2019, Hossain and Jayadeep 2018).

The sphingolipids are a class of lipids containing a long-chain base (commonly sphingosine), fatty acids, and various other compounds, such as phosphate and monosaccharides. The lipid group includes ceramides and glycosylceramides, being 1-O-glucosyl-N-20- hydroxyeicosanoyl-4,8-sphingadienine and 1-O-glucosyl-N-20-hydroxy-tetracosanoyl-4-hydroxy-8-sphingenine the most representative sphingolipids in corn (Tanaka, Onishi, and Fujino 1984, Akoh and Min 2002).

Waxes are esters of fatty acids and long-chain alcohols. Simple waxes are esters of medium-chain fatty acids (16:0, 18:0, 18:1ω9), and long-chain aliphatic alcohols. Complex waxes are formed from diols or alcohol acids (Akoh and Min 2002). The kernel wax is composed mainly of esters (76%), being C46 (21%), C48 (22%), C50 (9%), C52 (10%), and C54 (17%) the dominant chain lengths of the esters. Respect to alcohols, it has been identified C22 (16%), C24 (30%), C26 (10%), C32 (15%) as the major components. In the case of acids, C22 (48%) and C24 (37%) has been the foremost recognized (Wang and White 2019). Corn oil contains about 0.05% of waxes, especially myricil esters. Cutin is a polyester matrix composed mainly of glycerol and long-chain (C16 and C18) fatty acid monomers (Bourgault et al. 2019). According to the study of Espelie, Dean, and Kolattukudy (1979) ω-hydroxylated acids constituted the major class of monomers being ω-OH-18:1 (51%), diOH-16:0 (20%), 9,10,18-triOH-18:0 (7%), and ω-OH-16:0 (6%). Waxes and cutin provide mechanical strength and protect the kernel from environmental damages as extreme temperatures, UV radiation, and pathogens (e.g., *Aspergillus flavus*).

Nitrogenous Compounds

The nitrogenous compounds in maize kernel can be divided mainly into protein and non-protein nitrogenous compounds. Amino acids and amides, found mainly in the aleurone layer and the germ, represent the most non-protein nitrogenous compounds. Among free amino acids, the most abundant has been identified as asparagine, glutamine, proline, and alanine. Proteins play important roles in germination, maturation, and storage for the kernel. Albumins, globulins, prolamins, and glutelins represent the main categories of proteins in corn kernel, which are mainly found in the endosperm (70%-79%) and the germ (18%-28%). Average protein content in normal corn kernel ranges between 6% and 12% (dry-weight basis) (Győri 2017, Ai and Jane 2016, Nuss and Tanumihardjo 2010).

Albumins are found in minor amounts. They contribute about 7% of the kernel nitrogen, being leucosin the most important albumin in corn grain (Nuss and Tanumihardjo 2010, Győri 2017). Globulins contribute about 5% nitrogen of the kernel nitrogen. In germ, they represent 10 – 20% of embryo proteins, and in the starchy endosperm, only small quantities of storage globulins have been identified. The main identified globulins of corn kernel have been mayzin, legumin-1, and α-globulin (Győri 2017, Yamagata et al. 2003, Gu et al. 2010, Woo et al. 2001, Kriz 1999). Glutelins are endosperm-specific proteins. They account for 35%-40% of the protein in the mature corn kernel, being zeanin the most abundant (Nuss and Tanumihardjo 2010, Larkins 2019). Prolamins are the primary storage proteins of most cereals. In the case of corn, they accumulate mainly in the starchy endosperm, and they provide about half of the total kernel nitrogen, (Kriz 1999).

Among prolamins, zeins are by far the most abundant storage proteins of the endosperm (60%–70%). Structurally, zeins are divided into three types: α, β-, and γ-zeins. Zein proteins are low in lysine and tryptophan, which are essential amino acids for human survival. Albumins, globulins, and glutelins are higher in lysine (5%–6%), compared to zeins (0.1%), but they are less abundant (Larkins 2019). Therefore, in general, it is considered that the common corn kernel has poor quality proteins as staple food or feed. Fortunately, a natural mutation called *opaque2* alters the ratio of zeins to nonzeins in the seed, resulting in kernels with increased concentrations of lysine and tryptophan, and the resulting maize varieties and hybrids are called Quality Protein Maize (QPM). QPM contains almost twice as much lysine and tryptophan than regular counterparts (Prasanna et al. 2001).

MICRONUTRIENTS

Maize kernels supply many micronutrients necessary for human metabolic needs, being the pericarp, germ, and aleurone layer the anatomical parts that contain the highest concentration of most of them (Acosta-Estrada, Gutiérrez-Uribe, and Serna-Saldivar 2019).

Vitamins

Vitamins are vital micronutrients that cannot be synthesized endogenously or in insufficient amounts. Commonly classified as water-soluble or fat-soluble, they act as cofactors for enzymes that play fundamental roles in human metabolism. The water-soluble vitamins include ascorbic acid (vitamin C), thiamin (B_1), riboflavin (B_2), niacin (B_3), pantothenic acid (B_5), pyridoxine (B_6), biotin (B_8), folic acid (B_9) and cobalamine (B_{12}). And the fat-soluble vitamins include vitamins A, D, E, and K (Pinchen and Finglas 2019, Syeda Juveriya, Ilaiyaraja, and Farhath 2017).

Water-soluble vitamins are found mainly in the aleurone layer, followed by the endosperm and germ (Nuss and Tanumihardjo 2010). Corn provides significant quantities of most B-vitamins; however, mature corn kernel lacks vitamins B_{12} and C. Moreover, specifically in the case of niacin (B_3), if the corn kernel is not properly processed (e.g., as nixtamalization), this vitamin is biologically unavailable (Nuss and Tanumihardjo 2010). Table 4 shows the average composition of water-soluble vitamins of common maize.

Fat-soluble vitamins play integral roles in a multitude of physiological processes such as vision, immune function, bone health, and coagulation (Herrero et al. 2017). In corn kernel, fat-soluble vitamins are mainly found in the aleurone layer and the germ. Provitamin A (carotenoids) are mainly located in the endosperm in an average of 2 μg/g (Egesel et al. 2003, Acosta-Estrada, Gutiérrez-Uribe, and Serna-Saldivar 2019). Vitamin E, which is exclusively acquired from the diet, occurs in significant amounts in the germ of corn kernels. Native corn contains around 3.11 μg/g of vitamin E. Respect to vitamins D and K; corn contains a meager amount (0–10 μg/100 g) (Acosta-Estrada, Gutiérrez-Uribe, and Serna-Saldivar 2019).

Table 4. Water-soluble vitamins of common maize[a]

Vitamin	Content/ 100 g	Physiological functions
Thiamin (B1)	0.385 mg	Serves as a coenzyme for enzymes like ketoglutarate dehydrogenase and transketolase, helps to maintain proper functioning of the heart and nervous and digestive systems
Riboflavin (B_2)	0.201 mg	Acts as a coenzyme in oxidation and reduction reactions, as prosthetic group of flavoproteins
Niacin (B_3)	3.63 mg	Acts as a coenzyme in oxidation and reduction reactions, as a functional part of NAD+ and NADP+
Pantothenic acid (B_5)	0.424 mg	The functional part of coenzyme A and acyl carrier protein, synthesis and β-oxidation of fatty acids, synthesis of cholesterol, steroid hormones, vitamins A and D, synthesis of several amino acids and δ-aminolevulinic acid.
Pyridoxine (B_6)	0.622 mg	Acts as a coenzyme in transamination and decarboxylation of amino acids and glycogen phosphorylase, synthesis of niacin from tryptophan, synthesis of several neurotransmitters, and synthesis of δ-aminolevulinic acid, inhibits receptor affinity to DNA and thereby modulates steroid activity.
Folic acid (B_9)	19 μg	Acts as a coenzyme in the transfer of one-carbon groups, essential in the de novo synthesis of nucleotides, metabolism of several amino acids, as an integral component for the regeneration of the universal methyl donor

[a] Adapted from Nuss and Tanumihardjo (2010) and Syeda Juveriya, Ilaiyaraja, and Farhath (2017).

Minerals

Minerals are inorganic elements found in small or trace amounts in dietary constituents. Based on the amount present in the human body, they are classified as major and trace minerals. Major minerals are those present in the body in amounts greater of 5 g (e.g., calcium, phosphorus, magnesium, and potassium) and trace minerals are those daily required in few milligrams or even micrograms (e.g., iron, zinc, copper, manganese, selenium, iodine, among others) (Bock 2000). In the corn kernel, the highest concentration of minerals is found in the pericarp, germ, and aleurone layer, and they constitute <2% of the corn kernel. The germ contains nearly 80% of the total minerals of the kernel, and the endosperm has less than 1% (Nuss and Tanumihardjo 2010, Acosta-Estrada, Gutiérrez-Uribe, and Serna-Saldivar 2019).

The most prevalent major minerals found in maize are phosphorus (P), potassium (K), and magnesium (Mg), providing nearly 85% of kernel mineral content. Respect to calcium (Ca), corn is a poor source of this mineral (1–17mg/100g) (Suri and Tanumihardjo 2016). Among the major minerals, phosphorus is the most abundant (300mg/100g). However, it is mainly found in the form of phytate. About 90% of phytates are located in the germ, 3% in the endosperm, and about 10% in the outer layers (Nuss and Tanumihardjo 2010, Noureddini et al. 2009). Even if phytic acid has an important physiological role in seeds as a storage compound for phosphate, in food and feed, it is considered as a non-nutritional factor because it chelates minerals rendering them unavailable to the digestive system (Raboy et al. 2000). To make the phytate-bound minerals biologically available, phytate must be hydrolyzed by

60%–70% (Brinch-Pedersen et al. 2007). Germination and nixtamalization had shown good results in hydrolyzed phytic acid content in corn kernels (Sokrab, Mohamed Ahmed, and Babiker 2012, Bressani et al. 2004). Potassium (K) and magnesium (Mg) content in corn is about 287 mg/100g and 127 mg/100g, respectively. Magnesium is mainly located in the aleurone layer of corn (87%) as Ca5Mg phytate or as potassium-magnesium phytate (Bock 2000, Gwirtz and Garcia-Casal 2014).

Among trace minerals, iron (Fe) and copper (Cu) are mostly found in the bran. Fe content average 2.5 mg/100 g, and Cu ranging from 0.2 to 0.31 mg/100g (White and Broadley 2005). Zinc (Zn) content average 20 μg/g, and it is mainly found in the endosperm, aleurone, and the scutellum (Nuss and Tanumihardjo 2010, Brinch-Pedersen et al. 2007). Selenium (Se) content an average of 9.5 μg/100 g, and it is mainly concentrated in the germ (Suri and Tanumihardjo 2016). Manganese (Mn) content average of 3.65 mg/100g.

Even though maize kernels supply many vitamins and minerals necessary for human metabolic needs, the amounts of some of them are inadequate for consumers that use maize as a staple food. Therefore, the enrichment of cereal-based food is aimed toward the replacement of essential minerals (Fe and Zn) and vitamins (thiamine, riboflavin, niacin, and folic acid) (Serna-Saldivar and Perez Carrillo 2019, Nuss and Tanumihardjo 2010).

Bioactive Compounds in Maize

Whole grain has a wide range of phytochemicals exhibiting health benefits. Corn phytochemicals include phenolics, flavonoids, carotenoids, anthocyanins, phytosterols, phospholipids, policosanols, arabinoxylans, among others.

Phenolics Compounds

Plant phenolics include simple phenols, phenolic acids, flavonoids, anthocyanins, stilbenes, tannins, lignans, and lignins. Phenolics are characterized by possessing one common structural feature, a phenol (an aromatic ring bearing at least one hydroxyl substituent). According to their chemical structure, phenolics have been divided into polyphenols and simple phenols (phenolic acids) (Robbins 2003). Phenolics compounds are the most widely distributed secondary metabolites present in corn, mainly found in the pericarp and the monolayered aleurone cell walls (Adom and Liu 2002, Gutiérrez-Uribe et al. 2010).

Corn contains an average concentration of 255 mg/100g simple phenolics, including caffeic, *cis-p*-coumaric, *trans-p*-coumaric, *p*-hydroxybenzoic, (p-hydroxyphenyl) acetic, protocatechuic, syringic, vanillic, *trans*-sinapic, syringic, gallic, and ferulic acids (*cis*-ferulic, *trans*-ferulic). The major phenolic compounds from maize are ferulic acids, representing about 70% of the total phenolics compounds (Ndolo and Beta 2014, Nile and Park 2014, Del Pozo-Insfran et al. 2006).

Pigments

Pigments are chemical compounds that absorb light in the wavelength range of the visible region. Natural pigments of foods can be classified into four groups: the green chlorophylls, the yellow-orange-red carotenoids, the red-blue-purple anthocyanins, and the red betanin (Delgado-Vargas, Jiménez, and Paredes-López 2000). Corn kernels can be found in a wide range of colors as white, yellow, orange, red, purple, blue, and brown. All these variations may result from genetic differences in pericarp, germ, and endosperm structures. The pericarp can be colorless, orange, cherry red, red, dark red, brown, or multicolored. The germ can be colorless, yellow, orange-red, or purple. And the endosperm can be either colorless, yellow, orange, or orange-red (Coe Jr, Neuffer, and Hoisington 1988). In mature corn kernel, both carotenoids and anthocyanins are the natural pigments that occur in the largest amounts.

Carotenoids

Carotenoids, which are a provitamin A, are lipid-soluble compounds comprised of eight isoprenoid units whose order is inverted at the molecule center. They are classified by their chemical structure in two groups: (1) carotenes that are constituted by carbon and hydrogen and (2) oxycarotenoids or xanthophylls that have carbon, hydrogen, and, additionally, oxygen (Delgado-Vargas, Jiménez, and Paredes-López 2000). In corn kernel, the predominant carotenes are α-carotene and β-carotenes, and respect to xanthophylls, lutein, zeaxanthin, and β-Cryptoxanthin are the most prevalent.

White corn has low levels of total carotenoids, with only 3 mg/g of total carotenoids. In contrast, yellow genotypes contain significant amounts of both carotenes and xanthophylls, especially high-carotenoid kernels contain up to 66 and 95 mg/g of total carotenoids (Burt et al. 2010, Taleon et al. 2017, Díaz-Gómez, Ramos, et al. 2017).

In common yellow dent corn, the total carotenoid content range of 5–80 mg/kg (FAO 1992). The yellow color results from the content of carotenoid pigments in the horny endosperm (74%–86%), floury endosperm (9%–23%), germ (2%–4%), and pericarp (1%), being xanthophyll pigments the major responsibility of the color (Kurilich and Juvik 1999, Wang and White 2019). Lutein and zeaxanthin account for 90% of the total carotenoid pigments (Serna-Saldivar 2010). In red corn, according to Corrales-Bañuelos et al. (2016), total carotenoid content range from 1.49 to 3.49 mg of lutein equivalents per kg dry weight.

Anthocyanins

Anthocyanins are one of the major groups of pigments belonging to the secondary metabolite group of flavonoids. The basic C6-C3-C6 anthocyanin structure is the source of an infinity of colors, from scarlet to blue, produced by its chemical combination with glycosides and/or acyl groups and by its interaction with other molecules and/or media conditions (Delgado-Vargas, Jiménez, and Paredes-López 2000). In fact, according to Urias-Peraldí et al. (2013), anthocyanin concentrations in maize are sensitive to environmental conditions, growing location, and stresses such as drought and environmental temperatures, especially during grain development and maturation.

The term blue corn is used to describe corn varieties that may display colors ranging from light to dark blue as well as red or purple. Blue coloring in corn kernels is due mainly to the presence of anthocyanins (e.g., cyanidin, pelargonidin, and peonidin) in the aleurone layer or pericarp (Scott et al. 2019, Nuss and Tanumihardjo 2010).

According to various research studies, total monomeric anthocyanin content varied widely in colored corn, being reported values from 51 mg cyanidin-3-glucoside (C3G) equivalent/kg in red corn to 1300–16,400 mg e.q. C3G/kg in purple corn and 368 mg eq. C3G/kg in blue corn (Lao and Giusti 2016, Li et al. 2017, de la Parra, Serna Saldivar, and Liu 2007).

The main anthocyanins in colored corn are cyanidin, malvidin, pelargonidin, and peonidin based monoglucosides, malonyl and dimalonyl glucosides, and flavonol anthocyanin condensed forms. Among them, the cyanidin is considered the main anthocyanin in colored corn, being cyanidin 3-glucoside the major anthocyanin present in blue corn, representing 75% of the total anthocyanins content (Del Pozo-Insfran et al. 2006, Salinas-Moreno et al. 2003). Other predominant anthocyanins in colored corn, include peonidin-3-O-glucoside, peonidin-3-(60-malonylglucoside), peonidin-3-(dimalonylglucoside), peonidin-3-(600- ethyl malonyl glucoside), pelargonidin-3-glucoside, pelargonidin-3-(600- malonylglucoside), cyanidin-3-(600-malonylglucoside), cyanidin-3-O-glucoside-2-malonylglucoside, cyanidin-3-(600-succinylglucoside) and cyanidin-3-(600-disuccinylglucoside) (Mora-Rochín et al. 2016, Li et al. 2017, Camelo-Méndez et al. 2016).

Policosanols

Policosanols are nutraceuticals that consist of long-chain primary aliphatic alcohols derived from plant wax. Total policosanol content of corn kernels varied from 15.2 to 20.5mg/kg, being dotriacontanol, triacontanol, and tetracosanol, the major identified components in corn kernels. Corn pericarp presented higher contents of policosanols (72.7–110.9 mg/kg) than the endosperm (4.0–16.2 mg/kg) and germ (19.3–37.1 mg/kg) fractions (Harrabi et al. 2009). Recent studies have shown anti-hypercholesterolemia and antioxidant activity (Shen, Luo, and Lin 2019).

Arabinoxylans

Arabinoxylans are the main soluble fiber components of corn. Arabinoylans have a β-(1→4) xylan backbone, and, in its simplest form, β xylopyranosyl units of the backbone are branched with α-arabinofuranose units at their O-2 and/or O-3 positions. The corn fiber arabinoxylans yield ranges from 19% to 39%, and they are mainly found in the cell wall material from the endosperm and the bran (Hamaker, Tuncil, and Shen 2019, Acosta-Estrada, Gutiérrez-Uribe, and Serna-Saldivar 2019).

Functional Properties of Maize

Maize is a rich source of phytochemicals, for instance, phenolics acids, and carotenoids. The phytochemical compounds derived from maize and their health properties have recently become the major focus of studies. Table 5 describes several properties attributed to phytochemicals present in maize or its derived products.

Antioxidant Activity

The antioxidant activity in maize is one of the most reported biological activities. This activity is related to phytochemicals, such as carotenoids and phenolic compounds. For instance, Beta and Hwang (2018) reported the presence of β-carotene, β-cryptoxanthin, zeaxanthin, and lutein in orange corn flour. Also, identify the presence of ferulic acid. Methanolic extracts of orange corn flour demonstrated the capacity to scavenge free radicals such as ABTS and DPPH.

On the other hand, blue maize flour extracts could scavenge ABTS free radical and showed ferric reducing power. The antioxidant activity of blue maize was related to its anthocyanin content (Camelo-Méndez et al. 2017). Moreover, Salinas-Moreno et al. (2012) analyzed 18 blue corn varieties, for their DPPH scavenging capacity, finding high discoloration activity in all samples, attributing the antioxidant capacity to anthocyanins.

Maize beverage enriched with 9% of nejayote solids (the liquid resulting from alkaline cooking of maize) presented antioxidant activity, as measured by nitric oxide inhibition. The activity was attributed not only to the phytochemicals in maize but also to the ferulic acid contained in nejayote (Buitimea-Cantúa et al. 2019).

Zavala-López et al. (2018) reported that teosinte, an ancient corn variety, is rich in phenolic compounds, and present high antioxidant activity.

Derived products of maize, such as tortillas and protein concentrates, have also shown antioxidant capacity (Díaz-Gómez, Castorena-Torres, et al. 2017, Herrera-Sotero et al. 2017). Corn tortillas from yellow and red kernels had antioxidant activity related to the presence of carotenoids. Although yellow grains had more quantity of carotenoids than red ones, both tortillas presented high antioxidant capacity as measured by the ORAC method (Corrales-Bañuelos et al. 2016). Blue corn tortillas contain a variety of phenolic compounds, then their antioxidant capacity is high against DPPH free radical and also can inhibit lipid peroxidation (Herrera-Sotero et al. 2017).

Anticancer Activity

Due to the variety of phytochemicals present in corn kernels, anticancer activity was evaluated by several authors (Díaz-Gómez, Castorena-Torres, et al. 2017, Díaz-Gómez et al. 2018, Ortiz-Martinez, Gonzalez de Mejia, et al. 2017). Herrera-Sotero et al. (2017) evaluated the antiproliferative capacity of blue corn and tortilla extracts against hepatocellular carcinoma (HePG2), lung carcinoma (H-460), mammary adenocarcinoma (MCF-7) and prostate cancer androgen-dependent (PC-3) cells. The effective concentration tested was 1000 μg/mL. The presence of anthocyanins and proanthocyanidins were detected in blue corn and tortilla preparations (Herrera-Sotero et al. 2017).

Not only phenolic compounds in maize are potential anticarcinogenic agents but also bioactive peptides. Díaz-Gómez, Castorena-Torres, et al. (2017) suggested several mechanisms through bioactive peptides could act, such as apoptosis induction, caspase modulation, p53 restoration, regulation of cellular mechanisms related to cell proliferation, among others. Moreover, Díaz-Gómez et al. (2018) evaluated the cytotoxicity of zein hydrolysates on HepG2 cells, suggesting that peptides contained in the hydrolysates could induce apoptosis. HepG2 cells were also used by Ortiz-Martinez, Gonzalez de Mejia, et al. (2017) to evaluate maize protein hydrolysates, achieving similar conclusions, as the mechanisms responsible for anticancer activity was apoptosis induction.

ANTIHYPERTENSIVE ACTIVITY

Hypertension is one of the most common diseases in the world. Chinese and North American natives used corn silk tea to low blood pressure. Corn silk has been claimed to have several benefits to human health, such as anti-inflammatory, reducing edema, and obesity. Several authors had studied corn silk to evaluate its antihypertensive activity and look for its mechanism of action (Li et al. 2019, Shi et al. 2019, Puchalska, Luisa Marina, and Concepción García 2013, Puchalska, Marina, and García 2012).

Li et al. (2019) studied corn silk extract, prepared in boiling water, finding a significant reduction of blood pressure. These authors used spontaneously hypertensive rats. Moreover, no secondary effects were observed under the subchronic usage of corn silk tea. An interesting finding was a phyto-peptide present in corn silk that could be responsible for the antihypertensive effect through inhibition of the angiotensin-converting enzyme.

More studies on peptides found in corn proteins have been done by Puchalska, Marina, and García (2012) and Puchalska, Luisa Marina, and Concepción García (2013). After extracting zein, it was digested to produce several peptides, including three highly antihypertensive peptides (LRP, LQP, and LSP), which were identified by liquid chromate-graphy coupled to mass spectrometry (Puchalska, Luisa Marina, and Concepción García 2013). Furthermore, corn germ meal protein was hydrolyzed to produce peptides with angiotensin-converting enzyme inhibitory activity (Guo et al. 2020). These peptides also had antihypertensive activity *in vivo*, as tested in spontaneously hypertensive rats.

OTHER BIOLOGICAL ACTIVITIES

There are some other activities reported for corn, corn silk, and other corn products, such as anti-diabetic, antiobesogenic, and anti-inflammatory (Table 5).

Nessa, Ismail, and Mohamed (2012) tested polar and non- polar extracts against twelve pathogenic bacteria. Their findings demonstrated that chloroform extract was active against five out of twelve bacteria, while methanolic extract was effective against *Bacillus* and *Salmonella*. Similarly, Morshed and Islam (2015) demonstrated that the ethanol extract of corn silk was effective against Gram-positive and Gram-negative bacteria.

Table 5. Biological properties of phytochemicals present in maize and its derivatives

Functional property	Part plant	Phytochemical	Reference
Antioxidant	Flour	Carotenoids, phenolics	(Beta and Hwang 2018)
	Roasted grain	Phenolic acids	(Buitimea-Cantúa et al. 2019)
	Grain flour	Phenolics, anthocyanins	(Camelo-Méndez et al. 2017, Salinas-Moreno et al. 2012, Zavala-López et al. 2018)
	Grain tortillas	Carotenoids	(Corrales-Bañuelos et al. 2016, Herrera-Sotero et al. 2017)
	Grain protein	Peptides	(Díaz-Gómez et al. 2018, Ortiz-Martinez, Otero-Pappatheodorou, et al. 2017)
	Wastewater	Phenolics	(Ramírez, Quintero-Soto, and Rochín-Medina 2020)
Anticancer	Leaf	Phenolics	(Balasubramanian and Padma 2013)
	Grain protein	Peptides	(Díaz-Gómez, Castorena-Torres, et al. 2017, Díaz-Gómez et al. 2018, Ortiz-Martinez, Gonzalez de Mejia, et al. 2017)
	Grain and tortillas	Anthocyanins	(Herrera-Sotero et al. 2017)
	Corn silk	Methanolic extract	(Al-Oqail et al. 2019)
Anti-diabetic	Seed	Anthocyanins	(Bhaswant et al. 2017, Zhang et al. 2019)
Antimicrobial	Grain	Methanol and chloroform extracts	(Guo et al. 2016, Morshed and Islam 2015)
	Corn silk	Ethanol extract	(Nessa, Ismail, and Mohamed 2012)
	Wastewater	Phenolics	(Ramírez, Quintero-Soto, and Rochín-Medina 2020)
Antiobesogenic	Grain	Phenolics	(Luna-Vital et al. 2020, Zhang et al. 2019)
Antihypertensive	Grain protein Corn silk	Peptides	(Puchalska, Luisa Marina, and Concepción García 2013, Puchalska, Marina, and García 2012)
Anti-inflammatory	Purple maize pericarp	Anthocyanins quercetin, luteolin, and rutin	(Zhang et al. 2019)

On the other hand, Ramírez, Quintero-Soto, and Rochín-Medina (2020) evaluated the antimicrobial activity of phenolics contained in nejayote, the remaining liquid of nixtamalization, against several bacterial strains. Their findings showed high activity against nejayote against *Salmonella typhimurium* and *Enterococcus faecalis*.

Maize is a rich source of simple phenolics such as ferulic acid, which could prevent the accumulation of fat in adipocytes. In this sense, Luna-Vital et al. (2020) showed that maize extract and ferulic acid at different concentrations could modulate genes related to gluconeogenesis, adipogenesis, and inflammation, suggesting that not only ferulic acid but also anthocyanins could modulate obesity by reducing adipogenesis and adipose inflammation, and promoting energy expenditure.

Regarding inflammation Zhang et al. (2019) and Zhang, Luna-Vital, and Gonzalez de Mejia (2019) reported that phenolic compounds such as quercetin, luteolin and rutin had anti-inflammatory and anti-diabetic compounds, which could modulate the production of inflammation mediators.

Finally, corn and its diverse phytochemicals had several biological activities, which could be beneficial to human health.

ISSUES RELATED TO MAIZE CONSUMPTION: MYCOTOXINS

Even if maize contains various phytochemicals linked to human health benefits, many studies have shown that the consumption of fungal contaminated maize could be associated to mycotoxicosis and health risks in both humans (e.g., esophageal cancer, neural tube defects) and farm animals (e.g., leukoencephalomalacia in horses and pulmonary edema syndrome in swine) (Ponce-García, Serna-Saldivar, and Garcia-Lara 2018). In fact, the maize plant is easily infected via fungal attack during their development, harvest, or after their harvest, which in most of the cases triggers the production of mycotoxins (Sun, Su, and Shan 2017).

The word mycotoxin is derived from 2 words: "mukes" referring to "fungi" (Greek) and "toxicum" referring to "poison" (Latin). Mycotoxin is a broad term that applies to secondary metabolites of fungi that grow in able substrates, especially those with special humidity content and available nutrients, such as vegetables and grains (Bhat, Rai, and Karim 2010). Mycotoxins found in corn can be grouped according to the fungi that produce them, being *Fusarium*, *Aspergillus,* and *Penicillium* species, the major toxigenic fungi (Miller 1995). The *Fusarium* species includes the most widespread toxigenic fungi in corn. The *Aspergillus* species are associated with the most serious human health impacts for mycotoxin contaminated corn, and *Penicillium* species are secondarily important because they are not aggressive pathogens (Munkvold et al. 2019). Table 6 shows the major fungi species reported in the corn kernel.

One special fact about fungi and their metabolites is that even though swift growth of a particular mold can occur on a substrate, it is not a prerequisite that the mold should produce a mycotoxin. For example, *Fusarium* molds grow prolifically at temperatures of 25 to 30°C without producing any mycotoxin, but at near-freezing temperatures, large quantities of mycotoxins are produced without much mold growth (Bhat, Rai, and Karim 2010). In corn, the most important mycotoxins are aflatoxins (AFs), ochratoxins (OTs), fumonisins (FBs), trichothecenes (especially T-2 toxin and deoxynivalenol-DON), and zearalenone (ZEN). When talking about incidence data of mycotoxins on different food matrices across the globe, in Africa and the Asian subcontinents, AFs are the major toxins; in Australia, they are AFs and FUM; in North America, they are AFs, OTs, ZEN, and DON; in South America, they are AFs, FUM, OTs, DON, and T-2 toxin; in Eastern European countries, they are ZEN and DON; and in Western European regions they are OTs, ZEN, and DON (Bhat, Rai, and Karim 2010).

Table 6. Major fungi species reported in the corn kernel. Adapted from Munkvold et al. (2019)

Fungi species in the corn kernel	
Genus	Species
Fusarium	*Fusarium graminearum, Fusarium proliferatum, Fusarium subglutinans, Fusarium temperatum* and *Fusarium verticillioides*
Aspergillus	*A. alliaceus, A. chevalieri, A. flavus, A. glaucus, A. nomius, A. ochraceus, A. parasiticus, A. restrictus, A. rubrobrunneus, A. sulphureus, A. sydowii, A. pseudotamarii, A. ustus* and *A. versicolor*
Penicillium	*P. brevi-compactum, P. chrysogenum, P. citrinum, P. cyclopium, P. expansum, P. frequentans, P. funiculosum, P. oxalicum, P. purpurogenum, P. urticae, P. variabile, P. verrucosum,* and *P. viridicatum*

Aflatoxins (AFs)

Aflatoxins possess a high acute and chronic toxicity due to their capacity to bind nucleic acids and cellular nucleoproteins, resulting in deleterious effects on protein synthesis and cellular integrity. They are considered potent hepatotoxins with high genotoxic activity, and they are responsible for hepatocarcinoma (Covarelli, Beccari, and Salvi 2011). They are mainly produced by *Aspergillus* species, especially *A. flavus* and *A. parasiticus*. In maize, the most common types of aflatoxins are B1, B2, G1, G2, and M1, being aflatoxin B1 (AFB1), the most toxic and carcinogenic (Schaarschmidt and Fauhl-Hassek 2019).

Fumonisins (FBs)

Fumonisins are the most recently isolated mycotoxins (first discovered in 1988) that are known to possess high cancer-inducing properties. These toxins were originally isolated from *Fusarium verticillioides* and *Fusarium proliferatum* (Bhat, Rai, and Karim 2010). Among the number of different types of fumonisins, B1, B2, and B3 fumonisins are the major forms found in food. In corn, fumonisin B1 (FB1), the diester of propane-1,2,3-tricarboxylic acid and 2-amino-12,16-dimethyl- 3,5,10,14,15-pentahydroxyeicosane, is the most prevalent. Moreover, in the specific case of FB1, in addition to its carcinogenicity, this compound has also revealed immunotoxic, hepatotoxic, pneumotoxic, and nephrotoxic activities (Covarelli, Beccari, and Salvi 2011).

Ochratoxins (OTs)

Ochratoxins (OTs) are a family of mycotoxins (OTA, OTB, and OTC) that can be produced by species of *Aspergillus* and *Penicillium.* They are characterized by containing a phenylalanine and dihydroisocumarin moiety joined by a peptide bond (Riley et al. 2011). Among ochratoxins family, ochratoxin A (OTA) is considered the most commonly occurring member and a potent toxin; it has presented nephrotoxic, hepatotoxic, teratogenic, and immunosuppressive activities (Bozzo et al. 2011). OTA is mainly produced by *Aspergillus ochraceus* and *Penicillium verrucosum* (Bhat, Rai, and Karim 2010). In cereals, ochratoxins generally appear during storage, and its degradation is particularly complex because they are specially heated stable. Under normal conditions of cooking OTA is partially degraded, and even at 250°C, its destruction is not completed (Gbashi et al. 2019, Bhat, Rai, and Karim 2010, Boudra, Le Bars, and Le Bars 1995).

Zearalenone (ZEN)

Zearalenone (ZEN) is a metabolite produced by *Fusarium* species, which commonly contaminate most of the cereals (Bhat, Rai, and Karim 2010). This molecule is a potent fungal estrogen, rather than a toxin. Because it has a special similarity to female hormones, in humans it has been associated with precocious pubertal changes, and in animals with feminizing syndromes. For example, in pigs, ZEN caused genital/urinatory syndromes and atrophy of the testes (Bhat, Rai, and Karim 2010). To avoid its effects, the Joint FAO/WHO Expert Committee on Food Additives (JECFA) has established a provisional maximum tolerable daily intake for ZEN of 0.5 μg/kg of body weight (CODEX 2011).

Trichothecenes

Trichothecenes are sesquiterpenoid mycotoxins produced by several genera of fungi, including *Fusarium*, *Trichoderma*, *Trichotecium*, *Myrothecium*, *Stachybotrys*. Comprising over 160 different trichothecenes compounds, they are classified into four groups (types A, B, C, and D) (Mbundi et al. 2014). In corn, trichothecenes are typical of type A (principally diacetoxyscirpenol -DAS, T-2, and HT-2 toxins) and type B (principally nivalenol -NIV, deoxynivalenol -DON and acetyldeoxynivalenol -Ac-DON) groups; being the group A the most toxic to human and animals (Bhat, Rai, and Karim 2010, Munkvold et al. 2019). Mycotoxicosis by trichothecenes, also known as scabby grain toxicosis, has been reported to occur within hours after ingestion of contaminated foods, showing symptoms in the range of abdominal pain, nausea, vomiting, diarrhea, dizziness, and headache (Bhat, Rai, and Karim 2010).

Food Processing Technologies to Reduce Mycotoxins Contamination in Maize

Even if good agricultural practices and adequate storage conditions limit mycotoxin levels in the maize food chain, they do not eliminate mycotoxins. Food processing can further reduce mycotoxin levels by physical removal methods, thermal processing, chemical treatments, among others. Therefore, in this part of the chapter, it will be considered the most common processing techniques to reduce mycotoxins in maize.

Physical Removal Methods

Food processing can further reduce mycotoxin levels by physical removal of contaminated material. In maize, the most applied methods are sorting, trimming cleaning, and milling. The sorting, trimming, and cleaning do not destroy mycotoxins, but they help to reduce its concentrations (Bullerman and Bianchini 2007). The specific cleaning operations include sieving, aspiration, separation by gravity or by indented cylinders or discs, and/or scouring. Generally, it is custom to apply a coarse cleaning before storage, followed by an intensive cleaning before milling (Schaarschmidt and Fauhl-Hassek 2018). In maize, fumonisin concentrations were reduced up to 69% by the removal of “fines” (<3 mm) (Sydenham et al. 1994).

Milling mainly refers to particle size reduction and fractionation of cereal grains. As for cleaning procedures, this operation does not degrade mycotoxins; however, as mycotoxins tend to be concentrated in germ and bran fractions, both wet and dry milling processes can redistribute them (Schaarschmidt and Fauhl-Hassek 2018, Bullerman and Bianchini 2007). In wet milling, to facilitate germ separation and to break down of protein matrix, the first step involves soaking of maize in water containing 0.1 to 0.2% SO_2 (36-50 h at 50°C) (Karlovsky et al. 2016). Therefore, depending on the polarity of the mycotoxins, they may be dissolved into the steep water or distributed among the byproducts of the process. By the end of the wet milling process, mycotoxins can be mainly found in gluten fiber and germ, while the starch tends to be relatively free of them (Bullerman and Bianchini 2007). In dry milling, the highest concentrations of mycotoxins remain in germ and bran fractions of maize; and fractions used for food production, including flaking grits and flour, had the least amount of mycotoxins contamination (Bullerman and Bianchini 2007).

Thermal Processing Methods

Even if most of the mycotoxins are chemically and thermally stable, thermal processing still the most important industrial treatment to reduce mycotoxins content in finished products. In maize, reduction of mycotoxin contamination has been mainly studied after cooking (Stoloff and Trucksess 1981), canning, roasting (Castelo, Sumner, and Bullerman 1998), flaking (De Girolamo, Solfrizzo, and Visconti 2001), nixtamalization (Schaarschmidt and Fauhl-Hassek 2019, Dombrink-Kurtzman et al. 2000), extrusion (Cazzaniga et al. 2001), baking and frying (Jackson et al. 1997).

Greater degradation of mycotoxins has been achieved under harsh thermal processing (high amounts of heat energy during a long heating time); however, these conditions have also triggered nutritional and organoleptic quality losses of food. Therefore, in order to achieve the maximum degradation of mycotoxins at the lowest possible temperature and time conditions, many studies have been targeting to optimize these parameters. In this respect, Gbashi et al. (2019) investigated the thermal degradation patterns of multiple mycotoxins (aflatoxins, fumonisins, zearalenone, ochratoxins, T-2 toxin, alternariol monomethyl ether - AME- and sterigmatocystin – STERIG-) in pure form and spiked into maize flour. Their study revealed that mycotoxins with compact molecular structural configuration were the most relatively stable to heat treatment, like Ochratoxin-A (OTA); whereas the more structurally spread-out mycotoxins were more thermolabile, like fumonisins B1, B2, and B3.

Moreover, the thermal stability of mycotoxins varied between pure standards and spiked maize flour, showing that matrix interaction strongly influenced the thermal stability of mycotoxins. Studied thermal processing conditions ranged from 103.43 to 216.57°C and from 6.72 to 63.28 min. And, the optimal processing conditions to degrade pure multi-standards mycotoxins and spiked maize were 216.57°C during 63.28 min, and 210.85°C during 54.71 min respectively. Table 7 shows the optimum degradation conditions for each studied mycotoxin.

Table 7. Optimal thermal treatment conditions for mycotoxin degradation of pure standards and spiked maize matrix. Adapted from Gbashi et al. (2019)

Mycotoxin	Pure mycotoxin standard		Spiked maize matrix	
	Optimal Temperature and time conditions	% of degradation	Optimal Temperature and time conditions	% of degradation
Aflatoxin B1	216.57°C/35 min	100.00 ± 0.00	216.57°C/35 min	100.00 ± 0.00
Aflatoxin B2		85.88 ± 5.59		100.00 ± 0.00
Aflatoxin G1		100.00 ± 0.00		100.00 ± 0.00
Aflatoxin G2		100.00 ± 0.00		97.55 ± 0.22
AME		93.86 ± 0.40		62.66 ± 2.91
Fumonisin B1	200°C/15 min	100.00 ± 0.00	200°C/15 min	100.00 ± 0.00
Fumonisin B2		100.00 ± 0.00		100.00 ± 0.00
Fumonisin B3	216.57°C/35 min	100.00 ± 0.00		100.00 ± 0.00
Ochratoxin A		58.78 ± 0.63	216.57°C/35 min	93.84 ± 0.18
Ochratoxin B		50.50 ± 2.96		92.36 ± 0.13
STERIG		100.00 ± 0.00		98.09 ± 0.14
T2		97.53 ± 0.89		95.61 ± 2.19
ZEA		100.00 ± 0.00		93.13 ± 0.30
α-ZEA		100.00 ± 0.00		96.90 ± 0.61
β-ZEA		100.00 ± 0.00		97.51 ± 0.37

Canning, also known as retorting, is the action of adding the food materials to a metallic container (generally of aluminum), followed by high-temperature processing (Dewanto, Wu, and Liu 2002). In the case of maize mycotoxins, Castelo, Sumner, and Bullerman (1998) showed that canning (121°C for 60 min) of whole-kernel com and spiked cream-style corn resulted in a decrease in fumonisins of 15% and 9% respectively.

On the other hand, corn flaking has also outcome the reduction of aflatoxins, fumonisins, and ochratoxin content (Bullerman and Bianchini 2007). After cooking the corn grits with and without sugars and toasting, the aflatoxins reduction varied between 77 to 85% (Castelo 1999). And when flaking corn was coupled to extrusion, a 60 to 70% of fumonisins reduction was found, being identified that 30% of the overall reduction was linked to extrusion stage (De Girolamo, Solfrizzo, and Visconti 2001). Extrusion cooking considers the application of high temperature and pressure by forcing a dough-like mixture through a stationary metal tube or barrel by a rotating screw shaft. Through the generated high temperature, pressure, and mechanical shear force, the food matrix is transformed by chemical reactions and molecular modifications, being found the greatest reductions of mycotoxin concentrations at temperatures ≥ 160°C and longer residence times (Bullerman and Bianchini 2007). Elias-

Orozco et al. (2002) showed that the extrusion process of white corn naturally contaminated with aflatoxins, reduced the levels of AFB1, AFM1, and AFB1- dihydrodiol by 46%, 20%, and 53% respectively.

Roasting, frying, and baking also reduced mycotoxins concentration on spiked corn (5 μg of FB1/g). In the case of roasting, it has been reported that at 218°C for 15 min, fumonisins of naturally and artificially contaminated cornmeal were reduced up to 98% (Castelo, Sumner, and Bullerman 1998). Frying (190°C for 15 min) also performed a 67% loss of fumonisins (Jackson et al. 1997). Contrary, baking (200°C for 20 min) of corn muffins spiked with FB1 resulted in only 27.6% loss, being highlighted that baking caused significantly greater losses of FB1 at the surface than at the core (Jackson et al. 1997).

Nixtamalization of maize is an ancient thermal-alkaline process consisting of the three-step process. First, corn kernels are cooking (90-100°C) in a saturated calcium hydroxide solution (approximately 45 min), followed by steeping for 8–24 h and by a final washing to remove the excess of calcium and organic matter (Valderrama-Bravo et al. 2012). According to Schaarschmidt and Fauhl-Hassek (2019), nixtamalization reduces the maize mycotoxins amounts in three ways. The first one by leaching of water-soluble mycotoxins into the liquid fractions during cooking, steeping, and washing. The second one by the removal of the pericarp, tip cap, and germ during final washing. And the last one, linked to chemical degradation, caused by the high pH and elevated temperature, which can result in degradation, modification, and/or binding or release of mycotoxins. In the study of De Girolamo et al. (2016), the fumonisins (FBs) and partially hydrolyzed fumonisins (PHFBs) of maize naturally contaminated, were reduced by nixtamalization, troughs their conversion into hydrolyzed fumonisins (HFBs). Dombrink-Kurtzman et al. (2000) also showed that nixtamalization reduced the initial amount of FB1 into 81.5%, being identified that an average of 72.5% of FB1 was converted to HFB1 and remained in the steeping and washing water. Elias-Orozco et al. (2002) exposed that nixtamalization reduced aflatoxin B1, aflatoxin M1, and aflatoxin B1-8,9-dihydrodiol content in 94%, 90%, and 93% respectively. Moreover, Schaarschmidt and Fauhl-Hassek (2019) also reported a reduction of zearalenone (60 to 100%) and vomitoxin (70% to 80%) after maize nixtamalization.

Chemical Processing Methods

Chemical food processing methods destroy or inactivate mycotoxins. Among chemical treatments, in maize, it has been applied treatment with acids, bases, oxidizing, and reducing agents. For example, lactic acid bacteria fermentation in spiked maize meal reduced the concentration of fumonisin B1 (75%) and zearalenone (68%) (Mokoena, Chelule, and Gqaleni 2005). On the other hand, corn treated with phosphates (tetrasodium pyrophosphate, sodium hexametaphosphate, and sodium tripolyphosphate) prevented or reduced mold growth when added in powder form to corn at 1.0% or 2.0% (w/w). Moreover, whenever mold growth occurred, treatment of corn with tetrasodium pyrophosphate and sodium hexametaphosphate inhibited aflatoxin production (Lebron et al. 1989). Ammoniation decreased aflatoxin levels in maize by more than 75% and completely decomposed OTA. However, due to incomplete knowledge about composition and toxicities of end products, this technique has not been approved by the USA Food and Drug Administration (Park et al. 1988, Chelkowski et al. 1981). Ozonation also significantly affected aflatoxins in cornflour, after treatment with 75 mg/L ozone for 60 min, the contents of AFB1, AFG1, and AFB2 were decreased in 78.7%, 72.1% and 70.6% respectively (Luo et al. 2014). Sodium bisulfite also performed the greatest reductions of DON (up to 95%) when contaminated corn was autoclaved for 1h at 121°C in the presence of 8.33% aqueous sodium bisulfite (Young et al. 1987).

CONCLUSION

Maize represents a significant source of macro and micronutrients. Its physicochemical composition has shown that it is rich in carbohydrates, vitamins, minerals, carotenoids, phenolic compounds, phytosterols, policosanols, phospholipids, arabinoxylans, dietary fiber, among other phytochemicals. This composition varies in function of various parameters such as environmental growth conditions, genetic background, plant age, and geographic location. Moreover, in recent years, studies around health benefits of maize consumption have shown its antioxidant, anticancer, antihypertensive, anti-diabetic, antiobesogenic, and anti-inflammatory activities. However, one of the main issues of maize consumption is mycotoxins. Because the maize plant is easily infected via fungal attack during their development, harvest, or after their harvest, it is crucial to ensure the safety and quality of final maize-based products. Even if good agricultural practices and adequate storage conditions limit mycotoxin levels in the maize food chain, these practices are not enough to eliminate mycotoxins. Therefore, food processing becomes a key tool to reduce or even eliminate mycotoxins in maize. Among the most applied food processes, thermal processing is still the most important industrial treatment to reduce mycotoxins content in finished products. Even though there are various studies about the effect of food processing conditions into mycotoxins, to take full advantage of maize as a superfood, it is still necessary to carry out a multicriteria optimization of processing parameters, that assure the preservation of maize phytochemicals, and the elimination or at least the reduction at safe levels of mycotoxins.

REFERENCES

Acosta-Estrada, Beatriz A., Janet A. Gutiérrez-Uribe, and Sergio O. Serna-Saldivar. 2019. "Chapter 14 - Minor Constituents and Phytochemicals of the Kernel." In *Corn* (Third Edition), edited by Sergio O. Serna-Saldivar, 369-403. Oxford: AACC International Press.

Adom, Kafui Kwami, and Rui Hai Liu. 2002. "Antioxidant Activity of Grains." *Journal of Agricultural and Food Chemistry* 50 (21):6182-6187. doi: https://doi.org/10.1021/jf0205099.

Ai, Yongfeng, and Jay-lin Jane. 2016. "Macronutrients in Corn and Human Nutrition." *Comprehensive Reviews in Food Science and Food Safety* 15 (3):581-598. doi: https://doi.org/10.1111/1541-4337.12192.

Akoh, Casimir, and David Min. 2002. *Food Lipids: Chemistry, Nutrition, Biotechnology*. Edited by Food Science and Technology. New York, United States: Marcel Dekker, Inc.

Al-Oqail, Mai M., Ebtesam S. Al-Sheddi, Nida N. Farshori, Shaza M. Al-Massarani, Eman A. Al-Turki, Javed Ahmad, Abdulaziz A. Al-Khedhairy, and Maqsood A. Siddiqui. 2019. "Corn Silk (*Zea mays* L.) Induced Apoptosis in Human Breast Cancer (MCF-7) Cells via the ROS-Mediated Mitochondrial Pathway." *Oxidative Medicine and Cellular Longevity* 2019. doi: https://doi.org/10.1155/2019/9789241.

Arendt, Elke K., and Emanuele Zannini. 2013. "2 - Maizes." In *Cereal Grains for the Food and Beverage Industries*, edited by Elke K. Arendt and Emanuele Zannini, 67-115e. Cambridge, UK: Woodhead Publishing.

Bailoni, Lucia, Stefano Schiavon, Giovanna Pagnin, Franco Tagliapietra, and Mario Bonsembiante. 2005. "Quanti-qualitative evaluation of pectins in the dietary fibre of 24 foods." *Italian Journal of Animal Science* 4 (1):49-58. doi: https://doi.org/10.4081/ijas.2005.49.

Balasubramanian, Kiruthika, and Palghat Raghunathan Padma. 2013. "Anticancer Activity of *Zea mays* Leaf Extracts on Oxidative Stress-induced Hep2 Cells." *Journal of Acupuncture and Meridian Studies* 6 (3):149-158. doi: https://doi.org/10.1016/j.jams.2013.01.015.

Barrera-Arellano, Daniel, Ana Paula Badan-Ribeiro, and Sergio O. Serna-Saldivar. 2019. "Chapter 21 - Corn Oil: Composition, Processing, and Utilization." In *Corn* (Third Edition), edited by Sergio O. Serna-Saldivar, 593-613. Oxford: AACC International Press.

Beta, Trust, and Taeyoung Hwang. 2018. "Influence of heat and moisture treatment on carotenoids, phenolic content, and antioxidant capacity of orange maize flour." *Food Chemistry* 246:58-64. doi: https://doi.org/10.1016/j.foodchem.2017.10.150.

Bhaswant, Maharshi, Siti Raihanah Shafie, Michael L. Mathai, Peter Mouatt, and Lindsay Brown. 2017. "Anthocyanins in chokeberry and purple maize attenuate diet-induced metabolic syndrome in rats." *Nutrition* 41:24-31. doi: https://doi.org/10.1016/j.nut.2016.12.009.

Bhat, Rajeev, Ravishankar V. Rai, and A.A. Karim. 2010. "Mycotoxins in Food and Feed: Present Status and Future Concerns." *Comprehensive Reviews in Food Science and Food Safety* 9 (1):57-81. doi: https://doi.org/10.1111/j.1541-4337.2009.00094.x.

Bock, Margaret Ann 2000. "Minor Constituents of Cereals." In *Handbook of Cereal Science and Technology, Revised and Expanded*, edited by Karel Kulp and Joseph G. Jr Ponte. New York: CRC Press.

Boudra, Hamid, Pierrette Le Bars, and Joseph Le Bars. 1995. "Thermostability of Ochratoxin A in wheat under two moisture conditions." *Applied and Environmental Microbiology* 61 (3):1156-1158.

Bourgault, Richard, Susanne Matschi, Miguel Vasquez, Pengfei Qiao, Annika Sonntag, Caleb Charlebois, Marc Mohammadi, Michael J Scanlon, Laurie G Smith, and Isabel Molina. 2019. "Constructing functional cuticles: analysis of relationships between cuticle lipid composition, ultrastructure and water barrier function in developing adult maize leaves." *Annals of Botany* 125 (1):79-91. doi: https://doi.org/10.1093/aob/mcz143.

Bozzo, Giancarlo, Elisabetta Bonerba, Edmondo Ceci, and Colao Valeriana. 2011. "Determination of ochratoxin A in eggs and target tissues of experimentally drugged hens using HPLC–FLD." *Food Chemistry - FOOD CHEM* 126:1278-1282. doi: https://doi.org/10.1016/j.foodchem.2010.11.070.

Bressani, Ricardo, Juan Carlos Turcios, Ana Silvia Colmenares de Ruiz, and Patricia Palocios de Palomo. 2004. "Effect of Processing Conditions on Phytic Acid, Calcium, Iron, and Zinc Contents of Lime-Cooked Maize." *Journal of Agricultural and Food Chemistry* 52 (5):1157-1162. doi: https://doi.org/10.1021/jf030636k.

Brinch-Pedersen, Henrik, Søren Borg, Birgitte Tauris, and Preben B. Holm. 2007. "Molecular genetic approaches to increasing mineral availability and vitamin content of cereals." *Journal of Cereal Science* 46 (3):308-326. doi: https://doi.org/10.1016/j.jcs.2007.02.004.

Brown, W.L., M.S. Zuber, and L.L. Darrah. 1985. "Origin, Adaptation, and Types of Corn." In *The National Corn Handbook*, edited by University of Wisconsin. Ames, Iowa: Iowa State University, Cooperative Extension Service.

Buitimea-Cantúa, Nydia E., Marilena Antunes-Ricardo, Javier Villela-Castrejón, and Janet A. Gutiérrez-Uribe. 2019. "Changes in cellular antioxidant and anti-inflammatory activity after 12 months storage of roasted maize-based beverages supplemented with nejayote solids." *Journal of Cereal Science* 89:102807. doi: https://doi.org/10.1016/j.jcs.2019.102807.

Bullerman, Lloyd B., and Andreia Bianchini. 2007. "Stability of mycotoxins during food processing." *International Journal of Food Microbiology* 119 (1):140-146. doi: https://doi.org/10.1016/j.ijfoodmicro.2007.07.035.

Burt, Andrew J., Christopher M. Grainger, J. Christopher Young, Barry J. Shelp, and Elizabeth A. Lee. 2010. "Impact of Postharvest Handling on Carotenoid Concentration and Composition in High-Carotenoid Maize (*Zea mays* L.) Kernels." *Journal of Agricultural and Food Chemistry* 58 (14):8286-8292. doi: https://doi.org/10.1021/jf100161r.

Cabrera-Ponce, José Luis, Eliana Valencia-Lozano, and Diana Lilia Trejo-Saavedra. 2019. "Chapter 3 - Genetic Modifications of Corn." In *Corn* (Third Edition), edited by Sergio O. Serna-Saldivar, 43-85. Oxford: AACC International Press.

Camelo-Méndez, Gustavo A., Edith Agama-Acevedo, Mirna M. Sanchez-Rivera, and Luis A. Bello-Pérez. 2016. "Effect on in vitro starch digestibility of Mexican blue maize anthocyanins." *Food Chemistry* 211:281-284. doi: https://doi.org/10.1016/j.foodchem.2016.05.024.

Camelo-Méndez, Gustavo A., Edith Agama-Acevedo, Juscelino Tovar, and Luis A. Bello-Pérez. 2017. "Functional study of raw and cooked blue maize flour: Starch digestibility, total phenolic content and antioxidant activity." *Journal of Cereal Science* 76:179-185. doi: https://doi.org/10.1016/j.jcs.2017.06.009.

Castelo, M.M., S.S. Sumner, and L.B. Bullerman. 1998. "Stability of fumonisins in thermally processed corn products." *Journal of Food Protection* 61 (8):1030-1033. doi: https://doi.org/10.4315/0362-028X-61.8.1030.

Castelo, Mauricio Montoya. 1999. "*Stability of mycotoxins in thermally processed corn products*." PhD PhD Dissertation, Food Science and Technology, University of Nebraska.

Cazzaniga, D., J.C. Basílico, R.J. González, R.L. Torres, and D.M. De Greef. 2001. "Mycotoxins inactivation by extrusion cooking of corn flour." *Letters in Applied Microbiology* 33 (2):144-147. doi: https://doi.org/10.1046/j.1472-765x.2001.00968.x.

CODEX. 2011. *Working document for information and use in discussions related to contaminants and toxins in the GSCTFF, CF/5 INF/1*. edited by FAO/WHO. The Hague, The Netherlands: CODEX Alimentarius Commission.

Coe Jr., E., M. Neuffer, and D. Hoisington. 1988. "The genetics of corn." In *Corn and Corn Improvement*, edited by G Sprague and J Dudley. Madison, WI, USA: American Society of Agronomy, Inc. Crop Science Society of America, Inc. Soil Science Society of America, Inc.

CONABIO. 2011. *Recopilación, generación, actualización y análisis de información acerca de la diversidad genética de maíces y sus parientes silvestres en México.* [*Collection, generation, updating and analysis of information about the genetic diversity of maize and its wild relatives in Mexico*]. edited by Comisión Nacional para el Conocimiento y Uso de la Biodiversidad. Mexico: CONABIO.

Corrales-Bañuelos, Ana Belinda, Edith Oliva Cuevas-Rodríguez, Janet Alejandra Gutiérrez-Uribe, Evelia María Milán-Noris, Cuauhtémoc Reyes-Moreno, Jorge Milán-Carrillo, and Saraid Mora-Rochín. 2016. "Carotenoid composition and antioxidant activity of tortillas elaborated from pigmented maize landrace by traditional nixtamalization or lime cooking extrusion process." *Journal of Cereal Science* 69:64-70. doi: https://doi.org/10.1016/j.jcs.2016.02.009.

Covarelli, Lorenzo, Giovanni Beccari, and Selene Salvi. 2011. "Infection by mycotoxigenic fungal species and mycotoxin contamination of maize grain in Umbria, central Italy." *Food and Chemical Toxicology* 49 (9):2365-2369. doi: https://doi.org/10.1016/j.fct.2011.06.047.

Chelkowski, J., P. Goliński, B. Godlewska, W. Radomyska, K. Szebiotko, and M. Wiewiórowska. 1981. "Mycotoxins in cereal grain. Part IV. Inactivation of ochratoxin A and other mycotoxins during ammoniation." *Food/Nahrung* 25 (7):631-637. doi: https://doi.org/10.1002/food.19810250705.

Darrah, L.L., M.D. McMullen, and M.S. Zuber. 2019. "Chapter 2 - Breeding, Genetics and Seed Corn Production." In *Corn* (Third Edition), edited by Sergio O. Serna-Saldivar, 19-41. Oxford: AACC International Press.

De Girolamo, A., V.M.T. Lattanzio, R. Schena, A. Visconti, and M. Pascale. 2016. "Effect of alkaline cooking of maize on the content of fumonisins B1 and B2 and their hydrolysed forms." *Food Chemistry* 192:1083-1089. doi: https://doi.org/10.1016/j.foodchem.2015.07.059.

De Girolamo, Annalisa, Michele Solfrizzo, and Angelo Visconti. 2001. "Effect of Processing on Fumonisin Concentration in Corn Flakes." *Journal of food protection* 64:701-5. doi: https://doi.org/10.4315/0362-028X-64.5.701.

de la Parra, Columba, Sergio O. Serna Saldivar, and Rui Hai Liu. 2007. "Effect of Processing on the Phytochemical Profiles and Antioxidant Activity of Corn for Production of Masa, Tortillas, and Tortilla Chips." *Journal of Agricultural and Food Chemistry* 55 (10):4177-4183. doi: https://doi.org/10.1021/jf063487p.

Del Pozo-Insfran, David, Carmen H. Brenes, Sergio O. Serna Saldivar, and Stephen T. Talcott. 2006. "Polyphenolic and antioxidant content of white and blue corn (*Zea mays* L.) products." *Food Research International* 39 (6):696-703. doi: https://doi.org/10.1016/j.foodres.2006.01.014.

Delgado-Vargas, F., A.R. Jiménez, and O. Paredes-López. 2000. "Natural Pigments: Carotenoids, Anthocyanins, and Betalains — Characteristics, Biosynthesis, Processing, and Stability." *Critical Reviews in Food Science and Nutrition* 40 (3):173 - 289. doi: https://doi.org/10.1080/10408690091189257.

Dewanto, V., X. Wu, and R.H. Liu. 2002. "Processed sweet corn has higher antioxidant activity." *Journal of Agricultural and Food Chemistry* 50 (17):4959-4964. doi: https://doi.org/10.1021/jf0255937.

Díaz-Gómez, Joana, Antonio Javier Ramos, Changfu Zhu, Olga Martín-Belloso, and Robert Soliva-Fortuny. 2017. "Influence of Cooking Conditions on Carotenoid Content and Stability in Porridges Prepared from High-Carotenoid Maize." *Plant Foods for Human Nutrition* 72 (2):113-119. doi: https://doi.org/10.1007/s11130-017-0604-7.

Díaz-Gómez, Jorge L., Fabiola Castorena-Torres, Ricardo E. Preciado-Ortiz, and Silverio García-Lara. 2017. "Anti-Cancer Activity of Maize Bioactive Peptides." *Frontiers in chemistry* 5:44-44. doi: https://doi.org/10.3389/fchem.2017.00044.

Díaz-Gómez, Jorge L., Margarita Ortíz-Martínez, Oscar Aguilar, Silverio García-Lara, and Fabiola Castorena-Torres. 2018. "Antioxidant Activity of Zein Hydrolysates from Zea Species and Their Cytotoxic Effects in a Hepatic Cell Culture." *Molecules* 23 (2):312. doi: https://doi.org/10.3390/molecules23020312.

Dombrink-Kurtzman, Mary A., Timothy J. Dvorak, Marc E. Barron, and Lloyd W. Rooney. 2000. "Effect of Nixtamalization (Alkaline Cooking) on Fumonisin-Contaminated Corn for Production of Masa and Tortillas." *Journal of Agricultural and Food Chemistry* 48 (11):5781-5786. doi: https://doi.org/10.1021/jf000529f.

Eckhoff, Steven R., and Stanley A. Watson. 2009. "Chapter 9 - Corn and Sorghum Starches: Production." In *Starch* (Third Edition), edited by James BeMiller and Roy Whistler, 373-439. San Diego: Academic Press.

Egesel, C.O., J.C. Wong, R.J. Lambert, and T.R. Rocheford. 2003. "Combining Ability of Maize Inbreds for Carotenoids and Tocopherols." *Crop Science* 43 (3):818-823. doi: https://doi.org/10.2135/cropsci2003.8180.

Elias-Orozco, R., A. Castellanos-Nava, M. Gaytan-Martinez, J.D. Figueroa-Cardenas, and G. Loarca-Pina. 2002. "Comparison of nixtamalization and extrusion processes for a reduction in aflatoxin content." *Food Additives & Contaminants* 19 (9):878-85. doi: https://doi.org/10.1080/02652030210145054.

Espelie, K.E., B.B. Dean, and P.E. Kolattukudy. 1979. "Composition of Lipid-derived Polymers from Different Anatomical Regions of Several Plant Species." *Plant Physiology* 64 (6):1089-93. doi: https://doi.org/10.1104/pp.64.6.1089.

FAO. 1992. *Maize in human nutrition.* edited by Food and Agriculture Organization of the United Nations. Rome.

FAO. 2020a. *FAOSTAT.* Food and agriculture data Accessed March 27. http://www.fao.org/faostat/en/#home.

FAO. 2020b. "*Production/Yield quantities of Maize in World.*" Accessed March 27. http://www.fao.org/faostat/en/#data/QC/visualize/.

García-Lara, Silverio, Cristina Chuck-Hernandez, and Sergio O. Serna-Saldivar. 2019. "Chapter 6 - Development and Structure of the Corn Kernel." In *Corn* (Third Edition), edited by Sergio O. Serna-Saldivar, 147-163. Oxford: AACC International Press.

García-Lara, Silverio, and Sergio O. Serna-Saldivar. 2019. "Chapter 1 - Corn History and Culture." In *Corn* (Third Edition), edited by Sergio O. Serna-Saldivar, 1-18. Oxford: AACC International Press.

Gbashi, S., N.E. Madala, S. De Saeger, M. De Boevre, and P.B. Njobeh. 2019. "Numerical optimization of temperature-time degradation of multiple mycotoxins." *Food and Chemical Toxicology* 125:289-304. doi: https://doi.org/10.1016/j.fct.2019.01.009.

Goodman, Major M., and Walton C. Galinat. 1988. "The history and evolution of Maize." *Critical Reviews in Plant Sciences* 7 (3):197-220. doi: https://doi.org/10.1080/07352688809382264.

Gu, Yong Qiang, Humphrey Wanjugi, Devin Coleman-Derr, Xiuying Kong, and Olin D. Anderson. 2010. "Conserved globulin gene across eight grass genomes identify fundamental units of the loci encoding seed storage proteins." *Functional & Integrative Genomics* 10 (1):111-122. doi: https://doi.org/10.1007/s10142-009-0135-x.

Gunstone, F.D. 2005. "Vegetable oils." In *Bailey's Industrial Oil and Fat Products*, edited by F. Shahidi. New Jersey: John Wiley & Sons.

Guo, Bing, Yongqiang Zhang, Shili Li, Ting Lai, Liang Yang, Juanni Chen, and Wei Ding. 2016. "Extract from Maize (*Zea mays* L.): Antibacterial Activity of DIMBOA and Its Derivatives against Ralstonia solanacearum." *Molecules* 21 (10):1397. doi: https://doi.org/10.3390/molecules21101397.

Guo, Yiting, Ke Wang, Bengang Wu, Ping Wu, Yuqing Duan, and Haile Ma. 2020. "Production of ACE inhibitory peptides from corn germ meal by an enzymatic membrane reactor with a novel gradient diafiltration feeding working-mode and in vivo evaluation of antihypertensive effect." *Journal of Functional Foods* 64:103584. doi: https://doi.org/10.1016/j.jff.2019.103584.

Gutiérrez-Uribe, Janet A., Carlos Rojas-García, Silverio García-Lara, and Sergio O. Serna-Saldivar. 2010. "Phytochemical analysis of wastewater (nejayote) obtained after lime-cooking of different types of maize kernels processed into masa for tortillas." *Journal of Cereal Science* 52 (3):410-416. doi: https://doi.org/10.1016/j.jcs.2010.07.003.

Gwirtz, Jeffrey A., and Maria Nieves Garcia-Casal. 2014. "Processing maize flour and corn meal food products." *Annals of the New York Academy of Sciences* 1312 (1):66-75. doi: https://doi.org/10.1111/nyas.12299.

Győri, Zoltán. 2017. "Chapter 11 - Corn: Grain-Quality Characteristics and Management of Quality Requirements." In *Cereal Grains* (Second Edition), edited by Colin Wrigley, Ian Batey and Diane Miskelly, 257-290. Woodhead Publishing.

Hamaker, Bruce R., Yunus E. Tuncil, and Xinyu Shen. 2019. "Chapter 11 - Carbohydrates of the Kernel." In *Corn* (Third Edition), edited by Sergio O. Serna-Saldivar, 305-318. Oxford: AACC International Press.

Harrabi, Saoussem, Sadok Boukhchina, Paul M. Mayer, and Habib Kallel. 2009. "Policosanol distribution and accumulation in developing corn kernels." *Food Chemistry* 115 (3):918-923. doi: https://doi.org/10.1016/j.foodchem.2008.12.098.

Harrabi, Saoussem, Annik St-Amand, Faouzi Sakouhi, Khaled Sebei, Habib Kallel, Paul M. Mayer, and Sadok Boukhchina. 2008. "Phytostanols and phytosterols distributions in corn kernel." *Food Chemistry* 111 (1):115-120. doi: https://doi.org/10.1016/j.foodchem.2008.03.044.

Herrera-Sotero, Mónica Y., Carlos D. Cruz-Hernández, Carolina Trujillo-Carretero, Mauricio Rodríguez-Dorantes, Hugo S. García-Galindo, José L. Chávez-Servia, Rosa M. Oliart-Ros, and Rosa I. Guzmán-Gerónimo. 2017. "Antioxidant and antiproliferative activity of blue corn and tortilla from native maize." *Chemistry Central Journal* 11 (1):110. doi: https://doi.org/10.1186/s13065-017-0341-x.

Herrero, Miguel, María Castro-Puyana, Elena Ibáñez, and Alejandro Cifuentes. 2017. "Chapter 12 - Compositional analysis of foods." In *Liquid Chromatography* (Second

Edition*)*, edited by Salvatore Fanali, Paul R. Haddad, Colin F. Poole and Marja-Liisa Riekkola, 359-380. Elsevier.

Hossain, Ashrafi, and A. Jayadeep. 2018. "Determination of tocopherol and tocotrienol contents in maize by in vitro digestion and chemical methods." *Journal of Cereal Science* 83:90-95. doi: https://doi.org/10.1016/j.jcs.2018.08.003.

Jackson, Lauren S., Sarath K. Katta, David D. Fingerhut, Jonathan W. DeVries, and Lloyd B. Bullerman. 1997. "Effects of Baking and Frying on the Fumonisin B1 Content of Corn-Based Foods." *Journal of Agricultural and Food Chemistry* 45 (12):4800-4805. doi: https://doi.org/10.1021/jf970405b.

Jaworski, N.W., H.N. Lærke, K.E. Bach Knudsen, and H.H. Stein. 2015. "Carbohydrate composition and in vitro digestibility of dry matter and nonstarch polysaccharides in corn, sorghum, and wheat and coproducts from these grains." *Journal of Animal Science* 93 (3):1103-1113. doi: https://doi.org/10.2527/jas.2014-8147.

Karlovsky, Petr, Michele Suman, Franz Berthiller, Johan De Meester, Gerhard Eisenbrand, Irène Perrin, Isabelle P. Oswald, Gerrit Speijers, Alessandro Chiodini, Tobias Recker, and Pierre Dussort. 2016. "Impact of food processing and detoxification treatments on mycotoxin contamination." *Mycotoxin Research* 32 (4):179-205. doi: https://doi.org/10.1007/s12550-016-0257-7.

Kiesselbach, T.A. 1949. "The structure and reproduction of corn " *Bulletin of the Agricultural Experiment Station of Nebraska* No. 161.

Kriz, Alan L. 1999. "7S Globulins of Cereals." In *Seed Proteins*, edited by Peter R. Shewry and Rod Casey, 477-498. Dordrecht: Springer Netherlands.

Kurilich, Anne C., and John A. Juvik. 1999. "Quantification of Carotenoid and Tocopherol Antioxidants in *Zea mays*." *Journal of Agricultural and Food Chemistry* 47 (5):1948-1955. doi: https://doi.org/10.1021/jf981029d.

Lao, Fei, and M. Monica Giusti. 2016. "Quantification of Purple Corn (*Zea mays* L.) Anthocyanins Using Spectrophotometric and HPLC Approaches: Method Comparison and Correlation." *Food Analytical Methods* 9 (5):1367-1380. doi: https://doi.org/10.1007/s12161-015-0318-0.

Larkins, Brian A. 2019. "Chapter 12 - Proteins of the Kernel." In *Corn* (Third Edition), edited by Sergio O. Serna-Saldivar, 319-336. Oxford: AACC International Press.

Lebron, C.I., R.A. Molins, H.W. Walker, A.A. Kraft, and H.M. Stahr. 1989. "Inhibition of mold growth and mycotoxin production in high-moisture corn treated with phosphates." *Journal of food protection* 52 (5):329-336. doi: https://doi.org/10.4315/0362-028X-52.5.329.

Li, C.C., Y.C. Lee, H.Y. Lo, Y.W. Huang, C.Y. Hsiang, and T.Y. Ho. 2019. "Antihypertensive Effects of Corn Silk Extract and Its Novel Bioactive Constituent in Spontaneously Hypertensive Rats: The Involvement of Angiotensin-Converting Enzyme Inhibition." *Molecules* 24 (10). doi: https://doi.org/10.3390/molecules24101886.

Li, Qian, Pavel Somavat, Vijay Singh, Laura Chatham, and Elvira Gonzalez de Mejia. 2017. "A comparative study of anthocyanin distribution in purple and blue corn coproducts from three conventional fractionation processes." *Food Chemistry* 231:332-339. doi: https://doi.org/10.1016/j.foodchem.2017.03.146.

Listman, Mike, and Rodrigo Ordóñez. 2019. *Ten things you should know about maize and wheat.* Accessed March 27.

Luna-Vital, Diego, Iván Luzardo-Ocampo, M. Liceth Cuellar-Nuñez, Guadalupe Loarca-Piña, and Elvira Gonzalez de Mejia. 2020. "Maize extract rich in ferulic acid and anthocyanins prevents high-fat-induced obesity in mice by modulating SIRT1, AMPK and IL-6 associated metabolic and inflammatory pathways." *The Journal of Nutritional Biochemistry* 79:108343. doi: https://doi.org/10.1016/j.jnutbio.2020.108343.

Luo, Xiaohu, Ren Wang, Li Wang, Yongfu Li, Yong Wang, and Zhengxing Chen. 2014. "Detoxification of aflatoxin in corn flour by ozone." *Journal of the Science of Food and Agriculture* 94 (11):2253-2258. doi: https://doi.org/10.1002/jsfa.6550.

Mbundi, Lubinda, Hector Gallar-Ayala, Mohammad Rizwan Khan, Jonathan L. Barber, Sara Losada, and Rosa Busquets. 2014. "Chapter Two - Advances in the Analysis of Challenging Food Contaminants: Nanoparticles, Bisphenols, Mycotoxins, and Brominated Flame Retardants." In *Advances in Molecular Toxicology*, edited by James C. Fishbein and Jacqueline M. Heilman, 35-105. Elsevier.

Miller, J. David. 1995. "Fungi and mycotoxins in grain: Implications for stored product research." *Journal of Stored Products Research* 31 (1):1-16. doi: https://doi.org/10.1016/0022-474X(94)00039-V.

Mokoena, Mduduzi P, Paul K Chelule, and Nceba Gqaleni. 2005. "Reduction of fumonisin B1 and zearalenone by lactic acid bacteria in fermented maize meal." *Journal of food protection* 68 (10):2095-2099. doi: https://doi.org/10.4315/0362-028X-68.10.2095.

Mora-Rochín, Saraid, Nalleli Gaxiola-Cuevas, Janet Alejandra Gutiérrez-Uribe, Jorge Milán-Carrillo, Evelia María Milán-Noris, Cuauhtémoc Reyes-Moreno, Sergio Othon Serna-Saldivar, and Edith Oliva Cuevas-Rodríguez. 2016. "Effect of traditional nixtamalization on anthocyanin content and profile in Mexican blue maize (*Zea mays* L.) landraces." *LWT - Food Science and Technology* 68:563-569. doi: https://doi.org/10.1016/j.lwt.2016.01.009.

Moreau, R.A. 2005. "Corn Oil." In *Bailey's Industrial Oil and Fat Products*, edited by F. Shahidi. New Jersey: John Wiley & Sons.

Moreau, R.A., V. Singh, A. Nunez, and K.B. Hicks. 2000. "Phytosterols in the aleurone layer of corn kernels." *Biochemical Society Transactions* 28 (6):803-6.

Moreau, Robert A. 2011. "Corn Oil." In *Vegetable Oils in Food Technology*, edited by F.D. Gunstone, 273-289. Oxford: Wiley Blackwell.

Morshed, Selim, and S.M. Islam. 2015. "Antimicrobial Activity and Phytochemical Properties of Corn (*Zea mays* L.) Silk." *SKUAST Journal of Research* 17:8-14.

Munkvold, Gary P., Silvina Arias, Ines Taschl, and Christiane Gruber-Dorninger. 2019. "Chapter 9 - Mycotoxins in Corn: Occurrence, Impacts, and Management." In *Corn* (Third Edition), edited by Sergio O. Serna-Saldivar, 235-287. Oxford: AACC International Press.

Ndolo, Victoria U., and Trust Beta. 2014. "Comparative Studies on Composition and Distribution of Phenolic Acids in Cereal Grain Botanical Fractions." *Cereal Chemistry* 91 (5):522-530. doi: https://doi.org/10.1094/CCHEM-10-13-0225-R.

Nessa, Fazilatun, Zhari Ismail, and Nornisah Mohamed. 2012. "Antimicrobial activities of extracts and flavonoid glycosides of corn silk (*Zea mays* L)." *International Journal of Biotechnology for Wellness Industries* 1 (2):115-120.

Nile, Shivraj H., and Se W. Park. 2014. "Antioxidant, α-Glucosidase and Xanthine Oxidase Inhibitory Activity of Bioactive Compounds From Maize (*Zea mays* L.)." *Chemical Biology & Drug Design* 83 (1):119-125. doi: https://doi.org/10.1111/cbdd.12205.

Noureddini, H., M. Malik, J. Byun, and A.J. Ankeny. 2009. "Distribution of phosphorus compounds in corn processing." *Bioresource Technology* 100 (2):731-736. doi: https://doi.org/10.1016/j.biortech.2008.05.056.

Nuss, Emily T., and Sherry A. Tanumihardjo. 2010. "Maize: A Paramount Staple Crop in the Context of Global Nutrition." *Comprehensive Reviews in Food Science and Food Safety* 9 (4):417-436. doi: https://doi.org/10.1111/j.1541-4337.2010.00117.x.

Ortiz-Martinez, Margarita, Elvira Gonzalez de Mejia, Silverio García-Lara, Oscar Aguilar, L. Margarita Lopez-Castillo, and José T. Otero-Pappatheodorou. 2017. "Antiproliferative effect of peptide fractions isolated from a quality protein maize, a white hybrid maize, and their derived peptides on hepatocarcinoma human HepG2 cells." *Journal of Functional Foods* 34:36-48. doi: https://doi.org/10.1016/j.jff.2017.04.015.

Ortiz-Martinez, Margarita, José T. Otero-Pappatheodorou, Sergio O. Serna-Saldívar, and Silverio García-Lara. 2017. "Antioxidant activity and characterization of protein fractions and hydrolysates from normal and quality protein maize kernels." *Journal of Cereal Science* 76:85-91. doi: https://doi.org/10.1016/j.jcs.2017.05.021.

Paliwal, R.L., G. Granados, H.R. Lafitte, A.D. Violic, and J.P. Marathrée. 2000. *Tropical maize improvement and production*. Edited by R.L. Paliwal, G. Granados, H.R. Lafitte and A.D. Violic, *Plant Production and Protection Div. eng Granados*. Rome, Italy: FAO.

Park, D.L., L.S. Lee, R.L. Price, and A.E. Pohland. 1988. "Review of the decontamination of aflatoxins by ammoniation: current status and regulation." *Journal - Association of Official Analytical Chemists* 71 (4):685-703.

Pinchen, Hannah, and Paul Finglas. 2019. "Water-Soluble Vitamins." In *Encyclopedia of Food Chemistry*, edited by Laurence Melton, Fereidoon Shahidi and Peter Varelis, 305-311. Oxford: Academic Press.

Ponce-García, Nestor, Sergio O. Serna-Saldivar, and Silverio Garcia-Lara. 2018. "Fumonisins and their analogues in contaminated corn and its processed foods – a review." *Food Additives & Contaminants: Part A* 35 (11):2183-2203. doi: https://doi.org/10.1080/19440049.2018.1502476.

Prasanna, B., Shalley Vasal, Kassahun Bantte, and N. Singh. 2001. "Quality Protein Maize." *Current Science* 81 (10):25-2001.

Puchalska, Patrycja, M. Luisa Marina, and M. Concepción García. 2013. "Development of a high-performance liquid chromatography–electrospray ionization-quadrupole-time-of-flight-mass spectrometry methodology for the determination of three highly antihypertensive peptides in maize crops." *Journal of Chromatography A* 1285:69-77. doi: https://doi.org/10.1016/j.chroma.2013.02.015.

Puchalska, Patrycja, M. Luisa Marina, and M. Concepción García. 2012. "Development of a reversed-phase high-performance liquid chromatography analytical methodology for the determination of antihypertensive peptides in maize crops." *Journal of Chromatography A* 1234:64-71. doi: https://doi.org/10.1016/j.chroma.2011.12.079.

Raboy, Victor, Paola F. Gerbasi, Kevin A. Young, Sierra D. Stoneberg, Suewiya G. Pickett, Andrew T. Bauman, Pushpalatha P.N. Murthy, William F. Sheridan, and David S. Ertl. 2000. "Origin and Seed Phenotype of Maize *low phytic acid 1-1* and *low phytic acid 2-1*." *Plant Physiology* 124 (1):355-368. doi: https://doi.org/10.1104/pp.124.1.355.

Ramírez, Karina, María F. Quintero-Soto, and Jesús J. Rochín-Medina. 2020. "Enhancement of the antioxidant and antimicrobial activities of maize wastewater by an eco-friendly process." *Journal of Food Measurement and Characterization*. doi: https://doi.org/10.1007/s11694-020-00416-1.

Riley, R.T., K.A. Voss, R.A. Coulombe, J.J. Pestka, and D.E. Williams. 2011. "9 - Developing mechanism-based and exposure biomarkers for mycotoxins in animals." In *Determining Mycotoxins and Mycotoxigenic Fungi in Food and Feed*, edited by Sarah De Saeger, 245-275. Woodhead Publishing.

Robbins, Rebecca J. 2003. "Phenolic Acids in Foods: An Overview of Analytical Methodology." *Journal of Agricultural and Food Chemistry* 51 (10):2866-2887. doi: https://doi.org/10.1021/jf026182t.

Rouf Shah, Tajamul, Kamlesh Prasad, and Pradyuman Kumar. 2016. "Maize—A potential source of human nutrition and health: A review." *Cogent Food & Agriculture* 2 (1):1166995. doi: https://doi.org/10.1080/23311932.2016.1166995.

Salinas-Moreno, Yolanda, F. Martinez-Bustos, Marcos Soto-Hernandez, Rafael Paczka, and José Arellano-Vázquez. 2003. "Effect of alkaline cooking process on anthocyanins in pigmented maize grain." *Agrociencia* 37:617-628.

Salinas-Moreno, Yolanda, José J. Pérez-Alonso, Gricelda Vázquez-Carrillo, Flavio Aragón-Cuevas, and Gustavo A. Velázquez-Cardelas. 2012. "Antocianinas y actividad antioxidante en maíces (*Zea mays* L.) de las razas chalqueño, elotes cónicos y bolita. [Anthocyanins and antioxidant activity in maize grains (*Zea mays* L.) of chalqueño, elotes cónicos and bolita races]." *Agrociencia* 46:693-706.

Sanchez G., J.J., M.M. Goodman, and C.W. Stuber. 2000. "Isozymatic and Morphological Diversity in the Races of Maize of Mexico." *Economic Botany* 54 (1):43-59. doi: https://doi.org/10.1007/BF02866599.

Scott, Paul, Richard C. Pratt, Nicholas Hoffman, and Randall Montgomery. 2019. "Chapter 10 - Specialty Corns." In *Corn* (Third Edition), edited by Sergio O. Serna-Saldivar, 289-303. Oxford: AACC International Press.

Schaarschmidt, S., and C. Fauhl-Hassek. 2019. "Mycotoxins during the Processes of Nixtamalization and Tortilla Production." *Toxins (Basel)* 11 (4). doi: https://doi.org/10.3390/toxins11040227.

Schaarschmidt, Sara, and Carsten Fauhl-Hassek. 2018. "The Fate of Mycotoxins During the Processing of Wheat for Human Consumption." *Comprehensive Reviews in Food Science and Food Safety* 17 (3):556-593. doi: https://doi.org/10.1111/1541-4337.12338.

Serna-Saldivar, Sergio O. 2010. *Cereal Grains: Properties, Processing, and Nutritional Attributes*. Edited by CRC Press, *Food Preservation Technologies Series*. United States: CRC Press, Taylor & Francis Group, LLC.

Serna-Saldivar, Sergio O., and Esther Perez Carrillo. 2019. "Chapter 16 - Food Uses of Whole Corn and Dry-Milled Fractions." In *Corn* (Third Edition), edited by Sergio O. Serna-Saldivar, 435-467. Oxford: AACC International Press.

Shen, Junjun, Feijun Luo, and Qinlu Lin. 2019. "Policosanol: Extraction and biological functions." *Journal of Functional Foods* 57:351-360. doi: https://doi.org/10.1016/j.jff.2019.04.024.

Shi, Shihua, Siming Li, Weihao Li, and Hao Xu. 2019. "Corn Silk Tea for Hypertension: A Systematic Review and Meta-Analysis of Randomized Controlled Trials." *Evidence-Based Complementary and Alternative Medicine* 2019:2915498. doi: https://doi.org/10.1155/2019/2915498.

Sokrab, Awad M., Isam A. Mohamed Ahmed, and Elfadil E. Babiker. 2012. "Effect of germination on antinutritional factors, total, and extractable minerals of high and low phytate corn (*Zea mays* L.) genotypes." *Journal of the Saudi Society of Agricultural Sciences* 11 (2):123-128. doi: https://doi.org/10.1016/j.jssas.2012.02.002.

Srinivasan, Radhakrishnan, Vijay Singh, Ronald L. Belyea, Kent D. Rausch, Robert A. Moreau, and M.E. Tumbleson. 2006. "Economics of Fiber Separation from Distillers Dried Grains with Solubles (DDGS) Using Sieving and Elutriation." *Cereal Chemistry* 83 (4):324-330. doi: https://doi.org/10.1094/CC-83-0324.

Stoloff, L., and M.W. Trucksess. 1981. "Effect of boiling, frying, and baking on recovery of aflatoxin from naturally contaminated corn grits or cornmeal." *Journal - Association of Official Analytical Chemists* 64 (3):678-80.

Sun, Xiang Dong, Ping Su, and Hong Shan. 2017. "Mycotoxin Contamination of Maize in China." *Comprehensive Reviews in Food Science and Food Safety* 16 (5):835-849. doi: https://doi.org/10.1111/1541-4337.12286.

Suri, Devika J., and Sherry A. Tanumihardjo. 2016. "Effects of Different Processing Methods on the Micronutrient and Phytochemical Contents of Maize: From A to Z." *Comprehensive Reviews in Food Science and Food Safety* 15 (5):912-926. doi: https://doi.org/10.1111/1541-4337.12216.

Sydenham, E.W., L. Van der Westhuizen, S. Stockenstrom, G.S. Shephard, and P.G. Thiel. 1994. "Fumonisin-contaminated maize: physical treatment for the partial decontamination of bulk shipments." *Food Addit Contam* 11 (1):25-32. doi: https://doi.org/10.1080/02652039409374199.

Syeda Juveriya, Fathima, Nallamuthu Ilaiyaraja, and Khanum Farhath. 2017. "12 - Vitamins and minerals fortification using nanotechnology: bioavailability and Recommended Daily Allowances." In *Nutrient Delivery*, edited by Alexandru Mihai Grumezescu, 457-496. Academic Press.

Taleon, Víctor, Luke Mugode, Luisa Cabrera-Soto, and Natalia Palacios-Rojas. 2017. "Carotenoid retention in biofortified maize using different post-harvest storage and packaging methods." *Food Chemistry* 232:60-66. doi: https://doi.org/10.1016/j.foodchem.2017.03.158.

Tanaka, Hiroyuki, Masao Onishi, and Yasuhiko Fujino. 1984. "Glycolipids in corn seeds." *Journal of the Japanese Society of Agricultural Chemistry* 58:17-24. doi: https://doi.org/10.1271/nogeikagaku1924.58.17.

Urias-Peraldí, Marcos, Janet A. Gutiérrez-Uribe, Ricardo E. Preciado-Ortiz, Andrea S. Cruz-Morales, Sergio O. Serna-Saldívar, and Silverio García-Lara. 2013. "Nutraceutical profiles of improved blue maize (*Zea mays*) hybrids for subtropical regions." *Field Crops Research* 141:69-76. doi: https://doi.org/10.1016/j.fcr.2012.11.008.

Valderrama-Bravo, C., E. Gutiérrez-Cortez, M. Contreras-Padilla, I. Rojas-Molina, J.C. Mosquera, A. Rojas-Molina, F. Beristain, and M.E. Rodríguez-García. 2012. "Constant pressure filtration of lime water (nejayote) used to cook kernels in maize processing." *Journal of Food Engineering* 110 (3):478-486. doi: https://doi.org/10.1016/j.jfoodeng.2011.12.018.

Vázquez-Carrillo, María Gricelda, David Santiago-Ramos, and Juan de Dios Figueroa-Cárdenas. 2019. "Kernel properties and popping potential of Chapalote, a Mexican ancient native maize." *Journal of Cereal Science* 86:69-76. doi: https://doi.org/10.1016/j.jcs.2019.01.010.

Wang, Tong, and Pamela J. White. 2019. "Chapter 13 - Lipids of the Kernel." In *Corn* (Third Edition), edited by Sergio O. Serna-Saldivar, 337-368. Oxford: AACC International Press.

Watson, S.A. 2003. "Description, development, structure, and composition of the corn kernel." In *Corn: chemistry and technology* edited by P.J. White and L. Johnson, 69-106. St. Paul, MN: American Association of Cereal Chemists.

White, Philip J., and Martin R. Broadley. 2005. "Biofortifying crops with essential mineral elements." *Trends in Plant Science* 10 (12):586-593. doi: https://doi.org/10.1016/j.tplants.2005.10.001.

Winkler-Moser, Jill K., and Lotta Breyer. 2011. "Composition and oxidative stability of crude oil extracts of corn germ and distillers grains." *Industrial Crops and Products* 33 (3):572-578. doi: https://doi.org/10.1016/j.indcrop.2010.12.013.

Woo, Y.M., D.W. Hu, B.A. Larkins, and R. Jung. 2001. "Genomics analysis of genes expressed in maize endosperm identifies novel seed proteins and clarifies patterns of zein gene expression." *Plant Cell* 13 (10):2297-317. doi: https://doi.org/10.1105/tpc.010240.

Yamagata, T., H. Kato, S. Kuroda, S. Abe, and E. Davies. 2003. "Uncleaved legumin in developing maize endosperm: identification, accumulation and putative subcellular localization." *Journal of Experimental Botany* 54 (384):913-22. doi: https://doi.org/10.1093/jxb/erg090.

Young, J. Christopher, H. Locksley Trenholm, Douglas W. Friend, and Daniel B. Prelusky. 1987. "Detoxification of deoxynivalenol with sodium bisulfite and evaluation of the effects when pure mycotoxin or contaminated corn was treated and given to pigs." *Journal of Agricultural and Food Chemistry* 35 (2):259-261. doi: https://doi.org/10.1021/jf00074a023.

Yu, Xurun, Heng Yu, Jing Zhang, Shanshan Shao, Fei Xiong, and Zhong Wang. 2015. "Endosperm Structure and Physicochemical Properties of Starches from Normal, Waxy, and Super-Sweet Maize." *International Journal of Food Properties* 18 (12):2825-2839. doi: https://doi.org/10.1080/10942912.2015.1015732.

Zavala-López, Mariana, Esteban López-Tavera, Juan de Dios Figueroa-Cárdenas, Sergio O. Serna-Saldívar, and Silverio García-Lara. 2018. "Screening of major phenolics and antioxidant activities in teosinte populations and modern maize types." *Journal of Cereal Science* 79:276-285. doi: https://doi.org/10.1016/j.jcs.2017.11.007.

Zhang, Qiaozhi, Elvira Gonzalez de Mejia, Diego Luna-Vital, Tianyi Tao, Subhiksha Chandrasekaran, Laura Chatham, John Juvik, Vijay Singh, and Deepak Kumar. 2019. "Relationship of phenolic composition of selected purple maize (*Zea mays* L.) genotypes with their anti-inflammatory, anti-adipogenic and anti-diabetic potential." *Food Chemistry* 289:739-750. doi: https://doi.org/10.1016/j.foodchem.2019.03.116.

Zhang, Qiaozhi, Diego Luna-Vital, and Elvira Gonzalez de Mejia. 2019. "Anthocyanins from colored maize ameliorated the inflammatory paracrine interplay between macrophages and adipocytes through regulation of NF-κB and JNK-dependent MAPK pathways." *Journal of Functional Foods* 54:175-186. doi: https://doi.org/10.1016/j.jff.2019.01.016.

In: Super and Nutraceutical Foods
Editors: Sandra Teresita Martín-del-Campo et al.
ISBN: 978-1-53619-082-3

Chapter 10

QUINOA GRAINS PROCESSING: EFFECTS ON ITS PHYSICOCHEMICAL PROPERTIES

Nestor Ponce-García[1,*], Anayansi Escalante-Aburto[2,3], David Santiago-Ramos[4] and Rosa María Mariscal-Moreno[4,5,†]

[1]UAEMex Campus Universitario "El Cerrillo,"
Facultad de Ciencias Agrícolas, Toluca Edo. de México
[2]Universidad de Monterrey. Departamento de Nutrición,
Escuela de Ciencias de la Salud. San Pedro Garza García, NL
[3]Tecnologico de Monterrey, Escuela de Ingeniería y Ciencias, Monterrey, N.L., México
[4]INIFAP, Campo Experimental Valle de México, Coatlinchán, Texcoco, Edo. México
[4]Universidad Politécnica de Francisco I. Madero,
Departamento de Ingeniería Agroindustrial, Francisco I Madero, Hidalgo
[5]Universidad Iberoamericana. Departamento de Salud, Ciudad de México

ABSTRACT

Recently there is an interest in the production of foods with high nutritional quality because there is a global requirement for more efficient foods that help to decrease malnutrition and nutrient problems in general. Quinoa is a pseudocereal that possesses excellent nutritional characteristics. This grain contains a high percentage of protein, and the main protein fractions are albumin and globulins. Besides, it has a unique amino acid content that balances essential amino acids. Quinoa is an excellent source of dietary fiber; it contains almost 10-16%.

Additionally, quinoa contains a high concentration of minerals and vitamins. However, quinoa contains some anti-nutritional compounds such as saponins removed by washing, soaking, and boiling. Quinoa is usually submitted to different processes, mainly to increase palatability and sensory acceptance of quinoa. Methods applied for quinoa are milling, cooking, extrusion, roasted, microwave heating. Those processes usually change nutritional and physical quinoa characteristics. The purpose of this chapter is to describe

* Corresponding Author's E-mail: nponceg@uaemex.mx.
† Corresponding Author's E-mail: mariscal.rosa@gmail.com.

quinoa characteristics and explain how different methods affect nutritional quinoa properties.

Keywords: quinoa, thermal processing, physical processing, physicochemical properties, pseudo cereals.

INTRODUCTION

Quinoa (*Chenopodium quinoa* Willd) is a pseudo cereal botanically described for the first time by Willdenow in 1778 (FAO, 2011). The origin center is the Andean Region of Bolivia and Peru, where it has been cultivated for about 7000 years (Bhargava and Srivastava, 2013), mainly for the Incas. Due to its high climate and soil adaptability (Jacobsen, Mujica, and Jensen 2003), nutritional quality, genetic variability, and relatively low production cost, this crop recently has been tested and introduced in diverse climatic regions and countries around the world (Stikic et al. 2012; Jacobsen, Mujica, and Jensen, 2003; FAO, 2011). As a result of its nutritional benefits, quinoa is considered by FAO as one of the crops that can contribute to food security in the 21st century (Jacobsen, Mujica, and Jensen, 2003). The year 2013 was declared by the United Nations, the International year of quinoa (FAO, 2013).

SEED MORPHOLOGY AND NUTRITIONAL COMPONENTS

Mature Seed Morphology

The quinoa fruit is an indehiscent achene derived from a superior, unilocular ovary (Di Fabio and Parraga, 2017) covered by the perigonium, which can be removed (in a dry state). The pericarp (hull) is the outer layer of the fruit and includes one seed (Jacobsen and Stølen, 1993). The achene produces small (1.5 to 2.5 mm in diameter) and cylindrical-lenticular in shape seeds. The seeds correspond to the campylotrophus type (Valencia-Chamorro 2003; Di Fabio and Parraga, 2017), in which colors varied from white-grey-black or even yellow-red (Nowak, Du, and Charrondière, 2016). The fiber content in quinoa seed is high, and it is composed mainly of the seed coat (bran), which in turn surrounds the starchy-perisperm and a campylotropus embryo (Reguera and Haros, 2017).

In mature quinoa seeds, food reserves have three distinct segments: central perisperm, a peripheral embryo (formed by two cotyledons), and a micropylar endosperm region of one to two cell-layered tissues surrounding the hypocotyl–radicle axis of the embryo (Figure 1) (Prego, Maldonado, and Otegui, 1998). Pseudo cereal grains (quinoa, amaranth, and buckwheat) have shorter sizes as well as different physical and chemical properties compared with some cereal grains (e.g., wheat, corn) (Reguera and Haros, 2017).

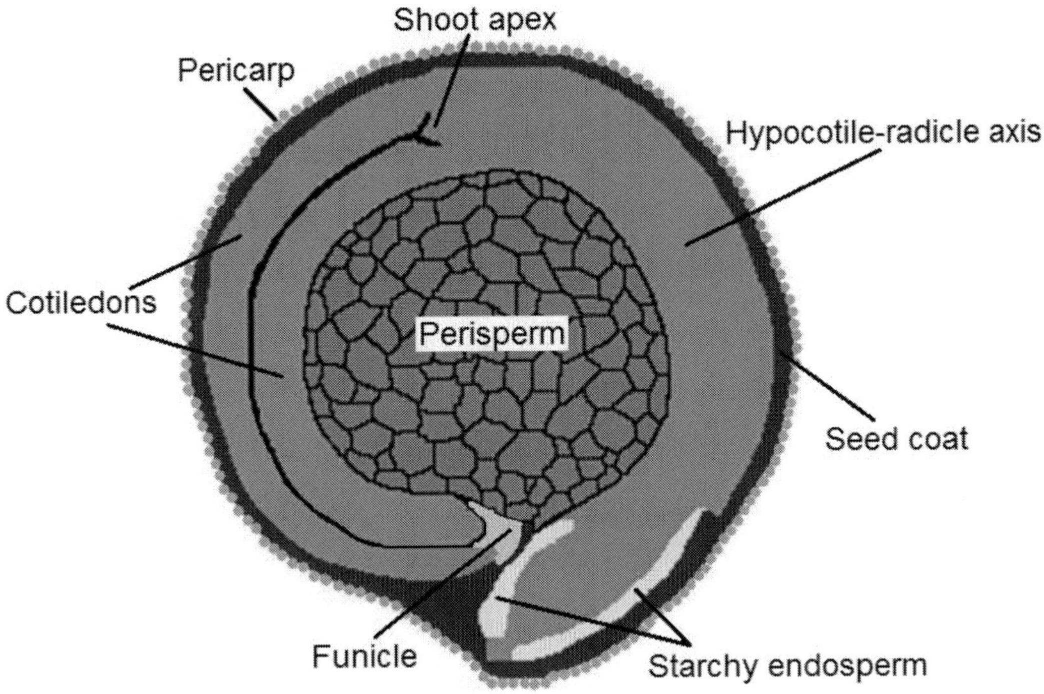

Figure 1. Morphology and structure of mature quinoa seed.

Nutritional Properties

According to Burrieza, López-Fernández, & Maldonado (2014), in a mature quinoa seed, the perisperm stores mainly starch and low sugar content. The embryo contains proteins, and lipids, while the micropylar cone (non-embryonic tissue) is a reservoir of proteins, lipids (fatty acids), and minerals. Quinoa seeds possess high protein content due to the embryo (about 60% of weight seed) (Valencia-Chamorro, 2003), which forms a ring around the endosperm that it is lost when the grain is cooked.

The main protein fractions include albumins and globulins (44–77% of the total protein) (Guerreo-Ochoa, Pedreschi, and Chirinos, 2015). Additionally, quinoa protein has an excellent balanced amino acid content with high levels of lysine, threonine, and methionine that are deficient in cereals (Stikic et al. 2012; Nowak, Du, and Charrondière, 2016). Quinoa grains have shown a low glycemic index when consumed; and also contains vitamins (B complex, E and C), trace elements (Ca, Mg, Fe, K, P, Mn, Zn, Cu, and Na) as well as other beneficial bioactive compounds (ferulic acid, quercetin, and anthocyanins) (Gordillo-Bastidas et al. 2016; Maradini-Filho, 2017). The range of dietary fiber in some varieties of quinoa is 10-16 g/100 g, and cooked quinoa consists of mostly insoluble fibers (80-90%) that act in the intestinal level as prebiotics (Singh, Singh, and Vikash Sing, 2016).

Anti-Nutritional Factors

Despite the excellent nutritional value than quinoa provides, its limited consumption is due to the presence of anti-nutritional or toxic compounds such as saponins (a water-soluble phytonutrient), which are mainly present in the pericarp (hull) and infer a bitter taste. Nowak, Du, and Charrondière (2016) point out the importance of dehulled/polished the seeds to reduce or remove the saponins content. The reduction of these compounds can also be achieved by washing, soaking, and boiling of quinoa grains (Singh, Singh, and Vikash Sing, 2016). Saponins content in quinoa is about 557 mg/100 g DW (dry weight), but after abrasive milling/pearling, its content is reduced up to 36 mg/100 g DW (Mattila et al. 2018).

Use of Quinoa in Foods and Beverages

Products contained quinoa seeds and flours are considered highly consumed and have an economic potential because they make use of its nutritional and physicochemical characteristics, reaching beyond the food industry and offering products to the chemical, cosmetic, and pharmaceutical industries (FAO, 2011).

The use of the seed is to make different food products, including flour, pasta (spaghetti and tagliatelle), breakfast cereal, bread (sourdough and non-sourdough), Chinese steamed bread, biscuits, cookies, crepes, muffins, pancakes, tortillas, popped like popcorn, sprouts, noodles, and even alcohol and edible films (Sánchez-Reséndiz et al. 2019; Haros and Wronkowska, 2017; Serna-Saldívar, 2010). More recently, attention has been given to quinoa for people with celiac disease (allergy to gluten), as an alternative to the consumption of certain cereals such as wheat, rye, and barley (Jacobsen, Mujica, and Jensen, 2003; Stikic et al. 2012). To obtain all that diversity of products, quinoa seeds components undergone several processing steps that change the properties of the final products.

Physical and Thermal Processing

During the protein extraction procedure to obtain defatted flours, the solubilization of carbohydrates to purchase concentrated products occurs. Besides, critical structural changes results (Sánchez-Reséndiz et al. 2019). Mainly, protein functionality and digestibility have been the most issues studied to improve the quinoa applications in foods.

Milling (Flour Production)

With the increase of the interest in grain diversification, quinoa is usually processed by milling to be used as raw material to produce several foods. Milling processes are classified as dry- or wet-milling. In dry milling, the anatomical part of the grain is separated to make mainly flour (Haros and Wronkowska, 2017). Dry milling of cereal grains is one of the oldest industries recorded, and the objective was to grind cereals into a more palatable, digestible, and convenient meal. The resulting fractions have particular particle-size distribution and particle-size index (granulation) and long shelf life due to their low moisture and fat contents due to the removal of the germ (Serna-Saldívar, 2010).

On the other hand, the purpose of wet-milling is to separate the chemical components of the grain (starch, proteins, fiber, and lipids) to obtain the pure fractions (Haros and Wronkowska, 2017). This process results in different types of milled fractions, such as bran (pericarp) and milled grain, both with different functionalities and properties. Although milling of quinoa improves nutritional properties by removing the antinutrients, it might also decrease the content of bioactive molecules in milled products (Hemalatha et al. 2016). The main objective of the starch milling industry is to obtain the highest yield of non-damaged starch that meets specifications for color, functional properties, and the maximum amount of protein (Serna-Saldivar, 2010).

There is not enough information about the specific effects of milling processing factors related to quinoa properties. There is a need to perform some experiments evaluating the quality of flours by the milling industry. Besides, the study of constituents such as protein, starch content, ashes, and moisture content its important since the grinding products have a specific end-use.

The type of milling and milling factors significantly impacts the functional properties of flours. In this sense, the effects of quinoa seed moisture content and temperature accomplished in a laboratory roller mill have been studied (Hassoon and Dziki, 2018). In summary, it was reported that a directly proportional increase of the specific grinding energy and the average particle size, as the seed moisture content was higher. On the contrary, the seed temperature did not significantly affect the specific grinding energy and had a sight influence on the particle size distribution. All those effects were related to the water-molecule interactions with the bran and endosperm components, causing changes in their mechanical properties.

Quinoa dry-milling is performed to produce flours to be used as a food ingredient. For example, to elaborate snacks. In this case, extruded snacks from pure quinoa flour are not suitable for consumption because of its hardness and unacceptable texture. Some strategies, such as the mixture of flour from different foods, are evaluated to produce snacks. Quinoa flours for this purpose have a median particle size of 575 μm, which is considered coarse when they are processed by a pin disc grinding device products with porosity up to 85.8% can be obtained with 35-50% of quinoa and corn flours mixture (Ramos Diaz et al. 2015).

The use of quinoa flours as an ingredient to produce wheat-based bread was studied by Park, Maeda, and Morita (2005). Two types of whole quinoa seeds, one stone-mortar milled and the other rolled-milled and roasted, were tested. Lipase addition was also evaluated to improve the rheological and breadmaking characteristics of the products. The addition of both flours reduced up to 30% the loaf volume of bread, but the addition of lipase distinctly improved the bread quality. The stone-mortar milling methods are useful to produce milder or more refined flours, this is, to obtain materials with smaller particle sizes than the roller-milling process.

Quinoa has uniform small starch granules, which have potential applications in the industry. In structural terms, quinoa starch showed a type A pattern when it is evaluated by X-ray dispersion, and the scanning electron microscopy shows aggregate starches (García-Salcedo, Torres-Vargas, and Ariza-Calderón, 2017).

Ballester-Sánchez et al. (2019) evaluated the effects of the wet-milling procedure on starch, germ, protein, and fiber components from quinoa. The wet-milling, with a NaOH solution at concentrations of 0.20-0.30%, was evaluated. As a result, they observed significant steeping time and temperature in fraction yields and starch recovery. Damaged starch was higher at longer steeping times (9 h) due to starch hydrolysis. However, starch recovery was higher when compared with other studies from other pseudo cereals (Calzetta Resio, Tolaba, and Suárez, 2009; Wronkowska and Haros, 2014; Haros and Wronkowska, 2017), and the material obtained could be used for a specific application.

Isolation of dietary fiber of red quinoa seeds using wet and dry-milling was assessed by Ballester-Sánchez, Fernández-Espinar, and Haros (2020). For wet milling, a plate mill was used to grind the quinoa seeds that were previously conditioned and steeped in a sodium bisulfite to obtain a concentration of 0.25% and pH 5. For the dry-milling procedure, quinoa seed was conditioned at 16% of moisture content and grinding in a CD1 laboratory mill. They concluded that the wet-milling system results in a higher yield of dietary fiber with lower lipid content. Besides, this process affected the particle size producing high granulometry dispersion in the fiber-rich fractions. More critical, extracts from bran and hull fractions of quinoa have shown remarkable antioxidant capacities.

Quinoa milled fractions such as bran and hulls also have shown functional activities. Potent inhibition of α-amylase with IC_{50} values of 108.6 and 148.2 μg/mL, respectively; and for α-glucosidase IC_{50} values of 62.1 and 68.4 μg/mL, respectively. Thus, quinoa grains and their milled fractions can be used as functional food ingredients in gluten-free products for health benefits (Hemalatha et al. 2016).

To protein enrichment of flour dry-milling approach is used. The continuous growth of the global population is creating a greater need to identify and develop sustainable solutions to provide high-quality food protein. Alonso-Miravalles and O'Mahony (2018) evaluated the nutrient composition, protein profile, morphology, and pasting properties of protein-rich pseudo cereal ingredients, including quinoa. The grains were milled using an impact mill equipped with different screens inserts to obtain flour and seed fragments. After this step, the protein-rich fractions were separated from the milled flours using a size-based dry sieve classification. They reported that protein-rich fractions of quinoa presented high levels of ash, fat, dietary fiber, and lower levels of starch. Besides, globulins and albumin fractions were detected under non-reducing and reducing conditions. These rich-protein fractions can be used as ingredients in the production of the vegetable protein beverages sector.

The development of more sustainable methods to obtain protein-rich fractions from quinoa are under study. A combination of dry and aqueous fractionation compared with wet-conventional fractionation was developed by Avila Ruiz et al. (2016). Quinoa seeds were milled using a 100 UPZ rotor mill with airflow and a built sieve of 2.0 mm. The ground grits were air classified, and the fine fraction was called "non-milled fraction." The generated embryo-rich fraction was milled further to facilitate the separation of proteins. An extra impact milling was applied to obtain protein-rich components and was identified as "milled-fraction." The two fine fractions were fractionated by aqueous phase separation with deionized water and NaCl at different concentrations to be ultra-filtrated. Protein concentrates with a purity of 59% w/DW, and 61% of protein yield were obtained, which similar values to conventional extraction.

The functionality of quinoa fractions obtained under different processing conditions did not guarantee the desired bioavailability in the organism. Therefore, the effect of *in vitro* gastric digestion of quinoa protein isolates obtained by wet and dry fractionation was studied by Opazo-Navarrete et al. (2018). Quinoa flours were obtained by grinding the seed in a laboratory-scale mill, and two fractionation methods were applied. Once again, the wet-milling procedure resulted in protein isolates with higher purity (87.1%) and protein yield of 54%. However, quinoa protein concentrates obtained by dry-milling showed higher digestibility rates than the isolates collected by the wet-milling process; this was explained by the prevention of the formation of large aggregates during pre-heating of the protein isolates.

Cooking

Quinoa is traditionally cooking like other whole-grain cereals (Bhathal, Kaur, and Gill, 2017; Brend et al. 2012). This process often is carried out with a double amount of water for 25-30 min at 100°C. Similar to other cereals and food products, heat treatment decreases the protein amount and increase moisture content (Bhathal, Kaur, and Gill, 2017). The protein digestibility or the bioavailability (true digestibility) of quinoa amino acids varies according to the grain variety and treatment. Digestibility increases considerably with cooking (Filho et al. 2017; Bertazzo et al. 2011; Flores-Silva et al. 2015). The *in vitro* digestibility of quinoa protein (from four different varieties), as studied by Repo-Carrasco-Valencia et al. (2010), varied between 76.3% and 80%. According to Bertazzo et al. (2011), the digestibility of quinoa protein is similar to quality protein-rich foods.

Mahoney, Lopez, and Hendricks (1975) reported the protein efficiency ratio (PER) values for QP (Protein quality) of cooked quinoa was like that of casein. According to these authors, the PER of the cooked quinoa was 30% greater than that of uncooked quinoa. Ranhotra et al. (1993) also concluded that the quality of protein in quinoa equals that of casein. Repo-Carrasco-Valencia et al. (2010) reported a significant decrease in iron content; wet processing procedures in general, cause loss of dry matter and iron. Also, boiling reduced the content of zinc in quinoa. Bhathal, Kaur, and Gill (2017) reported decreased lysine, methionine, tryptophan, and lipids and crude fiber contents.

Brend et al. (2012) reported that cooking increased TPC (total phenolics content) significantly, TFC (total flavonoid content), and FRAP AA (the ferric-reducing ability of plasma antioxidant activity in red-cooked quinoa seeds. In contrast, in yellow quinoa seeds, no significant differences with raw quinoa were reported.

Roasting

Roasting is a typical dry heating process used to cook or gelatinize, expand/pop/puff food materials. Moreover, this process inactivates anti-nutritional factors to obtain more amenable, palatable, and appealing food forms (Bhattacharya, 2014). Roasting alters texture, color, flavor, and appearance, resulting in product development with unique features of crispness and taste, compared with raw kernels (Carciochi, Galván D'Alessandro, and Manrique, 2016). High temperatures used in this process causes chemical and nutritional changes. Bhathal, Kaur, and Gill (2017) reported lower moisture content than other cooking methods. Roasting parameters used are 200°C for 10 min (Brady et al. 2007); 100-190°C for 15-120 min (Carciochi et al. 2016), and 190°C for 3 min (Repo-Carrasco-Valencia et al. 2010).

Previously, roasted quinoa seeds have been evaluated for different purposes. Brady et al. (2007) reported the degradation of saponins in quinoa seeds roasted at 200°C for 10 min.

Total phenolic content and flavonoid compounds increased with this process, and it is dependent on temperature; they increase when temperature and time rises. Total flavonoids content change with the process from 11.21 ± 0.34 quercetin equivalents (QE)/100 g DW to 29.96 ± 1.27 QE/100 g DW in roasted quinoa at 190°C 90 min, this pattern is like other roasting kernels (Carciochi, Galván D'Alessandro, and Marique, 2016). The increment in the TPC (total phenolic compounds) can be explained by glycosylation/esterification of phenolic compounds by roasting, making them more soluble during solvent extraction. Also, thermal treatment liberates phenolic acid aglycones and dissociates phenolic conjugated forms (Dewanto, Wu, and Adom, 2002). Additionally, Repo-Carrasco-Valencia et al. (2010) measured the mineral availability of seeds after roasting quinoa at 190°C for 3 min. A recent work evaluated the influence of roasted quinoa (177°C, 15–45 min) on sensory and physicochemical properties in the formulation of cakes (Rothschild et al. 2015).

Puffing

Young infants usually consume puffing cereals; the more common puffed grains are maize and rice. The puffing process involved a sudden release of water vapor, and the expansion of the pre-gelatinized kernel takes place. Superheated vapor is produced inside the grains by instantaneous heating, which cooks the grain and expands the endosperm while escaping with great force through the micropores of the grain structure (Hoke et al. 2007). Different technologies are used for puffing; however, for quinoa puffing, more recent studies have explorer microwave heating, extrusion expansion, gun puffing (Zapana et al. 2020), and expansion chamber (Paucar-Menacho et al. 2018). Puffing operations promote moisture loss in most of the quinoa snacks due to heating. Protein contents of quinoa samples varied between 11.06 and 13.28 g 100 g^{-1} DW (dry weight). A significant loss of proteins was found for quinoa treated by gun puffing compared to microwave and extrusion puffing due to the loss of cotyledon and grain embryo that possess the highest protein contents (Zapana et al. 2020). Popped quinoa grains showed a decrease in some non-essential and essential amino acids. Amino acid losses ranged from 17 to 31% and particularly affected to Arg, Ala, Pro, Val, Met, Cys, Ile, Phe, Tyr, Lys, and Trp. Reduction in amino acid content after puffing could be explained by thermally-induced chemical modifications of protein residues such as glycation, glycoxidation, and oxidation (Paucar-Menacho et al. 2018).

Extrusion puffing provokes degradation of lipids because of high temperature (110–120°C) and screw speed (500 rpm.). However, Paucar-Menacho et al. (2018) reported that fatty acid profile showed minimal differences after puffing; also, they detected small amounts of palmitic (0.27 g /100g oil) and erucic acid (1.3 g/100 g oil). On the other hand, popped quinoa showed a content of α-linolenic acid (C18:2 ω-3) 7-fold higher than popped kiwicha. Oleic acid (C18:1) was the second most abundant fatty acid in popped quinoa oils, although it is worth noting that the former was more abundant in this MUFA (monounsaturated fatty acid). Palmitic acid (C16:0) was the third most abundant fatty acid found.

Puffing decreased the content of total hydroxycinnamic acids in quinoa (35%). *Trans-p-*coumaric acid was the main compound notably reduced (57%) in quinoa. In contrast, a notable increase was observed for the contents of other phenolic compounds after puffing. Hydroxybenzoic acids presented a lower reduction (2.6%). Higher content of vanillic acid derivative (15% more), was found in popped quinoa than in raw grains. This denoted that dry heat puffing brought a general decrease in hydroxybenzoic and hydroxycinnamic acids contents in both pseudo cereal species. The latter is explained by the thermally-induced degradation of phenolic compounds (Ogrodowska et al. 2012). Also, a decrease in mineral content in gun puffed quinoa was detected because of the pericarp loss and processing conditions (Zapana et al., 2020).

Germination

Germination appears to be an effective method of achieving desirable changes in the nutritive value of seeds (Muzquiz, Rey, Cuadrado, and Fenwick, 1992). Germination starts when the dry seed begins to take up water, and is completed when the embryonic axis elongates. From the moment the seed breaks dormancy, protective responses emerge through the synthesis of phenolics and other compounds (Taiz and Zeiger, 2002). A fat reduction was found by Kumari and Srivastava (2000), reporting that during germination of seeds, hydrolysis of lipids and fatty acids oxidations occur. Another significant change was in crude fiber content, mainly due to the synthesis of more cell wall material to support the shoots and rootlets reported by Martín-Cabrejas et al. (2003). Germination of cereals/pseudo cereals has been suggested as an inexpensive and effective method to enhance the antioxidant capacity by increasing low-molecular-weight antioxidants (Alvarez-Jubete et al. 2010; Carciochi et al. 2016). Total phenol content (TPC) and antioxidant capacity were significantly higher after germination. The improvements observed of TPC during germination explains the action of endogenous esterases activated during germination, which can lead to the release of cell wall-bound phenolic compounds (Maillard et al. 1996).

The amount of vitamin C detected in unprocessed (raw) seeds was 0.38 mg/100 g. Germination showed a progressive increase of vitamin C during different days, reaching after 72 h of germination a level of almost 16 times the original content (Carciochi et al. 2016). The biosynthesis of vitamin C is a specific metabolic process during germination, which seems to be directly implicated in the modulation of initial embryo germination and plant growth, protecting other biological substances from oxidative damage (Fernandez-Orozco et al. 2009). Global germination provides a reduction between 20 and 38% in the content of γ-tocopherol. Thus, α-tocopherol was the most abundant isomer in germinated samples (Carciochi, Galván D'Alessandro, and Manrique, 2016).

Extrusion

Extrusion is a versatile process comprising operations like mixing, cooking, kneading, shearing, shaping, and forming (Xu et al. 2016). The variables that exert the most influence on the process concerning food material include moisture, composition, particle size, lipid content. On the other hand, the process considers other issues such as the compression die, pressure, temperature, residence time, output die geometry, and longitude/diameter (L/D) ratio of the screw. Some studies reported the influence of quinoa varieties, quinoa composition, and process conditions on quinoa extruded products. Those studies evaluated changes in the physical, mechanical, nutritional, and chemical structure of final products.

Tovar-Hernández, Perafan-Gil, Enríquez-Collazos, and Ceron-Fernandez (2017), and Repo-Carrasco-Valencia and Serna (2011) evaluated the extrusion quinoa process using a single screw extruder, evaluating samples with 20% and 30% of moisture. Kowalski et al. (2016) used a twin-screw extruder with temperatures of 40°C, 70°C and 100°C final temperature varied at 130°C, 140°C, and 160°C.

Regarding the chemical composition, some changes have been identified. During extrusion, a reduction in moisture content has been observed. This reduction is related to the temperature used during the extrusion process (Tovar-Hernández, Perafan-Gil, Enríquez-Collazos, and Ceron-Fernandez (2017) and Repo-Carrasco-Valencia and Serna (2011). Additionally, protein suffers a reduction after extrusion. Tovar-Hernández, Perafan-Gil, Enríquez-Collazos, and Ceron-Fernandez (2017) reported <3% of protein reduction in germinated and non-germinated quinoa. Similar research (Cerón-Fernández et al. 2016; Repo-Carrasco-Valencia and Serna 2011; Robin, Théoduloz, and Srichuwong 2015) reported the same pattern; this phenomenon can be attributed to protein denaturalization because protein is vulnerable at temperatures above 60°C. In contrast, Repo-Carrasco-Valencia and Serna (2011) and Zapana et al. (2020) reported an increase in protein digestibility, founding optimal conditions for protein digestibility at 100°C-150°C, and at a screw speed of 100 rpm. Moreover, Zapana et al. (2020) reported an increment from 23.67 to 87.16%.

Cerón-Fernández et al. (2016) reported a reduction of protein digestibility after extrusion, from 82.17% to 63.23% and 92.46% to 67.77% for *Blanca de Jericó* and *Tonkahuan* varieties, respectively. This decrement was related to protein modification structure and covalent bond formation. Besides, amino acid isomerization and lateral chain modifications change the structural protein conformation and enzyme recognition by adjusting the extrusion parameters. The fat content of extruded products presented a reduction associated with temperatures and speeds employed. Those conditions provoke lipids degradation and amylose-lipid complexation. Total dietary fiber and crude fiber suffer a slight modification; however, regarding soluble dietary fiber, there is an increase, this could be due to shear stress caused by high screw speed and high temperatures. Those conditions cause chemical bond brake and creating smaller particles that are soluble. Ash content presented a decrease by shear and high pressure (Cerón-Fernández et al. 2016). Maybe those minerals are interacting with other nutrients like protein and dietary fiber. The content of total phenolic compounds increased during extrusion. In the case of total polyphenols, this can be explained by the increase of phenolic compounds released by the thermal process. Extrusion can be beneficial for processing quinoa because it can significantly reduce saponin content due to heat and shear effects. Saponin content was decreased from 61.81 mg/g in raw quinoa flour to 1.23–39.59 mg/g in extrudates under the conditions studied (Kowalski et al. 2016). The most considerable saponin reduction was observed at the highest temperature (160°C) and lowest moisture content (15%, wet weight basis). Starch molecules suffer a molecular degradation by temperatures, moisture, and speed, obtaining dextrin and other interactions as amylose-lipid complexes and protein-starch complexes (Tovar-Hernández, Perafan-Gil, Enríquez-Collazos, and Ceron-Fernandez 2017).

Conclusion

Quinoa is a pseudo cereal that possesses excellent nutritional characteristics. It has a unique amino acid content that balances essential amino acids and is an excellent dietary fiber source. Quinoa seeds are usually submitted to different processes to increase palatability, digestibility, and sensory acceptance. The most important and usual methods applied in quinoa processing are milling, cooking, extrusion, roasted, and microwave heating. All these processes affected chemical and nutritional properties in different aspects, positive and negative. However, the impact of processing quinoa seeds is the increase of their properties and functionality. The main changes reported are diverse: different types of milled fractions, such as bran (pericarp) and milled grain, improves in the nutritional properties by removing anti-nutrient components, and by decreasing the content of bioactive molecules in milled products; a decrease of protein amount and increase moisture content and digestibility, total phenolics content, total flavonoid content, and the ferric reducing ability of plasma antioxidant activity in quinoa cooked products significantly increases. Roasting quinoa grains alter texture, color, flavor, and appearance, resulting in product development with unique crispness and taste features, compared with raw kernels. High temperatures used in this process mainly cause saponins degradation. Regarding puffed processing, quinoa grains showed a decrease in some non-essential and essential amino acids; additionally, it provokes lipids' degradation. Germination of quinoa causes fat reduction and changes crude fiber, and it is useful to enhance the antioxidant capacity by increasing low-molecular-weight antioxidant compounds. Extrusion of quinoa seeds causes protein reduction after extrusion, lipids degradation, amylose-lipid complexation, and an increase of the content of total phenolic compounds. As described earlier, constant evaluation and further research are crucial to optimize all the processes to achieve more novel food products with high nutritional quality and sensory characteristics. The exploitation of all the chemical and functional goodness of quinoa could be an excellent option to ensure food for the future population by improving its nutritional status and food security policies.

References

Alonso-Miravalles, Loreto, and James A O'Mahony. 2018. "Composition, Protein Profile and Rheological Properties of Pseudocereal-Based Protein-Rich Ingredients." *Foods (Basel, Switzerland)* 7 (5): 73. https://doi.org/10.3390/foods7050073.

Alvarez-Jubete, L., H. Wijngaard, E. K. Arendt, and E. Gallagher. 2010. "Polyphenol Composition and *in vitro* Antioxidant Activity of Amaranth, Quinoa Buckwheat and Wheat as Affected by Sprouting and Baking." *Food Chemistry* 119 (2): 770–78. https://doi.org/10.1016/j.foodchem.2009.07.032.

Avila Ruiz, Geraldine, Anke Arts, Marcel Minor, and Maarten Schutyser. 2016. "A Hybrid Dry and Aqueous Fractionation Method to Obtain Protein-Rich Fractions from Quinoa (*Chenopodium quinoa* Willd)." *Food and Bioprocess Technology* 9 (9): 1502–10. https://doi.org/10.1007/s11947-016-1731-0.

Ballester-Sánchez, J., M. T. Fernández-Espinar, and C. M. Haros. 2020. "Isolation of Red Quinoa Fibre by Wet and Dry Milling and Application as a Potential Functional Bakery

Ingredient." *Food Hydrocolloids* 101 (April): 105513. https://doi.org/10.1016/j.foodhyd.2019.105513.

Ballester-Sánchez, J., J. V. Gil, M. T. Fernández-Espinar, and C. M. Haros. 2019. "Quinoa Wet-Milling: Effect of Steeping Conditions on Starch Recovery and Quality." *Food Hydrocolloids* 89 (April): 837–43. https://doi.org/10.1016/j.foodhyd.2018.11.053.

Bertazzo, Antonella, Stefano Comai, Ilaria Brunato, Mirella Zancato, and Carlo V. L. Costa. 2011. "The Content of Protein and Non-Protein (Free and Protein-Bound) Tryptophan in *Theobroma cacao* Beans." *Food Chemistry* 124 (1): 93–96. https://doi.org/10.1016/j.foodchem.2010.05.110.

Bhargava, Atul, and Shilpi Srivastava. 2013. *Quinoa: Botany, Production and Uses*. CABI. http://books.google.com/books?id=Sxf4AgAAQBAJ&pgis=1.

Bhathal, Simranpreet Kaur, Navjot Kaur, and JPS Gill. 2017. "Effect of Processing on the Nutritional Composition of Quinoa (*Chenopodium quinoa* Willd)." *Agricultural Research Journal* 54 (1): 90. https://doi.org/10.5958/2395-146X.2017.00015.1.

Bhattacharya, Sila. 2014. "Roasting and Toasting Operations in Food: Process Engineering and Applications." *Conventional and Advanced Food Processing Technologies* 9781118406: 221–48. https://doi.org/10.1002/9781118406281.ch10.

Brady, Kevin, Chi-Tang Ho, Robert T. Rosen, Shengmin Sang, and Mukund V. Karwe. 2007. "Effects of Processing on the Nutraceutical Profile of Quinoa." *Food Chemistry* 100 (3): 1209–16. https://doi.org/10.1016/j.foodchem.2005.12.001.

Brend, Yael, Liel Galili, Hana Badani, Ran Hovav, and Shmuel Galili. 2012. "Total Phenolic Content and Antioxidant Activity of Red and Yellow Quinoa (*Chenopodium quinoa* Willd) Seeds as Affected by Baking and Cooking Conditions." *Food and Nutrition Sciences* 3 (August): 1150–55. https://doi.org/http://dx.doi.org/10.4236/fns.2012.38151.

Burrieza, H. P., M. P. López-Fernández, and Sara Maldonado. 2014. "Analogous Reserve Distribution and Tissue Characteristics in Quinoa and Grass Seeds Suggest Convergent Evolution." *Frontiers in Plant Science* 5 (October): 546. https://doi.org/10.3389/fpls.2014.00546.

Calzetta Resio, Andrea N., Marcela P. Tolaba, and Constantino Suárez. 2009. "Correlations Between Wet-Milling Characteristics of Amaranth Grain." *Journal of Food Engineering* 92 (3): 275–79. https://doi.org/10.1016/j.jfoodeng.2008.11.005.

Carciochi, Ramiro Ariel, Leandro Galván-D'Alessandro, Pierre Vandendriessche, and Sylvie Chollet. 2016. "Effect of Germination and Fermentation Process on the Antioxidant Compounds of Quinoa Seeds." *Plant Foods for Human Nutrition* 71 (4): 361–67. https://doi.org/10.1007/s11130-016-0567-0.

Carciochi, Ramiro Ariel, Leandro Galván D'Alessandro, and Guillermo Daniel Manrique. 2016. "Effect of Roasting Conditions on the Antioxidant Compounds of Quinoa Seeds." *International Journal of Food Science & Technology* 51 (4): 1018–25. https://doi.org/10.1111/ijfs.13061.

Cerón-Fernández, Claudia, Lina Guerra-Morcillo, Jorge Legarda-Quintero, Mario Enríquez-Collazos, and Yamid Pismag. 2016. "Effect of Extrusion in the Physicochemical Characterstics of Quinoa Flour (*Chenopodium quinoa* Willd)." *Biotecnología en el Sector Agropecuario y Agroindustrial* 14 (December): 92–99. https://doi.org/10.18684/BSAA(14)92-99.

Dewanto V, Wu X, Adom KK, Liu RH. 2002. "Thermal Processing Enhances the Nutritional Value of Tomatoes by Increasing Total Antioxidant Activity." *Journal of Agriculture and Food Chemistry* 8 (50): 3010–14.

Fabio, Amanda Di, and Gloria Parraga. 2017. "Origin, Production and Utilization of Pseudocereals." In *Pseudocereals: Chemistry and Technology*, edited by Claudia Monika Haros and Regine Schoenlechner, 1–24. UK: Wiley.

FAO. 2011. *Quinoa: An Ancient Crop to Contribute to World Food Security*. http://www.fao.org/3/aq287e/aq287e.pdf.

FAO. 2013. *International Year of Quinoa*. 2013. Accessed August 20. http://www.fao.org/quinoa-2013/en/.

Fernandez-Orozco, Rebeca, Juana Frias, Henryk Zielinski, Rosario Muñoz, MariuszK Piskula, Halina Kozlowska, and Concepción Vidal-Valverde. 2009. "Evaluation of Bioprocesses to Improve the Antioxidant Properties of Chickpeas." *Lebensmittel-Wissenschaft Und Technologie.* 42 (4): 885—892. https://doi.org/10.1016/j.lwt.2008.10.013.

Filho, Antonio Manoel Maradini, Mônica Ribeiro Pirozi, João Tomaz Da Silva Borges, Helena Maria Pinheiro Sant'Ana, José Benício Paes Chaves, and Jane Sélia Dos Reis Coimbra. 2017. "Quinoa: Nutritional, Functional, and Antinutritional Aspects." *Critical Reviews in Food Science and Nutrition* 57 (8): 1618–30. https://doi.org/10.1080/10408398.2014.1001811.

Flores-Silva, Pamela C., Jose De J. Berrios, James Pan, Edith Agama-Acevedo, Adelmo Monsalve-González, and Luis A. Bello-Pérez. 2015. "Gluten-Free Spaghetti with Unripe Plantain, Chickpea and Maize: Physicochemical, Texture and Sensory Properties." *CyTA - Journal of Food* 13 (2): 159–66. https://doi.org/10.1080/19476337.2014.929178.

García-Salcedo, Ángela Janet, Olga Lucía Torres-Vargas, and Hernando Ariza-Calderón. 2017. "Physical-Chemical Characterization of Quinoa (*Chenopodium quinoa* Willd), Amaranth (*Amaranthus caudatus* L.), and Chia (*Salvia hispanica* L.) Flours and Seeds." *Acta Agronomica* 67(2):215-222. https://doi.org/10.15446/acag.v67n2.63666.

Gordillo-Bastidas, E, DA Díaz-Rizzolo, E Roura, T Massanés, and R Gomis. 2016. "Quinoa (*Chenopodium quinoa* Willd), from Nutritional Value to Potential Health Benefits: An Integrative Review." *Journal of Nutrition & Food Sciences* 06 (03): 1–10. https://doi.org/10.4172/2155-9600.1000497.

Guerreo-Ochoa, Manuel R., Romina Pedreschi, and Rosana Chirinos. 2015. "Optimised Methodology for the Extraction of Protein from Quinoa (*Chenopodium quinoa* Willd)." *International Journal of Food Science and Technology* 50: 1815–22. https://doi.org/10.1111/ijfs.12834.

Haros, Claudia Monika, and Małgorzata Wronkowska. 2017. "Pseudocereal Dry and Wet Milling: Processes, Products and Applications." In *Pseudocereals*, edited by CM Haros and Regine Schoenlechner, 140–62. Chichester, UK: John Wiley & Sons, Ltd. https://doi.org/10.1002/9781118938256.ch7.

Hassoon, Waleed Hameed, and Dariusz Dziki. 2018. "The Effect of Seed Moisture and Temperature on Grinding Characteristics of Quinoa (*Chenopodium quinoa* Willd)." Edited by Anna Szeląg-Sikora. *BIO Web of Conferences* 10 (March): 01006. https://doi.org/10.1051/bioconf/20181001006.

Hemalatha, P., Dikki Pedenla Bomzan, B. V. Sathyendra Rao, and Yadahally N. Sreerama. 2016. "Distribution of Phenolic Antioxidants in Whole and Milled Fractions of Quinoa and Their Inhibitory Effects on α-Amylase and α-Glucosidase Activities." *Food Chemistry* 199 (May): 330–38. https://doi.org/10.1016/j.foodchem.2015.12.025.

Hoke, Karel, Milan Houška, Jiřina Průchová, Dana Gabrovská, Kateřina Vaculová, and Ivana Paulíčková. 2007. "Optimisation of Puffing Naked Barley." *Journal of Food Engineering* 80 (4): 1016–22. https://doi.org/10.1016/j.jfoodeng.2006.08.012.

Jacobsen, S. E., A. Mujica, and C. R. Jensen. 2003. "The Resistance of Quinoa (*Chenopodium quinoa* Willd) to Adverse Abiotic Factors." *Food Reviews International* 19 (1–2): 99–109. https://doi.org/10.1081/FRI-120018872.

Jacobsen, S.E., and O. Stølen. 1993. "Quinoa - Morphology, Phenology and Prospects for Its Production as a New Crop in Europe." *European Journal of Agronomy* 2 (1): 19–29. https://doi.org/10.1016/S1161-0301(14)80148-2.

Kowalski, Ryan J., Ilce Gabriela Medina-Meza, Bhim B. Thapa, Kevin M. Murphy, and Girish M. Ganjyal. 2016. "Extrusion Processing Characteristics of Quinoa (*Chenopodium quinoa* Willd.) Var. Cherry Vanilla." *Journal of Cereal Science* 70: 91–98. https://doi.org/10.1016/j.jcs.2016.05.024.

Kumari, S., and S. Srivastava. 2000. "Nutritive Value of Malted Flours of Finger Millet Genotypes and Their Use in the Preparation of Burfi." *Journal of Food Science and Technology* 37 (July): 419–22.

Mahoney, Arthur W., Javier G. Lopez, and Deloy G. Hendricks. 1975. "Evaluation of the Protein Quality of Quinoa." *Journal of Agricultural and Food Chemistry* 23 (2): 190–93. https://doi.org/10.1021/jf60198a035.

Maillard, M. N., M. H. Soum, P. Boivin, and C. Berset. 1996. "Antioxidant Activity of Barley and Malt : Relationship with Phenolic Content." *LWT—Food Science and Technology* 29 (3): 238–244.https://doi.org/10.1006/fstl.1996.0035Get.

Maradini-Filho, A M. 2017. "Quinoa: Nutritional Aspects." *Journal of Nutraceutical and Food Science* 3 (1): 1–5.

Martín-Cabrejas, María A., Nuria Ariza, Rosa Esteban, Esperanza Mollá, Keith Waldron, and Francisco J. López-Andréu. 2003. "Effect of Germination on the Carbohydrate Composition of the Dietary Fiber of Peas (*Pisum Sativum* L.)." *Journal of Agricultural and Food Chemistry* 51 (5): 1254–59. https://doi.org/10.1021/jf0207631.

Mattila, Pirjo H., Juha-Matti Pihlava, Jarkko Hellström, Markus Nurmi, Merja Eurola, Sari Mäkinen, Taina Jalava, and Anne Pihlanto. 2018. "Contents of Phytochemicals and Anti-nutritional Factors in Commercial Protein-Rich Plant Products." *Food Quality and Safety* 2 (4): 213–19. https://doi.org/10.1093/fqsafe/fyy021.

Muzquiz, M., Rey, C., Cuadrado, C., Fenwick, G.G. 1992. Effect of germination on the oligosaccharide content of lupin species. *Journal of Chromatography A* 607(2): 349-352. https://doi.org/10.1016/0021-9673(92)87094-O.

Nowak, Verena, Juan Du, and U. Ruth Charrondière. 2016. "Assessment of the Nutritional Composition of Quinoa (*Chenopodium quinoa* Willd)." *Food Chemistry* 193 (February): 47–54. https://doi.org/10.1016/j.foodchem.2015.02.111.

Ogrodowska, Dorota, Sylwester Czaplicki, Ryszard Zadernowski, Pirjo Mattila, Jarkko Hellström, and Marian Naczk. 2012. "Phenolic Acids in Seeds and Products Obtained from *Amaranthus cruentus*." *Journal of Food and Nutrition Research* 51 (2): 96–101.

Opazo-Navarrete, M., M. A. I. Schutyser, R. M. Boom, and A. E. M. Janssen. 2018. "Effect of Pre-Treatment on *in vitro* Gastric Digestion of Quinoa Protein (*Chenopodium quinoa*

Willd) Obtained by Wet and Dry Fractionation." *International Journal of Food Sciences and Nutrition* 69 (1): 1–11. https://doi.org/10.1080/09637486.2017.1332171.

Park, Sang Ha, Tomoko Maeda, and Naofumi Morita. 2005. "Effect of Whole Quinoa Flours and Lipase on the Chemical, Rheological and Breadmaking Characteristics of Wheat Flour." *Journal of Applied Glycoscience* 52: 337343.

Paucar-Menacho, Luz María, Montserrat Dueñas, Elena Peñas, Juana Frias, and Cristina Martínez-Villaluenga. 2018. "Effect of Dry Heat Puffing on Nutritional Composition, Fatty Acid, Amino Acid and Phenolic Profiles of Pseudocereals Grains." *Polish Journal of Food and Nutrition Sciences* 68 (4): 289–97. https://doi.org/10.1515/pjfns-2018-0005.

Prego, I., S. Maldonado, and M. Otegui. 1998. "Seed Structure and Localization of Reserves in *Chenopodium quinoa*." *Annals of Botany* 82 (4): 481–88. https://doi.org/10.1006/anbo.1998.0704.

Ramos Diaz, Jose Martin, Jussi-Petteri Suuronen, Kevin C. Deegan, Ritva Serimaa, Hely Tuorila, and Kirsi Jouppila. 2015. "Physical and Sensory Characteristics of Corn-Based Extruded Snacks Containing Amaranth, Quinoa and Kañiwa Flour." *LWT - Food Science and Technology* 64 (2): 1047–56. https://doi.org/10.1016/j.lwt.2015.07.011.

Ranhotra, G. S., B. K. Gelroth, K. J. Lorenz, and D. L. Johnson. 1993. "Composition and Protein Nutritional Quality of Quinoa." *Cereal Chemistry* 70 (3): 303–5.

Reguera, M., and CM Haros. 2017. "Structure and Composition of Kernels." In *Pseudocereals: Chemistry and Technology*, edited by Claudia Monika Haros and Regine Schoenlechner, 263. UK: Wiley.

Repo-Carrasco-Valencia, Ritva A. M., Christian R. Encina, Maria J. Binaghi, Carola B. Greco, and Patrícia A Ronayne de Ferrer. 2010. "Effects of Roasting and Boiling of Quinoa, Kiwicha and Kañiwa on Composition and Availability of Minerals *in vitro*." *Journal of the Science of Food and Agriculture* 90 (12): n/a-n/a. https://doi.org/10.1002/jsfa.4053.

Repo-Carrasco-Valencia, Ritva Ann-Mari, and Lesli Astuhuaman Serna. 2011. "Quinoa (*Chenopodium quinoa* Willd) as a Source of Dietary Fiber and Other Functional Components." *Ciência Tecnologia Alimentaria Campinas* 31 (1): 225–30.

Robin, Frédéric, Christine Théoduloz, and Sathaporn Srichuwong. 2015. "Properties of Extruded Whole Grain Cereals and Pseudocereals Flours." *International Journal of Food Science & Technology* 50 (10): 2152–59. https://doi.org/doi:10.1111/ijfs.12893.

Rothschild, Jeffrey, Kurt A. Rosentrater, Charles Onwulata, Mukti Singh, Lorena Menutti, Pera Jambazian, and María B. Omary. 2015. "Influence of Quinoa Roasting on Sensory and Physicochemical Properties of Allergen-Free, Gluten-Free Cakes." *International Journal of Food Science and Technology* 50 (8): 1873–81. https://doi.org/10.1111/ijfs.12837.

Sánchez-Reséndiz, Aidee I., Anayansi Escalante-Aburto, Vidalina Andía-Ayme, and Cristina Chuck-Hernández. 2019. "Structural Properties, Functional Evaluation, and *in vitro* Protein Digestibility of Black and Yellow Quinoa (*Chenopodium petiolare*) Protein Isolates." *CyTA - Journal of Food* 17 (1): 864–72. https://doi.org/10.1080/19476337.2019.1669714.

Serna-Saldivar, S. O. 2010. *Cereal Grains: Properties, Processing and Nutritional Attributes*. Boca Ratón, FL: CRC Press.

Serna-Saldívar, S. O. 2010. "Dry-Milling Operations." In *Cereal Grains. Properties, Processing and Nutritional Attributes*, edited by Sergio O. Serna-Saldívar, 1st ed., 747. CRC Press.

Singh, S, R Singh, and K Vikash Sing. 2016. "Quinoa (*Chenopodium quinoa* Willd), Functional Superfood for Today's World: A Review." *World Scientific News* 58: 84–96.

Stikic, Radmila, Djordje Glamoclija, Mirjana Demin, Biljana Vucelic-Radovic, Zorica Jovanovic, Dusanka Milojkovic-Opsenica, Sven-Erik Jacobsen, and Mirjana Milovanovic. 2012. "Agronomical and Nutritional Evaluation of Quinoa Seeds (*Chenopodium quinoa* Willd) as an Ingredient in Bread Formulations." *Journal of Cereal Science* 55 (2): 132–38. https://doi.org/10.1016/j.jcs.2011.10.010.

Taiz, Lincoln, and Eduardo Zeiger. 2002. *Plant Physiology 3rd Ed.* Edited by Lincoln Taiz and Eduardo Zeiger. 3rd Editio. Sinauer Associates;

Tovar-Hernández, C. E., Perafan-Gil, E. A. Enríquez-Collazos, M. G., Ceron-Fernandez, L. 2017. "Extrusion Process Effect Evaluation on Nornal and Germinated Flour Quinoa (*Chenopodium quinoa* Willd)." *Biotecnología en el Sector Agropecuario y Agroindustrial* 15 (2): 30–38. http://dx.doi.org/10.18684/BSAA(15)30-38. [*Biotechnology in the Agricultural and Agroindustrial Sector*]

Valencia-Chamorro, S. A. 2003. "Quinoa." In *Encyclopedia of Food Science and Nutrition*, edited by B. Caballero, 4895–4902. Amsterdam, The Netherlands: Academic Press.

Wronkowska, Małgorzata, and Monika Haros. 2014. "Wet-Milling of Buckwheat with Hull and Dehulled - The Properties of the Obtained Starch Fraction." *Journal of Cereal Science* 60 (3): 477–83. https://doi.org/10.1016/j.jcs.2014.09.004.

Xu, Enbo, Xiaowei Pan, Zhengzong Wu, Jie Long, Jingpeng Li, Xueming Xu, Zhengyu Jin, and Aiquan Jiao. 2016. "Response Surface Methodology for Evaluation and Optimization of Process Parameter and Antioxidant Capacity of Rice Flour Modified by Enzymatic Extrusion." *Food Chemistry* 212: 146–54. https://doi.org/10.1016/j.foodchem.2016.05.171.

Zapana, Franklyn, Johannes de Bruijn, Leslie Vidal, Pedro Melín, María Eugenia González, Gustavo Cabrera, Pamela Williams, and Rodrigo Bórquez. 2020. "Physical, Chemical and Nutritional Characteristics of Puffed Quinoa." *International Journal of Food Science and Technology* 55 (1): 313–22. https://doi.org/10.1111/ijfs.14290.

In: Super and Nutraceutical Foods
Editors: Sandra Teresita Martín-del-Campo et al.
ISBN: 978-1-53619-082-3

Chapter 11

RICE: COMPOSITION AND PROCESSING CHALLENGES

Elier Vélez-Jiménez, Carmen Téllez-Pérez, José G. Montejano and Anaberta Cardador-Martínez*
Tecnologico de Monterrey, Escuela de Ingeniería y Ciencias, Querétaro, México

ABSTRACT

Rice (Oryza sativa L. and O. glaberrima), being the second most global cultivated cereal, is a nutritional and nutraceutical food staple that complies with the caloric requirement of almost 50% of the world population. Rice must be dried to be a safe food, after harvesting to avoid microbiological and physical damage. Several drying technologies have been developed; however, some physical damage has been reported as a result of bad kernel handling during processing. Fissures and breakage are the main processing issues. Head Rice Yield (HRY) is the value referred to the number of whole grains after milling. The highest HRY produces the highest commercial worth. Parboiling, tempering, and emerging technologies condition rice kernels to improve both the moisture gradients and their structure. Glass transitions occurring during heating generate a stronger starchy matrix that can be considered as resistant starch. This physical characteristic develops a new taste, texture, nutraceutical, and cooking profile. On the other hand, foodborne illnesses caused by rice intake is still an important public health issue. Bacillus cereus is a pathogen that produces toxins when rice is not thoroughly cooked and kept under inadequate chilling conditions. New technologies for rice dying most consider nutraceutical, microbial, and morphological quality of rice kernel. The objective of this job was to review the literature about the problems associated with rice drying, evaluating the advantages and disadvantages, such as improving the physical as well as microbial qualities of rice.

Keywords: rice drying, drying emerging technologies, fissures, resistant starch, microbial rice quality

---

* Corresponding Author's E-mail: mcardador@tec.mx.

Introduction

Rice is considered a grass that belongs to the family Poaceae (Gramineae). Its scientific genus is *Oryza* L. Twenty-two species had been classified in the genus *Oryza*. Twenty species grow as wild species while the other two had been domesticated. Humans consume these only two species of rice. The most worldwide cultivated is *Oryza sativa*. *O. glaberrima* only grows in some areas of West Africa. There are more than 40,000 worldwide rice varieties, but not all of them have the quality to be commercialized. These varieties can be classified by size (long, medium, and short-grain) or according to their properties such as sweet or waxy rice, aromatic rice, basmati rice, black rice, Calrose, U.S. Della, Delrose, Delmont, Jasmine, red rice, Wehani, Arborio, and Carnaroli rice.

Oryza sativa has two subspecies: 1) *japonica* (Chinese variety), characterized by their rounded, medium or short grains cultivated by irrigation in tempered regions or submerged in tropical or sub-tropical zones; 2) *indica* (Indian variety), whose long, thin, and flat grains are cultivated by irrigation in tropical areas.

The Oryza plant grows as a self-pollinating perennial or semiaquatic long grass. At the end of the steam, there is a panicle with 50-300 florets (flower) or spikelets. These spikelets will contain future grains. The height of the plant is between 0.4-5 m according to the variety and the geographical zone (Maclean et al. 2002).

The seed or grain is the fruit of the plant. It is composed of the brown rice (caryopsis) covered by the husk (palea, lemmas, and rachilla constitute this hull; some varieties can include rudimentary glumes and some portions of pedicel). The brown rice has a multilayer surface of different tissues that keep the embryo and endosperm. Figure 1 shows the morphology of a rice grain. A single seed weights 10-45 mg at 0% moisture content. The husk represents about 20% of the total weight (Maclean et al. 2002).

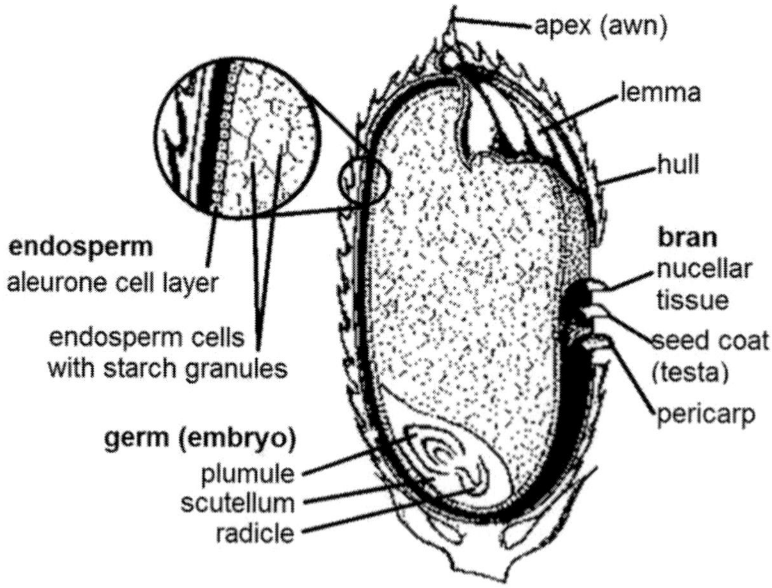

Figure 1. Rice kernel morphology.

Rice grows in warm temperatures. The higher temperatures and higher humidity levels, the faster plant growth. Nitrogen is the main nutrient that the plant needs because the number of filled spikelets depends directly on this source. Also, it improves the leaves' surface and contributes to general grain quality. Phosphorus stimulates the root development, promotes the plant precocity and maturity, making flowering uniformly, and improves the grain quality. This crop can be considered mature when its panicle has 95% light brown grains, and the 5% remaining is still yellow.

Rice is the second most cultivated cereal in the world (Hussaini et al. 2007). It is an important food crop because almost half of the world's population consumes rice (Cottyn et al. 2001), which contributes 20% of the diet calories (Aydin, Aksu, and Gunsen 2011). Comparing the number of calories that it can achieve per hectare against other cereals, it is ranked the first one on the list. That is why it has been cultivated in about 100 countries representing 154 million hectares planted annually (11% of the world's cultivated land) (Khush 2005).

Nutritional Aspects of Rice

Table 1 describes the nutritional content of two varieties of Rice. Carbohydrates are the main component, with an average content of 75%. The protein content is low, also fat content. But rice is rich in several minerals, B vitamins, and alfa-tocopherol.

Table 1. Nutrient content of brown and white rice (per 100 g of product) (Juliano 1993)

	Brown Rice	White Rice
Caloric Content (kcal)	360[a]-385[b]	351[a]-373[b]
Water (g)	12[a]	15.5[a]
Carbohydrates (g)	73-87[b]	76.9[a]-89[b]
Proteins (g)	7.1-8.3[b]	6.2[a]-7.1[b]
Fats (g)	1.6-2.8[b]	0.3[b]-0.8[a]
Ashes (g)	1.0-1.5[b]	0.6[a]
Crude Fiber (g)	0.6[b]	0.3[a]-0.6[b]
Phosphorus (mg)	170.0-430.0[b]	80[b]-150.0[a]
Potassium (mg)	214.0[a]	---
Calcium (mg)	10.0-50.0[b]	6.0[a]-30.0[b]
Zinc (mg)	0.6-2.8[b]	0.6-2.3[b]
Iron (mg)	0.2-5.2[b]	0.2-2.8[b]
Sodium (mg)	9.0[a]	2.0[a]
Niacin (Nicotinic Acid) (mg)	3.5-5.3[b]	1.3-2.4[b]
Vitamin B1 (Thiamine) (mg)	0.29-0.61[b]	0.02-0.11[b]
Vitamin B2 (Riboflavin) (mg)	0.05[a]-0.14[b]	0.02-0.06[b]
A-Tocopherol (mg)	0.90-2.50[b]	
Phytin P (g)	0.13-0.27[b]	0.02-0.07[b]

Quality and Commercial Value of Rice

Rice's appearance has a higher worth than as a nutritional staple food. Freshly harvested paddy must have higher quality standards to assure the final cooked rice quality, involving all losses during drying, storage, and milling steps (Xangsayasane et al. 2019). Rice, as a biomaterial, is very sensitive to heat and mold attacks resulting in fast deterioration. Test for rice quality includes external parameters such as head rice yield (HRY), fissures, and color; and internal characteristics such as taste, texture attributes, and physicochemical properties (Wiset et al. 2005). The number of broken kernels gives rice retail price after milling measured by the HRY. This quality parameter is directly dependent on variety, maturity when the grain is harvested, and drying conditions.

For quality improvement, parboiling is used to increase the kernel hardness: maximizing head rice yield, increasing yield percentage, reducing bran percentage, and slightly decreasing crust percentage (Cherati, Babatabar, and Nikzad 2012).

Texture profile analysis allows evaluating sensorial characteristics of cooked rice that can be appreciated by rice consumers, such as cohesiveness, hardness, chewiness, springiness, adhesiveness, adherence, gumminess, and stickiness. All these characteristics could be modified by parboiling (Zheng et al. 2011).

Rice Microbial Quality

In general, *Pseudomonadaceae,* coliforms, endospore-forming bacteria, yeast, and molds are the most common members of the rice microflora. Milling usually reduces aerobic bacteria counts plate and coliform counts being white rice, the one with the less contamination; however, viable bacteria are present throughout the kernel even in the endosperm itself. The necessity of treatments to reduce bacteria counts is essential to satisfy end-user requirements.

In general, the most common molds that can contaminate rice are from the genera *Aspergillus*, *Penicillium*, *Alternaria*, *Mucor*, *Cladosporium*, *Rhizopus,* and *Fusarium*. Several of these genera are mycotoxin-producing fungi (Table 2). Because of the effect on the health of aflatoxins, several studies have been done on rice to demonstrate the presence of them, for instance, total aflatoxin content of Spanish rice was 37.3 μg/kg, France rice 26.6 μg/kg, Pakistani rice 18.4 μg/kg, Mexican rice contained 16.9 μg/kg, United States rice 14.4 μg/kg and Uruguayan rice 15.6 μg/kg (Suarez-Bonnet et al. 2013). In rice from India, *A. flavus*, as well as other strains such as *Aspergillus niger*, *Aspergillus ochraceus,* and *Aspergillus parasiticus*, were detected, and a concentration of aflatoxin B_1 ranging from 0.1 to 308 μg/kg was measured by Reddy, Reddy, and Muralidharan (2009).

Table 2. Common mycotoxins found in rice and their effects on health

Mycotoxin	Mold producer	Effect on health	References
Aflatoxins	*Aspergillus flavus, Aspergillus parasiticus*	Carcinogenic	(Choi et al. 2015, Aydin, Aksu, and Gunsen 2011, Hussaini et al. 2007, Hussein and Brasel 2001)
Fumonisin	*Fusarium graminareum*	Carcinogenic	(Hussaini et al. 2007)
Ochratoxin	*Penicillium* spp, A. *ochraceus*,	Carcinogenic, nephrotoxic	(Aydin, Aksu, and Gunsen 2011, Hussaini et al. 2007)
Zearalenone	*Fusarium* spp.	Carcinogenic, phytoestrogenic, estrogenic	(Hussaini et al. 2007)
Deoxinivalenol	*Fusarium graminareum*	Carcinogenic	(Choi et al. 2015)

Other mycotoxins have been detected in rice, such as fumonisins, ochratoxins, zearalenone, and nivalenol (Rofiat et al. 2015, Nazari et al. 2014). Mycotoxins have various acute and chronic effects on humans and animals (especially monogastric) depending on the species and susceptibility of an animal within a species. Aflatoxins are classified as human carcinogens, while zearalenone, ochratoxins, trichothecenes, and fumonisins are possible carcinogens (Hussein and Brasel 2001). Hence, methods to remove either fungi or mycotoxins from rice are essential not to cause health consumer damages.

Since fungi spore contamination occurs in the field, efforts to prevent it have been made; however, application of triazole fungicides in crops do not avoid the presence of mycotoxins, such as deoxynivalenol, zearalenone, aflatoxin B_1 and ochratoxin in rice grain as studied by Dors et al. (2013).

Storage conditions are important for mycotoxin production. At a temperature of 21°C and 97% relative humidity (% RH), aflatoxins were produced by *A. flavus* in brown rice (Choi et al. 2015). Trucksess et al. (2011) reported that in healthy paddy rice, mycotoxins are produced during storage in poor conditions; however, milling of this contaminated rice causes a decrease in the concentration of aflatoxins depending on the fractions rice. Choi et al. (2015) evaluated the effect of the degree of milling on the growth of *Aspergillus flavus* and its production of aflatoxins, concluding that there were no differences among rough, brown, and white rice. On the contrary, Trucksess et al. (2011) demonstrated that white rice had less aflatoxin concentration than brown. Iqbal et al. (2012) reported that paddy has an aflatoxin content of 16.35 ± 1.67 µg/kg, while parboiled was 14.20 µg/kg, brown 9.85 µg/kg, white 7.10 µg/kg, and broken 8.5 ± 1.71 µg/kg, showing that processing can diminish mycotoxin content. However, it is still above the approved content (4 µg/kg) in the European Union.

Commercial parboiling provides ideal conditions for the growth of toxigenic *Aspergillus* species in rice because it involves steeping (a process where rice is soaked favoring fermentation). However, a low migration of AFB_1 into the starchy endosperm was observed when rice was soaked during 6 h, and autoclaved for 30 min; similar conditions were reported for deoxynivalenol by Dors, de Almeida Pinto, and Badiale-Furlong (2009). Therefore, parboiling is an effective treatment to reduce mycotoxin content in rice, making its consumption safer. On the other hand, the traditional 'cottage' method of parboiling rice, which does not involve steeping, appears to reduce *Aspergillus* growth even after long storage periods.

Bacillus Cereus in Rice Kernels and Treatments to Decrease Spore-Counts

Bacillus cereus is a Gram-positive, spore-forming bacterium that is widely distributed in the environment due to its ability to resist hostile conditions. It is a common food contaminant and can be found in raw and cooked rice. B. cereus in food is usually associated with food spoilage as well as food poisoning that usually occur in two types of illness: the emetic and diarrheal syndromes. The emetic syndrome is associated with the consumption of meat-containing foods, vegetables, and milk products, whereas diarrheal syndrome results from the consumption of rice and other farinaceous foods (Organji et al. 2015, Kim et al. 2014). Removal of B. cereus spores is crucial to avoid foodborne diseases.

Sarrías, Valero, and Salmerón (2002) described that after rice drying, husking, and polishing, the cell count of Bacillus cereus decreases. Ankolekar, Rahmati, and Labbé (2009) showed that brown rice contains higher *B. cereus* counts (average 52.9 CFU/g) than white rice (average 30.2 CFU/g), then polishing is an effective treatment to achieve B. cereus reduction.

The decimal reduction time, D-value, is referred to as the time required to destroy 90% of the organisms in one logarithmical cycle at a certain temperature. There is an exponential relationship between D-value and temperature but not with the initial cell concentration. If D-value is plotted against the corresponding temperatures, the reciprocal of the slope gives a Z-value. Z-value is referred to as increments in temperature to decrease 1-log in D-value (McKellar and Lu 2004).

Sarrías, Valero, and Salmerón (2002) calculated D-values at 90, 95 and 100°C for a heat-treated spores suspension obtaining the following ranges: 3.23-23.26, 0.69-5.17, and 0.43-1.09 min, respectively; demonstrating that there is a significant difference among D-value for different strains at each temperature. The Z-values ranged from 7.42 to 8.20°C. Desai and Varadaraj (2010) described the lethal behavior for vegetative cells of B. cereus as sigmoidal curves with D-values ranging from 3.45 min at 60°C to 10 min at 56°C in saline solution. Curvilinear plots were obtained for spore inactivation in saline solution with D-values of 19.45 min at 85°C. Z-values were 9.3°C in nutritive broth for vegetative cells and 16.6°C for spores in saline solution.

Daryaei, Balasubramanian, and Legan (2013) applied a pressure-heat treatment over samples of cooked rice inoculated with B. cereus. A viable spore reduction of 2.2-3.4 log occurred during 30 s. Holding this pressure during 4-8 min, the spore counts were reduced to less than 10 CFU/g. On the other hand, under atmospheric pressure (0.1 MPa) and 85°C, 180 min were required to reach <10 CFU/g. Thus, D-values for pressure-heat treatments were 1.08-2.36 min, while 34.6 min at 85°C was calculated for the atmospheric condition.

Once spores contaminate foods, toxin production occurs when foods are heated during periods that do not kill the bacteria or kept under ambient temperatures. The emetic type-toxin is produced rapidly on cooked rice and starchy foods at room temperature. In contrast, diarrheal toxins can be produced in the human intestine in 6-24 h after eating contaminated meals. B. cereus diarrheal enterotoxins can be inactivated at 56°C for 5 min, but emetic toxins are resistant to treatments at 121-126°C for 90 min (Ankolekar, Rahmati, and Labbé 2009). In any case, it is better to remove B. cereus spores from rice grain instead of looking for toxin removal.

RICE KERNEL PRESERVATION TECHNOLOGIES

Due to the susceptibility of rice to microbial contamination and growth, several technologies have been applied to preserve this grain.

Traditional Drying

The amount of water inside the kernel varies according to the growing stage or the trapped water that comes from the rain and dew. Paddy rice has a moisture content around 18 to 45% when harvested. Drying is a useful method to treat the grain and reach low values of water activity. When water is removed from food products, some microbial spoilage and deterioration reactions are stopped or become slower. Thus, dehydration must be done as quickly as possible. The recommended period of storage from harvesting to drying must not exceed 24 h to avoid deterioration by microorganism's activity and/or by the heat that is generated due to grain respiration. Removing foreign matter (as straw, soil, another kind of seeds, dust, broken grains, non-mature kernels, and weeds) improves the drying efficiency.

It is necessary to reach 12 to 15% moisture content (MC) on a dry basis in rough rice after artificial drying to meet the requirements for safe storage and human consumption (Zheng et al. 2011). When temperature ranges from 26.7 to 30°C, drying time takes from six to eight days to reach 15% moisture content. The following drying step to decrease it up to 12% can be delayed because the kernel is less susceptible to mold growth. However, this step is not easily achieved.

Physically, drying involves two stages. The first one is referred to as water removal from the surface. Outer layers will be deeply and sharply dried while the internal zone will not. The husk is widely dried at this period. The rate of drying diminishes during processing time and becomes limited because there is no more water to remove from the most external layers. Under this condition, the rice kernel tends to equilibrate its moisture content. Therefore, water from the core of the grain (being in a higher amount at that moment) migrates to the dried surfaces. The diffusion velocity that takes this water migration will allow the drying process to progress.

This drying kinetic is explained by two matter and energy transfer phenomena: external and internal. External water is converted and extracted as vapor. External water transference depends on the differential rate between the environmental conditions (pressure, % RH, and temperature) and those on the kernel surface. The vapor pressure in the rice kernel must be greater on the surface than that in the air that surrounds it. This condition is enhanced when humidity gradients are intensified by increasing the vapor partial pressure on the kernel surface. The vapor removal will reach a critical moisture point (when the MC on rice surface equals air RH), and the first stage will stop momentarily. If the humidity gradient is high, drying time is reduced. It can be easily achieved by increasing air temperature and decreasing air humidity (Al Haddad et al. 2008). These parameters affect the forced convective heat transfer coefficient h. The h value can be from 25 to 100 $W/m^2 \cdot °K$. This stage is mainly considered an isothermal operation (Al Haddad et al. 2008) because it is assumed that the drying air temperature is the same as the product that is being dried (Srikiatden and Roberts 2006). Heat transfer between air and rice kernels can take place by conduction and/or convection mechanisms.

The internal transfer occurs when the critical moisture point has been reached. At this point, the internal diffusion rate increases, but it is still so low that it does not keep the initial drying rate (Al Haddad et al. 2008). The water gradient is the driving force in this stage. Free water is moved in both liquid and vapor forms, while bounded water does as vapor from inside the kernel. In contrast with external water removal, energy transfer for internal migration is governed by conduction mechanisms according to the Fourier's laws. Matter properties (heat capacity, diffusivity, conductivity, and specific density) depend on factors such as material temperature, water content, matrix structure, and material porosity. Due to dehydration, the biological materials suffer shrinkage due to external water removal. As the structure becomes smaller and more compact, the internal water diffusivity, the thermal conductivity, and the heat transfer slow down.

Table 3. Optimal process conditions to obtain high Head Rice Yields (HRY)

Kind of Technology	Drying Method	*Oryza sativa* variety	Initial Rice % MC[1]	Final % MC	Parameters	Drying time (min)	% Breakage	% HRY	Reference
Traditional-mechanical	Pneumatic conveying	Supan 7	22-26 (wb[2])	18 (wb)	Hot air 50-60°C, 20-25 m/s	3-4 s		52-56	(Kaensup, Kulwong, and Wongwises 2006)
	Thin-layer with tempering	Bengal and Cypress	16-23		Tempering <60°C	Tempering 160-240 (Bengal), 80-160 (Cypress)		67.9 (Bengal), 64.5 (Cypress)	(Cnossen et al. 2001)
	Superheated-steam fluidized bed	Chainat 1	41-42.5 (db)	25 (db[3])	150-170°C			65-70	(Taechapairoj, Prachayawarakorn, and Soponronnarit 2004)
	Parboiling and tray-dryer	Tarom			Parboiling 65°C, 1 atm	Parboiling 15	7.3	75.93	(Cherati, Babatabar, and Nikzad 2012)
	Two stages: Fluidized bed + in-store	Amaroo	25 (wb)	14	Fluidized bed 40°C / In-store 18-30ªC, 60-70% RH	2660			(Jittanit, Srzednicki, and Driscoll 2010)
	Two stages: Spouted bed + in-store	Amaroo	25 (wb)	14	Spouted bed 40°C / In-store 18-30ªC, 60-70% RH	2740			(Jittanit, Srzednicki, and Driscoll 2010)
	Radial flow, circular bin with dehumidifier	IR64 PSB Rc10 PSB Rc34		14 (wb)	34-39°C	1440	6-7 6-7 3	51.6 38.4 36.5	(Regalado, Bekki, and Madamba 2000)
	Continuous bed (batch)	Khazar	20 (db)	16	45°C, 36% RH	35	Breakage susceptibility index at 10 mJ: 0.53±0.32		(Esehaghbeygi 2012)
	Natural on trays	Khazar	20 (db)	16	26°C, 36% RH	1440	Breakage susceptibility index at 10 mJ: 0.42±0.22		(Esehaghbeygi 2012)

Kind of Technology	Drying Method	*Oryza sativa* variety	Initial Rice % MC[1]	Final % MC	Parameters	Drying time (min)	% Breakage	% HRY	Reference
Traditional-mechanical	Three stages: Fluidized bed + Tempering + Thin layer	A10 OM2717	32 (wb) 24.5 (wb)	14 (wb)	Fluidized bed 80°C, tempering 75°C, thin-layer 35°C	Fluidized bed 3 (A10), 2.5 (OM2717). Tempering: 40 (A10), 60 (OM2717). Thin-layer: ¿960?	A10: 3.3 OM2717: 7.7	A10: 66.4 OM2717: 43.5	(Tuyen et al. 2009)
	Three stages: Fluidized bed + Tempering + Thin layer	Germinated Kao Dok Mali 105 Supan Buri 1	52 (db)	18 (db)	150°C. Air flow 4 m/s		20		(Srisang et al. 2010)
Emerging	Successive Pressure Drops (DDS)	Vietnamese	20-25 (db)	12-13	Hot air at 0.3 MPa	90-100	<3	70	(Cong et al. 2008)
	Instant Controlled Pressure Drop (DIC) + Oven	Vietnamese	32 (db)		DIC: 0.5 MPa, 16-30 s Oven at 50°C	180	<3	68-70	(Pilatowski et al. 2010)
	High voltage electrostatic field	Khazar	20 (db)	16	10 kV, 26°C, 36% RH	50	Breakage susceptibility index at 10 mJ: 0.41±0.35		(Esehaghbeygi 2012)
	Infrared radiation	Pathumthani 60	30 (wb)	18 (wb)	2100 W (radiation intensity 46.7 kW/m^2)	18		62.5	(Likitrattanaporn and Noomhorm 2011)
	Continuous fluidized bed microwave system		24% (wb)	14% (wb)	4.8 mW/cm^2 (60°C)	4		(+)7%	(Sangdao, Songsermpong, and Krairiksh 2010)

Moreover, those microstructural reductions obstruct the internal mass transfer (the limiting factor). Thus, molecular diffusivity is directly related to the specific surface area, porosity, and pore distribution (Al Haddad et al. 2008). Also, the temperature will play an important role. Arrhenius' law establishes that the higher temperature, the higher diffusivity. However, high temperatures have negative effects on rice grain quality, such as browning.

Fissure development is mainly caused by moisture loss. Moreover, a difference of 43°C between the rice kernel and the drying air generates cleavages. Drying air temperature below 53°C (rice transition temperature) avoids an excessive thermal expansion that could break the endosperm (Cnossen et al. 2001).

It is recommended to dry rice kernels intermittently to get a better rice quality. That means that cycles of drying and laying must be alternated to reach the final desired moisture. It avoids thermochemical degradations and physical damage. Commercially, those cycles operate from 15 to 20 h to keep costs low.

Table 3 contains information about treatments that have been applied to different varieties of rice. It shows the optimal process conditions that have been tested to achieve the highest HRY value and the shortest drying time.

Parboiling

Rice parboiling is a hydrothermal process that uses high temperature and pressure on rough rice kernels during short periods (2-5 min) (Serna-Saldivar 2012). Traditionally, parboiling is carried out by soaking unpolished rice in ambient temperature water for a few days. Afterward, rice is steamed or boiled at 100°C, followed by drying. Starch gelatinization takes place under these conditions. The starch modifies its molecular configuration and acts as glue by sealing micro-fissures and cracks (Serna-Saldivar 2012). So, the endosperm structure becomes harder, stiffer, and more resistant, which improves the HRY during milling because the number of broken kernels is decreased. Rice moisture content increases if temperature and pressure are high during steaming. This retention of hot water produces higher levels of swelling, partial melting, and gelatinization of the rice starch. Also, the final rice quality is improved due to the leach of the nutrients (essential minerals and B vitamins) from the aleurone cells to the endosperm, but keeping them in the new hard structure (Serna-Saldivar 2012, Dutta and Mahanta 2014). When the structure becomes more compact, mycotoxins and pesticides that could be present on the bran cannot reach the endosperm, but they are not eliminated (Oliveira et al. 2012). Moreover, rice kernels become glassy, translucent, and light yellow. Cooked parboiled rice keeps its shape, and it is firmer, fluffier, and less sticker than non-parboiled rice (Serna-Saldivar 2012).

The parboiling process is limited by the glass transition temperature, which is reached during heating rice kernels by boiling or steaming. Under 77°C water is only absorbed through the kernel matrix; on the other hand, above 77°C water diffusion is increased due to the change in starch structure. Thus, depending on the temperature, some parts of the starchy structure may stay glassy, rubbery, or combined. Then, the higher internal stress produced inside the kernel when its temperature and/or its moisture content changes, the higher the possibility of breakage (Cnossen et al. 2001).

Mechanical Drying Systems

Traditional drying includes sun and field drying, although they are low-cost, require large periods, high labor, and are useful just for small batches. Due to changes in environmental temperature, it is difficult to control the process. For that reason, mechanical drying systems have been developed to improve the efficiency of the drying process using heated air in continuous airflow, vacuum, rotation bins, and conveying systems to keep the kernel in the movement to dry large volume of rice grains (Cong et al. 2008). Therefore, rice batches can be kept in flat or inclined beds while hot air is injected through the basis of the granary or room. In the case of circular bins, hot air flows by a central column and contacts the rice kernel transversally (from the center to the external walls). It can be achieved using batch dryers, re-circulating batch dryers, or continuous flow dryers. According to the study of Regalado, Bekki, and Madamba (2000), attaching a dehumidifier can promote higher MC gradients. It can be the cause of the lowest HRY showed in Table 2.

Usually, drying processes must be carried out in two stages, including a tempering period of 4-12 h to stabilize the water content inside the grain (Cong et al. 2008, Jittanit, Srzednicki, and Driscoll 2010). Although drying systems are widely used, the damages on the products are very noticeable, such as structural, chemical, and physical changes (Aktas et al. 2007). Kaensup, Kulwong, and Wongwises (2006) reported that air temperatures from 50 to 60ºC keep the rice surface at 35ºC, avoiding damages, while temperatures over 70ºC develop breakage and a cloudy appearance in the kernel.

Other mechanical technologies hold the grain in flotation surrounded by drying air. This technology produces large amounts of broken rice (Cong et al. 2008) and losses of volatile components when air drying temperatures are over 100ºC. Other fluidized bed techniques use superheated steam at 150-170ºC to improve the starch gelatinization and then increase the HRY (Taechapairoj, Prachayawarakorn, and Soponronnarit 2004).

Using a combination of dying technologies with a fluidization system followed by tempering and finishing with a thin layer drying does not show a decrease in % breakage (Tuyen et al. 2009) in comparison with two stages of the same technique.

Emerging Drying Technologies

New technologies (based on irradiation, microwaves, ultrasound, etc.) have been applied over food samples to improve the drying process, reducing cost by using less energy and saving time and improving the final quality of the product. However, their high cost to scale them for industrial productions have become a limiting factor (Ratti 2008).

Infrared drying is a rapid method without excessive rice breakage. The drying rate is from 0.49 to 3.60% MC per minute. Pan et al. (2011) concluded that achieving 60ºC on the kernel during heating by applying high radiation intensity over thin rice layers or medium radiation intensity over deep rice beds, followed by tempering and natural cooling, rice with desirable milling and sensory qualities can be produced. Also, lower HRY, lighter yellowness, higher viscosity, better flavor, and higher starch gelatinization were obtained under IR drying as quality improvements on parboiled rice in comparison with steam-parboiled rice.

According to Ding et al. (2018), after 58 s of IR drying, rice was heated from 20°C to 60°C with corresponding moisture removal of 2.17 percentage points. IR drying has an additional advantage, during the natural cooling process, another 1.20 percentage points of moisture were removed without any additional energy input. Then, the rice moisture content was reduced to 15.92 ± 0.05% after 6h of ambient air ventilation.

Dielectric drying operates at 5-40 MHz radiofrequency with three stages of 5 min each. Thus, rice is heated and dried uniformly and rapidly. It represents a high cost for the industrial scale. Esehaghbeygi (2012) calculated the breakage susceptibility index after applying 10 kV on brown rice kernels that were dried at 45°C and received different levels of kinetic impact loadings. The higher kinetic impact loads the higher breakage susceptibility index. Although microwave technology works with 2,450 MHz and high temperature is achieved, the drying process occurs when rice is aerated after being heated. The final structure is too weak that reduces de HRY. Sangdao, Songsermpong, and Krairiksh (2010) tested a system that includes microwave treatment attached to fluidized bed equipment. The results show that a high drying capacity (3.1 kg/h using only 5.2 kJ of energy) can be reached while MC decreases from 24 to 14% on wb.

Instant Controlled Pressure Drop process (DIC from its name in French) is a thermo-mechanical treatment that consists of subjecting biological products to a short period of high pressure (0.1-1 MPa), followed by an instantaneous pressure drop close to 3 kPa. The sudden pressure drop, rate $\Delta P/\Delta t > 2.5 \times 10^5$ Pa/s, leads to auto vaporization of water and instantaneous cooling of the products, which stops thermal degradations (Mounir et al. 2012).

Focusing on rice drying, two DIC technologies have been studied on different varieties of paddy rice. In the study of Cong et al. (2008), DIC and DDS (Successive Pressure Drops) have been evaluated to obtain a fast and economical rice drying processing, also aiming for the prevention of grain fracture. The study was carried on 14 French and Vietnamese varieties of paddy rice. Just after harvesting, the paddy rice was processed under different DIC conditions (steam pressure: 1×10^5 - 8×10^5 Pa and time of treatment ≤ 30 s). MC of paddy rice just after DIC treatment was about 23-25% on wb. To get good milling and storage conditions, paddy rice was submitted to a traditional hot air drying (40-60°C, 0.12 m/s of air flux and 34% RH) and a DDS processing (highest level of gas pressure: 1.7-3.4 $\times 10^5$ Pa; average product temperature: 32-48.5°C; time at the highest air pressure: 17-33 s, and at the lowest pressure: 10 s; and lowest pressure: 1.0×10^5 Pa). Drying kinetics of husk, bran, and endosperm were obtained by continuous measurement of mass; breakage level was determined in all rice samples (Cong et al. 2008).

DIC treatment induced modifications on the microstructure level of paddy rice by developing a new micro alveoli structure. Else, by taking into account that the husk layer is considered to be a limiting process in rice drying (Pilatowski et al. 2010), the structural modification of DIC improved its drying kinetics; and its effective diffusivity (D_{eff} : 110,307 E^{-07} m^2/s) was increased. Also, the study showed that the greater the husk thickness, the longer the drying time. Furthermore, DDS reduced more than 4.8 times the drying time to achieve an MC of 12-14% on DB instead of 8 h for traditional hot air drying. DDS needed 90-100 min without any tempering period.

It is necessary to keep a low partial vapor pressure on the rice surface to allow the drying process. However, while heating the rice surface, the vapor pressure on the surface increases. As it is explained by the Al Haddad paradox, in traditional drying, due to these antagonist forces generated between the temperature gradient (higher in the surface than in the kernel core) and vapor pressure gradient (higher in the heart of the kernel toward the surface), the drying time becomes longer when MC on the kernel core is higher than on its surface. Improving porous structure by DIC makes total pressure becomes the driving force, and drying time becomes shorter (Cong et al. 2008).

The impact of DIC on starch (gelatinization and microporous structure) and the increasing of the effective diffusivity of water during DDS reduced the broken rice kernels to less than 3% instead of 12-25% from the traditional drying. The higher the diffusivity, the lower the internal constraints between the surface and the center of the grain (Cong et al. 2008).

Furthermore, the study of Pilatowski et al. (2010) showed that DIC treatment coupled to hot air drying could also improve the processing and the quality of rice. In the study, Vietnamese paddy rice (32% MC DB) was treated by DIC after post-harvesting (steam pressure: 0.40-0.60 MPa and thermal treatment time: 15-40 s), followed by a stage of hot air drying (50°C and 0.65 m/s of airflow). For all the samples, the final MC was 12.5% on DB. Drying time, the effective diffusivity of water within the grain husks, cooking time at boiling and ambient water temperature, and yield and broken rate were determined. Compared to the hot air drying, DIC treatment significantly reduced the drying time from 1,110 min to 205 min and increased the effective diffusivity from $0.22E^{-13}$ m^2/s to $1.18E^{-13}$ m^2/s. Cooking time of DIC rice at boiling temperature was reduced from 15 min to 5 min, and the broken rate was reduced from 21% to less than 3%. Storage time was increased from 9 to 18 months as well as HRY from 79 to 97%. Decontamination and elimination of insects by DIC treatment imply potentially long storage. THE optimized DIC process was defined at a steam pressure of 0.5 MPa for 16-30 s (Pilatowski et al. 2010). The emerging technologies improve rice quality as HRY

Final Storage of Rice

Storage conditions must be kept under low % RH and low temperatures. High % RH and high temperatures result in rapid grain deterioration with qualitative (nutrient losses, lower germination capacity, appearance changes, mold, and bacterial contamination) and quantitative losses.

Dried rice is a bio sensitive material that can reabsorb moisture from the environment. Most of the fissures observed in dried rice occur within 24-48 h after drying. Akowuah, Addo, and Bart-Plange (2012) reported an increment from 12 to 31% of fissured grains after the first month of storage. In addition, their study showed that the fissure rate diminishes during extended periods.

Shelf life for dried rice kernels will depend on storage conditions. They can be kept from six months (mixes of brown and white grains) to two years (white rice) in ambient temperature. Five years of storage can be achieved if they are kept under four °C or below. More extended periods (up to thirty years) require cooling temperatures and oxygen-free atmospheres.

Nutraceutical Importance of Rice: Resistant Starch

Starch is the major constituent of milled rice at about 70-90% of the dry matter. Starch is a polymer of D-glucose and usually consists of an essentially linear fraction (amylose) and a branched fraction (amylopectin). The ratio between both molecules varies according to the botanical origin of starch. Based on digestion rate, starch has been classified into three fractions, rapidly digestible starch (RDS) which is digested in vitro within 20 min, slowly digestible starch (SDS) which is digested between 20 and 120 min, and resistant starch (RS) which is the starch not hydrolyzed after 120 min of incubation (Ashwar et al. 2016). RS is further classified into four categories: Type 1 (RS1) is physically inaccessible starch granules, Type 2 (RS2) is native granular starch, Type 3 (RS3) is retrograded starch made by cooking/cooling processes on starchy materials, and Type 4 (RS4) is chemically modified starch (Chiu and Stewart 2013).

Retrogradation of starch occurs spontaneously in heat-processed starchy foods, reducing starch digestibility due to the reorganization of the crystalline structure of the starch molecules (Hsu et al. 2015) and could increase the content of both, RS and SDS.

Rice is considered a high glycemic index food because of its content of RDS. The glycemic index (GI), measures how a carbohydrate-containing food raises blood glucose after ingestion. GI of foods is ranked based on how they compare to a reference food, either glucose or white bread. However, depending on the content of RS, rice could be reclassified as low GI food. Therefore, several technological methods have been applied to rice to produce RS, such as oven baking, autoclaving, enzymatic modifications, acidification, among others (Hung, Vien, and Lan Phi 2016, Ashwar et al. 2016, Kim, Kim, and Yoo 2015, Takahama, Hirota, and Yanase 2019).

RS quantity depends not only on the cooking method but also on the variety of rice. Chiu and Stewart (2013) reported 0.37 and 0.58 g RS/100 g of freshly pressure cooked Short and Long grain rice, respectively. When the same varieties were oven cooked, RS content increased significantly. A combination of the use of a traditional rice cooker and refrigeration of cooked rice produces a two-fold increase of RS in Long grain rice. According to Sonia, Witjaksono, and Ridwan (2015), cooking white rice in a cooker and freezing for 24 and 48 h at 4°C, showed resistant starch contents of 0.64 g/100 g, 1.30 g/100 g, and 1.65 g/100 g, for freshly cooked, 24 h and 48 h refrigerated, respectively (Sonia, Witjaksono, and Ridwan 2015).

It is known that heat moisture treatment increases the RS content in normal rice by four-fold times. When this treatment is combined with organic acid treatments, the increase is six-fold times compared to native normal rice. Similar results were observed for high amylose rice (Hung, Vien, and Lan Phi 2016).

Autoclaving during heat-moisture treatment induces the formation of amylose-lipid and starch-lipid complexes decreasing the extraction yield of starch, suggesting an alteration in the structure of starch (Arns et al. 2015). On the other hand, Bian and Chung (2016) reported that heat-moisture treatment increased RDS and SDS contents by 3.6% and 6.0%, respectively, and decreased RS content by 9.5%, which could be due to the disruption of the double helices forming the starch crystallites. The consequence of these changes is related to an increase in digestibility and GI. However, Van Hung, Chau, and Phi (2016) reported an increase in RS and a decrease in SDS in amylose rich rice kinds, with a total RS content of 90%.

Starch digestibility of rice is an important parameter because of its relationship with postprandial blood glucose. RS is not digested or absorbed in the small intestinal lumen of healthy persons and, thus, its physiological effects include an inhibitory effect on the postprandial increase in blood glucose and insulin. A study of 4-week dietary treatment with rice containing resistant starch resulted in an improvement in glucose and insulin metabolism as well as a reduction in oxidative stress in subjects with impaired fasting glucose or newly diagnosed type 2 diabetes (Kwak et al. 2012). Consumption of waxy rice high in RS by diabetic rats reduces not only postprandial glucose levels but also helps to control plasma triacylglycerides and cholesterol levels (Matsumoto et al. 2016). Freshly cooked, 24 h-refrigerated, and reheated rice containing RS, diminished significantly plasma glucose levels after 120 min consumption (Sonia, Witjaksono, and Ridwan 2015). These authors recommend refrigerating and reheating cooked rice for diabetic patients. Heat-moisture treated rice administered to mice showed a low GI related to total amounts of SDS and RS (Van Hung, Chau, and Phi 2016), suggesting that during heat-moisture treatment, it is formed inherent amylose structures that reduce the accessibility of hydrolyzing enzymes and are responsible for low GI values.

Rice cultivars with high amylose content were tested for their effect on blood glucose and appetite after traditional cooking in a rice cooker 3-days previously to the study and reheated before consumption. Blood glucose concentration measured along time, and up to 120 min after consumption showed lower values (104 mg/dL) in subjects that consumed the high amylose rice than the control subjects (112 mg/dL) who consumed low amylose rice (Zenel and Stewart 2015).

Resistant starch is formed during successive steps of heating and cooling, then not only cooking and reheating could originate this kind of starch, but also the drying process before storage. Considering the effect of resistant starch on the health consumers, it is worth to study the relationship between different drying methods, with special emphasis on emerging technologies.

Recent Findings in Rice Derived Products

Since ancient times, rice has been used in many different products. Ray et al. (2016) reported that Indians had used rice to prepared fermented beverages from 7,000–8,000 BC. The same authors describe the interest in fermented rice products due to its low cholesterol content and high vitamin and mineral value. Moreover, fermented rice products confer additional benefits to health consumers because of the probiotics they contain.

Broken rice fermented with *Aspergillus niger* or *Rhizopus oligosporus*, produced various types of bioactive compounds such as phenolics and flavonoids with antioxidant activity. Also, fermented products inhibited the activity of elastase and tyrosinase. Those activities are of interest in the cosmetic industry, leading to a new trend called cosmeceutical (Abd Razak et al. 2019).

Puffed and popped rice are traditional breakfast cereals and snack foods. Raw rice is traditionally popped by heating the rough rice (13 to 17% moisture) at about 240°C for 30 to

35 seconds or at 275°C for 40 to 45 seconds or in an oil bath at 215 to 230°C. The hull contributes to pressure retention before popping, as evidenced by the lower popping percentage of brown rice. Good popping varieties have a tight hull, and a significant clearance between hull and brown rice and when freshly harvested, are free of grain. The tightness of hull, grain hardness, and degree of transparency? Could explain 80 percent of the variation in popping expansion among 25 rice varieties (Dalbhagat, Mahato, and Mishra 2019).

Rice flour is made from ground raw rice (glutinous or non-glutinous). Rice flour had been used to prepare gluten-free bread. This is important for people with celiac disease. However, gluten-free products are not the only goal in the use of rice but also getting food products that maintain and even enhance the nutraceutical potential of rice. In this sense, Pedron et al. (2016) studied the presence of essential elements in rice and rice-based products to evaluate the nutritional value and also, the associated risk to the consumption of metals present in rice.

Noodles is a product elaborated with rice flour, with a high glycemic index. Then, several types of research are focused on the addition of ingredients to decrease the glycemic index. Raungrusmee et al. (2020) evaluated the influence of resistant starch, xanthan gum, inulin, and defatted rice bran to rice noodles, getting a product with a lower glycemic index and acceptable sensorial characteristics. Similar results were achieved by Fu et al. (2020) adding buckwheat to rice. However, Srikaeo, Laothongsan, and Lerdluksamee (2018) reported that the addition of gums, for instance, carboxymethyl cellulose, xanthan gum, and guar gum to noodles increased the glycemic index, due to the formation of a more porous structure, which facilitated starch digestion.

Rice and byproducts such as rice bran, the main byproduct of polishing, have been used as a source of nutraceuticals (Huang, Benchamas, and Huang 2020, Peanparkdee and Iwamoto 2019, Sen, Chakraborty, and Kalita 2020). Aktaş and Akın (2020) added rice bran to nutritional composition and functionality of tarhana, a fermented food product, to improve the content of fiber, resulting in an increase on the content of phenolic compounds, flavonoids, and antioxidant capacity.

Germinated brown rice was evaluated as an ingredient to prepare yogurt-like products resulting in a product rich in phenolics, γ-orizanol, and γ-aminobutyric acid. Furthermore, yogurt-like products showed high antioxidant capacity and a 60% inhibition of the angiotensin-converting enzyme activity (Cáceres et al. 2019). Furthermore, rice bran was associated with preventing diabetes, fatty liver, and diabetic nephropathy. Rats fed with rice bran showed higher weight but lower fat deposits than rats with a standard diet. Moreover, there was an increase in metabolites related to glycolysis and lipolysis (Kubota et al. 2020).

The application or rice as food has many options, and more are being created, not only because of its nutritional value, but also its nutraceutical potential.

Although the sensory quality of the rice kernel is an important standard that processors must accomplish, several experiments have been carried out for a specific aim: increase HRY index to maximize economic profits for rice producers. Many drying techniques (probed in pilot plants or laboratory-scale) using different temperatures (hot air or superheated steam), % RH, electric kinetic pulses, microwaves, irradiation, and machinery design, have been tested to identify the optimal parameters to obtain high-quality rice kernels and efficient processes.

Physical appearance is directly related to structural modifications occurring during the glass transition step. Moreover, that transition modifies the thermomechanical properties of rice. Drying treatments rearrange the starchy matrix to allow water movement uniformly by increasing the water diffusivity coefficient. Harder or weaker structures can be achieved when amylose chains are reorganized inside the rice kernel. One example of this desirable phenomenon occurs during parboiling by producing a starch retrogradation while rice is cooled after its gelatinization. However, MC gradients during harvesting, drying, and storage must be perfectly controlled to avoid grain damages.

DIC process appears to be one important step in preconditioning rice. Better texture, color, flavor, nutrient retention, and drying and cooking times have been reported as advantages of using that technology. Also, due to the use of high pressure, it is highly possible that rice pathogen content could be eliminated. In other words, there is a noticeable profit if DIC technology is used. It will also contribute to the production of a larger volume of safety rice to attend recent and future food demand. There are few studies about positive and negative DIC effects on rice. It is convenient to investigate deeply more about improvements have given through its application.

Drying processes of rice have several advantages in both aspects, physically and over the health consumers. On the one hand, HRY is reduced, maximizing economic benefits, and on the other hand, retrograded starch induced by drying processes gives nutraceutical properties to rice, modifying blood glucose levels in consumers.

Acknowledgments

The first author, Elier Vélez-Jiménez, is grateful for the financial support of the Consejo Nacional de Ciencia y Tecnología (CONACYT) through doctoral scholarship 289574.

References

Abd Razak, Dang Lelamurni, Anisah Jamaluddin, Nur Yuhasliza Abd Rashid, Amsal Abd Ghani, and Musaalbakri Abdul Manan. 2019. "Assessment of fermented broken rice extracts for their potential as functional ingredients in cosmeceutical products." *Annals of Agricultural Sciences* 64 (2):176-182. doi: https://doi.org/10.1016/j.aoas.2019.11.003.

Akowuah, J. O., A. Addo, and A. Bart-Plange. 2012. "Influence of drying temperature and storage duration on fissuring and milling quality of Jasmine 85 rice variety." *Journal of Science and Technology* 32 (2):26-33.

Aktaş, K., and N. Akın. 2020. "Influence of rice bran and corn bran addition on the selected properties of tarhana, a fermented cereal based food product." *LWT* 129:109574. doi: https://doi.org/10.1016/j.lwt.2020.109574.

Aktas, T., S. Fujii, Y. Kawano, and S. Yamamoto. 2007. "Effects of pretreatments of sliced vegetables with trehalose on drying characteristics and quality of dried products." *Food and Bioproducts Processing* 85 (3):178-183. doi: 10.1205/fbp07037.

Al Haddad, M., S. Mounir, V. Sobolik, and K. Allaf. 2008. "Fruits & vegetables drying combining hot air, D.I.C. technology and microwaves." *International Journal of Food Engineering* 4 (6):1-6. doi: 10.2202/1556-3758.1491.

Ankolekar, Chandrakant, Talat Rahmati, and Ronald G. Labbé. 2009. "Detection of toxigenic *Bacillus cereus* and *Bacillus thuringiensis* spores in U.S. rice." *International Journal of*

Food Microbiology 128 (3):460-466. doi: http://dx.doi.org/10.1016/j.ijfoodmicro.2008.10.006.

Arns, Bruna, Josiane Bartz, Marjana Radunz, Jarine Amaral do Evangelho, Vânia Zanella Pinto, Elessandra da Rosa Zavareze, and Alvaro Renato Guerra Dias. 2015. "Impact of heat-moisture treatment on rice starch, applied directly in grain paddy rice or in isolated starch." *LWT - Food Science and Technology* 60 (2, Part 1):708-713. doi: http://dx.doi.org/10.1016/j.lwt.2014.10.059.

Ashwar, Bilal Ahmad, Adil Gani, Idrees Ahmed Wani, Asima Shah, Farooq Ahmad Masoodi, and Dharmesh Chandra Saxena. 2016. "Production of resistant starch from rice by dual autoclaving-retrogradation treatment: Invitro digestibility, thermal and structural characterization." *Food Hydrocolloids* 56:108-117. doi: http://dx.doi.org/10.1016/j.foodhyd.2015.12.004.

Aydin, Ali, Harun Aksu, and Ugur Gunsen. 2011. "Mycotoxin levels and incidence of mould in Turkish rice." *Environmental Monitoring and Assessment* 178:271-280.

Bian, Linlin, and Hyun-Jung Chung. 2016. "Molecular structure and physicochemical properties of starch isolated from hydrothermally treated brown rice flour." *Food Hydrocolloids* 60:345-352. doi: http://dx.doi.org/10.1016/j.foodhyd.2016.04.008.

Cáceres, Patricio J., Elena Peñas, Cristina Martínez-Villaluenga, Patricia García-Mora, and Juana Frías. 2019. "Development of a multifunctional yogurt-like product from germinated brown rice." *LWT* 99:306-312. doi: https://doi.org/10.1016/j.lwt.2018.10.008.

Cherati, Fazlollah E., Ramzam Babatabar, and Fariborz Nikzad. 2012. "Analysis and study of parboiling method, and the following impact on waste reduction and yield increase of Iranian rice in paddy conversion phase." *World Academy of Science, Engineering and Technology* 63:772-775.

Chiu, Y. T., and M. L. Stewart. 2013. "Effect of variety and cooking method on resistant starch content of white rice and subsequent postprandial glucose response and appetite in humans." *Asia Pacific Journal of Clinical Nutrition* 22 (3):372-9. doi: 10.6133/apjcn.2013.22.3.08.

Choi, S., H. Jun, J. Bang, S. H. Chung, Y. Kim, B. S. Kim, H. Kim, L. R. Beuchat, and J. H. Ryu. 2015. "Behaviour of *Aspergillus flavus* and *Fusarium graminearum* on rice as affected by degree of milling, temperature, and relative humidity during storage." *Food Microbiology* 46:307-13. doi: 10.1016/j.fm.2014.08.019.

Cnossen, Auke G., Terry J. Siebenmorgen, Wade Yang, and Rustico C. Bautista. 2001. "An application of glass transition temperature to explain rice kernel fissure occurrence during the drying process." *Drying Technology* 19 (8):1661-1682. doi: 10.1081/DRT-100107265.

Cong, Duong Thai, Mazen Al-Haddad, Zoulikha Rezzoug, Laurent Lefevre, and Karim Allaf. 2008. "Dehydration by successive pressure drops for drying paddy rice treated by Instant Controlled Pressure Drop." *Drying Technology* 26 (4):443-451. doi: 10.1080/07373930801929300.

Cottyn, B., E. Regalado, B. Lanoot, M. De Cleene, T. W. Mew, and J. Swings. 2001. "Bacterial populations associated with rice seed in the tropical environment." *Phytopathology* 91 (3):282-292. doi: 10.1094/PHYTO.2001.91.3.282.

Dalbhagat, Chandrakant Genu, Dipendra Kumar Mahato, and Hari Niwas Mishra. 2019. "Effect of extrusion processing on physicochemical, functional and nutritional characteristics of rice and rice-based products: A review." *Trends in Food Science & Technology* 85:226-240. doi: https://doi.org/10.1016/j.tifs.2019.01.001.

Daryaei, Hossein, V. M. Balasubramanian, and J. David Legan. 2013. "Kinetics of *Bacillus cereus* spore inactivation in cooked rice by combined pressure-heat treatment." *Journal of Food Protection* 76 (4):616-623. doi: 10.4315/0362-028X.JFP-12-447.

Desai, Shivalingsarj Vijaykumar, and Mandyam Chakravarathy Varadaraj. 2010. "Behavioural pattern of vegetative cells and spores of *Bacillus cereus* as affected by time-temperature combinations used in processing of Indian traditional foods." *Journal of Food Science and Technology* 47 (5):549-556. doi: 10.1007/s13197-010-0099-9.

Ding, Chao, Ragab Khir, Zhongli Pan, Delilah F. Wood, Chandrasekar Venkitasamy, Kang Tu, Hamed El-Mashad, and Jose Berrios. 2018. "Influence of infrared drying on storage characteristics of brown rice." *Food Chemistry* 264:149-156. doi: https://doi.org/10.1016/j.foodchem.2018.05.042.

Dors, Giniani Carla, Sergiane Souza Caldas, Helen Cristina dos Santos Hackbart, Ednei Gilberto Primel, Carlos Alberto Alves Fagundes, and Eliana Badiale-Furlong. 2013. "Fungicides and the effects of mycotoxins on milling fractions of irrigated rice." *Journal of Agricultural and Food Chemistry* 61 (8):1985-1990. doi: 10.1021/jf305144t.

Dors, Giniani Carla, Luiz Antônio de Almeida Pinto, and Eliana Badiale-Furlong. 2009. "Migration of mycotoxins into rice starchy endosperm during the parboiling process." *LWT-Food Science and Technology* 42 (1):433-437.

Dutta, Himjyoti, and Charu Lata Mahanta. 2014. "Traditional Parboiled Rice-Based Products Revisited: Current Status and Future Research Challenges." *Rice Science* 21 (4):187-200. doi: https://doi.org/10.1016/S1672-6308(13)60191-2.

Esehaghbeygi, Ali. 2012. "Effect of electrohydrodynamic and batch drying on rice fissuring." *Drying Technology* 30 (14):1644-1648. doi: 10.1080/07373937.2012.701262.

Fu, Meixia, Xuyang Sun, Di Wu, Linghan Meng, Xiao Feng, Weiwei Cheng, Chengcheng Gao, Yuling Yang, Xinchun Shen, and Xiaozhi Tang. 2020. "Effect of partial substitution of buckwheat on cooking characteristics, nutritional composition, and in vitro starch digestibility of extruded gluten-free rice noodles." *LWT* 126:109332. doi: https://doi.org/10.1016/j.lwt.2020.109332.

Hsu, Rachel Jui-cheng, Hong-Jhang Chen, Shin Lu, and Wenchang Chiang. 2015. "Effects of cooking, retrogradation and drying on starch digestibility in instant rice making." *Journal of Cereal Science* 65:154-161. doi: http://dx.doi.org/10.1016/j.jcs.2015.05.015.

Huang, Shiyu, Gunsriwiang Benchamas, and Gangliang Huang. 2020. "Whole processing and use of rice polishings." *Innovative Food Science & Emerging Technologies* 63:102373. doi: https://doi.org/10.1016/j.ifset.2020.102373.

Hung, Pham Van, Ngo Lam Vien, and Nguyen Thi Lan Phi. 2016. "Resistant starch improvement of rice starches under a combination of acid and heat-moisture treatments." *Food Chemistry* 191:67-73. doi: 10.1016/j.foodchem.2015.02.002.

Hussaini, A. Makun, A. Gbodi Timothy, H. Akanya Olufunmilayo, A. Salako Ezekiel, and H.Ogbadu Godwin. 2007. "Fungi and some mycotoxins contaminating rice (*Oryza sativa*) in Niger State, Nigeria." *African Journal of Biotechnology* 6 (2):099-108.

Hussein, Hussein S., and Jeffrey M. Brasel. 2001. "Toxicity, metabolism, and impact of mycotoxins on humans and animals." *Toxicology* 167 (2):101-134. doi: http://dx.doi.org/10.1016/S0300-483X(01)00471-1.

Iqbal, Shahzad Z., Muhammad R. Asi, Agustin Ariño, Noreen Akram, and Muhammad Zuber. 2012. "Aflatoxin contamination in different fractions of rice from Pakistan and estimation of dietary intakes." *Mycotoxin Research* 28 (3):175-180. doi: 10.1007/s12550-012-0131-1.

Jittanit, Weerachet, George Srzednicki, and Robert Driscoll. 2010. "Corn, rice, and wheat seed drying by two-stage concept." *Drying Technology* 28 (6):807-815. doi: 10.1080/07373937.2010.485081.

Juliano, Bienvenido. 1993. *Rice in human nutrition*. Rome: FAO.

Kaensup, Weerachai, Sittidej Kulwong, and Somchai Wongwises. 2006. "A small-scale pneumatic conveying dryer of rough rice." *Drying Technology* 24 (1):105-113. doi: 10.1080/07373930500538899.

Khush, Gurdev S. 2005. "What it will take to feed 5.0 billion rice consumers in 2030." *Plant Molecular Biology* 59 (1):1-6. doi: 10.1007/s11103-005-2159-5.

Kim, Booyoung, Jihyun Bang, Hoikyung Kim, Yoonsook Kim, Byeong-sam Kim, Larry R. Beuchat, and Jee-Hoon Ryu. 2014. "*Bacillus cereus* and *Bacillus thuringiensis* spores in Korean rice: Prevalence and toxin production as affected by production area and degree of milling." *Food Microbiology* 42:89-94.

Kim, Bum-Su, Hyun-Seok Kim, and Sang-Ho Yoo. 2015. "Characterization of enzymatically modified rice and barley starches with amylosucrase at scale-up production." *Carbohydrate Polymers* 125:61-68. doi: 10.1016/j.carbpol.2015.02.048.

Kubota, Masatoshi, Reiko Watanabe, Michihiro Hosojima, Akihiko Saito, Ai Sasou, Takehiro Masumura, Yukikazu Harada, Hiroyuki Hashimoto, Shinobu Fujimura, and Motoni Kadowaki. 2020. "Rice bran protein ameliorates diabetes, reduces fatty liver, and has renoprotective effects in Zucker Diabetic Fatty rats." *Journal of Functional Foods* 70:103981. doi: 10.1016/j.jff.2020.103981.

Kwak, Jung Hyun, Jean Kyung Paik, Hye In Kim, Oh Yoen Kim, Dong Yeob Shin, Hye-Jin Kim, Jin Hee Lee, and Jong Ho Lee. 2012. "Dietary treatment with rice containing resistant starch improves markers of endothelial function with reduction of postprandial blood glucose and oxidative stress in patients with prediabetes or newly diagnosed type 2 diabetes." *Atherosclerosis* 224 (2):457-464. doi: http://dx.doi.org/10.1016/j.atherosclerosis.2012.08.003.

Likitrattanaporn, Charun, and Athapol Noomhorm. 2011. "Effects of simultaneous parboiling and drying by infrared radiation heating on parboiled rice quality." *Drying Technology* 29 (9):1066-1075. doi: 10.1080/07373937.2011.566967.

Maclean, J. L., D. C. Dawe, B. Hardy, and G. P. Hettel, eds. 2002. *Rice almanac: source book for the most important economic activity on earth*. 3rd ed. China: CABI Publishing.

Matsumoto, Kenji, Eri Yasuyoshi, Kurumi Nishi, Yuji Honda, Makoto Nakaya, and Shinichi Kitamura. 2016. "Resistant starch-rich wx/ae brown rice prevents insulin resistance and hypertriglyceridaemia in type 2 diabetic NSY mice." *Journal of Functional Foods* 23:556-564. doi: http://dx.doi.org/10.1016/j.jff.2016.01.046.

McKellar, Robin C., and Xuewen Lu. 2004. "Primary Models." In *Modeling Microbial Responses in Food*, edited by Robin C. McKellar and Xuewen Lu, 21-62. USA: CRC Press.

Mounir, Sabah, Tamara Allaf, Arun S. Mujumdar, and Karim Allaf. 2012. "Swell drying: Coupling Instant Controlled Pressure Drop DIC to standard convection drying processes to intensify transfer phenomena and improve quality—an overview." *Drying Technology* 30 (14):1508-1531. doi: 10.1080/07373937.2012.693145.

Nazari, F., M. Sulyok, H. Yazdanpanah, F. Kobarfard, and R. Krska. 2014. "A survey of mycotoxins in domestic rice in Iran by liquid chromatography tandem mass spectrometry." *Toxicology Mechanisms Methods* 24 (1):37-41. doi: 10.3109/15376516.2013.844752.

Oliveira, Márcia Gonzaga de Castro, Priscila Zaczuk Bassinello, Valácia Lemes da Silva Lobo, and Maria Madalena Rinaldi. 2012. "Stability and microbiological quality of rice bran subjected to different heat treatments." *Food Science and Technology* 32:725-733.

Organji, Sameer Rushdi, Hussein Hasan Abulreesh, Khaled Elbanna, Gamal Ebrahim Haridy Osman, and Manal Khider. 2015. "Occurrence and characterization of toxigenic *Bacillus cereus* in food and infant feces." *Asian Pacific Journal of Tropical Biomedicine* 5 (7):515-520. doi: 10.1016/j.apjtb.2015.04.004.

Pan, Zhongli, Ragab Khir, K. L. Bett-Garber, E. T. Champagne, J. F. Thompson, Adel Salim, B.R. Hartsough, and S. Mohamed. 2011. "Drying characteristics and quality of rough rice under infrared radiation heating." *Transactions of the ASABE* 54 (1):203-210.

Peanparkdee, Methavee, and Satoshi Iwamoto. 2019. "Bioactive compounds from byproducts of rice cultivation and rice processing: Extraction and application in the food and pharmaceutical industries." *Trends in Food Science & Technology* 86:109-117. doi: https://doi.org/10.1016/j.tifs.2019.02.041.

Pedron, Tatiana, Fabiana Roberta Segura, Fabio Ferreira da Silva, Alexandre Luiz de Souza, Heloisa França Maltez, and Bruno Lemos Batista. 2016. "Essential and non-essential elements in Brazilian infant food and other rice-based products frequently consumed by children and celiac population." *Journal of Food Composition and Analysis* 49:78-86. doi: https://doi.org/10.1016/j.jfca.2016.04.005.

Pilatowski, Isaac, Sabah Mounir, Joseph Haddad, Duong Thai Cong, and Karim Allaf. 2010. "The Instant Controlled Pressure Drop process as a new post-harvesting treatment of paddy rice: Impacts on drying kinetics and end product attributes." *Food and Bioprocess Technology* 3 (6):901-907. doi: 10.1007/s11947-010-0332-6.

Ratti, Cristina. 2008. *Advances in Food Dehydration*. Edited by Da-Wen Sun, *Contemporary Food Engineering*. USA: CRC Press.

Raungrusmee, Sujitta, Smriti Shrestha, Muhammad Bilal Sadiq, and Anil Kumar Anal. 2020. "Influence of resistant starch, xanthan gum, inulin and defatted rice bran on the physicochemical, functional and sensory properties of low glycemic gluten-free noodles." *LWT* 126:109279. doi: https://doi.org/10.1016/j.lwt.2020.109279.

Ray, Mousumi, Kuntal Ghosh, Somnath Singh, and Keshab Chandra Mondal. 2016. "Folk to functional: An explorative overview of rice-based fermented foods and beverages in India." *Journal of Ethnic Foods* 3 (1):5-18. doi: https://doi.org/10.1016/j.jef.2016.02.002.

Reddy, K. R., C. S. Reddy, and K. Muralidharan. 2009. "Detection of *Aspergillus* spp. and aflatoxin B1 in rice in India." *Food Microbiology* 26 (1):27-31. doi: 10.1016/j.fm.2008.07.013.

Regalado, M. J. C., E. Bekki, and P. S. Madamba. 2000. "Dehumidification drying of high-moisutre rough rice in a radial-flow, circular bin." *Drying Technology* 18 (10):2307-2326. doi: 10.1080/07373930008917844.

Rofiat, A. S., F. Fanelli, O. Atanda, M. Sulyok, G. Cozzi, S. Bavaro, R. Krska, A. F. Logrieco, and C. N. Ezekiel. 2015. "Fungal and bacterial metabolites associated with natural contamination of locally processed rice (*Oryza sativa* L.) in Nigeria." *Food Additives and Contaminants Part A Chemical Analisys Control Exposition and Risk Assessesment* 32 (6):950-9. doi: 10.1080/19440049.2015.1027880.

Sangdao, Chokchai, Sirichai Songsermpong, and Monai Krairiksh. 2010. "A continuous fluidized bed microwave paddy drying system using applicators with perpendicular slots on a concentric cylindrical cavity." *Drying Technology* 29 (1):35-46. doi: 10.1080/07373937.2010.482721.

Sarrías, J. A., M. Valero, and M. C. Salmerón. 2002. "Enumeration, isolation and characterization of *Bacillus cereus* strains from Spanish raw rice." *Food Microbiology* 19 (6):589-595. doi: 10.1006/fmic.2002.0514.

Sen, Saikat, Raja Chakraborty, and Pratap Kalita. 2020. "Rice - not just a staple food: A comprehensive review on its phytochemicals and therapeutic potential." *Trends in Food Science & Technology* 97:265-285. doi: 10.1016/j.tifs.2020.01.022.

Serna-Saldivar, Sergio O. 2012. "Dry-milling processes and quality of dry-milled products." In *Cereal grains. Laboratory reference and procedures manual*, 125-. USA: CRC Press.

Sonia, Steffi, Fiastuti Witjaksono, and Rahmawati Ridwan. 2015. "Effect of cooling of cooked white rice on resistant starch content and glycemic response." *Asia Pacific Journal of Clinical Nutrition* 24 (4):620-625.

Srikaeo, Khongsak, Pronthip Laothongsan, and Chutima Lerdluksamee. 2018. "Effects of gums on physical properties, microstructure and starch digestibility of dried-natural fermented rice noodles." *International Journal of Biological Macromolecules* 109:517-523. doi: https://doi.org/10.1016/j.ijbiomac.2017.12.121.

Srikiatden, Jaruk, and John S. Roberts. 2006. "Measuring moisture diffusivity of potato and carrot (core and cortex) during convective hot air and isothermal drying." *Journal of Food Engineering* 74 (1):143-152. doi: http://dx.doi.org/10.1016/j.jfoodeng.2005.02.026.

Srisang, Naruebodee, Somkiat Prachayawarakorn, Warunee Varanyanond, and Somchart Soponronnarit. 2010. "Germinated brown rice drying by hot air fluidization technique." *Drying Technology* 29 (1):55-63. doi: 10.1080/07373937.2010.482691.

Suarez-Bonnet, E., M. Carvajal, I. Mendez-Ramirez, P. Castillo-Urueta, J. Cortes-Eslava, S. Gomez-Arroyo, and J. M. Melero-Vara. 2013. "Aflatoxin (B1, B2, G1, and G2) contamination in rice of Mexico and Spain, from local sources or imported." *Journal of Food Science* 78 (11):T1822-9. doi: 10.1111/1750-3841.12291.

Taechapairoj, Chaiyong, Somkiat Prachayawarakorn, and Somchart Soponronnarit. 2004. "Characteristics of rice dried in superheated-steam fluidized-bed." *Drying Technology* 22 (4):719-743. doi: 10.1081/DRT-120034259.

Takahama, Umeo, Sachiko Hirota, and Emiko Yanase. 2019. "Slow starch digestion in the rice cooked with adzuki bean: Contribution of procyanidins and the oxidation products." *Food Research International* 119:187-195. doi: https://doi.org/10.1016/j.foodres.2019.01.062.

Trucksess, M. W., H. K. Abbas, C. M. Weaver, and W. T. Shier. 2011. "Distribution of aflatoxins in shelling and milling fractions of naturally contaminated rice." *Food Additives and Contaminants Part A Chemical Analisys and Control Exposition Risk Assessessment* 28 (8):1076-82. doi: 10.1080/19440049.2011.576441.

Tuyen, Truong Thuc, Vinh Truong, Shu Fukai, and Bhesh Bhandari. 2009. "Effects of high-temperature fluidized bed drying and tempering on kernel cracking and milling quality of

Vietnamese rice varieties." *Drying Technology* 27 (3):486-494. doi: 10.1080/07373930802686099.

Van Hung, Pham, Huynh Thi Chau, and Nguyen Thi Lan Phi. 2016. "In vitro digestibility and in vivo glucose response of native and physically modified rice starches varying amylose contents." *Food Chemistry* 191:74-80. doi: http://dx.doi.org/10.1016/j.foodchem.2015.02.118.

Wiset, Lamul, George Srzednicki, Michael Wootton, Robert H Driscoll, and Anthony B Blakeney. 2005. "Effects of high-temperature drying on physicochemical properties of various cultivars of rice." *Drying technology* 23 (9-11):2227-2237.

Xangsayasane, Phetmanyseng, Khamtai Vongxayya, Senthong Phongchanmisai, Jaquie Mitchell, and Shu Fukai. 2019. "Rice milling quality as affected by drying method and harvesting time during ripening in wet and dry seasons." *Plant Production Science* 22 (1):98-106. doi: 10.1080/1343943X.2018.1544463.

Zenel, Alison M, and Maria L Stewart. 2015. "High amylose white rice reduces post-prandial glycemic response but not appetite in humans." *Nutrients* 7 (7):5362-5374.

Zheng, Xian-Zhe, Cheng-Hai Liu, Zhi-Ying Chen, Ning-Ye Ding, and Chang-Jiang Jin. 2011. "Effect of drying conditions on the texture and taste characteristics of rough rice." *Drying Technology* 29 (11):1297-1305. doi: 10.1080/07373937.2011.592032.

SECTION 3. ANIMAL ORIGIN AND DIVERSE

In: Super and Nutraceutical Foods
Editors: Sandra Teresita Martín-del-Campo et al.
ISBN: 978-1-53619-082-3

Chapter 12

CHLORELLA AS A VALUABLE INGREDIENT FOR FUNCTIONAL FOODS

Beatriz Martín-García[1,2], María del Carmen Razola-Díaz[2], Giovanni Luca Russo[3], Elisa Giambanelli[1], Antonio Luca Langellotti[3], Ana Mª Gómez-Caravaca[1,4] and Vito Verardo[2,4,*]

[1]Department of Analytical Chemistry, Faculty of Sciences, University of Granada, Granada, Spain
[2]Department of Nutrition and Food Science, University of Granada, Granada, Spain
[3]CAISIAL Centre and Department of Agricultural Sciences, University of Naples "Federico II," Portici, Naples, Italy
[4]Institute of Nutrition and Food Technology 'José Mataix,' Biomedical Research Center, University of Granada, Armilla, Granada, Spain

ABSTRACT

Chlorella sp. is a genus of green microalgae rich in proteins, carbohydrates, lipids and other valuable compounds such as pigments, vitamins, phenols and minerals. Many studies showed beneficial properties of *Chlorella* on human health such as antihypertensive, antidiabetic, anti-inflammatory, anticarcinogenic, neuroprotective, immunomodulatory, gastroprotective, anti-anemic and antimicrobial activities. Moreover, the extracts of chlorella showed the capacity to modulate and attenuate the progression of liver fibrosis and to promote muscle regeneration and ameliorate sarcopenia. For these reasons, *Chlorella* biomass is extensively used as a nutraceutical, and it is included among the most promising superfoods.

Keywords: *Chlorella*, functional foods, nutritional supplements, microalgae, bioactive compounds

---

* Corresponding Author's E-mail: vitoverardo@ugr.es.

INTRODUCTION

It is well known that 70% of the earth is the oceans, hosts to immense biodiversity, which includes the aquatic animals, plants and algae (Rani, Sandal, and Sahoo 2018). Algae, a multicellular or unicellular form of living organisms can be classified into macro or microalgae based on its size (Sathasivam et al. 2019). Of these, the microalgae comprise the lowest level of the aquatic food chain and provide oxygen for many species in the aquatic environment. Microalgae are a large, diverse group of microorganism consisting of prokaryotic cyanobacteria and eukaryotic microalgae (Stern et al. 2010). They are a rich source in compounds such as protein, carbohydrates, and lipids. Microalgae are photosynthetic microorganisms, and they do not have cell organelles in contrast with land plants (Hemantkumar and Rahimbhai 2019). They use the carbon from air for energy production. There are more than 300,000 species of microalgae; among them, only 30,000 have been studied. They live in complicated natural habitats and can adapt fast in extreme conditions (e.g., variable salinity, temperature, nutrients, UV–irradiation). This ability makes them capable of generating secondary metabolites with biological activities (Hemantkumar and Rahimbhai 2019). Microalgae produce some useful bio-products including carotenoids such as β-carotene, astaxanthin; lipids as docosahexaenoic acid (DHA) and eicosapentaenoic acid (EPA); bioactive and functional pigments, natural dyes, polysaccharides, antioxidants and algal extracts (Saleh Mobin and Alam 2017). Cultivation of microalgae for commercial use has begun five decades ago and was introduced for the first time in Japan in 1960s (Varfolomeev and Wasserman 2011).

Microalgae have an extensive range of industrial applications in food industries, wastewater purification, and pharmaceutical formulations (S Mobin and Alam 2014). Microalgae have also been used for high-value food, health food for human, polysaccharides, food and fodder additives, cosmetics, antioxidants, anti-inflammatory objects, dyes and feed for aquaculture, and preparation of biofilms (Saleh Mobin and Alam 2017). In the last 20 years, biotechnological and nutraceutical application of microalgae has focused on four important microalgae: *Spirulina*, *Chlorella*, *Dunaliella salina*, and *Haematococcus pluvialis*. These microalgae are extensively used in the food industry. Their biomass is also used for providing health product including tablets and capsules. Besides, these are used as supplements in noodles, bread, biscuits, candies, ice cream, bean curd and other common foods as a food additive to improve their nutritional and health values (Liang et al. 2004). Their tablets of dried biomass contain β-Carotene which works as a vitamin A precursor. *Spirulina*, *Chlorella*, *Dunaliella* and *H. pluvialis* are also used as dyes for foodstuffs (e.g., orange juice production), lutein, zeaxanthin and canthaxanthin are used for chicken skin coloration or for pharmaceutical purposes (Saleh Mobin and Alam 2017).

BIOCHEMICAL COMPOSITION OF *CHLORELLA*

Chlorella is a genus of eukaryotic unicellular green algae (phylum Chlorophyta) with an excellent capacity for photosynthesis, able to reproduce in few hours and requires only sunlight, carbon dioxide, water, and a small quantity of nutrients (Silva et al. 2019; Mata, Martins, and Caetano 2010). Chlorella is one of the most promising biomass for proteins production (Heredia-Arroyo et al. 2011), because more than half of its composition consists of proteins, the rest are carbohydrates, lipids and other valuable nutraceuticals such as pigments, vitamins and phenolic compounds (Panahi et al. 2019). Commercially, few species are cultured among them the main are: C. vulgaris, C. protothecoides, C. ellipsoidea, C. sorokiniana and C. pyrenoidosa.

Like many other algae, the proximal composition varies according to culture conditions. Generally, lipid increases, and biomass productivity ceases or decreases during unfavorable growth conditions such as nitrogen and phosphorus limitation, high CO_2 concentration, excessive exposure to light, excess of iron in the medium or increase in temperature. On the other hand, high protein content can be obtained using well-managed culture conditions.

Proteins

Proteins are an essential element for microalgae life cycle, contributing to growth, repair and maintenance mechanisms of the cells. Moreover, proteins can act as cellular motors, chemical messengers, regulators of cellular activities and defense (Safi et al. 2014). Proteins are components of the membranes; 20% of the total proteins are bound to the cell wall and are also involved in photosynthesis through catalytic enzymes. Compared to other sources of proteins, including the vegetable ones, microalgae contain a higher amount of proteins (Plaza et al. 2009). Protein content in *Chlorella* represents 42-58% of the total biomass on a dry weight basis. Microalgae growth conditions affect protein content. Higher protein contents are obtained in semi-continuous photobioreactor systems by means of constant irradiance (Becker 2007). On the one hand, it has been observed a positive correlation between the protein content of *Chlorella vulgaris* and light intensity (Baiee and Salman 2016), whereas it has been found that the protein content decrease with higher light intensity (He et al. 2015). However, in most recent studies, it has been found that a light intensity of 120-520 $\mu mol/m^2s^1$ does not significantly affect the protein content of *C. vulgaris* biomass (Metsoviti et al. 2020). Notwithstanding, the protein content has been demonstrated to be significantly higher with white or green light (Sánchez-Saavedra et al. 2020). On the other hand, protein content is highly affected by nitrogen sources and concentration that is required for their synthesis and used during exponential phase (Metsoviti, Katsoulas, et al. 2019). In *C. vulgaris* and *C. sorokininana* increasing amount of nitrogen especially in the form of urea and nitrate in the media, showed an increasing protein content and productivity of the cells (Lai et al. 2019).

Protein content in heterotrophic cultivation is usually lower than in autothrophic and mixotrophic conditions, although a protein content increase from 15.5% to 20.2% has been observed after heterotrophic batch cultivation of *Chlorella* under controlled conditions (Canelli et al. 2020). This result is mainly explained by the conversion of intracellular storage of nitrogen into protein (Li et al. 2008). Another possible explanation of this increase could be attributed to the fact that cells in starvation consume intracellular carbohydrates depositories, with a consequent protein content increase. Optimized culture strategies in heterotrophy have shown protein increase until 44.3% of dry weight (Xie et al. 2017).

And last, growing conditions, like autotrophic, heterotrophic or mixotrophic cultivation, affect also molecular protein weight. The particular interest for some microalgae species is due to the high percentage of protein and even to their amino acid profile (Buono et al. 2014). Several biological activities of microalgae are associated with the total protein content and also with protein hydrolyzates or peptides. Protein hydrolysates or biopeptides represent convenient sources of protein for human nutrition because they are absorbed more efficiently through the gastrointestinal tract compared with intact protein or free amino acids (Vaz et al. 2016). In that way, the amino acid profile is highly important to determine protein nutritional quality of microalgae. *Chlorella* cells can synthesize both essential and non-essential amino acids. The ratio of essential to total amino acids is almost 2/3 for *C. vulgaris*, which is clearly above the 1/3 index ratio recommended by WHO/FAO for a nitrogenous nutritional equilibrium (WHO/FAO 1985). The content of amino acids in *C. vulgaris* is reported in Table 1. The most abundant amino acids are aspartic acid, glutamic acid, glycine, leucine, alanine and tyrosine, although in the bibliography wide ranges have been reported.

Table 1. Amino acid content in *Chorella vulgaris*

Amino acid	Content (g/100g)	References
Aspartic acid	9.30-10.94	(Safi et al. 2014)
Threonine	5.15-6.09	(Safi et al. 2014)
Serine	4.32-7.77	(Safi et al. 2014)
Glutamic acid	8.37-13.70	(Safi et al. 2013; Safi et al. 2014)
Glycine	6.07-8.60	(Safi et al. 2014)
Alanine	8.33-10.90	(Safi et al. 2014)
Cysteine	0.18-1.28	(Safi et al. 2013; Safi et al. 2014)
Valine	2.85-7.00	(Safi et al. 2013; Safi et al. 2014)
Methionine	0.60-1.30	(Safi et al. 2013; Safi et al. 2014)
Isoleucine	0.09-4.44	(Safi et al. 2013; Safi et al. 2014)
Leucine	6.91-17	(Ferreira et al. 2020; Safi et al. 2013; Safi et al. 2014)
Tyrosine	2.80-13	(Ferreira et al. 2020; Safi et al. 2014)
Phenylalanine	5.36-5.81	(Safi et al. 2013; Safi et al. 2014)
Histidine	1.16-2.00	(Safi et al. 2013; Safi et al. 2014)
Lysine	6.30-6.83	(Safi et al. 2013; Safi et al. 2014)
Arginine	6.22-7.38	(Safi et al. 2014)
Tryptophan	2.21-2.30	(Safi et al. 2014)
Ornithine	0.12-0.13	(Safi et al. 2014)
Proline	2.74-5.00	(Safi et al. 2013; Safi et al. 2014)

Lipids

Microalgal lipids have two major functions directly derived from their hydrophobic property. Firstly, they structure lipid bilayers in membranes, forming barriers between extra- and intracellular media. These membrane lipids, also called polar lipids, are therefore structural lipids. Neutral lipids (triglycerides or triacylglycerols) are the second class. They represent a way to store carbon and energy in the cells (Cvejic et al. 2020). In the literature, a wide range of lipid content of *Chlorella* can be found with a mean usually around 5-20%, and the highest lipid content could be obtained using optimized culture condition (Heidari, Kariminia, and Shayegan 2016). This variation is strongly dependent on the cultivation conditions, as the biomass composition can be modified by adjusting temperature, nutrients (i.e., nitrogen and phosphorous), light and salt concentration. Lipids in *Chlorella* are mainly composed of glycolipids, waxes, hydrocarbons, phospholipids, and small amounts of free fatty acids. These components are synthesised by chloroplast and located on the cell membrane and in the membranes of organelles (chloroplast and mitochondrion). Neutral lipids are instead mainly located in lipid droplets in the cytoplasm or plastids. In general, microalgae tend to accumulate lipids under varied stress conditions such as light intensity, temperature, nitrogen starvation, salinity and metal stress (Zhu, Li, and Hiltunen 2016). Nitrogen is the critical nutrient which affecting microalgal growth and lipid accumulation.

Nitrogen starvation has been intensively proofed to affect lipid content. It has been demonstrated that under nitrate, urea or phosphate deprivation, the lipid content is major than with usual nitrate content (Anto, Pugazhendhi, and Mathimani 2019). Briefly, the increased lipid content is due to the alteration in the lipid biosynthesis pathway, where the polar membrane lipids are converted to a storage form of neutral lipids under stress conditions. Besides, the lipid accumulation under N stress in the metabolic pathway, brings out many possibilities such as inhibition of gluconeogenesis and redirecting carbon towards triacylglycerol accumulation, up-regulation of the glycolytic pathway for biosynthesis of fatty acids, or up-regulation of Kennedy pathway (Yang et al. 2013). On the other hand, phosphorus deprivation affects too the lipid content of biomass. It is a vital compound present in various biomolecules such as DNA, proteins, RNA, phospholipids, etc. However, under phosphate limitation, cell division ceases, and carbon gets accumulated in the form of triacylglycerol (Mühlroth et al. 2017). Moreover, it has been shown that the lipid content is maximum with light intensities of 560-600 $\mu mol/m^2s^1$ (Liao et al. 2017). An explanation for this phenomenon is that more energy was allocated to the synthesis of lipid serving as the storage material with the rising light intensity. About the light wavelength, it has been demonstrated that the lipid content of the cells is significantly higher, especially under blue light (Sánchez-Saavedra et al. 2020). Therefore, a two-stage cultivation strategy based on a first stage of high biomass productivity, followed by a stage of metabolic stress triggering the flux of uptaken carbon (CO_2 or organic carbon) toward lipids synthesis, can help to solve the problem of growth rate reduction under the stress conditions which enhance lipid content. (Cvejic et al. 2020).

Nowadays, some innovative techniques are being investigated to enhance lipid accumulation in *Chlorella*. Recently, lipid productivity from *Chlorella* mutant via physical approach by atmospheric and room temperature plasma has successfully improved by 16.8% of lipid productivity (Cao et al. 2017), meanwhile by Clustered Regularly Interspaced Short Palindromic Repeats (CRISPR/Cas) has increased lipid productivity by 46% (Lin and Ng 2020). *Chlorella* cultivated in autotrophy is not an important polyunsaturated fatty acid (PUFA) producer. However, in some studies it has been shown that the cultivation of *Chlorella sp.* in heterotrophy has increased the production of PUFAs, proving the possibility to use this green alga as a producer of important nutraceutical fatty acids (S. P. Ferreira et al. 2017).

A typical fatty acids composition for *C. vulgaris* is reported in Table 2. During the cultivation, linoleic acid decrease changing into oleic acid. These fractions yielded a ω3:ω6 ratio of approximately 1:2 ratio, which fit within the optimal range (from 1:1 to 1:4) according to nutrition studies regarding preventing effects towards chronic diseases (Canelli et al. 2020). The ratio of unsaturated to saturated fatty acids is 3:1, showing the good balance of fatty acids profile reported for *C. vulgaris* (Busi et al. 2015).

Table 2. Fatty acid composition of *Chlorella vulgaris*

Fatty acid	Content (g/100 g)	References
SFA	21.02-29.12	(Metsoviti, Katsoulas, et al. 2019)
Caproic acid (C6:0)	0.06-0.42	(Metsoviti, Katsoulas, et al. 2019)
Caprylic acid (C8:0)	0.23-0.78	(Metsoviti, Katsoulas, et al. 2019)
Capric acid (C10:0)	1.42-4.46	(Metsoviti, Katsoulas, et al. 2019)
Undecylic acid (C11:0)	0.55-1.49	(Metsoviti, Katsoulas, et al. 2019)
Lauric acid (C12:0)	0.05-0.12	(Metsoviti, Katsoulas, et al. 2019)
Tridecylic acid (C13:0)	0.87-1.97	(Metsoviti, Katsoulas, et al. 2019)
Myristic acid (C14:0)	0.29-0.88	(Metsoviti, Katsoulas, et al. 2019)
Pentadecanoic acid (C15:0)	0.14-0.31	(Metsoviti, Katsoulas, et al. 2019)
Palmitic acid (C16:0)	10.26-22.8	(Gao et al. 2019; Busi et al. 2015; Metsoviti, Katsoulas, et al. 2019)
Margaric acid (C17:0)	0.17-0.36	(Metsoviti, Katsoulas, et al. 2019)
Stearic acid (C18:0)	1.04-2.80	(Metsoviti, Katsoulas, et al. 2019)
Arachidic acid (C20:0)	0.16-0.34	(Metsoviti, Katsoulas, et al. 2019)
Heneicosylic acid (C21:0)	0.25	(Metsoviti, Katsoulas, et al. 2019)
Lignoceric acid (C24:0)	0.07-0.18	(Metsoviti, Katsoulas, et al. 2019)
MUFA	24.53-37.11	(Metsoviti, Katsoulas, et al. 2019)
Myristoleic acid (C14:1)	0.23-0.43	(Metsoviti, Katsoulas, et al. 2019)
Palmitoleic acid (C16:1)	6.80-11.46	(Metsoviti, Katsoulas, et al. 2019)
Nervonic acid (C24:1)	0.07-0.12	(Metsoviti, Katsoulas, et al. 2019)
Oleic acid (C18:1n9)	9.66-42.70	(Vishwakarma, Dhar, and Saxena 2019; Metsoviti, Katsoulas, et al. 2019; Busi et al. 2015)
Eicosenoic acid (C20:1n9)	0.07-0.46	(Metsoviti, Katsoulas, et al. 2019)
Erucic acid (C22:1n9)	0.14-0.50	(Metsoviti, Katsoulas, et al. 2019)
PUFA	34.13-52.93	(Metsoviti, Katsoulas, et al. 2019)
Linoleic acid (C18:2n6 Cis)	0.04-0.18	(Metsoviti, Katsoulas, et al. 2019)
Linoleaidic acid (C18:2n6 Trans)	8.38-29.4	(Gao et al. 2019; Metsoviti, Katsoulas, et al. 2019)
γ-Linolenic acid (C18:3n6)	0.05-2.20	(Metsoviti, Katsoulas, et al. 2019; Busi et al. 2015)
α-Linolenic acid (C18:3n3)	25.49-41.41	(Metsoviti, Katsoulas, et al. 2019; Busi et al. 2015)
Eicosadienoic acid (C20:2)	0.04-0.15	(Metsoviti, Katsoulas, et al. 2019)
Eicosatrienoic acid (C20:3n3)	0.07-0.47	(Metsoviti, Katsoulas, et al. 2019)
Arachidonic acid (C20:4n6)	0.20	(Metsoviti, Katsoulas, et al. 2019)
Eciosapentanoic acid (C20:5n3)	0.06-0.36	(Metsoviti, Katsoulas, et al. 2019)

MUFA, PUFA, SFA: Monounsaturated fatty acid, polyunsaturated fatty acid, saturated fatty acid, respectively.

Carbohydrates

Carbohydrates can be present in microalgae as mono-, oligo-, and polysaccharides. Microalgae can synthesize many structurally different carbohydrates with different types of glycosidic bonds. Therefore, they are also classified as homo- and hetero- or sulfated polysaccharides, depending on the constitutive monosaccharides (pentoses or hexoses), the non-sugar substitutes and the glycosidic linkages.

Moreover, carbohydrates can be found attached to proteins (glycoproteins) or lipids (glycolipids) (Nguyen 2016). Complex polysaccharides have different roles in the microalgae physiology: 1) they have structural functions, generally composing the major structural features of the cell wall; 2) they have metabolic and energetic functions; 3) and they are implicated in cellular communication and recognition sites (Arad and Levy-Ontman 2010). Approximately 10% of the total carbohydrates comprise cell wall polysaccharides (Bernaerts et al. 2018), while the remaining carbohydrates are mainly storage polysaccharides, such as starch and chrysolaminarin (Safi et al. 2014).

Carbohydrates in *C. vulgaris* are mainly composed by glucose (73%) mostly in form of starch, which account from 22 to 55% of the total biomass (A. S. Ferreira et al. 2020). *C. vulgaris* as Chlorophyta has the characteristic to synthesize starch in the form of amylopectin and amylose that are two glucose polymers (Busi et al. 2015). *C. vulgaris* has been reported to contain between 30-40% of the linear α-glucan amylose, which is similar to commercial potato (35%) and cereal starch (21–32%). In vitro digestibility of *C. vulgaris* starch is higher than that of the commercial potato starch, becoming more available for enzymes degradation. Nevertheless, the glycemic index of both starches is similar.

Non-starch carbohydrates in *Chlorella* are formed by galactose (10.3%), uronic acids (8%), and rhamnose (4%), with little amounts of xylose (1.7%), arabinose (0.4%) and fucose (0.3%) (Ferreira et al. 2020). The cell wall can be composed of glucosamine residues and glucose-mannose polysaccharides. *C. vulgaris* contains glucosamine as a chitin-like glycan which matrix is dominated by neutral sugar residues rhamnose and galactose. *Chlorella* can also produce extracellular polysaccharides mainly composed by galactose, followed by uronic acids, mannose, rhamnose, glucose, xylose, residues of glucosamine and sulphate esters, from higher to lower content (A. S. Ferreira et al. 2020), as products from photosynthesis (Williams and Laurens 2010). For this reason, the production of exopolysaccharides has been suggested as a way to monitor the photosynthetic activity of microalgae (Pignolet et al. 2013).

As mentioned for lipid and proteins, microalgae carbohydrate content can be regulated by their growth conditions. Among these cultivation and environmental factors such as nutrient starvation, salt stress, light intensity, and temperature could affect the carbohydrate content of microalgae (Buono et al. 2014). Moreover, sugar content is highly determined by the type of carbon source and metabolism process, autotrophic, heterotrophic or mixotrophic. Carbohydrate content is the main cellular component of *C. vulgaris.* Batch cultivated heterotrophically in a stirred tank reactor showed a decrease of carbohydrate content overgrowth (Canelli et al. 2020). The reduction in the carbohydrate content is probably due to conversion to other nutrients, such as lipids and proteins and to catabolic functions that occur during starvation. Carbohydrates productivity in *C. sorokiniana* was enhanced by semi-continuous mixotrophic cultivation using acetate as an organic carbon source reaching 695 mg/L/day.

Carbohydrates also increased with increased CO_2 (v/v %) concentration (Madhubalaji et al. 2019) in gas sparkling. About the color of light, it has been proved that the carbohydrate content of the cells is significantly higher, especially under yellow light (Sánchez-Saavedra et al. 2020).

Minerals and Vitamins

Chlorella contains up to 9% of total minerals such as Ca, K, P, Cu, Zn, Mg, Mn, and Fe. The most abundant minerals are calcium, potassium and phosphorus. Whereas the most abundant vitamins in *C. vulgaris* are vitamin C and E followed by retinol and niacin. *Chlorella* contains both water-soluble (C and B) and fat soluble (A, E) vitamins.

Table 3. Content of minerals in *C. vulgaris*

	Content (mg/100g)	References
Minerals		
Ca	33.04-2464	Kumar P, 2017; Prabakaran et al., 2019; Tokuşoglu & Ünal, 2003
K	49.92-1909	Kumar P, 2017; Tokuşoglu & Ünal, 2003
P	1761.5	Tokuşoglu & Ünal, 2003
Cu	0.06-1.86	Prabakaran et al., 2019; Tokuşoglu & Ünal, 2003
Zn	1.19-5.44	Prabakaran et al., 2019; Tokuşoglu & Ünal, 2003
Mg	37.3-344.3	Prabakaran et al., 2019; Tokuşoglu & Ünal, 2003
Mn	2.09-11.32	Prabakaran et al., 2019; Tokuşoglu & Ünal, 2003
Na	1308-1346.4	Kumar P, 2017; Tokuşoglu & Ünal, 2003
Fe	31.88-259.1	Prabakaran et al., 2019; Tokuşoglu & Ünal, 2003

Table 4. Content of vitamins in *C. vulgaris*

	Content (mg/100g)	References
Vitamins		
B_1 (Thiamine)	0.62-1.77	Liu & Hu, 2013; Prabakaran et al., 2019(Liu and Hu 2013; Prabakaran et al. 2019)(Liu and Hu 2013; Prabakaran et al. 2019)
B_2(Riboflavin)	2.0-5.68	Edelmann et al., 2019; J Liu & Hu, 2013; Prabakaran et al., 2019
B_3 (Niacin)	13.3-30.0	Edelmann et al., 2019; Prabakaran et al., 2019
B_6 (Pyridoxine)	1.4-2.3	Liu & Hu, 2013; Prabakaran et al., 2019
B_9 (Folate)	1.39-2.59	Edelmann et al., 2019
B_{12} (Cobalamin)	0.021-0.24	Edelmann et al., 2019; Jalilian, Najafpour, & Khajouei, 2019; Prabakaran et al., 2019
C (Ascorbic acid)	26-412	J Liu & Hu, 2013; Salvia, Mirzah, Marlida, & Purwati, 2014
E (Tocopherol)	23-132	J Liu & Hu, 2013; Salvia et al., 2014
A (Retinol)	2.08-11.8	J Liu & Hu, 2013; Prabakaran et al., 2019

Comparing four commercial *Chlorella* powder (three *C. vulgaris* and one *C. protechoides*), it has been reported that riboflavin ranges from 20 to 35 μg/g, total niacin from 0.15 to 0.30 mg/g, folate from 13.9 to 25.9 μg/g and B12 from 0 to 2.4 μg/g (Edelmann et al. 2019). Therefore, in one of the four products analysed the active B12 content in 5 g was able to provide more than the RDA. Although light and temperature greatly influence biomass production, they do not affect the ash content (minerals) of the algal biomass. (Metsoviti, Papapolymerou, et al. 2019). The effect of the light on vitamins production has been studied under photoautotrophic conditions in airlift photobioreactor with a transparent draft tube. It has been demonstrated that the highest vitamin B_{12} content was achieved under high light intensity (400 $\mu mol/m^2s^1$). Vitamin B_{12} content decrease when increase gas phase CO_2

concentration (Madhubalaji et al. 2019). In addition, it has been reported that vitamin E content is higher in *Chlorella* at pH 10 than at pH 6, whereas, there is a decline of vitamin C content at both pH (Khalil et al. 2010). Mineral and vitamin contents in *Chlorella* biomass are shown in Table 3 and Table 4.

Bioactive Compounds

Pigments

The pigment composition in *Chlorella* species includes carotenoids and chlorophylls. Chlorophylls are situated in the thylakoids (Guedes, Amaro, and Malcata 2011; Varela et al. 2015). Carotenoids comprise pigments, derived from tetraterpenes which consist of eight isoprene units comprising 40-carbon polyene chain (Guedes, Amaro, and Malcata 2011). The characteristic of carotenoids is light absorption features in the high-energy zone of the visible region (400–500 nm). Carotenoids are essential for photosynthesis, overall for life in the presence of oxygen (Zaghdoudi et al. 2017; Guedes, Amaro, and Malcata 2011). These hydrocarbons are divided into two groups: hydrocarbon carotenoids, known as carotenes (α-carotene, β-carotene, γ-carotene, phytoene, and lycopene) and xanthophylls with oxygen, carbon, hydrogen in their molecular structure. Oxygen may be present in the form of hydroxyl (lutein and zeaxanthin), as an oxy group (canthaxanthin and echinenone) or a mixture of both. In the case of astaxanthin and violaxanthin, the second group are carotenes with no oxygen, containing only carbon and hydrogen atoms (α-carotene, β-carotene, γ-carotene, phytoene, and lycopene) (Novoveská et al. 2019). Table 5. shows the content of these pigments extracted from *C. vulgaris*.

According to Table 5, the most abundant pigment in *C. vulgaris* is chlorophyll a, which comprises of 0.3-3% of dry weight. It could be observed a wide range in pigments content for *C. vulgaris* mainly due to different growth conditions and extraction techniques utilized in these analyses. A lot of studies have optimized some extraction conditions of pigments from *C. vulgaris* by using different extraction techniques such as ultrasound assisted extraction, pressurized liquid extraction (PLE), maceration, Soxhlet extraction, and different solvent (acetone, ethanol, 2-methyltetrahydrofuran and methanol) (Damergi et al. 2017; Safafar et al. 2016; Hynstova et al. 2018; Kwang et al. 2010). It has been reported PLE as the most efficient extraction than others. Temperature is an influent parameter on the extraction of carotenoids and chlorophylls, and at temperatures over 110°C, PLE minimized the formation of pheophorbide a harmful chlorophyll derivative. These results indicate that PLE could be a useful extraction method for the simultaneous extraction of carotenoids and chlorophylls (Kwang et al. 2010).

The most important factors that affect the pigment content in *Chlorella* are temperature, irradiance, pH, availability and source of nitrogen, salinity (or ionic strength) and presence of oxidizing substances (or redox potential). However, the specific growth rate also plays a crucial role. It has been reported a higher content of carotenoids in continuous light at temperatures of 30-35°C. Nevertheless, chlorophyll is reduced due to photooxidation, whereas carotenoids content increases to protect photooxidative damage of the cell (Sharma 2012).

Table 5. Content of pigments in Chlorella

Pigments	Content (mg/100 g dry weight (d.w.))	References
β-carotene	0.7-101.3	(Damergi et al. 2017; Kwang et al. 2010; Safafar et al. 2016; Hynstova et al. 2018)
Astaxanthin	6.0-80.1	(Damergi et al. 2017; Hynstova et al. 2018)
Violaxanthin	4	(Damergi et al. 2017)
Zeaxanthin	7.4-128.3	(Hynstova et al. 2018)
Lutein	8.0–714.1	(Derwenskus et al. 2019; Safafar et al. 2016; Damergi et al. 2017; Hynstova et al. 2018; Kwang et al. 2010)
Canthaxanthin	2.8-116.9	(Damergi et al. 2017; Hynstova et al. 2018)
Lycopene	1.8-2.2	(Damergi et al. 2017)
Chlorophyll a	332.0-3244.4	(Safafar et al. 2016; Kwang et al. 2010)
Chlorophyll b	258.0-577.0	(Kwang et al. 2010)
Pheophorbide a	1.0-515.0	(Kwang et al. 2010)
Pheophytin a	231.0-564.0	(Kwang et al. 2010)

For the irradiance effect, the highest content of chlorophyll a (13.25 μg/mL) is obtained at low irradiance of 37.5 $\mu mol/m^2s^1$, whereas the maximum content of β-carotene (0.07 μg/mL) is obtained at high irradiance of 100 $\mu mol/m^2s^1$ (Seyfabadi, Ramezanpour, and Khoeyi 2011). It has been reported the highest content of chlorophyll and biomass of *C.vulgaris* at 6% of CO_2 level, which is 60 and 20 times more than the obtained at ambient CO_2 (0.036%) (Singh and Singh 2014). Nevertheless, the production of lutein in a pilot-scale in *C. vulgaris* has shown a decrease of 58.11% with the increase of light intensity per cell from 160 $\mu mol/m^2s^1$ to 440 $\mu mol/m^2s^1$ (McClure et al. 2019). However, the irradiance effect depends on indoor or outdoor cultivation due to its difficult to work in the outdoor operation with the solar cycle and temperature fluctuation (Guedes, Amaro, and Malcata 2011). Besides, the color of light influences the pigments content. It has been reported maximum chlorophyll content with green light grown cultures, followed by blue and red lights. Carotenoids increase with all colored light, but green light has shown the lowest increase (El-Sayed and Olfat 2016).

At the optimum pH for biomass production, the highest lutein content is obtained, whereas, the concentration of nitrogen in the form of nitrate should be supplied to a moderate excess, the biomass growth continued even after extracellular nitrogen depletion. It has been reported an increase in total carotenoid content from 0.05 mg/g in N replete to 4.5 mg/g after 14 days of N depletion (Mulders et al. 2014). In respect of the salinity, among the various NaCl concentrations tested, 20 g/L solution has given the best results for total carotenoid content (0.80 g/100g d.w.) (Campenni' et al. 2013). In addition, lutein synthesis is improved via the addition of chemicals such as hydrogen peroxide and sodium chlorite, in the presence of Fe^{2+}; they affect the redox state and create stress-inducing chemical species. Moreover, a mutant strain of *C. sorokiniana* showed a promising opportunity to produce lutein in outdoor large-scale cultivation (J. H. Chen et al. 2019). Lutein provides a protective role in the induction of oxidative stress, particularly under heterotrophic growth, where spontaneous oxidative stress is normally absent (McClure et al. 2019; Guedes, Amaro, and Malcata 2011). *Chlorella zofingiensis* has been considered as a potential alternative for astaxanthin production in heterotrophy respect to the consolidated production using *Haematococcus pluvialis* (Liu et al. 2014).

Phenolic Compounds

Phenolic compounds in *Chlorella* are major contributors to the microalgae antioxidant capacity, preventing oxidative stress of cells (Hajimahmoodi et al. 2010). *C.vulgaris* contains small amounts of simple phenols such as phloroglucinol (74000 ng/g), ferulic acid (0.63 - 190000 ng/g), p-coumaric acid (540-190000 ng/g), gallic acid (32000-670000 ng/g), caffeic acid (110000-200000 ng/g) and flavonoids such as apigenin (9.9 ng/g) (Wan Mahmood et al. 2019; Goiris et al. 2014). Whereas, the phenolic compound quantified in *C. sorokiana* are caffeic acid (3.12-3.81 µg/g), ferulic acid (2.8-2.81 µg/g), p-coumaric acid (1.16-1.97 µg/g), cinnamic acid (0.13-0.47 µg/g) and its total content ranged from 7.26 to 9.06 µg/g d.w (Safafar et al. 2015). This content is lower than the obtained in *C. vulgaris* (75-860 µg/g d.w.) (Wan Mahmood et al. 2019; Goiris et al. 2014). The phenolic content depends mainly on light intensity and temperature. It has been reported the maximum content of caffeic acid at 18°C and light intensity of 112 µmol/m^{2}s; whereas, the highest content of ferulic acid is at 18°C and light intensity of 275 µmol /m^{2}s (Sozmen et al. 2018). Moreover, comparing three strains of *Chlorella* grown in different nitrogen concentration, no differences were found among the strains, but lower phenols were detected at higher N concentrations (Aremu et al. 2016).

PRODUCTION TECHNOLOGY OF *CHLORELLA*

Culture Conditions

The biomass productivity during algal cultivation is the result of two competitive processes: photosynthesis and respiration. The photosynthesis rate depends on many parameters, among which the most important are light intensity, temperature, nutrients concentration, CO_2 concentration, and pH. While respiration rates depend mostly on temperature and O_2. Generally, microalgae can be cultivated in any place where enough nutrients, tolerable temperature, and adequate sunlight are available. Even in places where edible crops cannot be cultivated like non-fertile lands, rocks, hilly areas and on rooftops, microalgae can be cultivated (Sarker and Salam 2019). But, to enhance biomass productivity, specific culture conditions and technology must be taken into account. *Chlorella* has a rapid growth rate and responds to each set of growth condition by modifying the yield of a specific component. It is an ideal organism for production because it is remarkably resistant against harsh conditions and invaders. Therefore, many growth techniques have been tested to increase biomass, lipids, proteins, carbohydrates or pigments productivity (Safi et al. 2014). The growing interest in microalgae has led to a diversification of cultivation technologies tailored to the species and the targeted molecules of interest. Open ponds and closed photobioreactors for autotrophic cultivation, and industrial fermenters for heterotrophic productions are the main available technologies (Cvejic et al. 2020). Culture conditions are briefly summarized in Table 6. Harvesting methods are necessary too to dewater and separate *Chlorella* biomass. It is important to remark that cultivation conditions and techniques depend on the target product (Daliry et al. 2017).

Table 6. Culture conditions of microalgae

Culture conditions	Energy source	Carbon source
Autotrophic	Light	Inorganic carbon
Heterotrophic	Organic carbon	Organic carbon
Mixotrophic	Light/organic carbon	Inorganic/organic carbon

Autotrophic Growth

In autotrophic conditions, algae use the energy of light, water, and carbon dioxide to produce their biomass. There are two main technologies used to cultivate microalgae in autotrophy: open pond and closed photobioreactor. The open pond system depends strongly from the environmental parameter (temperature, rain, sunlight, etc.); it is less productive and can be easily contaminated. The photobioreactors instead, provides an easy system of controlling cultivation parameters such as temperature, dissolved CO_2 and pH, and to prevent contaminations. However, photobioreactors have a high initial cost and are very specific to the physiology of microalgae strain being cultivated open pond system is the main technology used for commercial cultivation of microalgae in autotrophy. Open ponds are designed in many different shapes and forms, each having certain advantages and drawbacks. The location in which the pond is situated is a critical factor in determining the type of pond selected, microalgal strain and the amount of light for photosynthesis. Commercially, the open pond configurations for *Chlorella* are mainly circular mixed and raceway-type ponds. Open ponds are limited by key growth parameters, including light intensity, temperature, pH and CO_2 concentration. The construction, operation and maintenance costs of cultivation ponds are lower than photobioreactors. They are less technological in design and are more scalable, although the ponds contain a relatively lower biomass concentration, there is evidence that it has been cultivated in open plate pond (El-Sheekh et al. 2019) and raceway ponds system (Ramírez-López, Perales-Vela, and Fernández-Linares 2019) with good performance results. One of the major issues related to the open ponds system is the contamination by bacteria, protozoa, and other unwanted organisms. In fact, only a few microalgae can be cultivated in open ponds by using extreme culture conditions (i.e., high alkalinity and pH for *Arthrospira (Spirulina) spp*, high salinity for *Dunaliella spp*). Usually, for *Chlorella,* the cultivation strategy to reduce contamination in open systems is to increase the amount of nutrition salts (mainly nitrate) (Lee 2001).

Photobioreactor gives better control on cultivation parameters and until 10-fold higher surface productivity can be achieved compared to open pond systems. One of the most important factors in this type of cultivation is the design especially in high-cell density cultures (like *Chlorella*), due to the difficulties of dealing with cellular self-shading within it. The main objective in the design is the maximization of light penetration in the culture. It is obtained by increasing the surface to volume ratio, and although very different shapes have been designed, they are in general tubular or plate-types. Tubular reactors are considered to be more appropriate for outdoor cultivation, and they can be configured in a vertical, horizontal or inclined plane. The major difference between the configurations is the vertical design that allows the greater mass transfer and a decrease in energy usage. The horizontal reactor, instead, is more scalable but requires a large area of land. On the other hand, flat-plate photobioreactor is favourable due to low power energy consumption and high mass transfer capacity, reduction in oxygen build up, reduced dark volumes and high photosynthetic efficiency. Various designs of photobioreactors to cultivate *Chlorella* have been evaluated with good results using airlift photobioreactor (Zhu et al. 2019; Madhubalaji et al. 2019), or flat plate photobioreactor system (Cheah et al. 2020; Aremu et al. 2016).

Heterotrophic Growth

Heterotrophic cultivation of microalgae has the potential to overcome or reduce the problem of light penetration associated with autotrophic cultivation. This technique does not require light, and the biomass is fed with an organic carbon source such as simple sugars or organic acids (León-Vaz et al. 2019). Heterotrophic growth of microalgae is usually faster than the autotrophic growth. Thus, microalgae are grown in a stirred tank bioreactor or fermenter where higher growth rates and very high biomass concentrations are expected with a consequent lower harvesting costs respect to the autotrophic systems. The carbon sources commonly used for *Chlorella* cultivation are glucose, acetate, glycerol or glutamate with maximum specific growth rate obtained with glucose. Nevertheless, the major disadvantage of this system is the price and availability of sugars, which compete with feedstocks for other uses such as food and biofuel productions. *C. vulgaris* has been cultivated in heterotrophic batch conditions satisfactorily (Canelli et al. 2020), and some companies are selling depigmented *Chlorella* flour obtained in full heterotrophic conditions.

Chlorella is capable of combining both autotrophic and heterotrophic techniques by performing photosynthesis coupled with the metabolization of organic carbon sources. In fact, *Chlorella* is a facultative heterotrophic organism that means that could switch from autotrophic metabolism to a heterotrophic one or using both the metabolic pathways. The mixotrophic cultivation showed higher dry biomass productivity and lipids productivity if compared to the autotrophic mode. The main advantages of mixotrophic metabolism are the following: a lower energy cost due to its relatively lower requirements for light intensities, limiting the impact of biomass loss during dark respiration and reducing the number of organic substrates used for biomass growth. In summary, growth in mixotrophy embraces the positive factors of autotrophic and heterotrophic cultivation, mitigating their negative aspects or at least reducing its impact on the production process (Abreu et al. 2012).

The mixotrophic growth conditions of *C. vulgaris, C. pyrenoidosa and C. protethecoides* have been optimized with good results (Ward and Rehmann 2019; Liu et al. 2018; Sim et al. 2019).

Sustainable Production Methods

Heterotrophic and mixotrophic cultivation of algae is promising way to produce a high quantity of biomass. Nevertheless, the high cost for obtaining organic carbon sources is still a great concern. In recent years, there has been a great focus to study a more sustainable way to produce algae biomass. First of all, it has been studied the capacity of many algae, in particular *Chlorella*, to recycle organic components from the food industry. In fact, many food industry by-products and agro-industrial food waste are rich in sugars, proteins and other nutrients that could be used after simple pre-treatments as media for mixotrophic and heterotrophic microalgae growth preserving the safety of *Chlorella* biomass also for human consumption.

Agricultural waste molasses, rich in simple sugars and minerals, were successfully used as an organic carbon source for the growth of *Chlorella vulgaris* reducing the production costs (Mohammad Mirzaie et al. 2016). Cheese whey, a dairy by-product, is also a great source of nutrient for algae cultivation due to the quantity of lactose, protein and mineral salts present. *Chlorella protothecoides* has been successfully grown on a media based on cheese whey (Girard et al. 2017), or hydrolyzate of cassava bagasse (Chen et al. 2015). The mixotrophic cultivation of *C. sorokininana* was demonstrated effective using the acetate-rich wine waste lees (León-Vaz et al. 2019) Some research groups are currently working to enhance the nutrient recovery from food by-products and to test the optimal pre-treatments to increase the bioavailability of nutrients by microalgae.

From an economic point of view, this kind of approach in cultivation could lead to a more sustainable biomass production, reducing the feed cost of bioreactor up to 30% (Wang et al. 2020).

Harvesting Methods

Floculation

Chlorella is spherical shaped, and it is a very thin eukaryotic photosynthetic organism (~2 to 10 μm in diameter); therefore, harvesting is difficult and represents one of the main costs in production. Flocculation can be used as an initial dewatering step in order to reduce the cost of further processing. During the exponential growth phase, the algal cells have high negative surface charge, and they are difficult to neutralize, and thus the cells remain dispersed. After reaching the stationary or the declining phase, the negative charge decreases, allowing the cells to aggregate and to form lumps, thereby resulting in a process called auto-flocculation. This phenomenon is associated with elevated pH associated with photosynthesis, nitrate and phosphate assimilation. Moreover, auto-flocculation can occur by interactions between microalgae and bacteria or excreted organic molecules or by simply cutting CO_2 supply during photosynthesis; the latter method is less expensive but time consuming. In general, the culture of microalgae is very stable, and auto-flocculation probability is negligible and sometimes misleading. To accelerate coagulation, it is necessary to increase the pH by adding a base. The most effective is sodium hydroxide, which induces more than 90% of flocculation at pH 11. But on an industrial scale, lime seems to be the most costly-efficient. Recent studies have shown good *Chlorella* flocculation with cationic biopolymers as ε-poly-l-lysine (Noh et al. 2019), with calcium phosphate precipitates (Leite et al. 2019), metal sulfates (Zhu et al. 2020) and the most novels bioflocculants as waste chicken eggshells and clam shells (Suparmaniam et al. 2020).

Centrifugation

The most common harvesting technique for *Chlorella* biomass is centrifugation because it is highly efficient (95% recovery), not time consuming, and it is possible to treat large volumes. In addition, the morphology of *Chlorella* permits high centrifugal stress without damaging its structure during the process. Centrifugation can be combined with flocculation, flotation and filtration to maximise recovery of the biomass (Dassey and Theegala 2013).

Filtration

This method involves continuous passing of the microalgae culture across a filter on which algal cells will constantly concentrate until it reaches a certain thickness. Due to the small size of *Chlorella,* conventional filtration is not an adequate method to be applied. Instead, ultrafiltration or microfiltration are more efficient. Moreover, microfiltration and ultrafiltration are affected by different parameters such as filter type; transmembrane pressure, flow velocity, turbulent cross-flow and growth phase, and therefore a compromise that takes into consideration these parameters should be made. Low capital, energy and maintenance costs coupled with high harvesting efficiency are the main advantage of membrane filtration for *Chlorella* (Ahmad et al. 2014).

Nevertheless, there are recent studies that have shown dewatering effectively using dynamic filtration with acrylonitrile butadiene styrene membranes (Hapońska et al. 2018).

Flotation

There is very limited evidence of its feasibility, but this method consists of trapping the cells using dispersed micro-air bubbles and can be further helped by using surfactant-aided dispersed air (Alhattab and Brooks 2020). Flotation can also occur naturally when the lipid content in microalgae increases. It has been induced effective flotation on *C. vulgaris* by using dispersed ozone gas (Cheng et al. 2010), and electro-coagulation-flotation (Wong et al. 2017). Thus, unlike flocculation, this method does not require synthetic chemicals, but its economic viability is not yet well known, especially on an industrial scale. After harvesting the biomass is spray or drum-dried and the powder sold directly or used to make tablets or capsules.

Cell Disruption

Chlorella biomass, because of its rigid cellulosic cell wall, must be subjected to cell disruption after harvesting, to increase the digestibility of the whole product or to extract compounds of interest from the cells. Mechanical disruption methods of *Chorella* cells, such as bead milling, are reported to be quite efficient to increase protein digestibility and protein or carotenoids extraction efficiency, but sometimes they create cell debris that is hard to separate in industrial processes. As alternative cell disruption methods, pulsed-electric fields (PEF) (Scherer et al. 2019), ultrasound assisted extraction (Skorupskaite et al. 2019) and pressurized liquid extraction (Kwang et al. 2010), have also been investigated and have already been applied in *Chlorella* with satisfactory results.

It has been shown in sequential PEF and high pressure homogenization treatments of *C. vulgaris*, a high permeabilization degree and extractability of intracellular compounds (Carullo et al. 2018). Commercially it is possible to find *Chlorella* powder with cell wall disrupted by patented low-pressure flash expansion and or by high impact jet-spray drying with increased digestibility (Muys et al. 2019).

SCIENTIFIC EVIDENCE OF HEALTH BENEFITS OF *CHLORELLA VULGARIS*

Administration of *Chlorella* or *Chlorella* extracts has been shown to play positive roles in health care and disease prevention, such as lowering blood pressure (Rani, Sandal, and Sahoo 2018), cholesterol levels and the risk of heart diseases, accelerating wound healing, enhancing the immune system (Bishop and M. Zubeck 2012) and having a positive effect in reducing gastric ulcers, constipation, anemia, hypertension and diabetes (de Mello-Sampayo et al. 2013). It also possesses anti-inflammatory (Zhang et al. 2019a; Prabakaran et al. 2019), anti-cancer, and neuroprotective effects (Chan, Mong, and Yin 2009; Arnal et al. 2010; Novoveská et al. 2019).

Chlorella is an important source of antioxidant peptides. A bioactive peptide with free radical scavenging activity has been isolated after hydrolysis with pepsin. This kind of bioactive peptide has the potential to be a good dietary supplement for the prevention of oxidative stress-related diseases such as atherosclerosis, coronary heart disease and cancer (Sheih, Fang, and Wu 2009). Recently, a decapeptide has also been obtained after enzymatic hydrolysis of *C. pyrenoidosa*. That peptide showed inhibitory activity on pancreatic lipase, which could slow fatty acid synthesis and fat accumulation in the adipocyte (Zhang et al. 2019b).

This microalga is also able to modulate and attenuate the progression of liver fibrosis (Mohseni et al. 2019; Jaafar, Durani, and Makpol 2020), to promote muscle regeneration and ameliorate sarcopenia (Zainul Azlan et al. 2019), and also antimicrobial activity and makes protection against oxidative stress (Sikiru et al. 2019). Specifically, the use of carotenoids from *Chorella* as dietary supplements could neutralize free radicals and thus could prevent or decrease chronic diseases, cellular damage and aging (Griffiths et al. 2016). These pigments have multiple therapeutic properties, such as a protective effect against age-related macular degeneration and senile cataract (Chew 2013). β-carotene is one of the most important carotenoids because it has an active form of provitamin A (Gammone, Riccioni, and D'Orazio 2015). In addition, astaxanthin has the second most important carotenoid that has been sold as a nutraceutical and has a great antioxidant activity, which is 100 times higher than α-tocopherol (Sathasivam et al. 2019). Astaxanthin suppresses reactive oxygen species (ROS) formation. It modulates the expression of oxidative stress-responsive enzymes such as heme oxygenase-1 (HO-1), which is a marker of oxidative stress and a regulatory mechanism implicated in the cell adaptation against oxidative damage (Davinelli, Nielsen, and Scapagnini 2018).

There are some polysaccharides from *Chlorella* that showed significant antioxidant properties, tumor activities, and immunomodulatory properties. In particular a high molecular weight polysaccharide has been identified in *C. pyrenoidosa*. Named Immurella, it showed higher activity against cancer, when compared to the fungal polysaccharides such as schizoplyllan (Andrade et al., 2018).

Another component of the carbohydrate family is β-(1,3)-glucan. This compound has been isolated from cell wall of *Chlorella vulgaris* and it has been widely studied for its antitumoral potential action in human health, and also of other positive properties, such as immunostimulator, free radical scavenger and blood lipids reducer (Laroche and Michaud 2008). Randomized controlled trials on 797 subjects indicated that diet supplementation with

Chlorella, significantly decrease the levels of total cholesterol (from 9.09 to 5.26 mg/dl) lowering cardiovascular risk (Fallah et al. 2018).

FOOD APPLICATIONS OF *CHLORELLA VULGARIS*

Chlorella is the sunlight-driven single-cell factory for protein, lipids, carbohydrates, pigments, vitamins, and minerals. It has long been used as health food and additives for human consumption. Nowadays, Japan and China are the main Chlorella-producing countries. The produced *Chlorella* is commercialized mainly in the form of dried powder, tablets, liquid extracts, capsules or *Chlorella* growth factor (water-soluble extract that contain, among other nutrients, polysaccharides, glycoproteins and β-glucans (Barkia, Saari, and Manning 2019)) for human consumption (Lin 2005). In addition, extracts from these microalgae have also been incorporated in the form of pasta, snacks, sweets, and beans and marketed as a food supplement in the form of food additives, dyes, and emulsifiers, improving the nutritive properties of these products (Vaz et al. 2016). It has been added extract of *Chlorella* in fresh spaghetti and compared the quality of the pasta and the cooking, colour, texture, and sensorial parameters discovering that it originated a product with better chemical composition without affecting the cooking quality (Fradique et al. 2010).

The high protein content of *Chlorella* makes it an ideal candidate for a techno-functional use in the food industry. Protein extracts from *Chlorella sorokiniana* have shown good emulsifying properties, better than other food proteins (Ebert et al. 2019). In particular, soluble proteins of *Chlorella* are interesting as emulsifier thanks to their potential to stabilize acidified emulsions.

A study evaluated the possibility to use protein extract from *C. protothecoides* in beverage formulations, proving that it could be applied in acidified protein-rich beverages because of their high solubility and promising sensory properties (Grossmann et al. 2020). Another study from the same author remarked that protein aggregates from *C. protothecoides* might be useful for the formulation of foods with higher viscosities, or foods that are gelled, acting as a filler to strengthen the textures (Grossmann, Hinrichs, and Weiss 2019). A positive impact on viscoelastic characteristics of bread has been observed in a recent study using *C. vulgaris*, strengthening the gluten network and bread texture (Graça et al. 2018).

Also, some authors have used *Chlorella* for other technological uses like giving a greenish colour to buttery biscuits (Gouveia et al. 2007) and emulsions (Gouveia et al. 2006), or extending the shelf life of rainbow trout fillets during refrigerated storage (Taghavi Takyar, Haghighat Khajavi, and Safari 2019).

OTHER USES

Chlorella can be used for animal nutrition (Kotrbáček, Doubek, and Doucha 2015) and as a supplement in broiler feed (Abdelnour et al. 2019). It has been estimated that 30% of algae produced worldwide are used for animal feed. Great results are achieved with fish feed, were *Chlorella* addition in gibel carp diet proved to increase the growth performances, and to promote immune-response and activity of digestive enzymes (Xu et al. 2014).

Supplementation of rabbits, broilers and chicks diet with a small amount of *Chlorella* showed better growth performances and health than using the control diets (Abdelnour et al. 2019).

In addition, the green alga proves to be beneficial to environmental clean-up, such as bioremediation of industrial flue gases and wastewaters. Recently, due to the blooming of renewable energy, *Chlorella* has attracted unprecedented interest as a feedstock as a source of biofuels. Moreover, microalgae are raising interest as platforms for gene expression, although they do not reach a stage as mature as bacteria, yeast, mammalian cell, it shows substantial advantages and has also been proved feasible in *Chlorella* species (Williams and Laurens 2010).

Funding

This study is part of ERA-Net SUSFOOD2 project SUSPUFA, ID 145, funding provided by Spanish Ministry of Science, Education and Universities/ Spanish State Research Agency (PCI2018-093178) and co-funding by the European Union's Horizon 2020 research and innovation program.

Acknowledgments

Vito Verardo and Elisa Giambanelli thank the Spanish Ministry of Economy and Competitiveness (MINECO) for "Ramon y Cajal" contract (RYC-2015-18795) and "Juan de la Cierva-formación" contract (FJCI-2017-33216), respectively. Maria del Carmen Razola-Diaz would like to thank to the University of Granada for the "Beca de Iniciación a la Investigación para estudiantes de másteres oficiales" granted by the "Plan Propio 2019" program.

References

Abdelnour, S A, M E Abd El-Hack, M Arif, A F Khafaga, and A E Taha. 2019. "The Application of the Microalgae Chlorella Spp. as a Supplement in Broiler Feed." *World's Poultry Science Journal* 75 (2). Taylor & Francis: 305–318. doi:10.1017/S0043933919000047.

Abreu, Ana P., Bruno Fernandes, António A. Vicente, José Teixeira, and Giuliano Dragone. 2012. "Mixotrophic Cultivation of Chlorella Vulgaris Using Industrial Dairy Waste as Organic Carbon Source." *Bioresource Technology* 118. Elsevier Ltd: 61–66. doi:10.1016/j.biortech.2012.05.055.

Ahmad, A L, N H Mat Yasin, C J C Derek, and J K Lim. 2014. "Comparison of Harvesting Methods for Microalgae Chlorella Sp. and Its Potential Use as a Biodiesel Feedstock." *Environmental Technology* 35 (17). Taylor & Francis: 2244–2253. doi:10.1080/09593330.2014.900117.

Alhattab, Mariam, and Marianne Su Ling Brooks. 2020. "Optimization of Chlorella Saccharophila Harvesting by Surfactant-Aided Dispersed Air Flotation for Biodiesel

Production Processes." *Biomass and Bioenergy* 134 (January). Elsevier Ltd: 105472. doi:10.1016/j.biombioe.2020.105472.

Anto, Susaimanickam, Arivalagan Pugazhendhi, and Thangavel Mathimani. 2019. "Lipid Enhancement through Nutrient Starvation in Chlorella Sp. and Its Fatty Acid Profiling for Appropriate Bioenergy Feedstock." *Biocatalysis and Agricultural Biotechnology* 20 (May). Elsevier Ltd: 101179. doi:10.1016/j.bcab.2019.101179.

Arad, Shoshana, and Oshrat Levy-Ontman. 2010. "Red Microalgal Cell-Wall Polysaccharides: Biotechnological Aspects." *Current Opinion in Biotechnology* 21 (3). Elsevier Ltd: 358–364. doi:10.1016/j.copbio.2010.02.008.

Aremu, Adeyemi O., Nqobile A. Masondo, Zoltan Molnár, Wendy A. Stirk, Vince Ördög, and Johannes Van Staden. 2016. "Changes in Phytochemical Content and Pharmacological Activities of Three Chlorella Strains Grown in Different Nitrogen Conditions." *Journal of Applied Phycology* 28 (1): 149–159. doi:10.1007/s10811-015-0568-7.

Arnal, E., M. Miranda, J. Barcia, F. Bosch-Morell, and F. J. Romero. 2010. "Lutein and Docosahexaenoic Acid Prevent Cortex Lipid Peroxidation in Streptozotocin-Induced Diabetic Rat Cerebral Cortex." *Neuroscience* 166 (1). Elsevier Inc.: 271–278. doi:10.1016/j.neuroscience.2009.12.028.

Baiee, M. A., and J. M. Salman. 2016. "Effect of Phosphorus Concentration and Light Intensity on Protein Content of Microalga Chlorella Vulgaris." *Mesopotamia Environmental Journal* 2 (2): 75–86. doi:10.1039/c6ta09665d.

Barkia, Ines, Nazamid Saari, and Schonna R. Manning. 2019. "Microalgae for High-Value Products towards Human Health and Nutrition." *Marine Drugs* 17 (5): 1–29. doi:10. 3390/md17050304.

Becker, E. W. 2007. "Micro-Algae as a Source of Protein." *Biotechnology Advances* 25 (2): 207–210. doi:10.1016/j.biotechadv.2006.11.002.

Bernaerts, Tom M. M., Lore Gheysen, Clare Kyomugasho, Zahra Jamsazzadeh Kermani, Stéphanie Vandionant, Imogen Foubert, Marc E. Hendrickx, and Ann M. Van Loey. 2018. "Comparison of Microalgal Biomasses as Functional Food Ingredients: Focus on the Composition of Cell Wall Related Polysaccharides." *Algal Research* 32 (March). Elsevier: 150–161. doi:10.1016/j.algal.2018.03.017.

Buono, Silvia, Antonio Luca Langellotti, Anna Martello, Francesca Rinna, and Vincenzo Fogliano. 2014. "Functional Ingredients from Microalgae." *Food and Function* 5 (8): 1669–1685. doi:10.1039/c4fo00125g.

Busi, Maria V, Diego F Gomez-Casati, Mariana Martín, Julieta Barchiesi, Mauricio J Grisolía, Nicolás Hedín, and Julieta B Carrillo. 2015. "Starch Metabolism in Green Plants." In *Polysaccharides: Bioactivity and Biotechnology*, edited by Kishan Gopal Ramawat and Jean-Michel Mérillon, 329–376. Cham: Springer International Publishing. doi:10.1007/978-3-319-16298-0_78.

Campenni,' L., B. P. Nobre, C. A. Santos, A. C. Oliveira, M. R. Aires-Barros, A. M. F. Palavra, and L. Gouveia. 2013. "Carotenoid and Lipid Production by the Autotrophic Microalga Chlorella Protothecoides under Nutritional, Salinity, and Luminosity Stress Conditions." *Applied Microbiology and Biotechnology* 97 (3): 1383–1393. doi:10.1007/s00253-012-4570-6.

Canelli, Greta, Lukas Neutsch, Roberta Carpine, Sabrina Tevere, Francesca Giuffrida, Zhen Rohfritsch, Fabiola Dionisi, Christoph J. Bolten, and Alexander Mathys. 2020. "Chlorella

Vulgaris in a Heterotrophic Bioprocess: Study of the Lipid Bioaccessibility and Oxidative Stability." *Algal Research* 45 (October 2019). doi:10.1016/j.algal.2019.101754.

Cao, Song, Xu Zhou, Wenbiao Jin, Feng Wang, Renjie Tu, Songfang Han, Hongyi Chen, Chuan Chen, Guo Jun Xie, and Fang Ma. 2017. "Improving of Lipid Productivity of the Oleaginous Microalgae Chlorella Pyrenoidosa via Atmospheric and Room Temperature Plasma (ARTP)." *Bioresource Technology* 244. Elsevier Ltd: 1400–1406. doi:10.1016/j.biortech.2017.05.039.

Carullo, Daniele, Biresaw Demelash Abera, Alessandro Alberto Casazza, Francesco Donsì, Patrizia Perego, Giovanna Ferrari, and Gianpiero Pataro. 2018. "Effect of Pulsed Electric Fields and High Pressure Homogenization on the Aqueous Extraction of Intracellular Compounds from the Microalgae Chlorella Vulgaris." *Algal Research* 31 (February). Elsevier: 60–69. doi:10.1016/j.algal.2018.01.017.

Chan, Kung Chi, Mei Chin Mong, and Mei Chin Yin. 2009. "Antioxidative and Anti-Inflammatory Neuroprotective Effects of Astaxanthin and Canthaxanthin in Nerve Growth Factor Differentiated PC12 Cells." *Journal of Food Science* 74 (7): 225–231. doi:10.1111/j.1750-3841.2009.01274.x.

Cheah, Wai Yan, Pau Loke Show, Yee Jiun Yap, Hayyiratul Fatimah Mohd Zaid, Man Kee Lam, Jun Wei Lim, Yeek Chia Ho, and Yang Tao. 2020. "Enhancing Microalga Chlorella Sorokiniana CY-1 Biomass and Lipid Production in Palm Oil Mill Effluent (POME) Using Novel-Designed Photobioreactor." *Bioengineered* 11 (1). Taylor & Francis: 61–69. doi:10.1080/21655979.2019.1704536.

Chen, Jih Heng, Chun Yen Chen, Tomohisa Hasunuma, Akihiko Kondo, Chien Hsiang Chang, I. Son Ng, and Jo Shu Chang. 2019. "Enhancing Lutein Production with Mixotrophic Cultivation of Chlorella Sorokiniana MB-1-M12 Using Different Bioprocess Operation Strategies." *Bioresource Technology* 278 (January). Elsevier: 17–25. doi:10.1016/j.biortech.2019.01.041.

Chen, Jin Liu and Feng. 2016. "Biology and Industrial Applications of Chlorella: Advances and Prospects." *Adv Biochem Eng Biotechnol (2016) 153: 1–35* 123: 1–35. doi:10.1007/10_2014_286.

Chen, Junhui, Xiaoguang Liu, Dong Wei, and Gu Chen. 2015. "High Yields of Fatty Acid and Neutral Lipid Production from Cassava Bagasse Hydrolysate (CBH) by Heterotrophic Chlorella Protothecoides." *Bioresource Technology* 191. Elsevier Ltd: 281–290. doi:10.1016/j.biortech.2015.04.116.

Cheng, Ya Ling, Yu Chuan Juang, Guan Yu Liao, Shih Hsin Ho, Kuei Ling Yeh, Chun Yen Chen, Jo Shu Chang, Jhy Chern Liu, and Duu Jong Lee. 2010. "Dispersed Ozone Flotation of Chlorella Vulgaris." *Bioresource Technology* 101 (23). Elsevier Ltd: 9092–9096. doi:10.1016/j.biortech.2010.07.016.

Chew, Emily Y. 2013. "Lutein/Zeaxanthin for the Treatment of Age-Related Cataract." *JAMA Ophthalmology* 131 (7): 843. doi:10.1001/jamaophthalmol.2013.4412.

Cvejic, Jelena Helene, Antonio Luca Langellotti, Hubert Bonnefond, Vito Verardo, and Olivier Bernard. 2020. "*Lipids and Edible Oils. Properties, Processing and Applications*." In, edited by CHARIS M. GALANAKIS, 175–210. London: Elsevier Inc.

Daliry, S., A. Hallajisani, J. Mohammadi Roshandeh, H. Nouri, and A. Golzary. 2017. "Investigation of Optimal Condition for Chlorella Vulgaris Microalgae Growth." *Global*

Journal of Environmental Science and Management 3 (2): 217–230. doi:10.22034/gjesm.2017.03.02.010.

Damergi, Eya, Jean Paul Schwitzguébel, Dominik Refardt, Shivom Sharma, Christof Holliger, and Christian Ludwig. 2017. "Extraction of Carotenoids from Chlorella Vulgaris Using Green Solvents and Syngas Production from Residual Biomass." *Algal Research* 25 (June). Elsevier: 488–495. doi:10.1016/j.algal.2017.05.003.

Dassey, Adam J., and Chandra S. Theegala. 2013. "Harvesting Economics and Strategies Using Centrifugation for Cost Effective Separation of Microalgae Cells for Biodiesel Applications." *Bioresource Technology* 128. Elsevier Ltd: 241–245. doi:10.1016/j.biortech.2012.10.061.

Davinelli, Sergio, Michael E. Nielsen, and Giovanni Scapagnini. 2018. "Astaxanthin in Skin Health, Repair, and Disease: A Comprehensive Review." *Nutrients* 10 (4): 1–12. doi:10.3390/nu10040522.

de Mello-Sampayo, Cristina, M. Luisa Corvo, Rogério Mendes, Diana Duarte, Joana Lucas, Rui Pinto, Ana Paula Batista, et al. 2013. "Insights on the Safety of Carotenogenic Chlorella Vulgaris in Rodents." *Algal Research* 2 (4). Elsevier B.V.: 409–415. doi:10.1016/j.algal.2013.07.005.

Derwenskus, Felix, Felix Metz, Andrea Gille, Ulrike Schmid, Karlis Briviba, Ursula Schließmann, and Thomas Hirth. 2019. "Pressurized Extraction of Unsaturated Fatty Acids and Carotenoids from Wet Chlorella Vulgaris and Phaeodactylum Tricornutum Biomass Using Subcritical Liquids," no. July 2018: 335–344. doi:10.1111/gcbb.12563.

Ebert, Sandra, Lutz Grossmann, Jörg Hinrichs, and Jochen Weiss. 2019. "Emulsifying Properties of Water-Soluble Proteins Extracted from the Microalgae: Chlorella Sorokiniana and Phaeodactylum Tricornutum." *Food and Function* 10 (2). Royal Society of Chemistry: 754–764. doi:10.1039/c8fo02197j.

Edelmann, Minnamari, Sanni Aalto, Bhawani Chamlagain, Susanna Kariluoto, and Vieno Piironen. 2019. "Riboflavin, Niacin, Folate and Vitamin B12 in Commercial Microalgae Powders." *Journal of Food Composition and Analysis* 82 (January). Elsevier: 103226. doi:10.1016/j.jfca.2019.05.009.

El-Sayed, A. B, and M. A. Olfat. 2016. "Effect of Different Light Wave Lengths on the Growth and Oil Content of Chlorella Vulgaris." *Egyptian J. of Phycol* 17.

El-Sheekh, Mostafa M., Saly F. Gheda, Abo El Khair B. El-Sayed, Atef M. Abo Shady, Mai E. El-Sheikh, and Michael Schagerl. 2019. "Outdoor Cultivation of the Green Microalga Chlorella Vulgaris under Stress Conditions as a Feedstock for Biofuel." *Environmental Science and Pollution Research* 26 (18). Environmental Science and Pollution Research: 18520–18532. doi:10.1007/s11356-019-05108-y.

Fallah, Aziz A., Elham Sarmast, Saied Habibian Dehkordi, Javad Engardeh, Leila Mahmoodnia, Arsalan Khaledifar, and Tina Jafari. 2018. "Effect of Chlorella Supplementation on Cardiovascular Risk Factors: A Meta-Analysis of Randomized Controlled Trials." *Clinical Nutrition* 37 (6). Elsevier Ltd: 1892–1901. doi:10.1016/j.clnu.2017.09.019.

Ferreira, Andreia S., Sónia S. Ferreira, Alexandra Correia, Manuel Vilanova, Tiago H. Silva, Manuel A. Coimbra, and Cláudia Nunes. 2020. "Reserve, Structural and Extracellular Polysaccharides of Chlorella Vulgaris: A Holistic Approach." *Algal Research* 45 (July 2019). Elsevier: 101757. doi:10.1016/j.algal.2019.101757.

Ferreira, S. P., J. C. P. Holz, C. R. Lisboa, and J. A. V. Costa. 2017. "Fatty Acid Profile of Chlorella Biomass Obtained by Fed Batch Heterotrophic Cultivation." *International Food Research Journal* 24 (1): 284–291.

Fradique, Mónica, Ana Paula Batista, M. Cristiana Nunes, Lúisa Gouveia, Narcisa M. Bandarra, and Anabela Raymundo. 2010. "Incorporation of Chlorella Vulgaris and Spirulina Maxima Biomass in Pasta Products. Part 1: Preparation and Evaluation." *Journal of the Science of Food and Agriculture* 90 (10): 1656–1664. doi:10.1002/jsfa.3999.

Gammone, Maria Alessandra, Graziano Riccioni, and Nicolantonio D'Orazio. 2015. "Carotenoids: Potential Allies of Cardiovascular Health?" *Food and Nutrition Research* 59 (7): 1–11. doi:10.3402/fnr.v59.26762.

Gao, Feng, Hong Li Yang, Chen Li, Yuan Yuan Peng, Miao Miao Lu, Wei Hong Jin, Jing Jiao Bao, and Yuan Ming Guo. 2019. "Effect of Organic Carbon to Nitrogen Ratio in Wastewater on Growth, Nutrient Uptake and Lipid Accumulation of a Mixotrophic Microalgae Chlorella Sp." *Bioresource Technology* 282 (January). Elsevier: 118–124. doi:10.1016/j.biortech.2019.03.011.

Girard, Jean Michel, Réjean Tremblay, Nathalie Faucheux, Michèle Heitz, and Jean Sébastien Deschênes. 2017. "Phycoremediation of Cheese Whey Permeate Using Directed Commensalism between Scenedesmus Obliquus and Chlorella Protothecoides." *Algal Research* 22. Elsevier B.V.: 122–126. doi:10.1016/j.algal.2016.12.013.

Goiris, Koen, Koenraad Muylaert, Stefan Voorspoels, Bart Noten, Domien De Paepe, Gino J. E Baart, and Luc De Cooman. 2014. "Detection of Flavonoids in Microalgae from Different Evolutionary Lineages." *Journal of Phycology* 50 (3): 483–492. doi:10.1111/jpy.12180.

Gouveia, Luísa, Ana Paula Batista, Ana Miranda, José Empis, and Anabela Raymundo. 2007. "Chlorella Vulgaris Biomass Used as Colouring Source in Traditional Butter Cookies." *Innovative Food Science and Emerging Technologies* 8 (3): 433–436. doi:10.1016/j.ifset.2007.03.026.

Gouveia, Luísa, Anabela Raymundo, Ana Paula Batista, Isabel Sousa, and José Empis. 2006. "Chlorella Vulgaris and Haematococcus Pluvialis Biomass as Colouring and Antioxidant in Food Emulsions." *European Food Research and Technology* 222 (3–4): 362–367. doi:10.1007/s00217-005-0105-z.

Graça, C., P. Fradinho, I. Sousa, and A. Raymundo. 2018. "Impact of Chlorella Vulgaris on the Rheology of Wheat Flour Dough and Bread Texture." *LWT - Food Science and Technology* 89 (November 2017). Elsevier: 466–474. doi:10.1016/j.lwt.2017.11.024.

Griffiths, Keith, Bharat Aggarwal, Ram Singh, Harpal Buttar, Douglas Wilson, and Fabien De Meester. 2016. "Food Antioxidants and Their Anti-Inflammatory Properties: A Potential Role in Cardiovascular Diseases and Cancer Prevention." *Diseases* 4 (4): 28. doi:10.3390/diseases4030028.

Grossmann, Lutz, Jörg Hinrichs, and Jochen Weiss. 2019. "Solubility and Aggregation Behavior of Protein Fractions from the Heterotrophically Cultivated Microalga Chlorella Protothecoides." *Food Research International* 116 (May 2018). Elsevier: 283–290. doi:10.1016/j.foodres.2018.08.037.

Grossmann, Lutz, Vera Wörner, Jörg Hinrichs, and Jochen Weiss. 2020. "Sensory Properties of Aqueous Dispersions of Protein-Rich Extracts from Chlorella Protothecoides at

Neutral and Acidic pH." *Journal of the Science of Food and Agriculture* 100 (3): 1344–1349. doi:10.1002/jsfa.10082.

Guedes, Ana Catarina, Helena M. Amaro, and Francisco Xavier Malcata. 2011. "Microalgae as Sources of Carotenoids." *Marine Drugs* 9 (4): 625–644. doi:10.3390/md9040625.

Hajimahmoodi, Mannan, Mohammad Ali Faramarzi, Najmeh Mohammadi, Neda Soltani, Mohammad Reza Oveisi, and Nastaran Nafissi-Varcheh. 2010. "Evaluation of Antioxidant Properties and Total Phenolic Contents of Some Strains of Microalgae." *Journal of Applied Phycology* 22 (1): 43–50. doi:10.1007/s10811-009-9424-y.

Hapońska, M., E. Clavero, J. Salvadó, and C. Torras. 2018. "Application of ABS Membranes in Dynamic Filtration for Chlorella Sorokiniana Dewatering." *Biomass and Bioenergy* 111: 224–231. doi:10.1016/j.biombioe.2017.03.013.

He, Qiaoning, Haijian Yang, Lei Wu, and Chunxiang Hu. 2015. "Effect of Light Intensity on Physiological Changes, Carbon Allocation and Neutral Lipid Accumulation in Oleaginous Microalgae." *Bioresource Technology* 191. Elsevier Ltd: 219–228. doi:10.1016/j.biortech.2015.05.021.

Heidari, Malikeh, Hamid Reza Kariminia, and Jalal Shayegan. 2016. "Effect of Culture Age and Initial Inoculum Size on Lipid Accumulation and Productivity in a Hybrid Cultivation System of Chlorella Vulgaris." *Process Safety and Environmental Protection* 104. Institution of Chemical Engineers: 111–122. doi:10.1016/j.psep.2016.07.012.

Hemantkumar, Joshi Nilesh, and Mor Ilza Rahimbhai. 2019. "Microalgae and Its Use in Nutraceuticals and Food Supplements." In *Microalgae - From Physiology to Application*, edited by Milada Vítová, 1–11. IntechOpen.

Heredia-Arroyo, Tamarys, Wei Wei, Roger Ruan, and Bo Hu. 2011. "Mixotrophic Cultivation of Chlorella Vulgaris and Its Potential Application for the Oil Accumulation from Non-Sugar Materials." *Biomass and Bioenergy* 35 (5): 2245–2253. doi:10.1016/j.biombioe.2011.02.036.

Hynstova, Veronika, Dagmar Sterbova, Borivoj Klejdus, Josef Hedbavny, Dalibor Huska, and Vojtech Adam. 2018. "Separation, Identification and Quantification of Carotenoids and Chlorophylls in Dietary Supplements Containing Chlorella Vulgaris and Spirulina Platensis Using High Performance Thin Layer Chromatography." *Journal of Pharmaceutical and Biomedical Analysis* 148. Elsevier B.V.: 108–118. doi:10.1016/j.jpba.2017.09.018.

Jaafar, Faizul, Lina Wati Durani, and Suzana Makpol. 2020. "Chlorella Vulgaris Modulates the Expression of Senescence-Associated Genes in Replicative Senescence of Human Diploid Fibroblasts." *Molecular Biology Reports* 47 (1). Springer Netherlands: 369–379. doi:10.1007/s11033-019-05140-8.

Khalil, Zeinab I., Mohsen M. S. Asker, Salwa El-Sayed, and Imam A. Kobbia. 2010. "Effect of pH on Growth and Biochemical Responses of Dunaliella Bardawil and Chlorella Ellipsoidea." *World Journal of Microbiology and Biotechnology* 26 (7): 1225–1231. doi:10.1007/s11274-009-0292-z.

Kotrbáček, Václav, Jaroslav Doubek, and Jiří Doucha. 2015. "The Chlorococcalean Alga Chlorella in Animal Nutrition: A Review." *Journal of Applied Phycology* 27 (6): 2173–2180. doi:10.1007/s10811-014-0516-y.

Kwang, Hyun Cha, Hee Ju Lee, Song Yi Koo, Dae Geun Song, Dong Un Lee, and Cheol Ho Pan. 2010. "Optimization of Pressurized Liquid Extraction of Carotenoids and

Chlorophylls from Chlorella Vulgaris." *Journal of Agricultural and Food Chemistry* 58 (2): 793–797. doi:10.1021/jf902628j.

Lai, Yu Cheng, Chien Hsiang Chang, Chun Yen Chen, Jo Shu Chang, and I. Son Ng. 2019. "Towards Protein Production and Application by Using Chlorella Species as Circular Economy." *Bioresource Technology* 289 (June). Elsevier: 121625. doi:10.1016/j.biortech.2019.121625.

Laroche, Celine, and Philippe Michaud. 2008. "New Developments and Prospective Applications for Beta (1,3) Glucans." *Recent Patents on Biotechnology* 1 (1): 59–73. doi:10.2174/187220807779813938.

Lee, Yuan Kun. 2001. "Microalgal Mass Culture Systems and Methods: Their Limitation and Potential." *Journal of Applied Phycology* 13 (4): 307–315. doi:10.1023/A:1017560006941.

Leite, Luan de Souza, Luiz Antonio Daniel, Martin Pivokonsky, Katerina Novotna, Irena Branyikova, and Tomas Branyik. 2019. "Interference of Model Wastewater Components with Flocculation of Chlorella Sorokiniana Induced by Calcium Phosphate Precipitates." *Bioresource Technology* 286 (April). Elsevier: 121352. doi:10.1016/j.biortech.2019.121352.

León-Vaz, Antonio, Rosa León, Encarnación Díaz-Santos, Javier Vigara, and Sara Raposo. 2019. "Using Agro-Industrial Wastes for Mixotrophic Growth and Lipids Production by the Green Microalga Chlorella Sorokiniana." *New Biotechnology* 51 (February 2018). Elsevier: 31–38. doi:10.1016/j.nbt.2019.02.001.

Li, Yanqun, Mark Horsman, Bei Wang, Nan Wu, and Christopher Q. Lan. 2008. "Effects of Nitrogen Sources on Cell Growth and Lipid Accumulation of Green Alga Neochloris Oleoabundans." *Applied Microbiology and Biotechnology* 81 (4): 629–636. doi:10.1007/s00253-008-1681-1.

Liang, Shizhong, Xueming Liu, Feng Chen, and Zijian Chen. 2004. "Current Microalgal Health Food R & D Activities in China." *Hydrobiologia* 512: 45–48.

Liao, Qiang, Yahui Sun, Yun Huang, Ao Xia, Qian Fu, and Xun Zhu. 2017. "Simultaneous Enhancement of Chlorella Vulgaris Growth and Lipid Accumulation through the Synergy Effect between Light and Nitrate in a Planar Waveguide Flat-Plate Photobioreactor." *Bioresource Technology* 243. Elsevier Ltd: 528–538. doi:10.1016/j.biortech.2017.06.091.

Lin, Liang-ping. 2005. *Chlorella: Its Ecology, Structure, Cultivation, Bioprocess and Application.* Yi Hsien P. Taipei.

Lin, Way Rong, and I. Son Ng. 2020. "Development of CRISPR/Cas9 System in Chlorella Vulgaris FSP-E to Enhance Lipid Accumulation." *Enzyme and Microbial Technology* 133 (August 2019). Elsevier: 109458. doi:10.1016/j.enzmictec.2019.109458.

Liu, J, and Q Hu. 2013. "Chlorella: Industrial Production of Cell Mass and Chemicals." In *Handbook of Microalgal Culture: Applied Phycology and Biotechnology*, 329–338.

Liu, Jin, Zheng Sun, Henri Gerken, Zheng Liu, Yue Jiang, and Feng Chen. 2014. "Chlorella Zofingiensis as an Alternative Microalgal Producer of Astaxanthin: Biology and Industrial Potential." *Marine Drugs* 12 (6): 3487–3515. doi:10.3390/md12063487.

Liu, Lijun, Yuan Zhao, Xiaoxue Jiang, Xiaoyu Wang, and Wenyan Liang. 2018. "Lipid Accumulation of Chlorella Pyrenoidosa under Mixotrophic Cultivation Using Acetate and Ammonium." *Bioresource Technology* 262 (April). Elsevier: 342–346. doi:10.1016/j.biortech.2018.04.092.

M. Bishop, West, and Heidi M. Zubeck. 2012. "Evaluation of Microalgae for Use as Nutraceuticals and Nutritional Supplements." *Journal of Nutrition & Food Sciences* 2 (5). doi:10.4172/2155-9600.1000147.

Ma, C. Y., J. M. Zhao, L. H. Liu, and L. Zhang. 2019. "Growth-Dependent Radiative Properties of Chlorella Vulgaris and Its Influence on Prediction of Light Fluence Rate in Photobioreactor." *Journal of Applied Phycology* 31 (1): 235–247. doi:10.1007/s10811-018-1499-x.

Madhubalaji, C. K., T. Sarat Chandra, V. S. Chauhan, R. Sarada, and Sandeep N. Mudliar. 2019. "Chlorella Vulgaris Cultivation in Airlift Photobioreactor with Transparent Draft Tube: Effect of Hydrodynamics, Light and Carbon Dioxide on Biochemical Profile Particularly ω-6/ω-3 Fatty Acid Ratio." *Journal of Food Science and Technology*. Springer India. doi:10.1007/s13197-019-04118-5.

Mata, Teresa M., António A. Martins, and Nidia S. Caetano. 2010. "Microalgae for Biodiesel Production and Other Applications: A Review." *Renewable and Sustainable Energy Reviews* 14 (1): 217–232. doi:10.1016/j.rser.2009.07.020.

McClure, Dale D., Jonathan K. Nightingale, Audrey Luiz, Sachin Black, Jingyuan Zhu, and John M. Kavanagh. 2019. "Pilot-Scale Production of Lutein Using Chlorella Vulgaris." *Algal Research* 44 (October). Elsevier: 101707. doi:10.1016/j.algal.2019.101707.

Metsoviti, Maria N., Nikolaos Katsoulas, Ioannis T. Karapanagiotidis, and Georgios Papapolymerou. 2019. "Effect of Nitrogen Concentration, Two-Stage and Prolonged Cultivation on Growth Rate, Lipid and Protein Content of Chlorella Vulgaris." *Journal of Chemical Technology and Biotechnology* 94 (5): 1466–1473. doi:10.1002/jctb.5899.

Metsoviti, Maria N., George Papapolymerou, Ioannis T. Karapanagiotidis, and Nikolaos Katsoulas. 2019. "Comparison of Growth Rate and Nutrient Content of Five Microalgae Species Cultivated in Greenhouses." *Plants* 8 (8): 1–13. doi:10.3390/plants8080279.

Metsoviti, Maria N., George Papapolymerou, Ioannis T. Karapanagiotidis, and Nikolaos Katsoulas. 2020. "Effect of Light Intensity and Quality on Growth Rate and Composition of Chlorella Vulgaris." *Plants* 9 (1): 1–17. doi:10.3390/plants9010031.

Mobin, S, and F Alam. 2014. "Biofuel Production from Algae Utilizing Wastewater." In *19th Australasian Fluid Mechanics Conference*. Melbourne, Australia.

Mobin, Saleh, and Firoz Alam. 2017. "Some Promising Microalgal Species for Commercial Applications : A Review." *Energy Procedia* 110. The Author(s): 510–517. doi:10.1016/j.egypro.2017.03.177.

Mohammad Mirzaie, M. A., M. Kalbasi, S. M. Mousavi, and B. Ghobadian. 2016. "Investigation of Mixotrophic, Heterotrophic, and Autotrophic Growth of Chlorella Vulgaris under Agricultural Waste Medium." *Preparative Biochemistry and Biotechnology* 46 (2): 150–156. doi:10.1080/10826068.2014.995812.

Mohseni, Roohollah, Zahra Arab Sadeghabadi, Jamshid Karimi, Hamid Gholami, Hadi Ghasemi, Hamid Reza Ghadimipour, and Nejat Kheiripour. 2019. "Chlorella Vulgaris Supplementation Attenuates the Progression of Liver Fibrosis through Targeting TGF-β-Signaling Pathway in the CCl4-Induced Liver Fibrosis in Rats." *Toxin Reviews* 0 (0). Taylor & Francis: 1–9. doi:10.1080/15569543.2019.1700525.

Mühlroth, Alice, Per Winge, Aimen El Assimi, Juliette Jouhet, Eric Maréchal, Martin F. Hohmann-Marriott, Olav Vadstein, and Atle M. Bonesa. 2017. "Mechanisms of Phosphorus Acquisition and Lipid Class Remodeling under P Limitation in a Marine Microalga." *Plant Physiology* 175 (4): 1543–1559. doi:10.1104/pp.17.00621.

Mulders, Kim J. M., Jorijn H. Janssen, Dirk E. Martens, René H. Wijffels, and Packo P. Lamers. 2014. "Effect of Biomass Concentration on Secondary Carotenoids and Triacylglycerol (TAG) Accumulation in Nitrogen-Depleted Chlorella Zofingiensis." *Algal Research* 6 (PA). Elsevier B.V.: 8–16. doi:10.1016/j.algal.2014.08.006.

Muys, Maarten, Yixing Sui, Barbara Schwaiger, Céline Lesueur, Dieter Vandenheuvel, Pieter Vermeir, and Siegfried E. Vlaeminck. 2019. "High Variability in Nutritional Value and Safety of Commercially Available Chlorella and Spirulina Biomass Indicates the Need for Smart Production Strategies." *Bioresource Technology* 275 (December 2018). Elsevier: 247–257. doi:10.1016/j.biortech.2018.12.059.

Nguyen, Van Duy. 2016. "Marine Glycans in Relationship with Probiotic Microorganisms to Improve Human and Animal Health: Principles and Applications." In *Marine Glycobiology: Principles and Applications*, 67–84. doi:10.1201/9781315371399-7.

Noh, Won, Seonghwan Park, Sang Jun Lee, Byung Gon Ryu, and Jungmin Kim. 2019. "The Potential of a Natural Biopolymeric Flocculant, ε-Poly-L-Lysine, for Harvesting Chlorella Ellipsoidea and Its Sustainability Perspectives for Cost and Toxicity." *Bioprocess and Biosystems Engineering* 42 (6). Springer Berlin Heidelberg: 971–978. doi:10.1007/s00449-019-02098-6.

Novoveská, Lucie, Michael E. Ross, Michele S. Stanley, Rémi Pradelles, Virginie Wasiolek, and Jean François Sassi. 2019. "Microalgal Carotenoids: A Review of Production, Current Markets, Regulations, and Future Direction." *Marine Drugs* 17 (11): 1–21. doi:10.3390/md17110640.

Panahi, Yunes, Ahmad Yari Khosroshahi, Amirhossein Sahebkar, and Hamid Reza Heidari. 2019. "Impact of Cultivation Condition and Media Content on Chlorella Vulgaris Composition." *Advanced Pharmaceutical Bulletin* 9 (2): 182–194. doi:10.15171/apb.2019.022.

Papapolymerou, G., V. Karayannis, A. Besios, A. Riga, N. Gougoulias, and X. Spiliotis. 2019. "Scaling-up Sustainable Chlorella Vulgaris Microalgal Biomass Cultivation from Laboratory to Pilot-Plant Photobioreactor, towards Biofuel." *Global Nest Journal* 21 (1): 37–42. doi:10.30955/gnj.002777.

Pignolet, Olivier, Sébastien Jubeau, Carlos Vaca-Garcia, and Philippe Michaud. 2013. "Highly Valuable Microalgae: Biochemical and Topological Aspects." *Journal of Industrial Microbiology and Biotechnology* 40 (8): 781–796. doi:10.1007/s10295-013-1281-7.

Plaza, Merichel, Miguel Herrero, And Alejandro Cifuentes, and Elena Ibáñez. 2009. "Innovative Natural Functional Ingredients from Microalgae." *Journal of Agricultural and Food Chemistry* 57 (16): 7159–7170. doi:10.1021/jf901070g.

Prabakaran, Gopal, Meivelu Moovendhan, A. Arumugam, A. Matharasi, R. Dineshkumar, and Pitchai Sampathkumar. 2019. "Evaluation of Chemical Composition and In Vitro Antiinflammatory Effect of Marine Microalgae Chlorella Vulgaris." *Waste and Biomass Valorization* 10 (11). Springer Netherlands: 3263–3270. doi:10.1007/s12649-018-0370-2.

Ramírez-López, Citlally, Hugo Virgilio Perales-Vela, and Luis Fernández-Linares. 2019. "Biomass and Lipid Production from Chlorella Vulgaris UTEX 26 Cultivated in 2 m3 Raceway Ponds under Semicontinuous Mode during the Spring Season." *Bioresource Technology* 274 (November 2018). Elsevier: 252–260. doi:10.1016/j.biortech.2018.11.096.

Rani, Komal, Nidhi Sandal, and P K Sahoo. 2018. "A Comprehensive Review on Chlorella-Its Composition, Health Benefits, Market and Regulatory Scenario." *~ 584 ~ The Pharma Innovation Journal* 7 (7): 584–589.

Safafar, Hamed, Patrick Uldall Nørregaard, Anita Ljubic, Per Møller, Susan Løvstad Holdt, and Charlotte Jacobsen. 2016. "Enhancement of Protein and Pigment Content in Two Chlorella Species Cultivated on Industrial Process Water." *Journal of Marine Science and Engineering* 4 (4). doi:10.3390/jmse4040084.

Safafar, Hamed, Jonathan Van Wagenen, Per Møller, and Charlotte Jacobsen. 2015. "Carotenoids, Phenolic Compounds and Tocopherols Contribute to the Antioxidative Properties of Some Microalgae Species Grown on Industrial Wastewater." *Marine Drugs* 13 (12): 7339–7356. doi:10.3390/md13127069.

Safi, Carl, Michaël Charton, Olivier Pignolet, Françoise Silvestre, Carlos Vaca-Garcia, and Pierre Yves Pontalier. 2013. "Influence of Microalgae Cell Wall Characteristics on Protein Extractability and Determination of Nitrogen-to-Protein Conversion Factors." *Journal of Applied Phycology* 25 (2): 523–529. doi:10.1007/s10811-012-9886-1.

Safi, Carl, Bachar Zebib, Othmane Merah, Pierre Yves Pontalier, and Carlos Vaca-Garcia. 2014. "Morphology, Composition, Production, Processing and Applications of Chlorella Vulgaris: A Review." *Renewable and Sustainable Energy Reviews* 35 (July): 265–278. doi:10.1016/j.rser.2014.04.007.

Sánchez-Saavedra, M. del Pilar, Daniel Sauceda-Carvajal, Fátima Y. Castro-Ochoa, and Ceres A. Molina-Cárdenas. 2020. "The Use of Light Spectra to Improve the Growth and Lipid Content of Chlorella Vulgaris for Biofuels Production." *Bioenergy Research*. BioEnergy Research. doi:10.1007/s12155-019-10070-1.

Sarker, Nilay Kumar, and P. Abdul Salam. 2019. "Indoor and Outdoor Cultivation of Chlorella Vulgaris and Its Application in Wastewater Treatment in a Tropical city—Bangkok, Thailand." *SN Applied Sciences* 1 (12). Springer International Publishing: 1–13. doi:10.1007/s42452-019-1704-9.

Sathasivam, Ramaraj, Ramalingam Radhakrishnan, Abeer Hashem, and Elsayed F. Abd_Allah. 2019. "Microalgae Metabolites: A Rich Source for Food and Medicine." *Saudi Journal of Biological Sciences* 26 (4). King Saud University: 709–722. doi:10.1016/j.sjbs.2017.11.003.

Scherer, Daniel, Damaris Krust, Wolfgang Frey, Georg Mueller, Peter Nick, and Christian Gusbeth. 2019. "Pulsed Electric Field (PEF)-Assisted Protein Recovery from Chlorella Vulgaris Is Mediated by an Enzymatic Process after Cell Death." *Algal Research* 41 (May). Elsevier: 101536. doi:10.1016/j.algal.2019.101536.

Seyfabadi, Jafar, Zohreh Ramezanpour, and Zahra Amini Khoeyi. 2011. "Protein, Fatty Acid, and Pigment Content of Chlorella Vulgaris under Different Light Regimes." *Journal of Applied Phycology* 23 (4): 721–726. doi:10.1007/s10811-010-9569-8.

Sharma, Rekha. 2012. "Effects of Culture Conditions on Growth and Biochemical Profile of Chlorella Vulgaris." *Journal of Plant Pathology & Microbiology* 3 (5). doi:10.4172/2157-7471.1000131.

Sheih, I. Chuan, Tony J. Fang, and Tung Kung Wu. 2009. "Isolation and Characterisation of a Novel Angiotensin I-Converting Enzyme (ACE) Inhibitory Peptide from the Algae Protein Waste." *Food Chemistry* 115 (1). Elsevier Ltd: 279–284. doi:10.1016/j.foodchem.2008.12.019.

Sikiru, A. B., A. Arangasamy, I. C. Alemede, P. R. Guvvala, S. S.A. Egena, J. R. Ippala, and R. Bhatta. 2019. "Chlorella Vulgaris Supplementation Effects on Performances, Oxidative Stress and Antioxidant Genes Expression in Liver and Ovaries of New Zealand White Rabbits." *Heliyon* 5 (9). Elsevier Ltd: e02470. doi:10.1016/j.heliyon.2019.e02470.

Silva, Joana, Celso Alves, Susete Pinteus, João Reboleira, Rui Pedrosa, and Susana Bernardino. 2019. "Chapter 3.10. Chlorella." In *Nonvitamin and Nonmineral Nutritional Supplements*, edited by Seyed Mohammad Nabavi Ana Sanches Silva, 187–191. Academic Press is an imprint of Elsevier.

Sim, Sang Jun, Jaemin Joun, Min Eui Hong, and Anil Kumar Patel. 2019. "Split Mixotrophy: A Novel Cultivation Strategy to Enhance the Mixotrophic Biomass and Lipid Yields of Chlorella Protothecoides." *Bioresource Technology* 291 (July). Elsevier: 121820. doi:10.1016/j.biortech.2019.121820.

Singh, S. P., and Priyanka Singh. 2014. "Effect of CO2 Concentration on Algal Growth: A Review." *Renewable and Sustainable Energy Reviews* 38: 172–179. doi:10.1016/j.rser. 2014.05.043.

Skorupskaite, Virginija, Violeta Makareviciene, Egle Sendzikiene, and Milda Gumbyte. 2019. "Microalgae Chlorella Sp. Cell Disruption Efficiency Utilising Ultrasonication and Ultrahomogenisation Methods." *Journal of Applied Phycology* 31 (4). Journal of Applied Phycology: 2349–2354. doi:10.1007/s10811-019-01761-5.

Sozmen, Alper Baran, Erhan Canbay, Eser Yildirim Sozmen, and Bikem Ovez. 2018. "The Effect of Temperature and Light Intensity during Cultivation of Chlorella Miniata on Antioxidant, Anti-Inflammatory Potentials and Phenolic Compound Accumulation." *Biocatalysis and Agricultural Biotechnology* 14 (March). Elsevier Ltd: 366–374. doi:10.1016/j.bcab.2018.03.023.

Stern, Rowena F, Ales Horak, Rose L Andrew, Mary-alice Coffroth, Robert A Andersen, C Frithjof, Fumai Kasai, Jerry Brand, and Erick R James. 2010. "Environmental Barcoding Reveals Massive Dinoflagellate Diversity in Marine Environments." *Plos One* 5 (11): e13991. doi:10.1371/journal.pone.0013991.

Suparmaniam, Uganeeswary, Man Kee Lam, Yoshimitsu Uemura, Siew Hoong Shuit, Jun Wei Lim, Pau Loke Show, Keat Teong Lee, Yukihiko Matsumura, and Phung Thi Kim Le. 2020. "Flocculation of Chlorella Vulgaris by Shell Waste-Derived Bioflocculants for Biodiesel Production: Process Optimization, Characterization and Kinetic Studies." *Science of the Total Environment* 702. Elsevier B.V.: 134995. doi:10.1016/j.scitotenv. 2019.134995.

Taghavi Takyar, Mir Babak, Shabnam Haghighat Khajavi, and Reza Safari. 2019. "Evaluation of Antioxidant Properties of Chlorella Vulgaris and Spirulina Platensis and Their Application in Order to Extend the Shelf Life of Rainbow Trout (*Oncorhynchus Mykiss*) Fillets during Refrigerated Storage." *Lwt* 100 (July 2018). Elsevier: 244–249. doi:10.1016/j.lwt.2018.10.079.

Varela, João C., Hugo Pereira, Marta Vila, and Rosa León. 2015. "Production of Carotenoids by Microalgae: Achievements and Challenges." *Photosynthesis Research* 125 (3): 423–436. doi:10.1007/s11120-015-0149-2.

Varfolomeev, S D, and L A Wasserman. 2011. "Microalgae as Source of Biofuel, Food, Fodder, and Medicines." *Applied Biochemistry and Microbiology* 47 (9): 789–807. doi:10.1134/S0003683811090079.

Vaz, Bruna da Silva, Juliana Botelho Moreira, Michele Greque de Morais, and Jorge Alberto Vieira Costa. 2016. "Microalgae as a New Source of Bioactive Compounds in Food Supplements." *Current Opinion in Food Science* 7. Elsevier Ltd: 73–77. doi:10.1016/j.cofs.2015.12.006.

Vishwakarma, Rashi, Dolly Wattal Dhar, and Sudhir Saxena. 2019. "Influence of Nutrient Formulations on Growth, Lipid Yield, Carbon Partitioning and Biodiesel Quality Potential of Botryococcus Sp. and Chlorella Sp." *Environmental Science and Pollution Research* 26 (8). Environmental Science and Pollution Research: 7589–7600. doi:10.1007/s11356-019-04213-2.

Wan Mahmood, Wan M. Asyraf, Atiwich Lorwirachsutee, Constantinos Theodoropoulos, and Maria Gonzalez-Miquel. 2019. "Polyol-Based Deep Eutectic Solvents for Extraction of Natural Polyphenolic Antioxidants from Chlorella Vulgaris." Research-article. *ACS Sustainable Chemistry and Engineering* 7 (5). American Chemical Society: 5018–5026. doi:10.1021/acssuschemeng.8b05642.

Wang, Shi Kai, Xu Wang, Yong Ting Tian, and Yue Hua Cui. 2020. "Nutrient Recovery from Tofu Whey Wastewater for the Economical Production of Docosahexaenoic Acid by Schizochytrium Sp. S31." *Science of the Total Environment* 710. Elsevier B.V.: 136448. doi:10.1016/j.scitotenv.2019.136448.

Ward, Valerie C. A., and Lars Rehmann. 2019. "Fast Media Optimization for Mixotrophic Cultivation of Chlorella Vulgaris." *Scientific Reports* 9 (1). Springer US: 19262. doi:10.1038/s41598-019-55870-9.

WHO/FAO. 1985. *Energy and Protein Requirements*. Génova.

Williams, Peter J. Le B., and Lieve M. L. Laurens. 2010. "Microalgae as Biodiesel & Biomass Feedstocks: Review & Analysis of the Biochemistry, Energetics & Economics." *Energy and Environmental Science* 3 (5): 554–590. doi:10.1039/b924978h.

Wong, Y. K., Y. H. Ho, H. M. Leung, K. C. Ho, Y. H. Yau, and K. K.L. Yung. 2017. "Enhancement of Chlorella Vulgaris Harvesting via the Electro-Coagulation-Flotation (ECF) Method." *Environmental Science and Pollution Research International* 24 (10): 9102–9110. doi:10.1007/s11356-016-7856-x.

Xie, Tonghui, Yun Xia, Yu Zeng, Xingrui Li, and Yongkui Zhang. 2017. "Nitrate Concentration-Shift Cultivation to Enhance Protein Content of Heterotrophic Microalga Chlorella Vulgaris: Over-Compensation Strategy." *Bioresource Technology* 233. Elsevier Ltd: 247–255. doi:10.1016/j.biortech.2017.02.099.

Xu, Wei, Zhen Gao, Zhitao Qi, Ming Qiu, Jian-qing Peng, and Rong Shao. 2014. "Effect of Dietary Chlorella on the Growth Performance and Physiological Parameters of Gibel Carp, Carassius Auratus Gibelio." *Turkish Journal of Fisheries and Aquatic Sciences* 14 (1). PO BOX 129, TRABZON, 61001, TURKEY: CENTRAL FISHERIES RESEARCH INST: 53–57. doi:10.4194/1303-2712-v14_1_07.

Yang, Zhi Kai, Ying Fang Niu, Yu Han Ma, Jiao Xue, Meng Han Zhang, Wei Dong Yang, Jie Sheng Liu, Song Hui Lu, Yuanfang Guan, and Hong Ye Li. 2013. "Molecular and Cellular Mechanisms of Neutral Lipid Accumulation in Diatom Following Nitrogen Deprivation." *Biotechnology for Biofuels* 6 (1): 1–14. doi:10.1186/1754-6834-6-67.

Zaghdoudi, Khalil, Orleans Ngomo, Régis Vanderesse, Philippe Arnoux, Bauyrzhan Myrzakhmetov, Céline Frochot, and Yann Guiavarc'h. 2017. "Extraction, Identification and Photo-Physical Characterization of Persimmon (*Diospyros Kaki* L.) Carotenoids." *Foods* 6 (1): 4. doi:10.3390/foods6010004.

Zainul Azlan, Nurhazirah, Yasmin Anum Mohd Yusof, Ekram Alias, and Suzana Makpol. 2019. "Chlorella Vulgaris Modulates Genes and Muscle-Specific microRNAs Expression to Promote Myoblast Differentiation in Culture." *Evidence-Based Complementary and Alternative Medicine*. doi:10.1155/2019/8394648.

Zhang, Ruilin, Jian Chen, Xinwu Mao, Ping Qi, and Xuewu Zhang. 2019a. "Anti-Inflammatory and Anti-Aging Evaluation of Pigment–protein Complex Extracted from Chlorella Pyrenoidosa." *Marine Drugs* 17 (10). doi:10.3390/md17100586.

Zhang, Ruilin, Jian Chen, Xinwu Mao, Ping Qi, and Xuewu Zhang. 2019b. "Separation and Lipid Inhibition Effects of a Novel Decapeptide from Chlorella Pyenoidose." *Molecules* 24 (19): 1–16. doi:10.3390/molecules24193527.

Zhu, L. D., Z. H. Li, and E. Hiltunen. 2016. "Strategies for Lipid Production Improvement in Microalgae as a Biodiesel Feedstock." *BioMed Research International* 2016: 7–9. doi:10.1155/2016/8792548.

Zhu, Liandong, Tianyi Hu, Shuangxi Li, Yohanes K. Nugroho, Baishan Li, Jun Cao, Pau Loke Show, and Erkki Hiltunen. 2020. "Effects of Operating Parameters on Algae Chlorella Vulgaris Biomass Harvesting and Lipid Extraction Using Metal Sulfates as Flocculants." *Biomass and Bioenergy* 132. Elsevier Ltd: 105433. doi:10.1016/j.biombioe.2019.105433.

Zhu, Shunni, Lei Qin, Pingzhong Feng, Changhua Shang, Zhongming Wang, and Zhenhong Yuan. 2019. "Treatment of Low C/N Ratio Wastewater and Biomass Production Using Co-Culture of Chlorella Vulgaris and Activated Sludge in a Batch Photobioreactor." *Bioresource Technology* 274 (August 2018). Elsevier: 313–320. doi:10.1016/j.biortech.2018.10.034.

In: Super and Nutraceutical Foods
Editors: Sandra Teresita Martín-del-Campo et al.
ISBN: 978-1-53619-082-3

Chapter 13

HONEY BEE (*APIS MELLIFERA*) HIVE PRODUCTS: THEIR USE AND BENEFITS TO HUMAN HEALTH AND NUTRITION

***Jose Octavio Macias-Macias*[1,*]*, Jose Maria Tapia-Gonzalez*[1]*, Francisca Contreras-Escareño*[2]*, José Carlos Tapia-Rivera*[3] *and Ernesto Guzman-Novoa* [1,4]**

[1]Departamento de Ciencias de la Naturaleza, Centro de Investigaciones en Abejas (CIABE). CUSUR. Universidad de Guadalajara, Jalisco, Mexico
[2]Departamento de Producción Agrícola, Centro de Investigaciones en Abejas (CIABE). Centro Universitario de la Costa Sur, Universidad de Guadalajara, Jalisco, Mexico
[3]Departamento de Ciencias Económico Administrativas. Centro de Investigaciones en Abejas (CIABE), CUSUR, Universidad de Guadalajara, Jalisco, Mexico
[4]School of Environmental Sciences, University of Guelph, Guelph, Ontario, Canada

ABSTRACT

Although honey bees (*Apis mellifera*) have been managed for thousands of years, most people do not know about the benefits of using their products in food. Honey is one of those products to which many uses and properties are attributed, but honey is not the only hive product; other products such as propolis, pollen, and royal jelly, can be obtained from bee hives. Currently, many studies are aimed at investigating the pharmacological properties and benefits of natural products to human health, leading to the growing development of nutraceuticals and functional foods from them. The concept of functional food refers to foods that have the capability of promoting better physiological or psychological health compared to traditional foods. In this capacity, it has been shown that hive products contribute positively to the maintenance of good health and well-being, as well as to ameliorate chronic diseases. Here we present a literature review of research conducted on the properties and benefits of hive products

* Corresponding Author's E-mail: joseoc@cusur.udg.mx.

(honey, propolis, pollen, and royal jelly) for human nutrition and health, some of which have been used in traditional medicine for the control of certain diseases. According to the reports analyzed, it can be concluded that hive products have therapeutic properties that may help prevent and control various illnesses, so it may be advisable to consume them routinely. It is also important to highlight that much more research with humans in clinical trials is needed to confirm the properties and benefits of hive products.

Keywords: honey bee, hive products, human health, nutrition

INTRODUCTION

A functional food may increase its benefits and value to human health when new, specific ingredients, such as nutraceuticals, are added to it (Proestos 2018). It has been claimed that a nutraceutical product, or nutritive food products, may provide medical and health benefits to consumers, including the prevention and treatment of certain diseases (Maghsoudlou et al. 2019).

Aging, health, and life expectancy of humans are closely related concepts. Although aging is a natural phenomenon, many factors such as stress, poor nutrition, and pollution are associated with increased internal production of free radicals, which favor chronic subclinical inflammation and accelerate aging. Age disorders can be prevented or reversed through dietary intervention since functional foods can promote healthy aging, reduce morbidity, and even increase life expectancy (Proestos 2018, Kunugi and Mohammed 2019). Within this context, products obtained from bees can play a very important role due to their diverse properties related to health and well-being.

Honey bees stand out in the group of insects because their domestication and exploitation have resulted in the development of an entire industry with systems designed to care, cultivate and exploit these insects, which has been called beekeeping (Kritsky 2017, Thakur et al. 2019). Humans have practiced beekeeping since ancient times. Evidence of this activity has been found in cave paintings dating from 15000 BC in Europe. The relationship between bees and humans has allowed a very close symbiosis where beekeepers take care of bees by providing everything necessary for them to have the right conditions to reproduce, and bees provide humans in return, a diversity of products with which they obtain food, health and economic benefits (Crane 2013, Long 2020).

Various products such as honey, propolis, pollen, royal jelly, bee venom, and wax can be obtained from honey bees through production systems that have been developed over time. These products and their mixtures are widely used in industry, cosmetology, alternative medicine, and human food. Honey, propolis, pollen, and royal jelly stand out, as they are attributed various properties and effects mentioned below.

HONEY

Honey is defined as a natural sweet substance produced by bees from the nectar of flowers or from secretions of living parts of plants, or excretions of plant-sucking insects that bees collect, transform, and combine with their own specific substances, deposit, dehydrate, store, and leave in the honeycomb to mature and age (DOF 2018). The taste of honey depends on the plant source from which the bees collect the nectar, and its color can vary from an almost translucent liquid to a dark syrup. Regardless of the type of honey, it can be used to improve the health and vitality of animals and humans (Wolfe 2009, Long 2020). According to Bogdanov (2016), honey has 46 Kcal per 15 g of product and contains the following carbohydrates shown in Table 1.

Table 1. Different types of carbohydrates contained in honey

Fructose	Glucose	Sucrose	Other sugars
38%	31%	0.7%	5%

The proportion of carbohydrates in honey may vary depending on the diversity of plant sources visited by the bees. Carbohydrates play a main role in conferring flavors and aromas to honey (Pattamayutanon et al. 2017). In its composition, there can also be small amounts of amino acids. It may have a large variety of vitamins, including vitamins B2, B3, B5, B6, and minerals such as Calcium (Ca), Iron (Fe), Magnesium (Mg), Phosphorus (P), Potasium (K), Sodium (Na), Zinc (Zn), Copper (Cu) and Selenium (Se), in addition to bioactive components, such as phenolic acid, flavonoids and α-tocopherol (Nguyen et al. 2019, USDA 2019).

In addition to being recognized for its nutritional components, honey has also been used in the traditional medicine of various cultures such as the Greek, Roman, and Chinese cultures, among others (Raoa et al. 2016). In modern times, the medicinal properties of honey have been related to its plant origin and composition, and specifically to the phenolic compounds in it (Cianciosi et al. 2018). Phenolic compounds have antioxidant, antimicrobial, antiviral, anti-inflammatory, antifungal, and cardioprotective properties (Pasupuleti et al. 2017). Clearly, honey is rich in beneficial nutrients and, thus, can be listed as a superfood. Here is some information about how honey can benefit human health.

Anti-Inflammatory Effect

One of the beneficial effects of honey is its anti-inflammatory capability. According to several authors, honey reduces inflammation in cell tissues of experimental animals (Samarghandian et al. 2017). The specific mechanisms resulting in this effect are poorly understood, but it is known that the anti-inflammatory effect of honey is modulated by two pathways, the mitogen-activated protein kinase (MAPK) pathway and the nuclear factor kappa B (NF-kB) pathway. The activation of one of these two pathways brings a cascade reaction of proteins and genes present in inflammation, and within these, we find the cyclooxygenase 2 (COX-2), lipoxygenase 2 (LOX-2), C-reactive protein (PCR), and various interleukins, among other inflammatory-related factors. Honey consumption attenuates the expression of these interleukins and genes, especially in the NF-kB pathway (Ahmed et al. 2018); more specifically, honey flavonoid compounds such as chrysin, quercetin, and galangin, decrease the inflammatory response (Murtaza et al. 2014).

Immunostimulator

Honey consumption induces monocytes to emit cytokines, tumor necrosis factor (TNF), and interleukins, modulating the synthesis of fibroblasts and collagen, which initiates the inflammatory process; studies also mention the Apa-1 glycoprotein component that stimulates macrophages to release TNF (Majtán et al. 2006). These effects propose that honey has an immunostimulatory effect, possibly based on its protein content, specifically in glycoproteins (Raynaud et al. 2013).

Honey Consumption and Cancer

Cancer is a disease that results from alterations in genes that control cell growth and reproduction functions. Although there are some cancer types that are heritable, most of them are due to interactions with the environment, mainly by having contact with physical factors such as UV light, chemicals such as tobacco smoke, or biological factors such as some types of viruses. These factors can cause mutations in the normal function of some genes, turning them into oncogenes that can activate signaling pathways that result in uncontrolled cell growth, which alters the functions of cells (Castillo-Ruiz et al. 2019).

Physicians can resort to various treatments to fight cancer, such as radiation, chemotherapy, surgery, and nutritional treatments, to eradicate cancer cells or promote apoptosis in the cells. Apoptosis or programmed cell death could be one of the best ways to fight cancer (Pfeffer and Amareshwar 2018). Studies conducted with mice that were fed 1 g of honey per kg of weight showed improved apoptosis of cells and positive regulation of the expression of proapoptotic proteins like caspase-9. Additionally, the clinical results showed smaller size, weight, and multiplicity of cancerous tumors compared to the untreated positive control (Ahmed et al. 2017). The same authors also suggested that the consumption of honey has an important influence on the apoptotic pathways of cancer cells, and proposed the theory that honey consumption stimulates caspase enzymes, which mediate the process of apoptosis and inflammation or stimulate the p53 gene pathway, which induces the Bax factor to express itself in the mitochondria and activates the cytochrome c oxidase enzyme, which connects to caspase 9 and to caspase 3, that favors the apoptosis process.

Cough and Its Treatment with Honey

Cough is a voluntary or involuntary reflex that keeps the throat clear; physiologically, it can be defined as forced breathing when there is a closed glottis. However, a prolonged cough can lead to disease. Acute coughing is often caused by the spread of viruses that affect the upper respiratory tract and tend to disappear after four weeks (Morice 2002). This physiological process is derived from irritation or obstruction of the respiratory tract due to the presence of excess secretions or foreign bodies in the tract. Allopathic treatments include expectorant and mucolytic drugs to treat cough symptoms; however, the World Health Organization (WHO) states that honey can be used in the treatment of cough (WHO 2001). One of the properties with which honey could improve the symptoms of this condition is by fighting the pathogen that generates mucus. Honey has various germicide mechanisms, such as its low pH, its osmotic effect derived from its high concentration of sugars and low water content, the presence of bacteriostatic and bactericidal factors such as hydrogen peroxide, antioxidants, phenolic compounds, flavonoids and peptides (Israili 2014, Almasaudi et al. 2017). This is why some authors consider honey as a broad-spectrum antibiotic. In addition to these effects, the consumption of honey leads to salivation to digest the sugars present in it, and salivation produces lubrication and a demulcent effect on the pharynx and larynx, reducing cough (Ahmed et al. 2013).

The dosages of honey recommended by some authors to reduce cough-derived symptoms are as follows: children 2 to 5 years old, 8.5 mg/kg (equivalent to 1/2 teaspoon), children 6 to 11 years old, 17 mg/kg (equivalent to 1 teaspoon), and children aged 12 to 18, 34 mg/kg (equivalent to 2 teaspoons). These doses allow a decrease in cough frequency and severity, while the quality of sleep improves (Paul et al. 2007, Ahmed et al. 2013, Eduwole et al. 2018). With these evidences, honey must be considered as a superfood, either because of its concentration of nutrients or because of the pharmacological effects that it possesses. However, we must clarify that industrialized honey undergoes pasteurization, a process that could limit its nutraceutical effects. Therefore, the use of raw honey is recommended in these cases.

PROPOLIS

The term propolis is derived from two Greek words, Pro, which means before, and Polis, which means city, and it is used in the sense of defense of the hive or "city of bees" (Martinez-Perez et al. 2019). Propolis is produced when honey bees mix the secretions of their hypopharyngeal glands with the digested product of resins collected from plants (Skoskiewicz-Malinowska et al. 2017). These are gummy and resinous substances that bees collect from the buds and exudates of various plant species, such as poplar, palm, pine, herbaceous plants, and shrubs, and transport them to their colonies to seal cracks and crevices in their hives (Becerra et al. 2019, Anjum et al. 2019).

Propolis is a natural antiseptic used by the bees to prevent microbial infections in the hive, as well as to prevent the decomposition of animals that they sacrifice inside the hive when animals dare to invade their colonies (Valenzuela-Barra et al. 2015). Propolis consumed in the diet may confer nutraceutical, antioxidant, anti-inflammatory, lipid-modulatory, and immunomodulatory properties, among others properties (Mujica et al. 2017, Camacho et al. 2019, Perez et al. 2020).

The chemical composition of propolis and its biological effects depend on the plant sources, geographical area, season, time of collection, etc. More than 300 compounds have been identified in propolis, including phenolic compounds, aromatic acids, essential oils, waxes, and amino acids. In terms of composition, propolis generally contains 50% resin, 30% wax, 10% essential oils, 5% pollen, and 5% other substances, including cinnamic acid derivatives, phenolic acids, substituted benzoic acids, amino acids, and flavonoids (Devequi-Nunes et al. 2018, Anjum et al. 2019).

Official standards for the production and specifications of propolis have been created based on its chemical composition and study of its compounds. Propolis standards exist in Mexico (Official Mexican Standard nom-003-sag/gan-2016) to set out the establishment of quality control tests that Mexican propolis must comply with for their use in human and animal health (Perez et al. 2020). There are three factors that influence the production and quality of propolis, (1) Those inherent in bees (the race and genotype of bees, the colony's propolinizing behavior), (2) The environment in which the hive is located, including the flora surrounding the apiary, (3) The harvesting technique used by the beekeeper (Martinez-Perez et al. 2019).

Beneficial effects on various conditions caused to humans and animals by parasites, bacteria, viruses, and other stressors, have been attributed to propolis (Valenzuela-Barra et al. 2015, Dantas et al. 2017, Skoskiewicz-Malinowska et al. 2017, Brezani, 2018). In relation to diseases caused by bacteria and fungi, multiple scientific investigations have been carried out on the use of propolis. Becerra et al. (2019) evaluated the antibacterial efficacy of a 5% propolis alcoholic solution on the growth of *Streptococcus mutans* colonies plated in Petri dishes. The results indicated an inhibition area of 26.4 ± 2.6 and 18.2 ± 1.8 mm on bacterial colonies. In a more extensive investigation, the antimicrobial properties of propolis from different origins were evaluated. Most of the extracts demonstrated antimicrobial activity against *Staphylococcus aureus* and an inhibitory effect of all the tested ethanolic extracts on the growth of the Y strain of *Trypanosoma cruzi* in the first 24 hours (Dantas et al. 2017). In a similar work, Devequi-Nunes et al. (2018) found an antimicrobial effect against *S. aureus* and *Escherichia coli*, even though they were more effective against gram (+) bacteria due to their flavonoids and aromatic compounds; gram (-) bacteria presented greater resistance (Przybylek and Karpinsky 2019). Research has also been conducted on effective treatments with alcoholic propolis extracts to reduce the multiplication of resistant bacteria and fungi, such as *Enterococcus faecalis* and *Candida albicans*, that are found in root canal surgery (root canals) (Skoskiewicz-Malinowska et al. 2017). When the alcohol extract of propolis at concentrations of 0.1%, 0.3%, and 0.5% is added to honey, this combination has demonstrated synergistic antimicrobial effects and is considered a promising functional food (Oses et al. 2016). Propolis added or mixed in a cream at a 20% concentration has also been used to treat mild to moderate acne problems. Compared with the effect of 3% erythromycin, the results were statistically different; the papular and scar lesions treated with the propolis cream mixture showed a high reduction of erythema compared to the treatment with erythromycin (Mazzarello et al. 2018).

The use of propolis formulations in the pharmaceutical, cosmetic, and food industries faces some challenges, including the standardization of its biological quality, chemical composition, and unpleasant taste for consumption. Research has been conducted regarding its use as an innovative food preservative and bioactive food supplement (Dantas et al. 2017). It has been shown that the alcoholic extract of propolis used at 0.1%, 0.3%, and 0.5%, added to honey, is better accepted for ingestion due to its antimicrobial, antioxidant, and inflammatory properties. Osés et al. (2016) concluded that honey enriched with small amounts of propolis extracts can be regarded as functional and promising food. Tests have been conducted to ameliorate the strong flavor of propolis using nanoemulsions, and it was observed that they did not lose their antimicrobial and antioxidant capacity and that they could be used as a food preservative, preventing food degradation (Seibert et al. 2019).

Concerning human health, there is evidence about the beneficial effects of flavonoids and phenolic compounds of propolis on various conditions caused by bacteria, fungi, viruses, and parasites. This is in addition to the inhibition of lipid peroxidation, platelet aggregation, capillary permeability and fragility, and enzymatic systems, including cyclooxygenase and lipoxygenase (Braakluis 2019). It has also been shown that propolis consumption increases insulin production to control the effects of diabetes (Rivera-Yañez et al. 2018). Moreover, propolis anti-inflammatory action has been proven in numerous investigations in vivo and in vitro (Valenzuela-Barra et al. 2015; Osés et al. 2016; Brezani 2018). Furthermore, propolis improves immune system responses by acting as an immunomodulator; it also controls blood pressure, allergies and skin conditions, and in cardiometabolic health, may have anti-tumor and anti-aging activity (Viuda-Martos et al. 2008). Additionally, propolis reduces oxidative stress, which is a precursor to neurovegetative diseases such as cancer and atherosclerosis (Mujica et al. 2017, Kocot et al. 2018, Braakhuis 2019).

Pollen

Pollen is a compound produced by flowering plants composed of microscopic grains, which contain male germ cells for the reproduction of plants. Bees collect pollen from flowers and use it to feed their larvae and the queen bee. Its high protein content is essential for the production of royal jelly by worker bees that feed young larvae and the queen and is also needed for the development of older bee larvae (Wrigth et al. 2018).

Pollen is a product highly appreciated by people who want to have a healthy diet because of its high nutrient content and beneficial elements for the functionality of several organs. Pollen contains large amounts of protein, between 16 and 30%, with an average of 25%. Also, it is rich in vitamins A, D, E, B1, C, K, rutin and choline, and also contains minerals, amino acids, and trace elements, such as calcium (Ca), sodium (Na), zinc (Zn), magnesium (Mg), potassium (K), aluminum (Al), iron (Fe) and many others. Thanks to this amount of nutrients, pollen is considered a very complete and beneficial superfood for well-functioning of the organism (Jean-Prost and Le Cont 2007, Yucel et al. 2017).

Pollen grains are coated with a strong wall composed of hard layers of exine and intin. Therefore, for the pollen to be better assimilated by the bee or human digestive system, it should undergo a fermentation process, so that the layers of exine and intin break down, to release the content of grains and make it more digestible (Utoiu et al. 2018). This fermentation process begins when the bees collect pollen, adding saliva to it, and forming pellets that are carried to the colony. At the colony, bees add honey or nectar, digestive enzymes, and microorganisms to pollen pellets, and place them inside comb cells. When a cell is full of pollen, bees cover it with a thin wax layer and keep it covered for two weeks for lactic acid fermentation to occur (Vasquez and Olofsson 2015, Tomas et al. 2017), after which it is used in the colony. Pollen used as a food in the human diet is usually consumed fresh, without undergoing most of the fermentation process described above, because it is collected from hives before the bees store it in their combs. Despite this, the nutritional value of pollen is high, and according to Markiewicz-Żukowska et al. (2013) and Urcan et al. (2017), its bromatological composition can be summarized in Table 2.

Table 2. Composition of pollen collected from hives

Elements	Percent content
Proteins	17.19-22.7
Amino acids	10.14
Carbohydrates	30.8
Reducing sugars	15.1-30.2
Lipids	4.1-13.37
Lactic acid	0.56
Moisture	14.14-27
Ashes	2.1-4
pH	3.8-6.3

Fermented or partially fermented pollen allows the growth of microorganisms, which may provide benefits to organisms that consume it. Among them are the group of bacillus, lactobacillus, and yeasts of the genus *Candida*, as well as bifidobacteria from the bees' microbiota. It is known that bifidobacteria and lactobacilli protect their hosts employing antimicrobial metabolites, immunomodulation and competition (Vasquez and Olofsson 2015, Urcan et al. 2017).

Pollen Use and Health Benefits

Pollen has been used as food since humans learned to manage and cultivate honey bees in their ancestral regions of origin (the old world). In the Americas, and specifically in Mexico, the Maya people used pollen collected from other bee species (not honey bees), stingless bees, which are known as melipona bees. There is evidence in the Maya codices about how pollen was used to treat the negative effects of various health afflictions, including problems of low fertility in humans (Quezada-Euan 2018).

Pollen is particularly appreciated because of its effect in controlling oxidative stress, which is caused by various factors such as the presence of toxins and oxidizing agents in the environment, which are precursors of the appearance of malignant cells in organs of the human body. Therefore, regular consumption of pollen, either alone or in combination with other hive products, counteracts the development of cancer, particularly prostate cancer (Hurd 2002; Leja et al. 2007, Negri et al. 2011, Premratanachai and Chanchao 2014, Al-Lawati et al. 2018).

The nutritional elements, the associated microorganisms, and the high amount of active bio components, carbohydrates, enzymes, essential amino acids and carotenoids, make pollen one of the hive products with more therapeutic properties, which include antibiotics, anti anemic, antitussive, diuretic, cardioprotective, hepatoprotective, and anti-inflammatory properties. Pollen is also beneficial for the control of colds and stomach conditions, such as colitis and gastric ulcers (Wolfe 2009, Bruneau 2017, Utoiu et al. 2018, Margaoan et al. 2019).

Royal Jelly

Royal jelly is a secretion of the mandibular and hypopharyngeal glands of young bees that is used to feed their larvae during early development (1-3 days), and the queen bee throughout her entire life. Royal jelly is a viscous, gelatinous substance of bright yellowish-white color, which has a sour and spicy smell and taste. Royal jelly influences the fertility and longevity of the queen bee that may live for up to five years (Proestos 2018, Flanjak et al. 2019, Yamaga et al. 2019). As with other bee products, royal jelly is also used as a nutritional supplement in human diets to provide nutrients that help prevent or treat some health ailments because it is rich in elements that contribute to protect and strengthen human health (Proestos 2018, Strant et al. 2019).

Regarding its conservation and use, royal jelly is a food that requires some care. If intended for fresh consumption, it must be kept refrigerated and packaged in dark glass containers to avoid its degradation, because air, light, and heat drastically modify its biological and organoleptic properties. Alternatively, one conservation technique that can be used to preserve royal jelly is the process of lyophilization, where the royal jelly is dehydrated to become a powder that retains all its properties (Maghsoudlou et al. 2019).

Regarding its characteristics and composition, fresh royal jelly is an acidic substance (pH of 3.6-4.2) that has 60-70% moisture, is rich in proteins, carbohydrates, vitamins, lipids, minerals, flavonoids, vitamins, polyphenols, and several biologically active substances, such as acetylcholine, peptides, the hormones testosterone, progesterone, prolactin, estradiol, (hydroxidecanoic acid) (HAD), adenosine monophosphate (AMP), N1 oxide, polyphenols, flavonoids, and adenosine (Kunugi and Mohammed 2019, Maghsoudlou et al. 2019).

Due to its bioactive components, royal jelly may be considered as a functional and nutraceutical food, known in the pharmacopeia and popular in traditional medicine. It is one of the bee products with germicide, antioxidant, anti-aging, and immunomodulatory effects that can be tested in laboratory animals, microbial organisms, farm animals, and clinical trials (Maghsoudlou et al. 2019, Pan et al. 2019). It has been suggested that royal jelly not only improves overall health, but it is of aid in the treatment of some diseases and health conditions such as diabetes, cardiovascular problems, Alzheimer's, and cancer (Kunugi and Mohammed 2019, Ahmad et al. 2020).

It has been claimed that royal jelly may play a primary role in modulating the mechanisms of aging, which can promote longevity in humans. The anti-aging properties of royal jelly are related to the essential amino acids contained in its proteins, as well as to 10-HDA (10-hydroxy-2-decenoic acid, a bioactive compound present in royal jelly) (Kunugi and Mohammed 2019).

Yamaga et al. (2019) observed that dicarboxylic acids and hydroxyl derivatives of royal jelly are absorbed into the circulatory system and then excreted in the urine; likewise, they observed that the absorption of fatty acids increased by treatment with royal jelly. Therefore, royal jelly is considered a promising food but also safe and effective. These authors mention and emphasize, for the first time, the potential benefits of using royal jelly in humans due to the absorption, metabolism, and pharmacokinetics of its fatty acids.

One study carried out with rats that were injected streptozotocin (diabetogenic drug) into the brain and were treated with ingested royal jelly, resulted in reduced neurogenesis and increased neuronal proliferation, which improved memory loss in the experimental animals, which in turn could help in cases of Alzheimer's disease (Zamani et al. 2012, e Silva et al. 2020). The proliferation of new neurons by oral treatment with royal jelly can also be considered a result of a neuroprotective property of the product. The newly generated neurons, still immature, exert a significant influence on the dynamics of the hippocampus circuits and cognitive and behavioral aspects related to learning and memory, pattern recognition, stress, and mood disorders, acting together with its other beneficial effects, reducing oxidative stress levels, and neuronal degeneration (e Silva et al. 2020).

Hadia et al. (2018) suggested that royal jelly reduces cholesterol blood levels and that the consumption of royal jelly could be effective in improving lipid parameters since poor control of dyslipidemia is related to a set of problems that reduce the quality of life, increase mortality and impose huge costs on social health systems.

Royal jelly consumed fresh induces vasorelaxation through the production of nitric oxide from the vascular endothelium in healthy rats, which leads to an increase in blood circulation, so it may improve peripheral circulation (Liang et al. 2018). Several other studies have shown a significant reduction in fasting blood glucose after supplementation with royal jelly (Khoshpey et al. 2016, Omer et al. 2019). Besides, it substantially improved serum triglyceride levels, reduced cholesterol levels, reduced oxidative stress indicators, and increased antioxidant enzyme levels. This is why royal jelly could improve the glycemic status, lipid profiles, and oxidative stress in patients with diabetes (Maleki et al. 2019).

Final Considerations

Nutritionists have focused in recent years on highlighting the properties of superfoods in several ways, including that they are ecological, functional, and primarily healthy. Hive products are attracting the attention of the food companies because they are considered as foods with high nutritional value and with a positive impact on health. Numerous studies have also shown that bee products are rich in bioactive and nutritive components such as proteins, minerals, nucleic acids, amino acids, polyphenols, carbohydrates, phytosterols, vitamins, and sugars, that together can promote the prevention and control of various diseases that afflict humans. However, the most specific studies have been conducted with laboratory animals due to ethical considerations and restrictions that exist in certain countries that do not allow testing in humans. Therefore, more research ought to be conducted with humans in countries where these restrictions are not a barrier. This strategy could generate more evidence to reliably confirm the properties of hive products as superfoods and their effects on the prevention and control of various diseases.

References

Ahmad, Saabor., Campos, Graca M., Fratini, Filippo, Altaye, Zewdu S., and Jianke Li. 2020. "New Insights into the Biological and Pharmaceutical Properties of Royal Jelly". *International Journal of Molecular Sciences*, 21, 382. Accessed February 18, 2020. doi:10.3390/ijms21020382.

Ahmed, Naveed, Sutcliffe, Alastair, and Claire Tipper. 2013. "Feasibility study: Honey for treatment of cough in children". *Pediatric Reports*, 5:31-34. doi10.4081/pr.2013.e8.

Ahmed, Sarfraz, Sulaiman, Amrah S., and Nor Hayaty, O. 2017. "Oral Administration of Tualang and Manuka Honeys Modulates Breast Cancer Progression in Sprague-Dawley Rats Model". *Evidence-Based Complementary and Alternative Medicine* 2017: ID 5904361,1-15 doi:10.1155/2017/5904361.

Ahmed, Sarfraz, Sulaiman, Amrah S., Baig, Amin. A., Ibrahim, Muhammad, Liaqat, Sana, Fatima, Saira, Jabeen, Sadia, Shamim, Nighat, and Nor Hayati O. 2018. "Honey as a Potential Natural Antioxidant Medicine: An Insight into Its Molecular Mechanisms of Action". *Oxidative Medicine and Cellular Longevity* ID 8367846, 1-19. doi:10.1155/2018/8367846.

Al-Lawati, Hassam. T., Al-Ajmi, Salim H. I., and Mostafa Waly, I. 2018. "Antioxidant and health properties of beehive products against oxidative stress-mediated carcinogenesis". In *Bioactive Components, Diet and Medical Treatment in Cancer Prevention,* edited by Mostafa I. Waly, Mohammad Shafiur Rahman, 97-103. Springer International Publishing.

Almasaudi, Saad, B., Al-Nahari, Alam. A. M., Abb El-Ghany, El Sayed. M., Barbour, Eli, Al Muhayawi, Sadd. M., Al-Jaouni, Soad A., Azhar, Esam, Qari, Mohamad, Qari, Yousef. A., and Steve Harakeh. 2017. "Antimicrobial effect of different types of honey on *Staphylococcus aureus*". *Saudi Journal of Biological Sciences* 24:1255-1261. Doi: 10.1016/j.sjbs.2016.08.007.

Anjum, Ishtiaq Syed, Ullah, Amjad, Khan, Ali K., Attaullah Mohammad, Khan, Hikmatullah, Ali, Hussain, Bashir, Amjad M., Tair, Muhammad, Ansari Javed Mohammad, Ghramh, A Hamed, Adgaba, Nuru, Dash Kanta C. 2019. "Composition and functional properties of propolis (bee glue): A review". *Saudi Journal of Biological Sciences* 26:1695-1703. doi:10.1016/j.sjbs.2018.08.013.

Becerra B, Thalia, Calla-Poma, Roger D., Requena-Mendizabal, Margarita F., and Pablo A. Millones-Gómez. 2019. "Antibacterial Effect of Peruvian Propolis Collected During Different Seasons on the Growth of *Streptococcus mutans*". *The Open Dentistry Journal* 13:328-331. doi: 10.2174/1874210601913010327.

Bogdanov, Stefan. 2016. "Honey Composition". In *The Honey Book.* Bee Product Science. 27-36. www.bee-hexagon.net. Accessed March 19.

Braakhuis, Andrea. 2019. "Evidence on the Health Benefits of Supplemental Propolis". *Nutrients* 11, 2705. doi:10.3390/nu11112705.

Brezani, Viliam, Smejkal, Karel, Hosek Jan, and Veronika Tomasova. 2018. "Anti-inflammatory Natural Prenylated Phenolic Compounds-Potential Lead Substances". *Current Medical Chemistry* 25:1094-1159. doi:10.2174/0929867324666170810161157.

Bruneau, Stephanie. 2017. *The benevolent bee. Capture the Bounty of the Hive Through Science, History, Remedies and Crafth*. Minneapolis. Quarry books.

Camacho, Bernal, Gloria I., Cruz-Cansino, Nelly, S., Alanís-García Ernesto, Calderón-Ramos, Zuli G y Arianna Omaña-Covarrubias. 2019. "Análisis químico proximal (AQP) de productos de la colmena de un apiario de Tulancingo, Hidalgo" [Proximal chemical analysis (PCA) of products from the hive of an apiary in Tulancingo, Hidalgo]: *Educación y Salud Boletín Científico de Ciencias de la Salud del ICSA* 7,14:65-69. doi:10.29057/icsa.v7i14.4441.

Canales-Martínez, M.M., Penieres-Castillo, J.G., y Rodríguez-Pérez, B. 2019. "Análisis de diferentes técnicas para la recolecta de propóleo apegándonos a la NOM-003-SAG/GAN 2017" [Analysis of different techniques for propolis collection according to NOM-003-SAG/GAN 2017] En *Agroecosistemas Tropicales:Conservación de recursos naturales y seguridad alimentaria*, editado por William Cetzal-Ix, Fernando Casanova-Lugo, Alfonso J. Chay-Canul y Jesús F. Martínez-Puc, 15-23. Tecnológico Nacional de México, Instituto Tecnológico de Chiná, Instituto Tecnológico de la Zona Maya.

Castillo-Ruiz, V., Uranga, Hernández, R. D., y Zafra de la Rosa, G. F. 2019. *Genética clínica* [*Clinical Genetics*]. Accesado Marzo 13. https://elibro.net/ereader/elibrodemo/39804.

Cianciosi, Danila, Forbes-Hernández, Tamara Y., Afrin, Sadia, Gasparrini, Massimiliano, Reboredo-Rodriguez, Patricia, Manna, Piera P., Zhang, Jiaojiao, Bravo, Lamas, Leire, Martínez Flórez, Susana, Agudo, Toyos, Pablo, Quiles, Jose L., Giampieri, Francesca, and Maurizio Battino. 2018. "Phenolic Compounds in Honey and Their Associated Health Benefits: A Review". *Molecules* 23, 2322. Doi:10.3390/molecules23092322.

Crane, Eva. 2013. *The World History of Beekeeping and Honey Hunting*. New York: Taylor and Francis Group.

Dantas, Silva, Rejane P., Machado, Souza, Bruna A., Barreto, Abreu, Gabriele A., Costa, Serra S., Andrade, Nalone L, Amaral Guimaraes R., Carvalho, Andrade A., Padilha, Ferreira F., Barbosa, Viana, Josiaba D., Umsza-Guez, Marcelo A. 2017. "Antioxidant, antimicrobial, antiparasitic, and cytotoxic properties of various Brazilian propolis extracts". *PLoS One* 12(3):e0172585. doi:10.1371/journal.pone.0172585.

Devequi-Nunes D., Souza, Machado Bruna A., Abreu, Barreto de G., Rebouças Silva J., da Silva Figueredo D., da Rocha Carneiro J.L., Brandão Neves H., Borges V.M., and Marcelo Andres Umsza-Guez. 2018. "Chemical characterization and biological activity of six different extracts of propolis through conventional methods and supercritical extraction". *PLoS One* 13(12):e0207676. doi:10.1371/journal.pone.0207676.

DOF. 2018. "Proyecto de Norma Oficial Mexicana PROY-NOM-004-SAG/GAN-2018, Producción de miel y especificaciones". *Secretaría de Agricultura, Ganadería, Desarrollo Rural, Pesca y Alimentación*. Accesado Abril 2020. https://dof.gob.mx/nota_detalle.php?codigo=5546962&fecha=20/12/2018.

e Silva, de Souza, Tiago. G., de Paulo do Val, Ferreira Maria E., da Silva Mamona Rubens J., Alves, da Silva A., Britto, G. Luiz R., Xavier, Fernando G., and Maria Regina Lopes. 2020. "Oral treatment with royal jelly improves memory and presents neuroprotective effects on icv-STZ rat model of sporadic Alzheimer's disease". *Heliyon* 6:(2), e03281. doi:10.1016/j.heliyon.2020.e03281.

Flanjak, Ivana, Primorac, Ljiljana, Vukadin, Ilijana, Kovačić, Marin, Puškadija, Zlatko, Bilić Raj, Blanca. 2019. "Physicochemical characteristics of Croatian royal jelly". *Croatian Journal of Food Science and Technology* 11: 266-271. doi: 10.17508/CJFST.2019.11.2.18.

Hadia, Amir, Najafgholizadeh, Ameneh, Aydenluc, Smadi E., Shafiei, Zahra, Pirivand, Fatemeh, Golpour, Sahar, and Makan Pourmasoumi. 2018. "Royal jelly is an effective and relatively safe alternative approach to blood lipid modulation: A meta-analysis". *Journal of Functional Foods* 41:202-209. doi:10.1016/j.jff.2017.12.005.

Hurd, Lyle. 2002. "Bee Pollen: Top rank antioxidant". *Total Health* 24: 40-43.

Israili, Zafar, H. 2014. "Antimicrobial Properties of Honey". American *Journal of Therapeutics* 21:304-323. doi:/10.1097/MJT.0b013e318293b09b.

Jean-Prost, Pierre, and Yves Le Cont. 2007. *Apicultura: conocimiento de la abeja. Manejo de la colmena*. España: Mundi Prensa.

Khoshpey, Basemeh, Djazayeri, Shima, Amiri, Fatemehsadat, Malek, Mojtaba, Hosseini, Fateme A., Hosseini, Sharieh, Shidfar, Shahrzad, and Farzad Shidfar. 2016. "Effect of royal jelly intake on serum glucose, apolipoprotein AI (ApoA-I), apolipoprotein B (ApoB) and ApoB/ApoA-I ratios in patients with type 2 diabetes: a randomized, double-blind clinical trial study". *Canadian Journal of Diabetes* 40: 324-328. doi:10.1016/j.jcjd.2016.01.003.

Kocot, Joanna, Kiełczykowska, Małgorzata, Luchowska-Kocot D., Kurzepa Jacek, and Irena Musik. 2018. "Antioxidant Potential of Propolis, Bee Pollen, and Royal Jelly: Possible Medical Application". *Oxidative Medicine and Cellular Longevity* ID 7074209, 29. Doi:10.1155/2018/7074209.

Kritsky, Gene. 2017. "Beekeeping from Antiquity Through the Middle Ages". *Annual Review of Entomology* 62:249-264. doi:10.1146/annurev-ento-031616-035115.

Kunugi, Hiroshi, and Amira Mohammed A. 2019. "Royal Jelly and Its Components Promote Healthy Aging and Longevity: From Animal Models to Humans". *International Journal of Molecular Sciences* 20:4662. doi:10.3390/ijms20194662.

Leja, M., Mareczek, A., Wyzgolik, G., Klepacz-Baniak, J., and K Czekońska. 2007. "Antioxidative properties of bee pollen in selected plant species". *Food Chemistry* 100: 237-240. doi:10.1016/j.foodchem.2005.09.047.

Liang, Yaoyue, Kagota, Satomi, Maruyama, Kana, Oonishi, Yuri, Miyauchi-Wakuda, S., Ito, Yoshihito, Yamada, Shizuo and Kazumasa Shinozuka. 2018. "Royal jelly increases peripheral circulation by inducing vasorelaxation through nitric oxide production under healthy conditions". *Biomedicine & Pharmacotherapy* 106:1210-1219. doi:10.1016/j.biopha.2018.07.047.

Long, L. (2020) *Honey: A Global History*. London: Reaktion Books.

Maghsoudlou, Atefe, Mahoonak, Sadeghi A., Mohebodini, Hossein, and Fidel Toldra. 2019. "Royal jelly: Chemistry, storage and bioactivities". *Journal of Apicultural Science* 63:17-40. doi:10.2478/jas-2019-0007.

Majtan, Juraj, Kovacova, Elena, Bılikova, Katarına, and Jozef Simuth. 2006. "The immunostimulatory effect of the recombinant apalbumin 1–major honeybee royal jelly protein–on TNFa release". *International Immunopharmacology* 6:269-278. doi:10.1016/j.intimp.2005.08.014.

Maleki, Vahid, Jafari-Vayghan, Hamed, Saleh-Ghadimi, Sevda, Adibian, Mahsa, Kheirouri, Sorayya, and Mohammad Alizadeh. 2019. "Effects of Royal jelly on metabolic variables in diabetes mellitus: A systematic review". *Complementary therapies in medicine* 43:20-27. doi:10.1016/j.ctim.2018.12.022.

Margaoan, Rodica, Stranț, Mirela, Varadi, Alina, Topal, Ercan, Yücel, Banu, Cornea-Cipcigan, Mihaiela, Campos, Maria. G., & Dan C. Vodnar. 2019. "Bee collected pollen and bee bread: Bioactive constituents and health benefits". *Antioxidants* 8,568. doi:10.3390/antiox8120568.

Markiewicz-Żukowska, Renata, Naliwajko, K. Sylwia, Bartosiuk, Emilia, Moskwa, Justyna, Isidorov, Valery, Soroczyńska, Jolanta, and Maria H. Borawska. 2013. "Chemical composition and antioxidant activity of beebread, and its influence on the glioblastoma cell line (U87MG)". *Journal Apicultural Research Science* 57:147-157. doi:10.2478/jas-2013-0025.

Martínez-Pérez de Ayala, R.L., Tucuch-Tun, J.R., Cruz-Sánchez, Tonatihuh A., Canales-Martinez M.M., Penieres-Castillo, J.G., Rodriguez-Perez, B. 2019. "Análisis de Diferentes Técnicas para la Recolecta de Propóleo Apegándonos a la NOM-003-SAG/GAN-2017" [Analysis of Different Techniques for Propolis Collection According to NOM-003-SAG/GAN-2017]. En *Agroecosistemas tropicales: conservación de recursos naturales y seguridad alimentaria*, editado por William Cetzal-Ix, Fernando Casanova Lugo, Alfonso J. Chay-Canul y Jesús F. Martínez-Puc. 15-24. México: Tecnológico Nacional de México, Instituto Tecnológico de Chiná, Instituto Tecnológico de la Zona Maya. [*Tropical agroecosystems: conservation of natural resources and food security*]

Mazzarello, V., Donadu, M.G., Ferrari, M., Piga, G., Usai, D., Zanetti, S., and M.A. Sotgiu. 2018. "Treatment of acne with a combination of propolis, tea tree oil, and Aloe vera compared to erythromycin cream: two double-blind investigations". *Clinical Pharmacology: Advances and Applications* 10:175-181. doi: 10.2147/CPAA.S180474.

Morice, H. Alyn. 2002. "Epidemiology of Cough". *Pulmonary Pharmacology and Therapeutics* 15:253-259. doi:10.1006/pupt.2002.0352.

Mujica, Veronica, Orrego, Roxana, Pérez, Jorge, Romero, Paula, Ovalle Paz, Zúñiga-Hernández Jessica, Arredondo, Miguel, and Elba Leiva. 2017. "The Role of Propolis in Oxidative Stress and Lipid Metabolism: A Randomized Controlled Trial". *Evidence-Based Complementary and Alternative Medicine* ID 4272940:11. doi:10.1155/2017/4272940.

Murtaza, Ghulam, Karim, Sabiha, Akram, Muhammad, R., Khan, Shujaat A., Azhar, Saira, Mumtaz, Amara, and Muhammad Hassham Hassan B. A. 2014. "Caffeic Acid Phenethyl Ester and Therapeutic Potentials". *BioMedi Research International* ID 145342,9. doi:10.1155/2014/145342.

Negri, Giuseppina, Teixeira, Weinstein, E., Florêncio Alves, Florencio, Marques, Teles, M.L., Moreti, Carmello de Camargo, A.C., Otsuk, Pozar I., Borguini, Galhardo, R., and Antonio Salatino. 2011. "Hydroxycinnamic Acid Amide Derivatives, Phenolic Compounds and Antioxidant Activities of Extracts of Pollen Samples from Southeast Brazil". *Journal of Agricultural and Food Chemistry* 59:5516-5522. doi:10.1021/jf200602k.

Nguyen, Huong Thi L., Panyoyai, Naksit, Kasapis, Stefan, Pang, Edwin, and Nitin Mantri. 2019. "Honey and Its Role in Relieving Multiple Facets of Atherosclerosis". *Nutrients* 11,167. doi:10.3390/nu11010167.

Oduwole, Olabisi, Udoh, Ekong E., Oyo-Ita, Angela, and Martin M. Meremikwu. 2018. "Honey for acute cough in children". *Cochrane Database of Systematic Reviews* IDCD007094. doi:10.1002/14651858.CD007094.pub5.

Omer, Kamel, Gelkopf, Maxwell J., and Genevieve Newton. 2019. "Effectiveness of royal jelly supplementation in glycemic regulation: A systematic review". *World Journal of Diabetes* 10: 96-113. doi: 10.4239/wjd.v10.i2.96.

Osés, S.M., Pascual-Maté, A., Fernández-Muiño, M.A., López-Díaz, T.M., and Sancho, M.T. 2016. "Bioactive properties of honey with propolis". *Food Chemestry* 196:1215-1223. doi:10.1016/j.foodchem.2015.10.050.

Pan, Yongming, Rong, Yili, You, Mengmeng, Ma, Quanxin, Chen, Minli, and Fuliang Hu. 2019. "Royal jelly causes hypotension and vasodilation induced by increasing nitric oxide production". *Food Science & Nutrition* 7:1361-1370. doi: 10.1002/fsn3.970.

Pasupuleti, Rao V., Sammugam, Lakhsmi, Ramesh, Nagesvari, and Siew Hua Gany Gan. 2017. "Honey, Propolis, and Royal Jelly: A Comprehensive Review of Their Biological Actions and Health Benefits". *Oxidative Medicine and Cellular Longevity* ID 1259510,21. doi:10.1155/2017/1259510.

Pasupuleti, Rao Visweswara, Sammugam, Lakhsmi, Ramesh, Nagesvari, and Siew Hua Gan. 2017. "Honey, Propolis, and Royal Jelly: A Comprehensive Review of Their Biological Actions and Health Benefits". *Oxidative Medicine and Cellular Longevity* ID 1259510. doi:10.1155/2017/1259510.

Pattamayutanon, Praetinee, Angeli, Sergio, Thakeow, Prodpran, Abraham, John, Disayathanoowat, Terd, and Panuwan Chantawannakul. 2017. "Volatile organic compounds of Thai honeys produced from several floral sources by different honey bee species". *Plos One* 12 e0172099. doi:10.1371/journal.pone.0172099.

Paul, Ian, M., Beiler, Jessica, Mc Monagle, Amyee, Shaffer L. Michele, Duda, Laura, and Cheston M. Berlin. 2007. "Effect of Honey, Dextromethorphan, and No Treatment on Nocturnal Cough and Sleep Quality for Coughing Children and Their Parents". *Archives of Pediatrics and Adolescent Medicine* 161:1140-1146. doi:10.1001/archpedi.161.12.1140.

Perez, Rodriguez, Betsabe, Martínez, Canales, Maria. M., Carrillo, Penieres, Jose G., and Tonatiuh Alejandro Sánchez Cruz. 2020. "Chemical composition, antioxidant properties and antimicrobial activity of Mexican propolis". *Acta Universitaria* 30, e2435. doi. http://doi.org/10.15174.au.2020.2435.

Pfeffer, M. Claire, and Amareshwar T. K. Singh. 2018. "Apoptosis: A Target for Anticancer Therapy". *International Journal of Molecular Sciences* 19:448. Doi:org/10.3390/ijms19020448.

Premratanachai, Pongsathon, and Chanpen Chanchao. 2014. "Review of the anticancer activities of bee products". *Asian Pacific Journal of Tropical Biomedicine* 4:337-344. Doi:org/10.12980/APJTB.4.2014C1262.

Proestos, Charalampos. 2018. "Superfoods: Recent data on their role in the prevention of diseases". *Current Research in Nutrition and Food Science Journal* 6:576-593. Doi:10.12944/CRNFSJ.6.3.02.

Przybyłek, Izabela and Tomasz Karpiński Z. 2019. "Antibacterial Properties of Propolis". *Molecules* 24:2047. doi: 10.3390/molecules24112047.

Quezada-Euan, J.J.G. 2018. *Stinglees bees of Mexico: The biology, management and conservation of an ancient heritage*. Switzerland; Springer International Publishing.

Raoa, Visweswara P., Krishnan, Thevan K., Salleh, Naguib, and Gan, Siew Hua. 2016. "Biological and therapeutic effects of honey produced by honey bees and stingless bees: A comparative review". *Revista Brasileira de Farmacognosia* 26:657-664. doi:10.1016/j.bjp.2016.01.012.

Raynaud, A., Ghezali, L., Gloaguen, V., Liagre, B., Quero, F., and Petit, J. M. 2013. "Honey-induced macrophage stimulation: AP-1 and NF-κB activation and cytokine production are unrelated to LPS content of honey". *International Immunopharmacology* 17:874-879. doi:10.1016/j.intimp.2013.09.014.

Rivera-Yañez, Nelly, Rodriguez-Canales, Mario, Nieto-Yañez Oscar, Jimenez-Estrada, Manuel, Ibarra-Barajas, Maximiliano, Canales-Martinez M.M., and M.A. Rodriguez-Monroy. 2018. "Hypoglycaemic and Antioxidant Effects of Propolis of Chihuahua in a

Model of Experimental Diabetes". *Evidence-Based Complementary and Alternative Medicine* ID 4360356,10. https://doi.org/10.1155/2018/4360356.

Samarghandian, Saeed, Farkhondeh, Tahereh, and Fariborz Samini. 2017. "Honey and Health: A Review of Recent Clinical Research". *Pharmacognosy Research* 9: 21-127. doi: 10.4103/0974-8490.204647.

Seibert, Brandao, A., Bautista-Silva Jessica P., Amparo Roquete Tatiane, Petit Alicia, Pervier, Pauline, dos Santos, Almeida, Juliana C., Azevedo, Costa M., Silveira, Maria B., Brandão, Celio G., de Souza, Bianco, Henrique G., Teixeira, de Medeiros Luis F., dos Santos, Henrique David O. 2019. "Development of propolis nanoemulsion with antioxidant and antimicrobial activity for use as a potential natural preservative". *Food Chemestry* 287:61-67. doi:10.1016/j.foodchem.2019.02.078.

Skoskiewicz-Malinowska, Katarzyna, Kaczmarek, Urszula, Malicka, Barbara, Walczak, Katarzyna, Zietek, Marek. 2017. "Application of Chitosan and Propolis in Endodontic Treatment: A Review". *Mini Reviews in Medicinal Chemestry* 17:410-434.

Strant, Mirela, Yücel, Banu, Topal, Erkan, Puscasu, Ana Maria, Margaoan, Rodica, and Alina Varad. 2019. "Use of Royal Jelly as Functional Food on Human and Animal Health". *Hayvansal Üretim* 60:131-144. doi:10.29185/hayuretim.513449.

Thakur, R. K., Manzoor, Uzma, and Bhadauriya, Amarpal Singh. 2019. "Apiculture for Sustainable Rural Development". *RASSA Journal of Science for Society* 1:46-48.

Tomás, Andreia, Falcão, I Soraia, Russo-Almeida, Paulo, Vilas-Boas, Miguel. 2017. "Potentialities of beebread as a food supplement and source of nutraceuticals: Botanical origin, nutritional composition and antioxidant activity". *Journal Apicultural Research* 56:219-230. doi:10.1080/00218839.2017.1294526.

Urcan, Adriana, Mărghitaş, Al Liviu, Dezmirean, S. Daniel, Bobiş, Otilia, Bonta, Victorita, Mureşan, I. Carmen, Mărgăoan, Rodica. 2017. "Chemical composition and biological activities of beebread-review". *Bulletin UASVM Animal Science and Biotechnologies* 74(1)/ 2017 doi:10.15835/buasvmcn-asb: 12646.

USDA. 2019. "Food Data Central-Honey". *United States Department of Agriculture.* Accessed March 2020. https://fdc.nal.usda.gov/fdc-app.html#/?query=ndb Number:19296.

Uțoiu, Elena, Matei, Florentina, Toma, Agnes, Diguță, Filofteia C., Ștefan, Mihaela L., Mănoiu, Sorin, Vrăjmașu, Valeriu V., Moraru, Ionut, Oancea, Anca, Israel-Roming, Florentina, Cornea, Petrut C., Constantinescu-Aruxandei, Diana, Moraru, Angela, and Florin Oancea. 2018. "Bee Collected Pollen with Enhanced Health Benefits, Produced by Fermentation with a Kombucha Consortium". *Nutrients* 10,1365. doi:10.3390/nu10101365.

Valenzuela-Barra, Gabriela, Castro, Consuelo, Figueroa, Catalina, Barriga, Andres, Silva Ximena, de Las Heras, Beatriz, Hortelano, Sonsoles, Delporte, Carla. 2015. "Anti-inflammatory activity and phenolic profile of propolis from two locations in Región Metropolitana de Santiago, Chile". *Journal of Ethnopharmacol* 168:37-44.doi:10.1016/j.jep.2015.03.050.

Vásquez, Alejandra, and Tobias C. Olofsson. 2015. "The lactic acid bacteria involved in the production of bee pollen and bee bread". *Journal of Apicultural Research* 48:189-195. doi:10.3896/IBRA.1.48.3.07.

Viuda-Martos, M., Ruiz-Navajas Y., Fernández-López J., and J.A. Pérez-Alvarez. 2008. "Functional Properties of Honey, Propolis, and Royal Jelly". *Journal of Food Science* 73:117-124. doi: 10.1111/j.1750-3841.2008.00966.

WHO. 2001. *Cough and cold remedies for the treatment of acute respiratory infections in young children. Maternal, newborn, child and adolescent health.* World Health Organization. Accessed March 2020. https://www.who.int/maternal_child_adolescent/documents/fch_cah_01_02/en/

Wolfe, David. 2009. *Superfoods: The Food and Medicine of the Future.* Californa: North Atlanctic Books.

Wright, A. Geraldine, Nicolson, W. Susan, and Sharoni Shafir. 2018. "Nutritional Physiology and Ecology of Honey Bees". *Annual Review of Entomology* 63:327-344. Doi; 10.1146/annurev-ento-020117-043423.

Yamaga, Masayuki, Tani, Hiroto, Yamaki, Ayanori, Tatefuji, Tomoki, and Ken Hashimoto. 2019. "Metabolism and pharmacokinetics of medium chain fatty acids after oral administration of royal jelly to healthy subjects". *RSC advances* 9:15392-15401. doi: 10.1039/C9RA02991E.

Yucel, Banu, Topal, Erkan, Kosoglu, Mustafa. 2017. "Bee products as functional food". In *Superfood and functional food-A overview on their processing and utilization*, edited by Waisundara V. and Shiomi N. 63-87. Croatia: InTech Open.

Zamani, Zohre, Reisi, Parham, Alaei, Hojjatallah, and Ali Asghar Pilehvarian. 2012. "Effect of Royal Jelly on spatial learning and memory in rat model of streptozotocin-induced sporadic Alzheimer's disease". *Advanced Biomedical Research* 12:1-26. doi: 10.4103/2277-9175.98150.

In: Super and Nutraceutical Foods
Editors: Sandra Teresita Martín-del-Campo et al.
ISBN: 978-1-53619-082-3

Chapter 14

MILK AND DAIRY PRODUCTS: NEW PERSPECTIVES FOR ANCIENT PRODUCTS

Rosa Vázquez-García, Cristina Chuck-Hernández and Sandra Teresita Martín-del-Campo*
Tecnologico de Monterrey, School of Engineering and Sciences, Monterrey, Mexico

ABSTRACT

Milk is an animal product consumed for human nutrition since the beginning of humankind. Depending on the geographical area and environmental characteristics, humans have used milk from different species for direct consumption or processed to increase shelf life. Milk from cow, goat, sheep, buffalo, mare, and camel is used to be drunk or produce cheeses, fermented milk, butter and other derivatives. The nutritional value of milk and dairy products is well-known and valued with most recent studies highlighting their health benefits associated with antihypertensive, immunomodulatory, antioxidant, and antibacterial activities linked to peptides obtained during dairy products' manufacture.

Keywords: milk, dairy products, ACE inhibition, antioxidant, bioactive peptides

INTRODUCTION

Functional foods or superfoods are understood as those products that can produce a specific effect on the health of the person who consumes it. In general, superfoods are food products that are minimally processed or fortified with some natural products to give added value (Proestos 2018, Gossard and Lipski 2019).

Milk is in itself a superfood with many biological benefits. Since the beginning of humankind, the milk traits have been well-known and used in all the world's cultures. Depending on the geographical area, weather, environmental characteristics, and diet, humans

* Corresponding Author's E-mail: smartinde@tec.mx.

have used the milk of different animals, where the cow, goat, sheep, buffalo, and camel stand out. Due to its animal origin and its nutritional components, milk has been used as a supplement for infant food, mainly in those unable to ingest breast milk from the earliest ages (Figure 1). Besides, millions of reports show that milk and dairy products are used as coadjuvants of all kinds of disorders, including gastrointestinal, cardiovascular, nutritional, sleep-related, and even as a palliative in skin diseases (Valenze 2011).

Figure 1. Legend of Rome foundation, Romulus and Remus suckled by a she-wolf.

On the other hand, milk-derived products have shown high physiological benefits. Products such as cheese, yogurt, fermented milk, whey, and others, have demonstrated antihypertensive, immunomodulatory, antioxidant, and antibacterial activities. Thanks to current analytical technologies, it is possible to determine which compounds provide a functional activity to milk and dairy derivatives under different production and processing conditions. Most of these are known to be due to peptides released after hydrolysis of the most important proteins in milk: caseins. These peptides obtained from dairy products could produce the mentioned effects and, when added to other foods, improve food technologies. The following lines describe the main bioactivities and the most current studies that make milk and its derivatives an example of superfoods.

Whole Milk as a Superfood

Milk is one of the most important and complete foods available. Its production reached in 2018 843 million tonnes, representing an increase of more than 59% compared with the 530 million tonnes produced worldwide in 1988 (Wu, Aluko, and Muir 2002). This product is an important source of balanced nutrients and has been shown to have many biological activities no matter which mammal is obtained. (Abdel-Hamid et al. 2017, Ahmed et al. 2015, Atanasova and Ivanova 2010).

Proteins are the milk component with more biological properties reported. From the whole protein fraction, caseins are the most important, representing 77% (Lynch and Barbano 1999). Its chemical structure includes micelles that can protect the internal protein from changes in the media, as pH and temperature fluctuations. This structure also helps to protect the protein chains from the presence of minerals, such as calcium, which can cause denaturation and chemical damage. Micelles are formed by αS_1, αS_2, β, and κ caseins, representing an important fraction of the structure. The micelle structure consists of a sphere presenting a hydrophobic center with α and β caseins, protected by a κ casein layer which function is to repel external water or another molecule, avoiding the denaturation of internal proteins. The presence of negative charges in the κ casein gives higher repulsion properties, increasing the structure's protection (Akuzawa, Miura, and Kawakami 2009, Park 2009).

Despite the excellent repulsion properties of casein micelles, some enzymes can hydrolyze this molecule, aiding in the digestion process when milk is ingested. These enzymes can be obtained from the same milk, from starter bacteria, and somatic cell membranes. Among the most important enzymes are plasmin, plasminogen, and lactocepin, producing small peptides and free amino acids (Ardö et al. 2017). The production of these peptides is crucial since they present synergy with several physiological functions, showing specific biological traits (Choi et al. 2012, McSweeney and Fox 2004).

Milk is one of the most studied sources of bioactive peptides. There are a large number of reports that shows antihypertensive properties due to the Angiotensin Converting Enzyme (ACE)-inhibitory effect (Abdel-Hamid et al. 2017, De Gobba et al. 2014, Espejo-Carpio et al. 2013, Huang et al. 2020, Shu et al. 2018), antioxidant activity (De Gobba et al. 2014, Zanutto-Elgui et al. 2019, Tonolo, Fiorese, et al. 2020, Huma et al. 2018), antimicrobial (Atanasova and Ivanova 2010, Esmaeilpour et al. 2016), hypocholesterolemic (Jiang et al. 2020, Mudgil, Baby, Ngoh, Vijayan, et al. 2019) and hemodynamic regulatory activities. These results, despite the milk source since their presence has been recorded in cow, goat, sheep, buffalo, camel, donkey and human milk (Ahmed et al. 2015, Mudgil, Baby, Ngoh, Vijayan, et al. 2019, Li et al. 2020, Nguyen et al. 2020, Aspri et al. 2018).

Although some authors have observed biological activities without hydrolysis, to study the qualities of whole milk in the production of bioactive peptides, it is necessary to carry out the *in vitro* proteolysis process using specific proteases looking to find specific activities. The human digestion process or the use of specific enzymes from animal, vegetal, or microorganisms are the most used methodologies for releasing bioactive peptides found in the casein structure. Once the peptides are produced, they are isolated using several analytical methods to analyze their specific activity (Choi et al. 2012, Espejo-Carpio et al. 2013, Abdel-Hamid et al. 2017, Skrzypczak et al. 2017).

Antihypertensive and ACEI Effects

One of the most studied functional activities in milk and its peptides is the Angiotensin I-Converting Enzyme Inhibitory Activity (ACEI). The angiotensin-converting-enzyme catalyzes the reaction that produces angiotensin II, a highly vasoconstrictive hormone, part of the renin-angiotensin-aldosterone system. If ACE is inhibited, blood pressure will decrease. This mechanism is one of the most used in drugs aiming to lower blood pressure (Soffer

1976, Wu, Aluko, and Muir 2002). Many authors consider that milk is the best source of ACEI peptides even after the digestion process.

Abdel-Hamid et al. (2017) found in buffalo milk hydrolyzed with papain, pepsin, and trypsin good ACEI inhibitory activities, identifying seven peptides from β-casein. Espejo-Carpio et al. (2013) hydrolyzed goat's milk under different temperatures using trypsin and subtilisin in several concentrations, finding the best functionality in the products of trypsin hydrolysis at 5 °C and 5% concentration. αS1 casein has been considered as the source of a high amount of bioactive peptides in goat milk. Tagliazucchi et al. (2017) observed the same activity in this type of milk, simulating gastrointestinal digestion. Li et al. (2020) and Giromini et al. (2019) determined that after a digestion process under stomach conditions, peptides related to ACEI activity are released, in addition to anticancer, antioxidant, and anti-anxiolytic counterparts.

Mudgil, Baby, Ngoh, Kamal, et al. (2019), Rahimi et al. (2016), and Mudgil, Baby, Ngoh, Vijayan, et al. (2019) tested the effect of papain, bromelin, alcalase, and proteinase K in camel's milk, detecting an increase in ACEI activity specifically with papain hydrolysis. Solanki, Hati, and Sakure (2017) used bacterial enzymes from *Lactobacillus bulgaricus* and *Lactobacillus fermentum*. Ugwu et al. (2019) observed the same effect using an in-silico model in camel's and horse's milk with pepsin and trypsin combined.

Antioxidant Effect

The production of substances with antioxidant effects from a natural source having the same qualities compared with a synthetic counterpart has been sought. Devi et al. (2020) observed an important antioxidant activity with cytoprotective potential in caseins obtained from buffalo milk and determined a bioactive peptide sequence. Sobral Bezerra et al. (2013), De Gobba et al. (2014), Ahmed et al. (2015) and Zanutto-Elgui et al. (2019) observed the activity with an *in vitro* model in goat milk hydrolyzed with papain followed by trypsinand subtilisin before being treated with pepsin, *Aspergillus oryzae* and *Aspergillus flavipes* enzymes. Tonolo, Folda, et al. (2020) studied the antioxidant properties of artificial peptides made after the sequence of those determined in milk.

Antimicrobial Effect

Milk has also demonstrated the presence of high amounts of peptides with antimicrobial effect, attributed to the presence of immunoglobulins in addition to non-immune proteins such as lactoferrins, lactoperoxidases, and lysozymes (Floris et al. 2003).

Milk proteins have encrypted peptides showing an antimicrobial effect when released by enzymatic reactions. Esmaeilpour et al. (2016) used goat milk hydrolyzates with trypsin and ficin to evaluate the antimicrobial activity, observing a significant effect over *Escherichia coli* and *Bacillus cereus*. Zanutto-Elgui et al. (2019) obtained antimicrobial peptides in goat's milk using *Aspergillus oryzae* and *Aspergillus flavipes* enzymes. Laiche, Khelef, and Daoudi (2019) observed the effect on camel's milk using *Lactococcus lactis* subsp. *cremoris* producing a bacteriocin against *Staphylococcus aureus*. All of these peptides have potential applications in the food and phrmaceutical industries.

Hypocholesterolemic Effect

Elevated cholesterol concentrations in plasma are associated with an increased risk for heart diseases in humans (Marcone, Belton, and Fitzgerald 2017). Many studies have shown that bioactive peptides derived from milk can exert a hypocholesterolemic effect. Nagaoka et al. (2001) used cow milk hydrolyzates with trypsin to obtain the peptide lactostatin and demonstrated a decrease in cholesterol levels in studies *in vivo*. Morikawa et al. (2007) described that lactostatin regulates the phosphorylation of extracellular signal-regulated kinase (ERK) and intracellular Ca^{2+} concentration, resulting in a lactostatin-mediated cholesterol degradation. They also showed that water-soluble lactostatin could activate the cholesterol 7α-hydroxylase (CYP7A1) gene, which increases cholesterol metabolism. Mudgil, Baby, Ngoh, Vijayan, et al. (2019) demonstrated that peptides are important agents against lipid oxidation and have a potential cholesterol esterase inhibitor (CEase), generated by proteins obtained after hydrolysis of camel milk with papain. Jafar et al. (2018) observed the inhibition of cholesteryl esterase (CE) and pancreatic lipase using bioactive peptides obtained from camel milk hydrolyzed with pepsin.

Mineral-Binding Effect

Milk has been determined to have peptides with a mineral-binding effect, which has been attributed to caseins' presence (Hartmann and Meisel 2007). Cross, Huq, and Reynolds (2007) worked with casein from cow's milk by tryptic digestion. They observed a property associated with localization and stabilization of calcium and phosphate ions at the tooth surface, promoting enamel subsurface lesions' remineralization. Korhonen and Pihlanto (2007) observed that casein phosphopeptides (CCPs) from αs1-, αs2- and β-caseins cow's milk can be released by digestive enzymes, forming soluble organophosphate salts and increasing Ca absorption by limiting the mineral precipitation within the ileum. Also, it has been detected phosphopeptides that withstand proteolysis in the gut, forming soluble complexes with calcium enhancing intestinal absorption and its retention in the body (Gobbetti et al. 2002).

Dairy Products

Cheese

Cheese is one of the most consumed dairy products in the world. Cheese characteristics depend on the production process and the milk origin giving a wide variety of cheeses (Figure 2). For all types of cheeses production, it is necessary to carry out a hydrolysis process of the casein micelles in the whole milk of any mammal to obtain a precipitate that will subsequently be processed to obtain one of the hundreds of varieties. The enzymes that carry out the process are generally exogenous, like trypsin, chymosin, and pepsin. Once the precipitation has been carried out, the appropriate microorganisms *Lactococcus lactis* and *Lactobacillus*, the most used for cheese production, will be added for the starting of the ripening process (Ardö et al. 2017, McAuliffe 2017, Tonolo, Folda, et al. 2020).

Figure 2. Variety of cheeses produced in Europe with different origin milk and the cheese-making process.

Proteolysis generally occurs during the cheese ripening stage, which depends on various external factors, such as moisture, temperature, salt content, and the ripening period applied in the cheese production process. The selection of microorganisms is crucial for the production of specific bioactive peptides. Yeasts are frequently observed in cheese, and they have an important relationship with the ripening process. Its growth characteristics such as lactose fermentation or proteolytic and lipolytic activities made them closely participate in the formation of different chemical compounds causing unique flavor characteristics in the final product (Barać et al. 2017, Liu et al. 2010, Ferreira and Viljoen 2003, McAuliffe 2017, Galli et al. 2019).

Once the protein denaturation has been carried out and other processes of bacterial elimination and decomposition after the cheese ripening, a great variety of bioactive compounds are released in the cheese matrix. They are obtained from ripened or digested *in vitro* or *in vivo* cheeses (Ardö et al. 2017). Also of products from proteins in cheese, others as lipid derivatives and metals are released during enzymatic digestion (Asledottir et al. 2017).

The ripening and bacterial enzymes in cheeses are known to develop different types of peptides in the same product. For example, sheep's milk cheeses have demonstrated the presence of immunomodulatory, antihypertensive, and antioxidant peptides (Meira et al. 2012, Pisanu et al. 2015, Galli et al. 2019). In the next lines, we will describe these activities in different cheese varieties.

Several antihypertensive and antithrombotic effects have also been found. In the casein case, there are reports that a prolonged ripening can lead to its hydrolysis with peptides' production depicting the ACE-I effect. This is the case of Val-Pro-Pro and Ile-Pro-Pro peptides, which have been detected in casein hydrolyzates from cow's milk. The antithrombotic effect has been studied in κ-casein, particularly the casoplateline peptides, which can inhibit the fibrinogen union and avoid the coagulation process. (Bütikofer et al. 2007, Akuzawa, Miura, and Kawakami 2009, Rafiq et al. 2017, Sagardia et al. 2013, Skrzypczak et al. 2017, Sahingil et al. 2014).

The antihypertensive activity has been found in ripened cheeses like Manchego, Norvegia, Jarlsberg, Cheddar, Blue, Emmental, Blue, Camembert, Edam, Havarti and Gouda (Bütikofer et al. 2007). Basiricò et al. (2015) determined eight different potent ACE-inhibitor in Parmigiano Reggiano cheese one year aged.

In cheeses with 180 days or more in ripening time, authors as Hernández-Galán et al. (2016) in Mexican Cotija cheese, Gonzalez-Gonzalez et al. (2019) in semi-ripened Chiapas cheese, and Dimitrov et al. (2015) in Bulgarian white brined cheese fermented with lactic acid bacterias observed antihypertensive peptides.

Hypotensive peptides have been identified in Roquefort cheese, Cheddar cheese from buffalo and cow and in Gamalost cheese (Meira et al. 2012; Rafiq et al. 2017 and Qureshi et al., 2012), with a significant increase of this peptides after proteolysis of up to 120 days. All peptides seem derived mostly from β-casein, with differences depending on cheese varieties and reduced concentration up to the end of the ripening process.

These characteristics were also observed by Erkaya and Şengul (2015), and Paul and Van Hekken (2011) in fresh and white cheeses. They found ACEI peptides released after at least eight weeks of ripening; however, in those studies, cheeses did not retain good sensory properties, being any longer suitable for human consumption. In contrast, it was determined by Kocak et al. (2020) that it is possible to modify the production of peptides with ACEI activity by modifying starter cultures. Torres-Llanez et al. (2011) used different starter cultures to produce ACEI peptides in Mexican Fresco cheese, concluding that the ripening is not necessary for the presence of the effect.

Another important property studied in cheeses is the antioxidant one. There are some antioxidant effects observed in ripened cheese like Gouda, Cheddar, Roquefort, Feta, and Parmigiano Reggiano. The ripening process induces the release of many peptides with this activity, both in caseins and whey proteins, particularly in cheese prepared with overheated milk. This effect is caused by the production of tridimensional peptides, the hydrophobicity of these, and Maillard derivatives production. Barać et al. (2016) observed the effect in their studies with white cheese prepared with overheated cow milk. Bottesini et al. (2013) found this bioactivity in 24 months aged Parmigiano Reggiano cheese. Taha et al. (2020) observed the effect during accelerated ripening in Domati cheese using a commercial enzyme and after nine months in Cheddar cheese.

Erkaya and Şengul (2015) determined antioxidant activity in vacuum packed white cheese, although it was no longer suitable for consumption. They observed the biological effect during the first 150 ripening days, and after this time, it decreases. Some authors have tested antioxidant activity *in vivo,* as Huma et al. (2018) observing an increase in the production of antioxidant peptides in extracts of Cheddar cheese from buffalo milk up to 180 ripening days. However, when tested in human colon adenocarcinoma cells, these peptides did not show significant antioxidant activity.

The immunomodulatory cheese effect has been demonstrated with α and β-caseins against phagocytic activity *in vitro* and their immunologic activity over a wide array of microorganisms. This effect is well-known as antimicrobial properties, and while there are many reports in milk and dairy products, there is not much information about peptides in cheese. After the proteolysis process during ripening, there are reports of proteins such as lactoferrin, lactoperoxidase, and lysozymes. They have been attributed to the antimicrobial capacity of any ripened cheese (Floris et al. 2003). There is evidence of peptides' presence and ability to eliminate certain microorganisms so that hydrolyzates could be used commercially as a preservative and food additive (Benkerroum 2010).

Like other peptides, those with antimicrobial activity are obtained from prolonged ripening times and then detected in aged cheeses, like Cheddar and Parmigiano Reggiano. Fialho et al. (2018) found antimicrobial peptides against *Escherichia coli* in Canastra artisanal minas cheese produced with cow milk after 30 ripening days. Lignitto et al. (2012) used Asiago dÁllevo cheese with different ripening times and determined antimicrobial activity against *Listeria innocua* in 18 months aged cheeses. In cheddar cheese it was observed an important antimicrobial activity against *Escherichia coli*, *Bacillus cereus* and *Staphylococcus aureus* and the ripening time was related to the peptides released (Pritchard, Phillips, and Kailasapathy 2010).

Mineralizing properties have been reported for phosphoserine and phosphothreonine, which are referred to as phosphopeptides. They can be easily obtained from caseins during ripening. This characteristic has the main quality that can limit calcium precipitation and plays an important role in bone mineralization. These peptides are among the best known in dry Italian cheeses such as Grana Padano and Parmigiano Reggiano, which can be found even after *in vitro* digestion (Cattaneo et al. 2017).

Whey

Whey is considered a waste product after milk has been processed into cheese. In general, caseins are precipitated thanks to external enzymes' action as previously mentioned, and the rest of the proteins remains in milk whey, which contains high amounts of water-soluble molecules such as lactose, minerals (calcium and magnesium), and proteins such as lactoglobulins and lactalbumin. In recent years, these proteins have been considered an important source of amino acids and peptides of great nutritional value. Many authors have used it as another important dairy product (Corrochano et al. 2019, D'Souza et al. 2020).

Abdel-Hamid et al. (2020) observed an important production of antimicrobial peptides that can inhibit the growth of *Pseudomona aeruginosa* and *Staphylococcus aureus* in camel whey hydrolyzed with papain. Luz et al. (2020) determined antifungal action against ten different fungi types from the genus *Penicillium*.

There is a record of the serum's ability to promote certain hormonal activities, such as the function of insulin. These proteins showed the ability to alter lipid metabolism via pepsin-pancreatin in adipocytes, although the peptide responsible for this activity was not determined (D'Souza et al. 2020). Mazorra-Manzano et al. (2020) used a fermentative whey method with native microflora. They obtained ACEI peptides from β-lactoglobulin and α-lactoalbumin.

Fermented Milk

Fermented milk is considered an important source of nutrients since the beginning of humanity. Although the fermentation process occurs on the carbohydrates, the bacteria that develop this process also can produce free peptides in an adequate environment (Pourjoula et al. 2020, Moreno-Fernandez et al. 2017).

Fermented milk produces antioxidant peptides, as studied by Tonolo, Fiorese, et al. (2020) and Sowmya et al. (2018). This effect was evaluated, producing *in vivo* activity against oxidative stress and inhibiting reactive oxygen species production in Caco-2 cells. The peptides were identified using mass spectrometry. Also, Rubak et al. (2020) produced fermented milk using *Lactobacillus delbrueckii, Lactococcus lactis ssp. lactis* and *Lactobacillus kefiri* during fermentation observed the production of ACEI activity. Aslam, Shoukat, et al. (2019) and Aslam, Aslam, et al. (2019) observed the same activity in goat and bovine fermented milk with *Lactobacillus casei*, and determined that peptides with high concentrations of proline, lysine, and aromatic amino acids get the higher activity, proposing it as a nutraceutical product. Parmar et al. (2019) studied the same effect in milk fermented with five different *Lactobacillus* cultures in different fermented conditions.

In cow milk kefir, mass spectrometry determined the presence of almost 35 peptides that are related to ACEI activity, concluding they can be used in antihypertensive drug development (Amorim et al. 2019). There have been observed the preservation of the peptides with this and other biological activities in fermented milk products with *Lactobacillus casei* after 21 storage days at 4°C (Abdel-Hamid et al. 2019). The effect was studied in donkey milk fermented with *Lactobacillus casei* DM214, analyzed by HPLC. The possible effects as antioxidant and antimicrobial using the same fermentative procedure were studied (Aspri et al. 2018).

On the other hand, some studies observed antithrombotic and hypocholesterolemic activities. The study elaborated by Rendon-Rosales et al. (2019) showed the effect in fermented milk with *Lactococcus lactis* followed by simulated digestion in an *in vitro* model. Similar results were observed by Guzmán-Rodríguez et al. (2019) with *Lactobacillus casei Shirota* and different fermentation conditions, as pH and temperature. The highest antithrombotic activity was detected at pH 6.0 and 39.5°C.

Pourjoula et al. (2020) observed phosphorylated peptides related to the bone mineralization function in kashk, a dairy product fermented in the Middle East and Iran. The author considered to use them as additives in other foods for this purpose.

NEW TRENDS

The presence and type of bioactive peptides in milk and dairy products depend on many external factors, as the type of milk or ripening conditions. The biological activity of peptides is based on its composition and sequence of amino acids. Its presence will depend entirely on balance between the degradation and production of the proteolytic process during manufacturing (Barać et al. 2017, Bernabucci et al. 2014, Bottari et al. 2017).

It has been shown that the use of milk and dairy products as a supplement in other products can potentiate each of the previously mentioned effects, as observed by Balthazar et al. (2019), who produced a beverage with sheep's milk and strawberry pulp, creating a product that can be an excellent prebiotic after a controlled digestion process, proving to be a good source of protein and minerals.

Other properties have been associated with the continuous consumption of milk. Sartorius et al. (2019) observed the milk's ability to decrease postprandial glucose using low concentrations over long periods. They conclude its use as a treatment to reduce the risk of developing type 2 diabetes mellitus. Kinariwala et al. (2019) observed the same effect in bovine milk fermented with *Lactobacillus fermentum*. In this sense, some studies determined that it can be used as antiosteopenic. In buffalo milk, the effect was observed with a restored bone microarchitectural in an *in vivo* model. These results can be used in the prevention of postmenopausal osteoporosis (Reddi et al. 2019).

In dairy products, there are also reports of properties that have been few studies, but they are known for the wide use. In buffalo mozzarella cheese, Tenore et al. (2019) analyzed the presence of anti-inflammatory properties in the small intestine thanks to the regulation of inflammatory cells reducing oxidative stress in chronic diseases such as colitis using *in vivo* and *in vitro* models. In camembert cheese, Schlienger et al. (2014) observed that ripening does not affect lipid concentration or blood pressure parameters in adults who consume it. In yogurt enriched with bioactive peptides obtained from milk, it was determined that its consumption could help treat cardiovascular diseases. However, the mechanism of action is not exact yet (Abd El-Fattah et al. 2018).

Acknowledgments

The first author, Vazquez-Garcia, R gratefully acknowledges Consejo Nacional de Ciencia y Tecnología (CONACyT) for granting scholarship No. 260794.

References

Abd El-Fattah, A., S. Sakr, S. El-Dieb, and H. Elkashef. 2018. "Developing functional yogurt rich in bioactive peptides and gamma-aminobutyric acid related to cardiovascular health." *LWT-Food Science and Technology* 98:390-397. doi: 10.1016/j.lwt.2018.09.022.

Abdel-Hamid, M., E. Romeih, R. R. Gamba, E. Nagai, T. Suzuki, T. Koyanagi, and T. Enomoto. 2019. "The biological activity of fermented milk produced by Lactobacillus casei ATCC 393 during cold storage." *International Dairy Journal* 91:1-8. doi: 10.1016/j.idairyj.2018.12.007.

Abdel-Hamid, M., E. Romeih, P. Saporito, A. Osman, R. V. Mateiu, B. Mojsoska, and H. Jenssen. 2020. "Camel milk whey hydrolysate inhibits growth and biofilm formation of Pseudomonas aeruginosa PAO1 and methicillin-resistant Staphylococcus aureus." *Food Control* 111. doi: 10.1016/j.foodcont.2019.107056.

Abdel-Hamid, Mahmoud, Jeanette Otte, Cristian De Gobba, Ali Osman, and Essam Hamad. 2017. "Angiotensin I-converting enzyme inhibitory activity and antioxidant capacity of bioactive peptides derived from enzymatic hydrolysis of buffalo milk proteins." *International dairy journal* 66:91-98. doi: 10.1016/j.idairyj.2016.11.006.

Ahmed, Ahmed S, Tawfik El-Bassiony, Laila M Elmalt, and Hisham R Ibrahim. 2015. "Identification of potent antioxidant bioactive peptides from goat milk proteins." *Food research international* 74:80-88. doi: 10.1016/j.foodres.2015.04.032.

Akuzawa, Ryozo, Takayuki Miura, and Hiroshi Kawakami. 2009. "Bioactive components in caseins, caseinates, and cheeses." In *Bioactive components in milk and dairy products*, edited by Young W Park, 217-234. Athens, GA.

Amorim, F. G., L. B. Coitinho, A. T. Dias, A. G. F. Friques, B. L. Monteiro, L. C. D. D. Rezende, T. D. M. C. Pereira, B. P. Campagnaro, E. De Pauw, E. C. Vasquez, and L. Quinton. 2019. "Identification of new bioactive peptides from Kefir milk through proteopeptidomics: Bioprospection of antihypertensive molecules." *Food Chemistry* 282:109-119. doi: 10.1016/j.foodchem.2019.01.010.

Ardö, Ylva, Paul L. H. McSweeney, Abdallah A. A. Magboul, Vivek K. Upadhyay, and Patrick F. Fox. 2017. "Biochemistry of cheese ripening: proteolysis." In *Cheese (Fourth Edition)*, 445-482. Elsevier.

Aslam, M. Z., M. S. Aslam, S. Firdos, G. Ghous, G. Firdos, Z. Hongfei, and Z. Bolin. 2019. "Role of Bioactive Peptides in Reducing the Severity of Hypertension with the Inhibition of ACE." *International Journal of Peptide Research and Therapeutics* 25 (4):1639-1649. doi: 10.1007/s10989-018-09806-y.

Aslam, M. Z., S. Shoukat, Z. Hongfei, and Z. Bolin. 2019. "Peptidomic Analysis of ACE Inhibitory Peptides Extracted from Fermented Goat Milk." *International Journal of Peptide Research and Therapeutics* 25 (4):1259-1270. doi: 10.1007/s10989-018-9771-0.

Asledottir, Tora, Thao T. Le, Bjørn Petrat-Melin, Tove G. Devold, Lotte B. Larsen, and Gerd E. Vegarud. 2017. "Identification of bioactive peptides and quantification of β-casomorphin-7 from bovine β-casein A1, A2 and I after ex vivo gastrointestinal digestion." *International Dairy Journal* 71:98-106. doi: 10.1016/j.idairyj.2017.03.008.

Aspri, M., G. Leni, G. Galaverna, and P. Papademas. 2018. "Bioactive properties of fermented donkey milk, before and after in vitro simulated gastrointestinal digestion." *Food Chemistry* 268:476-484. doi: 10.1016/j.foodchem.2018.06.119.

Atanasova, J., and I. Ivanova. 2010. "Antibacterial peptides from goat and sheep milk proteins." *Biotechnology & Biotechnological Equipment* 24 (2):1799-1803. doi: 10.2478/V10133-010-0049-8.

Balthazar, C. F., A. Santillo, J. T. Guimarães, V. Capozzi, P. Russo, M. Caroprese, R. Marino, E. A. Esmerino, R. S. L. Raices, M. C. Silva, H. L. A. Silva, M. Q. Freitas, D. Granato, A. G. Cruz, and M. Albenzio. 2019. "Novel milk–juice beverage with fermented sheep milk and strawberry (Fragaria ananassa): Nutritional and functional characterization." *Journal of Dairy Science* 102 (12):10724-10736. doi: 10.3168/jds.2019-16909.

Barać, Miroljub, Mirjana Pešić, Tanja Vučić, Milena Vasić, and Milenko Smiljanić. 2017. "White cheeses as a potential source of bioactive peptides." *Mljekarstvo/Dairy* 67 (1). doi: 10.2478/V10133-010-0049-8.

Barać, Miroljub, Milenko Smiljanić, Sladjana Žilić, Mirjana Pešić, Sladjana Stanojević, Milena Vasić, and Tanja Vučić. 2016. "Protein profiles and total antioxidant capacity of water soluble and insoluble protein fractions of white cow cheese at different stage of ripening." *Mljekarstvo/Dairy* 66 (3). doi: 10.15567/mljekarstvo.2016.0303.

Basiricò, L., E. Catalani, P. Morera, S. Cattaneo, M. Stuknytė, U. Bernabucci, I. De Noni, and A. Nardone. 2015. "Release of angiotensin converting enzyme-inhibitor peptides during in vitro gastrointestinal digestion of Parmigiano Reggiano PDO cheese and their absorption through an *in vitro* model of intestinal epithelium." *Journal of dairy science* 98 (11):7595-7601. doi: 10.3168/jds.2015-9801.

Benkerroum, Noreddine. 2010. "Antimicrobial peptides generated from milk proteins: a survey and prospects for application in the food industry. A review." *International Journal of Dairy Technology* 63 (3):320-338. doi: 10.1111/j.1471-0307.2010.00584.x.

Bernabucci, Umberto, Elisabetta Catalani, Loredana Basiricò, Patrizia Morera, and Alessandro Nardone. 2014. "In vitro ACE-inhibitory activity and in vivo antihypertensive effects of water-soluble extract by Parmigiano Reggiano and Grana Padano cheeses." *International Dairy Journal* 37 (1):16-19. doi: 10.1016/j.idairyj.2014.02.009.

Bottari, Benedetta, Andrea Quartieri, Barbara Prandi, Stefano Raimondi, Alan Leonardi, Maddalena Rossi, Alessandro Ulrici, Monica Gatti, Stefano Sforza, and Marco Nocetti. 2017. "Characterization of the peptide fraction from digested Parmigiano Reggiano cheese and its effect on growth of lactobacilli and bifidobacteria." *International journal of food microbiology* 255:32-41. doi: 10.1016/j.ijfoodmicro.2017.05.015.

Bottesini, Chiara, Sara Paolella, Francesca Lambertini, Gianni Galaverna, Tullia Tedeschi, Arnaldo Dossena, Rosangela Marchelli, and Stefano Sforza. 2013. "Antioxidant capacity of water soluble extracts from Parmigiano-Reggiano cheese." *International journal of food sciences and nutrition* 64 (8):953-958. doi: 10.3109/09637486.2013.821696.

Bütikofer, Ueli, Jacques Meyer, Robert Sieber, and Daniel Wechsler. 2007. "Quantification of the angiotensin-converting enzyme-inhibiting tripeptides Val-Pro-Pro and Ile-Pro-Pro in hard, semi-hard and soft cheeses." *International dairy journal* 17 (8):968-975. doi: 10.1016/j.idairyj.2006.11.003.

Cattaneo, Stefano, Milda Stuknytė, Anita Ferraretto, and Ivano De Noni. 2017. "Impact of the in vitro gastrointestinal digestion protocol on casein phosphopeptide profile of Grana Padano cheese digestates." *LWT-Food Science and Technology* 77 (1):356-361. doi: 10.1016/j.lwt.2016.11.069.

Choi, Jongwoo, Latha Sabikhi, Ashraf Hassan, and Sanjeev Anand. 2012. "Bioactive peptides in dairy products." *International Journal of Dairy Technology* 65 (1):1-12. doi: 10.1111/j.1471-0307.2011.00725.x.

Corrochano, A. R., A. Ferraretto, E. Arranz, M. Stuknytė, M. Bottani, P. M. O'Connor, P. M. Kelly, I. De Noni, V. Buckin, and L. Giblin. 2019. "Bovine whey peptides transit the intestinal barrier to reduce oxidative stress in muscle cells." *Food Chemistry* 288 (16):306-314. doi: 10.1016/j.foodchem.2019.03.009.

Cross, K. J., N. L. Huq, and E. C. Reynolds. 2007. "Casein Phosphopeptides in Oral Health - Chemistry and Clinical Applications." *Current Pharmaceutical Design* 13 (8):793-800. doi: 10.2174/138161207780363086.

D'Souza, K., A. Mercer, H. Mawhinney, T. Pulinilkunnil, C. C. Udenigwe, and P. C. Kienesberger. 2020. "Whey peptides stimulate differentiation and lipid metabolism in adipocytes and ameliorate lipotoxicity-induced insulin resistance in muscle cells." *Nutrients* 12 (2). doi: 10.3390/nu12020425.

De Gobba, Cristian, F Javier Espejo-Carpio, Leif H Skibsted, and Jeanette Otte. 2014. "Antioxidant peptides from goat milk protein fractions hydrolysed by two commercial proteases." *International dairy journal* 39 (1):28-40. doi: 10.1016/j.idairyj.2014.03.015.

Devi, S., G. K. Pal, R. Kapila, and S. Kapila. 2020. "C-terminal sequence deletion effect on antioxidative characteristics of VLPVPQK bioactive peptide from buffalo milk casein." *LWT-Food Science and Technology* 119. doi: 10.1016/j.lwt.2019.108816.

Dimitrov, Zhechko, Elena Chorbadjiyska, Irina Gotova, Kalinka Pashova, and Svetla Ilieva. 2015. "Selected adjunct cultures remarkably increase the content of bioactive peptides in Bulgarian white brined cheese." *Biotechnology & Biotechnological Equipment* 29 (1):78-83. doi: 10.1080/13102818.2014.969918.

Erkaya, Tuba, and Mustafa Şengul. 2015. "Bioactivity of water soluble extracts and some characteristics of white cheese during the ripening period as effected by packaging type and probiotic adjunct cultures." *Journal of Dairy Research* 82 (1):47-55. doi: 10.1017/S0022029914000703.

Esmaeilpour, Mehrnaz, Mohammad Reza Ehsani, Mahmoud Aminlari, Shahram Shekarforoush, and Ebrahim Hoseini. 2016. "Antimicrobial activity of peptides derived from enzymatic hydrolysis of goat milk caseins." *Comparative Clinical Pathology* 25 (3):599-605. doi: 10.1007/s00580-016-2237-x.

Espejo-Carpio, Francisco Javier, Raúl Pérez-Gálvez, Emilia M Guadix, and Antonio Guadix. 2013. "Optimisation of the hydrolysis of goat milk protein for the production of ACE-inhibitory peptides." *Journal of dairy research* 80 (2):214-222. doi: 10.1017/S0022029913000083.

Ferreira, A. D., and B. C. Viljoen. 2003. "Yeasts as adjunct starters in matured Cheddar cheese." *International journal of food microbiology* 86 (1-2):131-140. doi: 10.1016/S0168-1605(03)00252-6.

Fialho, Tatiana Lopes, Lanna Clícia Carrijo, Marcos Jorge Magalhães Júnior, Maria Cristina Baracat-Pereira, Roberta Hilsdorf Piccoli, and Luiz Ronaldo de Abreu. 2018. "Extraction and identification of antimicrobial peptides from the Canastra artisanal minas cheese." *Food Research International* 107:406-413. doi: 10.1016/j.foodres.2018.02.009.

Floris, René, Isidra Recio, Ben Berkhout, and Servaas Visser. 2003. "Antibacterial and antiviral effects of milk proteins and derivatives thereof." *Current Pharmaceutical Design* 9 (16):1257-1275. doi: 10.2174/1381612033454810.

Galli, B. D., D. P. Baptista, F. G. Cavalheiro, F. Negrão, M. N. Eberlin, and M. L. Gigante. 2019. "Peptide profile of Camembert-type cheese: Effect of heat treatment and adjunct culture Lactobacillus rhamnosus GG." *Food Research International* 123:393-402. doi: 10.1016/j.foodres.2019.05.009.

Giromini, C., J. A. Lovegrove, D. I. Givens, R. Rebucci, L. Pinotti, E. Maffioli, G. Tedeschi, T. S. Sundaram, and A. Baldi. 2019. "In vitro-digested milk proteins: Evaluation of angiotensin-1-converting enzyme inhibitory and antioxidant activities, peptidomic profile, and mucin gene expression in HT29-MTX cells." *Journal of Dairy Science* 102 (12):10760-10771. doi: 10.3168/jds.2019-16833.

Gobbetti, M., L. Stepaniak, M. De Angelis, A. Corsetti, and R. Di Cagno. 2002. "Latent Bioactive Peptides in Milk Proteins: Proteolytic Activation and Significance in Dairy Processing." *Critical Reviews in Food Science and Nutrition* 42 (3):223-239. doi: 10.1080/10408690290825538.

Gonzalez-Gonzalez, C. R., J. Machado, S. Correia, A. L. McCartney, J. Stephen Elmore, and P. Jauregi. 2019. "Highly proteolytic bacteria from semi-ripened Chiapas cheese elicit angiotensin-I converting enzyme inhibition and antioxidant activity." *LWT-Food Science and Technology* 111:449-456. doi: 10.1016/j.lwt.2019.05.039.

Gossard, Crystal M., and Elizabeth Lipski. 2019. "Superfoods for Digestive Health." In *Integrative Gastroenterology*, edited by Gerard E Mullin, Marvin Singh and Alyssa Parian, 141-174.

Guzmán-Rodríguez, F., L. Gómez-Ruizy, G. Rodríguez-Serrano, S. Alatorre-Santamaría, M. García-Garibay, and W. A. Cruz-Guerrero. 2019. "Iron binding and antithrombotic peptides released during the fermentation of milk by Lactobacillus casei shirota." *Revista Mexicana de Ingeniera Quimica* 18 (3):1161-1166. doi: 10.24275/uam/izt/dcbi/revmexingquim/2019v18n3/Guzman. [*Mexican Journal of Chemical Engineering*]

Hartmann, Rainer, and Hans Meisel. 2007. "Food-derived peptides with biological activity: from research to food applications." *Current opinion in biotechnology* 18 (2):163-169. doi: /10.1016/j.copbio.2007.01.013.

Hernández-Galán, Leticia, Anaberta Cardador-Martínez, Daniel Picque, Henry Eric Spinnler, Micloth López-del-Castillo Lozano, and Sandra Teresita Martín del Campo. 2016. "ACEI and antioxidant peptides release during ripening of Mexican Cotija hard cheese." *Journal of Food Research* 5 (3):85. doi: 10.1080/19476337.2016.1202325.

Huang, S., Y. Gong, Y. Li, S. Ruan, S. M. Roknul Azam, Y. Duan, X. Ye, and H. Ma. 2020. "Preparation of ACE-inhibitory peptides from milk protein in continuous enzyme membrane reactor with gradient dilution feeding substrate." *Process Biochemistry* 92:130-137. doi: 10.1016/j.procbio.2020.02.030.

Huma, Nuzhat, Saima Rafiq, Aysha Sameen, Imran Pasha, and Muhammad Issa Khan. 2018. "Antioxidant potential of buffalo and cow milk Cheddar cheeses to tackle human colon adenocarcinoma (Caco-2) cells." *Asian-Australasian Journal of Animal Sciences* 31 (2):287-292. doi: 10.5713/ajas.17.0031.

Jafar, Sabika, Hina Kamal, Priti Mudgil, Hassan Mohamed Hassan, and Sajid Maqsood. 2018. "Camel whey protein hydrolysates displayed enhanced cholesteryl esterase and lipase inhibitory, anti-hypertensive and anti-haemolytic properties." *LWT* 98:212-218. doi: 10.1016/j.lwt.2018.08.024.

Jiang, X., D. Pan, T. Zhang, C. Liu, J. Zhang, M. Su, Z. Wu, X. Zeng, Y. Sun, and Y. Guo. 2020. "Novel milk casein–derived peptides decrease cholesterol micellar solubility and cholesterol intestinal absorption in Caco-2 cells." *Journal of Dairy Science*. doi: 10.3168/jds.2019-17586.

Kinariwala, D., G. Panchal, A. Sakure, and S. Hati. 2019. "Exploring the Potentiality of Lactobacillus Cultures on the Production of Milk-Derived Bioactive Peptides with Antidiabetic Activity." *International Journal of Peptide Research and Therapeutics* 1-15. doi: 10.1007/s10989-019-09958-5.

Kocak, Ali, Tuba Sanli, Elif Ayse Anli, and Ali Adnan Hayaloglu. 2020. "Role of using adjunct cultures in release of bioactive peptides in white-brined goat-milk cheese." *LWT-Food Science and Technology* 123:109127. doi: 10.1016/j.lwt.2020.109127.

Korhonen, Hannu, and Anne Pihlanto. 2007. "Bioactive peptides from food proteins." *Handbook of food products manufacturing–Health, meat, milk, poultry, seafood, and vegetables* 5-37.

Laiche, A. T., C. Khelef, and H. Daoudi. 2019. "Study of the antimicrobial activity of bacteriocins produced by lactic bacteria isolated from camel milk in southern Algeria." *Journal of Pure and Applied Microbiology* 13 (2):1285-1292. doi: 10.22207/JPAM. 13.2.72.

Li, S., Q. Hu, C. Chen, J. Liu, G. He, L. Li, J. Wu, and D. Ren. 2020. "Formation of bioactive peptides during simulated gastrointestinal digestion is affected by αs1-casein polymorphism in buffalo milk." *Food Chemistry* 313. doi: 10.1016/j.foodchem. 2020.126159.

Lignitto, Laura, Severino Segato, Stefania Balzan, Valeria Cavatorta, Nadia Oulahal, Stefano Sforza, Pascal Degraeve, Gianni Galaverna, and Enrico Novelli. 2012. "Preliminary investigation on the presence of peptides inhibiting the growth of Listeria innocua and Listeria monocytogenes in Asiago d'Allevo cheese." *Dairy science & technology* 92 (3):297-308. doi: 10.1007/s13594-012-0057-6.

Liu, Mengjin, Jumamurat R. Bayjanov, Bernadet Renckens, Arjen Nauta, and Roland J. Siezen. 2010. "The proteolytic system of lactic acid bacteria revisited: a genomic comparison." *BMC genomics* 11 (1):36. doi: 10.1186/1471-2164-11-36.

Luz, C., L. Izzo, A. Ritieni, J. Mañes, and G. Meca. 2020. "Antifungal and antimycotoxigenic activity of hydrolyzed goat whey on Penicillium spp: An application as biopreservation agent in pita bread." *LWT* 118. doi: 10.1016/j.lwt.2019.108717.

Lynch, Joanna M., and David M. Barbano. 1999. "Kjeldahl nitrogen analysis as a reference method for protein determination in dairy products." *Journal-aoac international* 82 (6):1389-1398. doi: 10.1.1.459.1821.

Marcone, Simone, Orina Belton, and Desmond J. Fitzgerald. 2017. "Milk-derived bioactive peptides and their health promoting effects: a potential role in atherosclerosis." 83 (1):152-162. doi: 10.1111/bcp.13002.

Mazorra-Manzano, M. A., G. R. Robles-Porchas, D. A. González-Velázquez, M. J. Torres-Llanez, M. Martínez-Porchas, C. O. García-Sifuentes, A. F. González-Córdova, and B. Vallejo-Córdoba. 2020. "Cheese whey fermentation by its native microbiota: Proteolysis and bioactive peptides release with ACE-inhibitory activity." *Fermentation* 6 (1). doi: 10.3390/fermentation6010019.

McAuliffe, Olivia. 2017. "Genetics of Lactic Acid Bacteria." In *Cheese: Chemestry, physics and microbiology: general aspects*, 227-247.

McSweeney, P. L. H., and P. F. Fox. 2004. "Metabolism of residual lactose and of lactate and citrate." In *Cheese: chemistry, physics and microbiology*, 361-371.

Meira, Stela Maris Meister, Daniel Joner Daroit, Virginia Etges Helfer, Ana Paula Folmer Corrêa, Jéferson Segalin, Silvana Carro, and Adriano Brandelli. 2012. "Bioactive peptides in water-soluble extracts of ovine cheeses from Southern Brazil and Uruguay." *Food Research International* 48 (1):322-329. doi: 10.1016/j.foodres.2012.05.009.

Moreno-Fernandez, Jorge, Javier Diaz-Castro, María J. M. Alférez, Christine Boesch, Teresa Nestares, and Inmaculada López-Aliaga. 2017. "Fermented goat milk improves antioxidant status and protects from oxidative damage to biomolecules during anemia recovery." *Journal of the science of food and agriculture* 97 (5):1433-1442. doi: 10.1002/jsfa.7882.

Morikawa, Kensei, Itsuo Kondo, Yoshihiro Kanamaru, and Satoshi Nagaoka. 2007. "A novel regulatory pathway for cholesterol degradation via lactostatin." *Biochemical and Biophysical Research Communications* 352 (3):697-702. doi: 10.1016/j.bbrc. 2006.11.090.

Mudgil, P., B. Baby, Y. Y. Ngoh, R. Vijayan, C. Y. Gan, and S. Maqsood. 2019. "Identification and molecular docking study of novel cholesterol esterase inhibitory peptides from camel milk proteins." *Journal of Dairy Science* 102 (12):10748-10759. doi: 10.3168/jds.2019-16520.

Mudgil, Priti, Bincy Baby, Ying-Yuan Ngoh, Hina Kamal, Ranjit Vijayan, Chee-Yuen Gan, and Sajid Maqsood. 2019. "Molecular binding mechanism and identification of novel anti-hypertensive and anti-inflammatory bioactive peptides from camel milk protein hydrolysates." *LWT-Food Science and Technology* 112:108193. doi: 10.1016/j.lwt. 2019.05.091.

Nagaoka, Satoshi, Yu Futamura, Keiji Miwa, Takako Awano, Kouhei Yamauchi, Yoshihiro Kanamaru, Kojima Tadashi, and Tamotsu Kuwata. 2001. "Identification of Novel Hypocholesterolemic Peptides Derived from Bovine Milk β-Lactoglobulin." *Biochemical and Biophysical Research Communications* 281 (1):11-17. doi: 10.1006/bbrc.2001.4298.

Nguyen, H. T. H., J. L. Gathercole, L. Day, and J. E. Dalziel. 2020. "Differences in peptide generation following in vitro gastrointestinal digestion of yogurt and milk from cow, sheep and goat." *Food Chemistry* 317. doi: 10.1016/j.foodchem.2020.126419.

Park, Young W. 2009. "Bioactive components in goat milk." In *Bioactive components in milk and dairy products*, edited by Young W Park, 43-82. Athens, GA.

Parmar, H., S. Hati, G. Panchal, and A. A. Sakure. 2019. "Purification and Production of Novel Angiotensin I-Converting Enzyme (ACE) Inhibitory Bioactive Peptides Derived from Fermented Goat Milk." *International Journal of Peptide Research and Therapeutics*. doi: 10.1007/s10989-019-09902-7.

Paul, Moushumi, and D. L. Van Hekken. 2011. "Assessing antihypertensive activity in native and model Queso Fresco cheeses." *Journal of dairy science* 94 (5):2280-2284. doi: 10.3168/jds.2010-3852.

Pisanu, Salvatore, Daniela Pagnozzi, Massimo Pes, Antonio Pirisi, Tonina Roggio, Sergio Uzzau, and Maria Filippa Addis. 2015. "Differences in the peptide profile of raw and pasteurised ovine milk cheese and implications for its bioactive potential." *International Dairy Journal* 42:26-33. doi: 10.1016/j.idairyj.2014.10.007.

Pourjoula, M., G. Picariello, G. Garro, G. D'Auria, C. Nitride, A. Rheza Ghaisari, and P. Ferranti. 2020. "The protein and peptide fractions of kashk, a traditional Middle East fermented dairy product." *Food Research International* 132. doi: 10.1016/j.foodres.2020.109107.

Pritchard, Stephanie Rae, Michael Phillips, and Kasipathy Kailasapathy. 2010. "Identification of bioactive peptides in commercial Cheddar cheese." *Food research international* 43 (5):1545-1548. doi: 10.1016/j.foodres.2010.03.007.

Proestos, Charalampos. 2018. "Superfoods: Recent data on their role in the prevention of diseases." *Current Research in Nutrition and Food Science Journal* 6 (3):576-593. doi: 10.12944/CRNFSJ.6.3.02.

Rafiq, Saima, Nuzhat Huma, Imran Pasha, Muhammad Shahid, and Hang Xiao. 2017. "Angiotensin-converting enzyme-inhibitory and antithrombotic activities of soluble peptide extracts from buffalo and cow milk Cheddar cheeses." *International Journal of Dairy Technology* 70 (3):380-388. doi: 10.1111/1471-0307.12373.

Rahimi, Mahmood, Seyed Mahmood Ghaffari, Maryam Salami, Seyed Jafar Mousavy, Amir Niasari-Naslaji, Raheleh Jahanbani, Saeed Yousefinejad, Mohammadreza Khalesi, and Ali Akbar Moosavi-Movahedi. 2016. "ACE-inhibitory and radical scavenging activities of bioactive peptides obtained from camel milk casein hydrolysis with proteinase K." *Dairy Science & Technology* 96 (4):489-499. doi: 10.1007/s13594-016-0283-4.

Reddi, S., S. B. Mada, N. Kumar, R. Kumar, N. Ahmad, A. Karvande, S. Kapila, R. Kapila, and R. Trivedi. 2019. "Antiosteopenic Effect of Buffalo Milk Casein-Derived Peptide (NAVPITPTL) in Ovariectomized Rats." *International Journal of Peptide Research and Therapeutics* 25 (3):1147-1158. doi: 10.1007/s10989-018-9763-0.

Rendon-Rosales, M. Á, M. J. Torres-Llanez, A. F. González-Córdova, A. Hernández-Mendoza, M. A. Mazorra-Manzano, and B. Vallejo-Cordoba. 2019. "In vitro antithrombotic and hypocholesterolemic activities of milk fermented with specific strains of Lactococcus lactis." *Nutrients* 11 (9). doi: 10.3390/nu11092150.

Rubak, Y. T., L. Nuraida, D. Iswantini, and E. Prangdimurti. 2020. "Angiotensin-I-converting enzyme inhibitory peptides in milk fermented by indigenous lactic acid bacteria." *Veterinary World* 13 (2):345-353. doi: 10.14202/vetworld.2020.345-353.

Sagardia, Iban, Ibon Iloro, Felix Elortza, and Carlos Bald. 2013. "Quantitative structure–activity relationship based screening of bioactive peptides identified in ripened cheese." *International Dairy Journal* 33 (2):184-190. doi: 10.1016/j.idairyj.2012.12.006.

Sahingil, Didem, Ali A. Hayaloglu, Huseyin A. Kirmaci, Barbaros Özer, and Osman Simsek. 2014. "Changes of proteolysis and angiotensin-I converting enzyme-inhibitory activity in white-brined cheese as affected by adjunct culture and ripening temperature." *Journal of Dairy Research* 81 (4):394-402. doi: 10.1017/S0022029914000326.

Sartorius, T., A. Weidner, T. Dharsono, M. Wilhelm, A. Boulier, and L. Brown. 2019. "Postprandial effects of a proprietary milk protein hydrolysate containing bioactive peptides in prediabetic subjects." *Nutrients* 11 (7). doi: 10.3390/nu11071700.

Schlienger, Jean-Louis, Francois Paillard, Jean-Michel Lecerf, Monique Romon, Cécile Bonhomme, Bernard Schmitt, Yves Donazzolo, Catherine Defoort, Cécilia Mallmann, and Pascale Le Ruyet. 2014. "Effect on blood lipids of two daily servings of Camembert cheese. An intervention trial in mildly hypercholesterolemic subjects." *International journal of food sciences and nutrition* 65 (8):1013-1018. doi: 10.3109/09637486.2014.945156.

Shu, G., J. Huang, C. Bao, J. Meng, H. Chen, and J. Cao. 2018. "Effect of different proteases on the degree of hydrolysis and angiotensin I-converting enzyme-inhibitory activity in goat and cow milk." *Biomolecules* 8 (4). doi: 10.3390/biom8040101.

Skrzypczak, Katarzyna, Waldemar Gustaw, Dominik Szwajgier, Emilia Fornal, and Adam Waśko. 2017. "κ-Casein as a source of short-chain bioactive peptides generated by Lactobacillus helveticus." *Journal of food science and technology* 54 (11):3679-3688. doi: 10.1007/s13197-017-2830-2.

Sobral Bezerra, Vilma, Julia Furtado Campos, Roberto Afonso da Silva, Tatiana Souza Porto, José Luiz de Lima Filho, and Ana Lúcia Figueiredo Porto. 2013. "Biotechnological richness of the northeastern semi-arid region: antioxidant activity of casein hydrolysates from Moxotó goat milk." *Food Science and Technology* 33 (3). doi: 10.1590/S0101-20612013005000074.

Soffer, Richard L. 1976. "Angiotensin-converting enzyme and the regulation of vasoactive peptides." *Annual review of biochemistry* 45 (1):73-94. doi: 10.1146/annurev.bi.45.070176.000445.

Solanki, Divyangkumar, Subrota Hati, and Amar Sakure. 2017. "In silico and in vitro analysis of novel angiotensin i-converting enzyme (ACE) inhibitory bioactive peptides derived from fermented camel milk (Camelus dromedarius)." *International Journal of Peptide Research and Therapeutics* 23 (4):441-459. doi: 10.1007/s10989-017-9577-5.

Sowmya, K., D. Mala, M. I. Bhat, N. Kumar, R. K. Bajaj, S. Kapila, and R. Kapila. 2018. "Bio-accessible milk casein derived tripeptide (LLY) mediates overlapping anti-inflammatory and anti-oxidative effects under cellular (Caco-2) and in vivo milieu." *Journal of Nutritional Biochemistry* 62:167-180. doi: 10.1016/j.jnutbio.2018.09.002.

Tagliazucchi, Davide, Sherif Shamsia, Ahmed Helal, and Angela Conte. 2017. "Angiotensin-converting enzyme inhibitory peptides from goats' milk released by in vitro gastro-intestinal digestion." *International Dairy Journal* 71:6-16. doi: 10.1016/j.idairyj.2017.03.001.

Taha, S., M. El Abd, C. De Gobba, M. Abdel-Hamid, E. Khalil, F. U. Hassan, and D. Fathy. 2020. "The multifunctional activity of water-soluble peptides' extract of Domiati cheese during accelerated ripening by Neutrase." *Journal of Food Processing and Preservation*. doi: 10.1111/jfpp.14434.

Tenore, G. C., E. Pagano, S. Lama, D. Vanacore, S. Di Maro, M. Maisto, R. Capasso, F. Merlino, F. Borrelli, P. Stiuso, and E. Novellino. 2019. "Intestinal anti-inflammatory effect of a peptide derived from gastrointestinal digestion of buffalo (Bubalus bubalis) mozzarella cheese." *Nutrients* 11 (3). doi: 10.3390/nu11030610.

Tonolo, F., F. Fiorese, L. Moretto, A. Folda, V. Scalcon, A. Grinzato, S. Ferro, G. Arrigoni, A. Bindoli, E. Feller, M. Bellamio, O. Marin, and M. P. Rigobello. 2020. "Identification of new peptides from fermented milk showing antioxidant properties: Mechanism of action." *Antioxidants* 9 (2). doi: 10.3390/antiox9020117.

Tonolo, F., A. Folda, L. Cesaro, V. Scalcon, O. Marin, S. Ferro, A. Bindoli, and M. P. Rigobello. 2020. "Milk-derived bioactive peptides exhibit antioxidant activity through the Keap1-Nrf2 signaling pathway." *Journal of Functional Foods* 64. doi: 10.1016/j.jff. 2019.103696.

Torres-Llanez, M. J., A. F. González-Córdova, A. Hernandez-Mendoza, H. S. Garcia, and B. Vallejo-Cordoba. 2011. "Angiotensin-converting enzyme inhibitory activity in Mexican Fresco cheese." *Journal of dairy science* 94 (8):3794-3800. doi: 10.3168/jds.2011-4237.

Ugwu, C. P., M. M. Abarshi, S. B. Mada, B. Sanusi, and H. C. Nzelibe. 2019. "Camel and Horse Milk Casein Hydrolysates Exhibit Angiotensin Converting Enzyme Inhibitory and Antioxidative Effects In Vitro and In Silico." *International Journal of Peptide Research and Therapeutics* 25 (4):1595-1604. doi: 10.1007/s10989-018-09802-2.

Valenze, Deborah. 2011. *Milk: a local and global history*: Yale University Press.

Wu, Jianping, Rotimi E. Aluko, and Alister D. Muir. 2002. "Improved method for direct high-performance liquid chromatography assay of angiotensin-converting enzyme-catalyzed reactions." *Journal of Chromatography A* 950 (1-2):125-130. doi: 10.1016/S0021-9673(02)00052-3.

Zanutto-Elgui, M. R., J. C. S. Vieira, D. Z. D. Prado, M. A. R. Buzalaf, P. D. M. Padilha, D. Elgui de Oliveira, and L. F. Fleuri. 2019. "Production of milk peptides with antimicrobial and antioxidant properties through fungal proteases." *Food Chemistry* 278:823-831. doi: 10.1016/j.foodchem.2018.11.119.

In: Super and Nutraceutical Foods
Editors: Sandra Teresita Martín-del-Campo et al.
ISBN: 978-1-53619-082-3

Chapter 15

OLIVE OIL: COMPOSITION AND BIOACTIVITY

Caterina Pipino[1], Beatriz Martín-García[2,3], Ana Mª Gómez-Caravaca[2,*] and Vito Verardo[3,4]

[1]Department of Medical, Oral and Biotechnological Sciences, University "G. d'Annunzio" of Chieti-Pescara, Chieti, Italy
[2]Department of Analytical Chemistry, Faculty of Sciences, University of Granada, Granada, Spain
[3]Department of Nutrition and Food Science, University of Granada, Granada, Spain
[4]Institute of Nutrition and Food Technology' José Mataix', Biomedical Research Center, University of Granada, Armilla, Granada, Spain

ABSTRACT

Olive oil has mainly been produced in the Mediterranean area since ancient times; however, nowadays, several extra-Mediterranean countries are producing this oil due to its crucial role in the Mediterranean diet. Thanks to its characteristic chemical composition, olive oil consume is related to the prevention of several diseases. Different families of bioactive compounds have been described in olive oil apolar and polar fractions. Virgin olive oil (VOO) is a rich source of bioactive lipids (i.e., oleic acid, terpenes, squalene) and natural antioxidants (i.e., tocopherols, phenolic compounds). Therefore, the main focus of this chapter is to summarize data collected on olive oil composition and its relation with the bioactivity.

Keywords: olive oil, phenolic compounds, terpenes, bioactivity, oxidative stress, endothelial function, cardiovascular diseases (CVD), microbiota, cancer, neurodegenerative disease

* Corresponding Author's E-mail: anagomez@ugr.es.

Introduction

Among vegetable oils, olive oil (OO) is emphasized by its sensory and high nutritional properties which make it a valuable diet component in the Mediterranean region. OO has become popular in recent years, and now it is consumed in most of the world countries such as Japan, Indonesia, USA, and South Africa because of its unique flavor (Rohman et al. 2020). In fats and oils industry, OO has a higher price compared with other vegetable oils because of its greater quality in some aspects, including health benefits, sensory, and oxidative stability. These characteristics are not only due to its composition of fatty acids, but some minority compounds, including phenolic compounds (Gómez-Caravaca, Maggio, and Cerretani 2016).

Composition of Olive Oil

Virgin olive oils are classified into extra virgin olive oil (EVOO) with acidity up to 0.8%, calculated as oleic acid; virgin olive oil (VOO) acidity about 2.0% and lampante with acidity higher than 2.0%; the last one is not edible because of that it is submitted to refining processes. Refined olive oil presents an acidity of 0.3%; regular olive oil, which is a mixture of refined olive oil and virgin olive oil has a free acidity of 0.1%; other edible oil are refined pomace oil, and olive pomace oil, a blend of refined pomace oil and virgin olive oil. EVOO is the highest quality of olive oil and represents only 10% of the produced vegetable oil (Rohman et al. 2020).

Virgin olive oil is mainly constituted by triacylglycerols, besides other minor compounds, which include around 230 different compounds (Aiello et al. 2015). Glycerols of OO is the most characteristic fraction that represents about 90.0% to 99.0% of total OO weight, which includes a high proportion of mono and polyunsaturated fatty acids (Pérez-Jiménez et al. 2007). Also, virgin olive oils represent a valuable source of minor compounds that constitute its unsaponifiable fraction (tocopherols, sterols, pigments, and triterpenes) including a polar fraction containing phenolic acids, phenolic alcohols such as tyrosol and hydroxytyrosol, secoiridoids, flavonoids, lignans, and other minor constituents (elenolic acids) (Gouvinhas et al. 2017; Termentzi, Halabalaki, and Skaltsounis 2015). The nutritional composition of olive oil depends on the olives used (varying with agronomic factors, exposure to stress, environmental conditions, cultivar, processing system, etc.); thus not all virgin olive oils present similar composition (Gouvinhas et al. 2017).

Fatty Acids

Olive oil is composed mainly of monounsaturated fatty acids (MUFA), which comprise around 72–77% of total lipid content, being this content higher than other vegetable oils such as corn, soybean and sunflower oils. Besides, saturated fatty acids (SFA) comprise around 14% of OO content, which is lower than other edible oils. Olive oil has a low content in polyunsaturated fatty acids (PUFA) (Guo et al. 2018). The main MUFA, SFA and PUFA in OO are oleic acid, palmitic acid, and linoleic acid, respectively (Table 1). The fatty acid profile of OO varies based on the cultivar and maturation of olives. Oleic acid predominates the fatty acids in all the stages (unripe, semi-ripe, and ripe) of maturation. Palmitic acid and oleic acid are found in higher quantities in the unripe stage. However, palmitoleic acid, stearic acid, and linoleic acid are predominant in the ripe stage (Wani et al. 2018).

Table 1. Fatty acids content in virgin olive oil

Fatty acid	Lipid numbers	Content (%)	References
Myristic acid	C14:0	0.01	Guo et al. 2018
Palmitic acid	C16:0	6.6 - 15.34	El Riachy et al. 2019; Piscopo et al. 2016; Navajas-porras et al. 2020; Guo et al. 2018; Wani et al. 2018; Rohman et al. 2020
Hypogein acid	C16:1ω9	0.09 - 0.10	Wani et al. 2018
Palmitoleic acid	C16:1 ω7	0.31 - 1.67	Guo et al. 2018; Wani et al. 2018; El Riachy et al. 2019; Rohman et al. 2020; Piscopo et al. 2016; Navajas-Porras et al. 2020
Margaric acid	C17:0	0.04 - 0.16	Guo et al. 2018; Wani et al. 2018; Piscopo et al. 2016
Margaroleic acid	C17:1	0.05 - 0.30	Guo et al. 2018; Wani et al. 2018; Piscopo et al. 2016
Stearic acid	C18:0	1.90 - 15	Guo et al. 2018; Wani et al. 2018; El Riachy et al. 2019; Rohman et al. 2020; Piscopo et al. 2016; Navajas-Porras et al. 2020
Oleic acid	C18:1 ω9	59.6 - 83.1	Guo et al. 2018; Wani et al. 2018; El Riachy et al. 2019; Rohman et al. 2020; Piscopo et al. 2016; Navajas-Porras et al. 2020
cis–vaccenic acid	C18:1 ω7	2.03 - 3.42	Wani et al. 2018
Linoleic acid	C18:2	3.78 - 14.59	Guo et al. 2018; Wani et al. 2018; El Riachy et al. 2019; Rohman et al. 2020; Piscopo et al. 2016; Navajas-Porras et al. 2020
Linolenic acid	C18:3	0.42 - 1.00	Guo et al. 2018; Wani et al. 2018; El Riachy et al. 2019; Rohman et al. 2020; Piscopo et al. 2016; Navajas-Porras et al. 2020
Arachidic acid	C20:0	0.14 - 0.47	Guo et al. 2018; Wani et al. 2018; Piscopo et al. 2016; Navajas-Porras et al. 2020
Eicosenoic acid	C20:1	0.11 - 0.47	Guo et al. 2018; Wani et al. 2018; Piscopo et al. 2016
Behenic acid	C22:0	0.08 - 0.13	Guo et al. 2018; Wani et al. 2018; Piscopo et al. 2016
Lignoceric acid	C24:0	0.04 - 0.63	Guo et al. 2018; Wani et al. 2018; Piscopo et al. 2016

Phenolic Compounds

Olive oil contains phenolic compounds in low quantities. The total phenolic content in EVOO showed considerable variation, which is influenced mainly by different agronomic factors such as cultivar. In general, EVOO has the highest phenol content due to minimal and mechanical processing. By contrast, a refined OO loses its color, odor, flavor and large quantities of phenols (Kotsiou and Tasioula-Margari 2016). Phenolic compounds in extra virgin olive oil are classified in phenolic alcohols (hydroxytyrosol and tyrosol), phenolic acids (vanillic acid and p-coumaric acid), secoiridoids (mainly derivatives of oleuropein and ligstroside), lignans, flavonoids (luteolin and apigenin) (Laincer et al. 2016). In addition, elenolic acids, which are nonphenolic compound, and their derivatives and analogues constitute the iridoid part of the most important olive secondary metabolites (Termentzi, Halabalaki, and Skaltsounis 2015). Table 2 shows the range of phenolic compounds in extra virgin olive oils.

Table 2. Phenolic content in extra virgin olive oil

Phenolic compounds	Content (mg/kg)	References
Phenolic alcohols		
Hydroxytyrosol	0.4 - 299.5	Tuberoso et al. 2016; Dugo et al. 2020; El Riachy et al. 2019; Laincer et al. 2016
Tyrosol	0.25 - 297.6	El Riachy et al. 2019; Laincer et al. 2016; Tuberoso et al. 2016; Dugo et al. 2020
Phenolic acids		
Vanillic acid	0.3 - 3.9	Tuberoso et al. 2016
p-coumaric acid	0.1 - 3.20	El Riachy et al. 2019; Tuberoso et al. 2016
Secoiridoids		
Oleacein	15.2 - 762.9	El Riachy et al. 2019; Dugo et al. 2020
Oleocanthal	4.1 - 407.9	El Riachy et al. 2019; Laincer et al. 2016; Dugo et al. 2020
Oleuropein	0.2 - 1.2	Laincer et al. 2016
Oleuropein aglycone	0.3 - 3482.4	Tuberoso et al. 2016; Dugo et al. 2020
Decarboxymethyl oleuropein aglycone	0.4 - 67.6	Tuberoso et al. 2016
Ligstroside aglycone	17.6 - 1495.5	Dugo et al. 2020
Decarboxymethyl ligstroside aglycone	0.5-66.4	Tuberoso et al. 2016
Flavonoids		
Luteolin	0.2-149	El Riachy et al. 2019; Laincer et al. 2016; Tuberoso et al. 2016; Dugo et al. 2020
Apigenin	0.1-38.8	El Riachy et al. 2019; Laincer et al. 2016; Tuberoso et al. 2016; Dugo et al. 2020
Lignans		
Pinoresinol	4.1 - 32.3	Tuberoso et al. 2016
Total	8.8 - 70.7	Laincer et al. 2016
Other compounds		
Elenolic acid	17.3 - 252.6	Laincer et al. 2016
Decarboxymethylelenolic acid	0.7 - 116.8	Laincer et al. 2016

The most known phenolic compounds in olive phenols are the phenolic alcohols such as tyrosol and hydroxytyrosol. Both are also abundant as esters with various forms of elenolic acid, providing open aldehydic and decarboxymethyl forms such as oleocanthal and oleacein. Glycosilated derivatives of these compounds, such as ligstroside and oleuropein are the precursors of all other secoiridoids (Termentzi, Halabalaki, and Skaltsounis 2015).

Tocopherols

Tocopherols are important liposoluble antioxidants that represent only 0.02-0.06% of the OO content. Olive oil is composed of four tocopherol isomers, being α-tocopherol the most abundant in OO (≥ 90% of tocopherols in EVOO) (Table 3). Tocopherols have exhibited a strong protective effect in preventing tissue oxidation and inflammation by influencing the expression of homeostatic genes, thus reducing the progress of aging, degenerative diseases and cancer (Piroddi et al. 2017).

Sterols

An important part of the unsaponifiable fraction of virgin olive oil is composed by phytosterols derived from squalene. Their structures are similar to cholesterol, although each phytosterol possesses an additional side chain. Phytosterols are classified into two sub-groups: sterols with a double bond at position and stanols, which do not have a double bond in the sterol ring structure (Kyçyk et al. 2016). In olive oil the phytosterol concentration ranges from 800 to 2600 mg/kg, being β-sitosterol (75–90%) the most abundant, followed by Δ^5-avenasterol (5–36%) and campesterol (3%) (Boulkroune et al. 2017; Gu et al. 2016; Kyçyk et al. 2016). Table 4 shows the range of sterol content in virgin olive oil.

Olive oil also contains other phytosterols in low quantities including brassicasterol, 24-methylene-cholesterol, campestanol, stigmasterol, clerosterol, sitostanol, $\Delta^{5,24}$-stigmastadienol, Δ^7-stigmastenol, Δ^7-avenasterol, furgosterol and ergosterol. Sterol composition in OO depends on the cultivar, olive ripening, oil extraction and refining procedures and storage conditions. It has been reported an increase of three times in the total sterol content in OO with the progress of olive ripening from green to ripe (Yorulmaz and Konuskan 2017).

Table 3. Tocopherols content in extra virgin olive oil

Tocopherols	Content (mg/kg)	References
α-Tocopherol (Vitamin E)	76.4 - 573.1	Laincer et al. 2016; Rohman et al. 2020; Tuberoso et al. 2016; Dugo et al. 2020
α-Tocotrienol	1.3 - 6.7	Dugo et al. 2020
δ-Tocopherol	0.1 - 1.6	Laincer et al. 2016
γ-Tocopherol	0.7 - 59.1	Laincer et al. 2016; Tuberoso et al. 2016; Dugo et al. 2020
β-Tocopherol	1.1 - 9.2	Laincer et al. 2016; Dugo et al. 2020

Table 4. Sterol content in virgin olive oil

Sterols	Content (mg/kg)	References
Cholesterol	0.84 - 16.1	Gu et al. 2016; Boulkroune et al. 2017; Kyçyk et al. 2016; Rohman et al. 2020
Brassicasterol	0.24 - 1.36	Kyçyk et al. 2016
24-methylene-cholesterol	0.06 - 6.41	Kyçyk et al. 2016
Campesterol	13.64 - 88.23	Gu et al. 2016; Boulkroune et al. 2017; Kyçyk et al. 2016
Campestanol	0.5 - 5.48	Boulkroune et al. 2017; Kyçyk et al. 2016
Stigmasterol	1.283 - 38.47	Gu et al. 2016; Boulkroune et al. 2017; Kyçyk et al. 2016; Rohman et al. 2020
Clerosterol	7.3 - 26.61	Gu et al. 2004; Boulkroune et al. 2017; Kyçyk et al. 2016
β-sitosterol	672 - 2123.6	Gu et al. 2016; Boulkroune et al. 2017; Kyçyk et al. 2016
Sitostanol	0.06 - 25.98	Kyçyk et al. 2016; Rohman et al. 2020
Δ^5-avenasterol	11.07 - 343.6	Boulkroune et al. 2017; Kyçyk et al. 2016; Rohman et al. 2020
$\Delta^{5,24}$-stigmastadienol	2.02 - 32.07	Boulkroune et al. 2017; Kyçyk et al. 2016
Δ^7-stigmastenol	1.01 - 11.74	Boulkroune et al. 2017; Kyçyk et al. 2016
Δ^7-avenasterol	2.29 - 19.48	Boulkroune et al. 2017; Kyçyk et al. 2016
Fucosterol	5.24 - 8.98	Gu et al. 2016
Ergosterol	1.12 - 2.36	Heleno et al. 2016

Pigments

The green and yellow of virgin olive oil are due to the presence of chlorophylls and carotenoids. The main carotenoids of virgin olive oil are β-carotene and lutein. Xanthophylls, mainly violaxanthin and neoxanthin, also are found in small quantities. Chlorophyll compounds are found as pheophytins (Boskou 2015). The stage of ripening and cultivar mainly influences the total content of chlorophyll and carotenoid in OO. It has been reported a decline in the chlorophyll and carotenoid content in OO samples during olive ripening. Total chlorophyll content in extra virgin oil range from 2.8 to 25.1 mg pheophytin/kg, whereas the total carotenoid content range from 5.34 to 14.4 mg lutein/kg (Condelli et al. 2015; Yorulmaz and Konuskan 2017).

Triterpenes

Erythrodiol (homo-olestranol, 5α-olean-12-ene-3β,28-diol) and uvaol (Δ12-ursen-3β,28-diol) are the main triterpene dialcohols in OO. Oleanolic (3β-hydroxyolean-12-en-28-oic acid) and maslinic acid (2α,3β-dihydroxyolean-12-en-28-oic acid) are important pentacyclic hydroxyterpenic acids in olive fruits and are biologically active compounds (Boskou 2015). During the oil production procedure, part of these triterpenes is transferred to the oil. The concentration of triterpenoids in the oil depends on several factors such as olive cultivar, the malaxation time and temperature, and even the diameter of the sieves influences in the final triterpenic concentrations (Termentzi, Halabalaki, and Skaltsounis 2015). Table 5 shows the triterpenes content in different VOO.

Bioactivity of Olive Oil

In the last years, a growing number of scientific evidence have directed attention to the high nutritional quality of OO. It was suggested that its consumption has a protective role in the incidence of pathologies characterized by oxidative stress and inflammation, such as cardiovascular diseases and cancer, as well as neurodegenerative disorders, metabolic syndrome and related complications (Del Río et al. 2016; Parkinson and Cicerale 2016). The nutraceutical and nutritional value of OO is ascribed to its bioactive components. It is showing a positive effect on genes involved in the pathogenesis of most common age-related human diseases and modern lifestyle disorders, denoting a natural homeostatic role of these molecules as preventive agents against conditions of premature and pathologic aging associated to increased oxidative stress (Flori et al. 2019).

OO composition is the result of different variables, such as cultivar, climate, degree of maturation, ripening, and some agronomic and technological factors including harvesting and processing techniques adopted, thus leading to variation in the percentage of the components (Romani et al. 2019). Of note, linoleic acid, an essential fatty acid that must be provided with the diet, is a necessary component of cell structure and function. Moreover, a little unsaponifiable fraction (about 1–2%), including vitamins (α-, β-, γ- and δ- tocopherols) and polyphenols (mainly tyrosol, hydroxytyrosol, and oleuropein) with bioactive properties is comprised in olive oil (Flori et al. 2019).

Several pieces of evidence suggest that the healthy and biological properties of OO are associated to the presence of MUFA and other saturated and PUFA (Romani et al. 2019). Besides the well-known healthy properties of these compounds, recent studies have shed light on the beneficial effects of some constituents of the unsaponifiable fraction, such as phenolic compounds for their antioxidant properties and their positive impact on various physiological biomarkers as well as antimicrobial, antioxidant and anti-inflammatory activities. As previously mentioned, the main classes of phenols in OO are phenolic acids, phenolic alcohols, flavonoids, secoiridoids and lignans. To date, more than 36 phenolic compounds have been isolated from OO, at very different concentrations (0.02–600 mg/kg) (Parkinson and Cicerale 2016). The most important phenolic compounds in OO are oleuropein, hydroxytyrosol and tyrosol, representing about 90% of total phenol fraction and whose content changes with the variety of OO. They mainly act as chelating agents of metal ions and free radical scavengers. They are also considered to reduce the incidence of certain diseases, such as hypertension, atherosclerosis, hypercholesterolemia and obesity (Santangelo, et al. 2018). Oleuropein has been found effective in obesity control, as antiatherogenic and cardioprotective bioactive compound (Ahamad et al. 2019; Castejón et al. 2020). Besides its antioxidant and antiinflammatory properties, it is efficient as antibacterial and antiviral (Li et al. 2016; Castejón et al. 2020). Hydroxytyrosol was found particularly protective in the development of atherosclerosis (Wani et al. 2018).

Table 5. Triterpenes content in virgin olive oil

Triterpenes	Content (mg/kg)	References
Maslinic acid	47.3	Sanchez-Rodriguez et al. 2019
Oleanolic acid	39.2	Sanchez-Rodriguez et al. 2019
Ursolic acid	< 10	Sanchez-Rodriguez et al. 2019
Erythrodiol	5.421 - 44.86	Gu et al. 2016
Uvaol	1.50 - 18.93	Gu et al. 2016

Among tocopherols, also known as the precursor of vitamin E, α-tocopherol is the main bioactive which possess a strong antioxidant activity. α-tocopherol has a clearly defined role as free radical and inhibitor of lipid peroxidation of the cellular membrane and its integrity maintenance (Romani et al. 2019). Also, a class of organic molecules derived from an isoprene compound with five carbon atoms, the so called terpenoids, are found in OO. They have been found beneficial in the ocular system, skin and mucous, in improving the immune system and protect against cardiovascular diseases (Pandolfi 2013; DiTomo et al. 2012).

In summary, accumulated evidence indicates that bioactive compounds of OO possess properties that broadly explain the beneficial effect of the Mediterranean Diet, in which OO is the main source of fat. Further human studies are needed to validate the biological properties of these compounds better and whether the biological effect can be achieved following a proper diet which includes OO. However, findings regarding the biological effect of OO bioactive compounds in protecting against chronic diseases, such as cardiovascular disease (e.g., hypertension, coronary artery disease) are abundant. Thus OO consumption is strongly recommended as the main type of dietary fat to prevent such disorders.

Olive Oil vs. Oxidative Stress, Endothelial Dysfunction and Cardiovascular Disease

Oxidative stress is a physiologic process playing an essential role in the maintenance of intracellular homeostasis. In physiological conditions, several intracellular protective mechanisms, including reduced glutathione, superoxide dismutase and catalase, are activated against the production of reactive oxygen species (ROS). However, some disease conditions, such as obesity and type 2 diabetes, are characterized by excessive production of ROS overwhelming intracellular defences (Alkhatib, Tsang, and Tuomilehto 2018; Saibandith et al. 2017). Increased levels of ROS are mainly associated with oxidative stress in the pathogenesis of cardiovascular disease, the main causes of death worldwide. When ROS bioavailability overtakes the antioxidant defences, an alteration in the balance of ROS production and antioxidants occurs, leading to cellular damage, tissue injury, and inflammation. Altogether, these events have a fundamental role in vascular dysfunction and atherosclerosis progression, affecting several cells in the vascular wall and leading to endothelial dysfunction (Marchio et al. 2019).

Of note, consuming OO, as in a typical Mediterranean diet, is associated with reduced oxidative stress and lower risk of cardiovascular events, thus promoting the so called "healthy aging". Aging is a process involving several age-related physical changes characterized by genomic instability, epigenetic alterations, as well as some other challenges associated with cellular senescence (Del Río et al. 2016). Olive oil consumption has been demonstrated to have a beneficial impact on the aging process, thanks to the effects of bioactive compounds in modulating telomerase activity, epigenetics changes and nutrient-sensing pathways (Del Río et al. 2016). Although several epidemiological pieces of evidence support the beneficial health effects of the Mediterranean Diet on health and chronic disease prevention, clinical trials are still few.

Of interest, a randomized clinical trial in the PREDIMED project, where people at high cardiovascular risk were involved, demonstrated that the incidence of cardiovascular events was lower among people enrolled into a Mediterranean diet supplemented with extra-virgin OO compared to those assigned to a control diet (Estruch et al. 2018).

A recent meta-analysis reported that consuming OO for at least four weeks in an amount ranging from 1 to 50 mg daily had a beneficial effect on endothelial function (Schwingshackl, Christoph, and Hoffmann 2015). Indeed, the USA Food and Drug Administration and the European Food Safety Authority recommend consuming around 20 grams or two tablespoons per day of OO.

The right balance of monounsaturated, polyunsaturated and saturated fatty acids in a diet is involved in the prevention of cardiovascular pathologies. Indeed, a generally accepted concept is that a diet rich of saturated fatty acids increases serum cholesterol and induce inflammation and insulin resistance, thus contributing to the risk of atherosclerosis and cardiovascular diseases. Various translational studies have shown a protective role of unsaturated components, MUFA, and more widely for PUFA and bioactive phenolic compounds (Silva et al. 2015).

Overall, the endothelial protective effect of OO can be explained by its antiinflammatory activity. The first protective effect is the inhibition of monocytes and lymphocytes adhesion to the endothelium by reducing the expression of the monocyte chemotactic protein (MCP-1) and several cytokines (IL-1, IL-6, IL-8). Moreover, the expression of the adhesion molecule E-selectin, intercellular adhesion molecule-1 (ICAM-1) and vascular cell adhesion molecule-1 (VCAM-1), which are involved in the first step of the development of the atherosclerotic plaque was also noticed (Bourque, Davidge, and Adams 2011; Marchio et al. 2019). Indeed, VCAM-1 acts by inducing leukocyte migration and activation of nicotinamide adenine dinucleotide phosphate oxidase in endothelial cells, thus starting the production of ROS and vascular oxidative stress. Olive oil polyphenols also act by inhibiting the NF-kB (nuclear factor kappa-light-chain-enhancer of activated B cells) pathway, a key regulator of the atherosclerotic plaque formation (Storniolo et al. 2014; Incalza et al. 2018). Interestingly, it was found that oleuropein was able to reduce endothelial dysfunction by inhibiting the activity of signaling mitogen-activated protein kinase (MAPK), a pathway involved in cell proliferation and differentiation (Ryu et al. 2015). Another study showed the capability of polyphenols of OO to modulate some genes and miRNAs expression changes involved in inflammation, metabolism and cancer and evaluated on the peripheral blood mononuclear cells of both healthy subjects and patients with metabolic syndrome, following acute intake of OO rich in polyphenols (D'Amore et al. 2016).

Several shreds of evidence from epidemiological studies suggest that the beneficial effect of OO is attributable to oleic acid, the main MUFA capable of decreasing the incidence of cardiovascular and several chronic diseases (Buckland and Gonzalez 2015). Oleic acid is beneficial in lowering the low density lipoproteins (LDL) oxidation, maintaining a normal cholesterol efflux, thus reducing the circulating LDL, that becomes less susceptible to oxidation and limit to cross the endothelial vessel. Also, phenolic compounds of OO, such as hydroxytyrosol, tyrosol, and oleuropein, display a recognized lowered LDL oxidation effect (Perrone and Gualtieri 2019). Again, oleuropein was found effective in reducing the expression of VCAM-1 in human umbilical vein endothelial cells pre-stimulated with inflammatory cytokines (Nediani et al. 2019). Furthermore, OO bioactive compounds modulated endothelial nitric oxide synthase (eNOS)-dependent ROS generation, thus preventing oxidative stress and endothelial dysfunction (Varadharaj et al. 2017). Notably, oleic acid has been found to protect against cardiovascular events, to improve endothelial function and reduce inflammation, as well as proliferation and apoptosis in vascular smooth muscle cells (Perdomo et al. 2015). It was found helpful in reducing the formation of foamy cells and the accumulation of macrophages in the circulatory system, thus decreasing the development of atheromas (Quintero-Flórez et al. 2015).

Additionally, oleic acid was found beneficial in lowering blood pressure and in the maintenance of a normal cholesterol level; therefore it is considered a valuable compound in reducing the risk of heart attacks and in the prevention of cardiovascular pathologies (Marcelino et al. 2019; Nocella et al. 2017; Quintero-Flórez et al. 2015). Furthermore, oleic acid has been found protective in the postprandial state (Bermudez et al. 2011). Postprandial hypertriglyceridemia is an important risk factor for cardiovascular disease given a genetic predisposition or conditions such as obesity and insulin resistance. The increased production of triglyceride-rich lipoproteins (TRLs) leads to increased formation of LDL proteins with consequent oxidative stress, inflammation and endothelial dysfunction (DiPietro, Formoso, and Pandolfi 2016). Of interest, the number of TRLs was found reduced following OO or oleuropein intake (Carnevale et al. 2018; Perrone and Gualtieri 2019). All these activities may explain its role in counteracting the atherosclerotic process *in vivo*. Therefore, the health-promoting effect of oleic acid has been well established.

In general, the cardiovascular protective role of bioactive compounds of OO is demonstrated mainly by *in vitro* studies in the protection of cells against oxidative stress, as well as *in vivo* studies and cardiovascular prevention trials (Perona, Cabello-Moruno, and Ruiz-Gutierrez 2006; Marcelino et al. 2019). The worth of note is linoleic acid, a fatty acid of OO implicated in the metabolic regulation of diabetes and inflammation. Linoleic acid is an essential fatty acid, and its lack seems to be associated with skin disorders, inflammation, vasoconstriction problems and poor control of blood pressure (Quintero-Flórez et al. 2015).

However, several human and animal studies have demonstrated that the antioxidant activity of OO is mainly given by the phenolic compounds, especially hydroxytyrosol, the major OO phenolic compound (Bertelli et al. 2020). Of note, hydroxytyrosol was able to reduce adhesion molecule expression in human aortic endothelial cells pre-treated with the inflammatory stimulus TNF-α (Catalán et al. 2015). Again, OO polyphenols were found to ameliorate endothelial dysfunction induced by high glucose and free fatty acids through modulation of nitric oxide and endothelin-1 in endothelial cells (Storniolo et al. 2014). Moreover, hydroxytyrosol reduced VCAM-1 and IL-1β levels and abated cell proliferation in the vascular wall of streptozotocin-diabetic rats (López-Villodres et al. 2016).

In cell culture, hydroxytyrosol was effective in normalizing glutathione concentrations, increased glutathione peroxidase, glutathione reductase and glutathione-S-transferase protein expression, and lowered the reduced glutathione/oxidized glutathione ratio, suggesting a decrease in the oxidative status (Peyrol, Riva, and Amiot 2017). Oxidation of LDL, the hallmark of early atherosclerosis, which leads to a more permeability of monocytes/macrophage of the arterial wall, is significantly reduced by hydroxytyrosol, that has been demonstrated strongly protective against LDL oxidation (Perrone and Gualtieri 2019; Peyrol, Riva, and Amiot 2017).

Of interest, a recent study showed that men who consumed OO rich in polyphenols had a reduction in the expression of pro-atherogenic genes such as CD40 antigen ligand and oxLDL receptor-1 (Mateos et al. 2016). Again, the protection of LDL by OO seems due to the antioxidant effect of high density lipoprotein (HDL) by enhancing the plasma activity of several enzymes present on HDL. A clinical trial on both healthy people and cardiovascular risk individuals demonstrated that HDL oxidation was reduced following a diet with OO supplementation. In detail, HDL was capable of inducing the release of nitric oxide in endothelial cells through activation of eNOS (Hernáez et al. 2017). Besides LDL oxidation, hydroxytyrosol was able to ameliorate mitochondrial dysfunction, to prevent the increase of lipid peroxidation markers, and normalize the levels of liver glutathione, glutathione S-transferase, and superoxide dismutase (SOD) activity in high-fat-diet mice (Poudyal et al. 2017; Cao et al. 2014). The attenuation of lipid peroxidation through a regular diet which includes OO provides a significant clinical finding regarding the beneficial effect of OO on endothelial function.

Another interesting study demonstrated that consumption of a diet rich in extra virgin OO improved endothelial function in patients with prediabetes and diabetes (Torres-Peña et al. 2018). This study emphasizes the importance of a correct diet in the treatment of patients with diabetes at high cardiovascular risk with attention to personalized dietary recommendations. In this context, an individual approach can be offered by omics technologies, including genetics (Deoxyribonucleic acid (DNA) sequence), epigenetics (DNA modification), transcriptomics (gene expression), proteomics (protein products of coding genes), and metabolomics (metabolite products of metabolism) which provide unique opportunities to detect biomarkers in response to nutrients and bioactive compound intake. The use of these technologies in personalized and even predictive protocols may provide a nutritional and lifestyle personalized approach for preventing Type 2 diabetes and related cardiovascular complication (Alkhatib et al. 2017).

Altogether, this evidence shows that OO is a functional food with a healthy profile and the widely studied phenolic compounds, as well as tocopherols and oleic acid, can contribute in different ways acting on different types of molecular targets to ensure beneficial, healthy effects.

Olive Oil, Gastrointestinal Health and Gut Microbiota

Olive oil is widely recommended in a diet as a source of oleic acid for its gastrointestinal and metabolic function. Indeed, oleic acid acts by protecting the gastric mucosa through a reduced secretion of hydrochloric acid leading to a reduced risk of gastric-duodenal ulcers (Piccinin et al. 2019). Also, a beneficial effect in the secretory activity of pancreas and liver bile secretion was demonstrated (Piroddi et al. 2017).

In a recent study, OO supplementation in a mouse model of dextrane sodium sulfate (DSS)-induced colitis had positive effects on the prevention and development of colitis. In particular, the effect of four Apulian EVO cultivars (Ogliarola, Coratina, Peranzana and Cima di Mola) was evaluated. Olive oil supplementation in mice was able to reduce body weight loss and rectal bleeding and abates the NF-κB pathway, leading to a lower pro-inflammatory cytokines and consequently lower inflammation (Cariello 2020). These effects were mainly attributable to the bioactive compounds of OO, oleic acid and phenols.

Of note, few studies have demonstrated that OO consumption is associated with the improvement of intestinal health, leading to a higher richness and diversity of intestinal microbiota (Marcelino et al. 2019; Deiana and Serra, 2018). Indeed, a healthy balance of the gut microbiota, composed of several species of microorganisms, including bacteria should be maintained in order to sustain a general health, while variation in gut bacteria content is influenced by genetic, lifestyle and dietetic factors (Tomasello et al. 2016; Marcelino et al. 2019).

Interestingly, a randomized, double-blind human trial showed that the ingestion of OO enriched of phenolic compounds for three weeks was beneficial not only in lowering blood oxLDL in hypercholesterolemic humans, but also in increasing microbial metabolites with antioxidant activities (Martín-Peláez et al. 2017). Again, a study in eighteen overweight/obese subjects indicated that following a Mediterranean diet rich in OO for three months, increased gut bacteria and beneficial effects in the gastrointestinal tract were observed (Luisi et al. 2019).

In conclusion, Mediterranean and vegetarian diets that includes abundant fruits, vegetables and OO seems to prevent dysbiosis ascribed to the antiinflammatory properties of bioactive compounds and their ability of modulating the intestine oxidative status (Tomasello et al. 2016).

Olive Oil vs. Cancer

Tumorigenesis can be caused by somatic mutations or epigenetic alteration in cells given to the exposure to endogenous substances or environmental ones (Rajagopala et al. 2017). Several critical environmental factors, including nutrition, are involved in epigenetic modifications in cancer. Inadequate or excessive nutrient consumption leads to oxidative stress and among the endogenously produced genotoxic substances, the reactive oxygen species have been strongly associated to the onset of cancer (Saha et al. 2017). ROS may promote cancer progression either by playing as signaling elements or by provoking DNA damage.

Evidences from epidemiological studies correlating a lower incidence of certain cancers within the Mediterranean region together with several *in vitro* and *in vivo* studies suggest that OO intake is associated to a reduced risk of cancer, specifically breast, prostate, and colorectal cancer (Salazar, López-Cortés, and Salazar-García 2017; Yang et al. 2018). The association between OO consumption and its effect against tumorigenesis seems to be due to the scavenging properties of phenolic compounds together with their antiinflammatory and antioxidant activities. Olive oil phenolic compounds have been found helpful in counteracting the classical features of cancer, such as angiogenesis, apoptosis evasion, self-sufficiency in growth signals, insensitivity to growth-inhibitory and tissue invasion (Toric et al. 2019).

In vitro studies on several cancer cell types demonstrated that the chemopreventive ability of OO is mainly attributable to the minor phenolic bioactive fraction acting by reducing cell proliferation and promoting apoptosis in several tumor cell lines (Mansouri et al. 2019).

The inhibition of carcinogenesis process at initiation/progression phases was confirmed in animal studies demonstrating the beneficial role of phenolic compounds from OO in counteracting tumor cell growth, migration and invasion (Giner et al. 2016; Terzuoli et al. 2016). Most of the animal studies have confirmed the ability of these compounds to inhibit the carcinogenesis process at both initiation and promotion/progression phases. Moreover, some human intervention trials have been performed and are based on the evaluation of the effects of OO phenols on DNA damage (Fabiani 2016). Few of these studies have shown a significant preventive effect on DNA damage through a reduction of 8-oxo-7,8-dihydro-2'-deoxyguanosine, a product of DNA oxidation, in urine, mitochondria DNA of mononuclear cells and in lymphocyte DNA (Fabiani 2016).

Overall, several studies have demonstrated that bioactive compounds from OO are able to inhibit proliferation and induce apoptosis in different tumor cell lines. However, additional studies are necessary to clarify the real chemopreventive potential of OO phenols on humans, in particular by means of intervention studies on populations at high cancer risk. Moreover, further investigation is necessary to correlate the *in vitro* data with *in vivo* responsiveness to dietary supplementation determining the adequate dose-dependent efficacy of single compounds and their combinations with anti-cancer drugs.

Olive Oil vs. Neurodegenerative Disease

Brain tissue is characterized by high energy demand, high oxygen consumption and low antioxidant defenses (Niedzielska et al. 2016). Different neurodegenerative diseases share increased ROS level and induced oxidative stress as common features. In this regard, a large body of evidence supports the beneficial effects of OO polyphenols in preventing neurodegeneration, thanks to their antioxidant properties able to counteract oxidative stress in brain tissue.

Some studies have indicated a possible role of polyphenols in cerebral ischemia problems, Alzheimer's disease (AD), Huntington's disease and Parkinson's disease by targeting multiple pathological aspects of the diseases (Angeloni et al. 2017; Salis et al., 2018). The neuroprotective effect of Olive Oil seems to be a result of oleuropein and hydroxytyrosol behaving as direct free radical scavengers. Also, hydroxytyrosol acts as strong cyclooxygenases inhibitors while oleuropein was found effective against low density lipoprotein oxidations (Imran et al. 2018; Perrone and Gualtieri 2019). Again, oleuropein was able to ameliorate arsenic-induced oxidative stress in mice thus improving nitrogen oxide production (Ogun et al. 2016). Also, similar results were obtained with hydroxytyrosol, found protective on arsenic-induced oxidative stress and mitochondrial dysfunction in rat brains. Mice treated with hydroxytyrosol (10 mg/kg/day) for 8 weeks neutralized arsenic depletion of the antioxidant enzymes catalase (CAT) and manganese superoxide dismutase (MnSOD) (Soni et al. 2017). Moreover, it was recently found that the antioxidant effect in the brain of both hydroxytyrosol and oleuropein are mediated by the activation of the Kelch-like ECH-associated protein 1- Nuclear factor erythroid 2-related factor 2 (Keap1-Nrf2) pathway, thus attenuating oxidative stress and improving mitochondrial function (Sun et al. 2017; Soni et al. 2017).

Furthermore, oleuropein treatment exhibited neuroprotective activity against infarct volume and neuronal apoptosis in cerebral ischemia/reperfusion injured mice, suggesting its potential application in clinical trials for the treatment of stroke (Yu et al. 2016).

Interestingly, it was found that pretreatment with oleuropein counteracted oxidative stress and cognitive dysfunction induced by colchicine in the hippocampal area in rats (an animal model of senile dementia of Alzheimer's type) thus reducing lipid peroxidation, nitrite level, caspase 3 activity and increasing CAT and gluatione peroxidase (GSH) activities (Pourkhodadad et al. 2016). Again, six months of treatment with hydroxytyrosol (5 mg/kg/day) in APP/PS1 transgenic mice, a model of Alzheimer's disease, was found beneficial in reducing inflammation and mitochondrial oxidative stress (Peng et al. 2016). Furthermore, hydroxytyrosol increased SOD2 expression and GSH levels while decreased Heme oxygenase 1 (HO-1) and NAD(P)H quinone dehydrogenase 1 (NQO1) expression that were overexpressed in AD mice (Peng et al. 2016).

Of note, some evidences indicate a link between diabetes and AD. Interestingly, it was found that hydroxytyrosol exerts beneficial effects on insulin resistance associated with AD in an astrocytic cell line by improving insulin sensitivity and restoring appropriate insulin-signaling function (Crespo et al. 2017).

In addition, OO phenols have been found able to modulate different cellular mechanism involved in the onset and progression of Parkinson's disease. Indeed, dopamine neurons are susceptible to oxidative stress as dopamine can go through autoxidation with production of toxins, such as hydrogen peroxide and dopamine auto-oxidation (DA)-quinone molecules.

Most of the evidences regarding the beneficial effect of polyphenols from OO on Parkinson's disease come from *in vitro* cell culture studies (Yu et al. 2016; Mischley, Lau, and Bennett 2017). In the dopaminergic human cell line SH-SY5Y, hydroxytyrosol showed a strong protective effect against production of toxins and induced phase II detoxifying enzyme expression thus protecting dopaminergic cells against dopamine-induced cytotoxicity (Yu et al. 2016).

Moreover, the effect of hydroxytyrosol was evaluated in PC12 cells in the presence of different mono ammino oxidase (MAO) inhibitors, drugs used in Parkinson's disease, suggesting the beneficial effect of hydroxytyrosol by synergistically acting with these drugs (Goldstein et al. 2016).

Of note, the neuroprotective effects of OO phenols especially related to the modulation of inflammatory mediators, suggest a potential role of these compounds in neurodegenerative diseases such as multiple sclerosis (MS) and amyotrophic lateral sclerosis (ALS). An interesting study found that phenolic extract from commercial Italian OO was effective in modulating toll-like receptor 4 (TLR4) signaling, known to be involved in ALS pathogenic mechanisms (De Paola et al. 2016).

Overall, OO phenols seem to have great potential for therapeutic success in counteracting neurological disorders and chronic neurodegenerative diseases. They act by direct activity to scavenge ROS *in vitro*, although their low concentration *in vivo* is not sufficient to justify their protective effect. Neurodegenerative diseases are complex disorders, thus as discussed herein, although several studies have indicated the beneficial health effect of OO bioactive compounds, clinical trials are needed to provide further insights on their preventive potential together with effective drugs.

ACKNOWLEDGMENTS

Vito Verardo thanks the Spanish Ministry of Economy and Competitiveness (MINECO) for "Ramon y Cajal" contract (RYC-2015-18795).

REFERENCES

Ahamad, Javed, Ibrahim Toufeeq, Mohammad Ahmed Khan, Muath Sh Mohammed Ameen, Esra T. Anwer, Subha Uthirapathy, Showkat R. Mir. & Javed Ahmad. (2019). "Oleuropein: A Natural Antioxidant Molecule in the Treatment of Metabolic Syndrome." *Phytotherapy Research*, *33* (12), 3112–3128. doi:10.1002/ptr.6511.

Aiello, Anna, Giovanni Dara Guccione, Giulia Accardi. & Calogero Caruso. (2015). "What Olive Oil for Healthy Ageing?" *Maturitas*, *80* (2), 117–118. doi:10.1016/ j.maturitas.2014.10.016.

Alkhatib, Ahmad, Catherine Tsang, Ali Tiss, Theeshan Bahorun, Hossein Arefanian, Roula Barake, Abdelkrim Khadir. & Jaakko Tuomilehto. (2017). "Functional Foods and Lifestyle Approaches for Diabetes Prevention and Management." *Nutrients*, *9* (12), 1–18. doi:10.3390/nu9121310.

Alkhatib, Ahmad, Catherine Tsang. & Jaakko Tuomilehto. (2018). "Olive Oil Nutraceuticals in the Prevention and Management of Diabetes: From Molecules to Lifestyle." *International Journal of Molecular Sciences*, *19* (7). doi:10.3390/ ijms19072024.

Angeloni, Cristina, Marco Malaguti, Maria Cristina Barbalace. & Silvana Hrelia. (2017). "Bioactivity of Olive Oil Phenols in Neuroprotection." *International Journal of Molecular Sciences*, *18* (11), 1–27. doi:10.3390/ijms18112230.

Bermudez, Beatriz, Sergio Lopez, Almudena Ortega, Lourdes M. Varela, Yolanda M. Pacheco, Rocio Abia. & Francisco J. G. Muriana. (2011). "Oleic Acid in Olive Oil: From a Metabolic Framework Toward a Clinical Perspective." *Current Pharmaceutical Design*, *17* (8), 831–843. doi:10.2174/138161211795428957.

Bertelli, Matteo, Aysha Karim Kiani, Stefano Paolacci, Elena Manara, Danjela Kurti, Kristjana Dhuli, Vilma Bushati., et al. (2020). "Hydroxytyrosol: A Natural Compound with Promising Pharmacological Activities." *Journal of Biotechnology*, *309*, 29–33. doi:10.1016/j.jbiotec.2019.12.016.

Boskou, D. (2015). *Olive Oil: Properties and Processing for Use in Food. Specialty Oils and Fats in Food and Nutrition*. doi:10.1016/B978-1-78242-376-8.00001-6.

Boulkroune, Hasna, Aida Lazzez, Mokhtar Guissous, Yuva Bellik, Sami Smaoui, Naziha Grati Kamoun. & Toufik Madani. (2017). "Characterization of Sterolic and Alcoholic Fractions of Some Algerian Olive Oils according to the Variety and Ripening Stage." *OCL - Oilseeds and Fats, Crops and Lipids*, *24* (5), doi:10.1051/ocl/2017026.

Bourque, Stephane L., Sandra T. Davidge. & Michael A. Adams. (2011). "The Interaction between Endothelin-1 and Nitric Oxide in the Vasculature: New Perspectives." *American Journal of Physiology - Regulatory Integrative and Comparative Physiology*, *300* (6), 1288–1295. doi:10.1152/ajpregu.00397.2010.

Buckland, Genevieve. & Carlos A. Gonzalez. (2015). "The Role of Olive Oil in Disease Prevention: A Focus on the Recent Epidemiological Evidence from Cohort Studies and

Dietary Intervention Trials." *British Journal of Nutrition*, *113* (S2), S94–S101. doi:10.1017/S0007114514003936.

Cao, Ke, Jie Xu, Xuan Zou, Yuan Li, Cong Chen, Adi Zheng, Hao Li., et al. (2014). "Hydroxytyrosol Prevents Diet-Induced Metabolic Syndrome and Attenuates Mitochondrial Abnormalities in Obese Mice." *Free Radical Biology and Medicine*, *67*, 396–407. doi:10.1016/j.freeradbiomed.2013.11.029.

Cariello, Marica, Annalisa Contursi, Raffaella Maria Gadaleta, Elena Piccinin, Stefania De Santis, Marilidia Piglionica, Ada Fiorenza Spaziante., et al. (2020). "Extra-Virgin Olive Oil from Apulian Cultivars and Intestinal Inflammation." *Nutrients*, *12*(4), 1084. doi: 10.3390/nu12041084.

Carnevale, Roberto, Romano Silvestri, Lorenzo Loffredo, Marta Novo, Vittoria Cammisotto, Valentina Castellani, Simona Bartimoccia, Cristina Nocella. & Francesco Violi. (2018). "Oleuropein, a Component of Extra Virgin Olive Oil, Lowers Postprandial Glycaemia in Healthy Subjects." *British Journal of Clinical Pharmacology*, *84* (7), 1566–1574. doi:10.1111/bcp.13589.

Castejón, María Luisa, Tatiana Montoya, Catalina Alarcón-de-la-lastra. & Marina Sánchez-hidalgo. (2020). "Potential Protective Role Exerted by Secoiridoids from *Olea Europaea* L. In Cancer, Cardiovascular, Neurodegenerative, Aging-Related, and Immuno-inflammatory Diseases." *Antioxidants*, *9* (2), doi:10.3390/antiox9020149.

Catalán, Úrsula, Maria Carmen López de las Hazas, Laura Rubió, Sara Fernández-Castillejo, Anna Pedret, Rafael de la Torre, Maria José Motilva. & Rosa Solà. (2015). "Protective Effect of Hydroxytyrosol and Its Predominant Plasmatic Human Metabolites against Endothelial Dysfunction in Human Aortic Endothelial Cells." *Molecular Nutrition and Food Research*, *59* (12), 2523–2536. doi:10.1002/ mnfr.201500361.

Condelli, Nicola, Marisa Carmela Caruso, Fernanda Galgano, Daniela Russo, Luigi Milella. & Fabio Favati. (2015). "Prediction of the Antioxidant Activity of Extra Virgin Olive Oils Produced in the Mediterranean Area." *Food Chemistry*, *177*, 233–239. doi:10.1016/j.foodchem.2015.01.001.

Crespo, M. Carmen, Joao Tomé-Carneiro, Cristina Pintado, Alberto Dávalos, Francesco Visioli. & Emma Burgos-Ramos. (2017). "Hydroxytyrosol Restores Proper Insulin Signaling in an Astrocytic Model of Alzheimer's Disease." *BioFactors*, *43* (4), 540–548. doi:10.1002/biof.1356.

D'Amore, Simona, Michele Vacca, Marica Cariello, Giusi Graziano, Andria D'Orazio, Roberto Salvia, Rosa Cinzia Sasso, Carlo Sabbà, Giuseppe Palasciano. & Antonio Moschetta. (2016). "Genes and miRNA Expression Signatures in Peripheral Blood Mononuclear Cells in Healthy Subjects and Patients with Metabolic Syndrome after Acute Intake of Extra Virgin Olive Oil." *Biochimica et Biophysica Acta - Molecular and Cell Biology of Lipids*, *1861* (11), 1671–1680. doi:10.1016/ j.bbalip.2016.07.003.

De Paola, Massimiliano, Stefania E. Sestito, Alessandro Mariani, Christian Memo, Roberto Fanelli, Mattia Freschi, Caterina Bendotti, Valentina Calabrese. & Francesco Peri. (2016). "Synthetic and Natural Small Molecule TLR4 Antagonists Inhibit Motoneuron Death in Cultures from ALS Mouse Model." *Pharmacological Research*, *103*, 180–187. doi:10.1016/j.phrs.2015.11.020.

Deiana, M., Serra, G. & Corona, G. (2018). "Modulation of Intestinal Epithelium Homeostasis by Extra Virgin Olive Oil Phenolic Compounds." *Food Funct.*, *9* (8), 4085–4099. doi:doi:10.1039/c8fo00354h.

Del Río, Lucía Fernández, Elena Gutiérrez-Casado, Alfonso Varela-López. & José M. Villalba. (2016). "Olive Oil and the Hallmarks of Aging." *Molecules*, *21* (2), 1–30. doi:10.3390/molecules21020163.

Di Pietro, Natalia, Gloria Formoso & Assunta Pandolfi. (2016). "Physiology and Pathophysiology of oxLDL Uptake by Vascular Wall Cells in Atherosclerosis." *Vascular Pharmacology*, *84*, 1–7. doi:10.1016/j.vph.2016.05.013.

Di Tomo, Pamela, Raffaella Canali, Domenico Ciavardelli, Sara Di Silvestre, Alessandro De Marco, Annalisa Giardinelli, Caterina Pipino, Natalia Di Pietro, Fabio Virgili. & Assunta Pandolfi. (2012). "β-Carotene and Lycopene Affect Endothelial Response to TNF-α Reducing Nitro-Oxidative Stress and Interaction with Monocytes." *Molecular Nutrition and Food Research*, *56* (2), 217–227. doi:10.1002/mnfr.201100500.

Dugo, Laura, Marina Russo, Francesco Cacciola, Filippo Mandolfino, Fabio Salafia, Alessandra Vilmercati, Chiara Fanali., et al. (2020). "Determination of the Phenol and Tocopherol Content in Italian High-Quality Extra-Virgin Olive Oils by Using LC-MS and Multivariate Data Analysis." *Food Analytical Methods*, *13*, 1027–1041.

El Riachy, Milad, Athar Hamade, Rabih Ayoub, Faten Dandachi. & Lamis Chalak. (2019). "Oil Content, Fatty Acid and Phenolic Profiles of Some Olive Varieties Growing in Lebanon." *Frontiers in Nutrition*, *6*, doi:10.3389/fnut.2019.00094.

Estruch, R., Ros, E., Salas-Salvadó, J., Covas, M. I., Corella, D., Arós, F., Gómez-Gracia, E., et al. (2018). "Primary Prevention of Cardiovascular Disease with a Mediterranean Diet Supplemented with Extra-Virgin Olive Oil or Nuts." *New England Journal of Medicine*, *378* (25), 1–14. doi:10.1056/NEJMoa1800389.

Fabiani, R. (2016). "Anti-Cancer Properties of Olive Oil Secoiridoid Phenols: A Systematic Review of: *In Vivo* Studies." *Food and Function*, *7* (10), Royal Society of Chemistry, 4145–4159. doi:10.1039/c6fo00958a.

Flori, Lorenzo, Sandra Donnini, Vincenzo Calderone, Angela Zinnai, Isabella Taglieri, Francesca Venturi. & Lara Testai. (2019). "The Nutraceutical Value of Olive Oil and Its Bioactive Constituents on the Cardiovascular System. Focusing on Main Strategies to Slow Down Its Quality Decay during Production and Storage." *Nutrients*, *11* (9), doi:10.3390/nu11091962.

Giner, Elisa, Carmen Recio, M., José Luis Ríos, José Miguel Cerdá-Nicolás. & Rosa María Giner. (2016). "Chemopreventive Effect of Oleuropein in Colitis-Associated Colorectal Cancer in c57bl/6 Mice." *Molecular Nutrition and Food Research*, *60* (2), 242–255. doi:10.1002/mnfr.201500605.

Goldstein, David S., Yunden Jinsmaa, Patti Sullivan, Courtney Holmes, Irwin J. Kopin, & Yehonatan Sharabi. (2016). "3,4-Dihydroxyphenylethanol (Hydroxytyrosol) Mitigates the Increase in Spontaneous Oxidation of Dopamine During Monoamine Oxidase Inhibition in PC12 Cells." *Neurochemical Research*, *41* (9), Springer US: 2173–2178. doi:10.1007/s11064-016-1959-0.

Gómez-Caravaca, Ana M., Rubén M. Maggio. & Lorenzo Cerretani. (2016). "Chemometric Applications to Assess Quality and Critical Parameters of Virgin and Extra-Virgin Olive Oil. A Review." *Analytica Chimica Acta*, *913*, doi:10.1016/ j.aca.2016.01.025.

Gouvinhas, Irene, Nelson Machado, Carla Sobreira, Raúl Domínguez-Perles, Sónia Gomes, Eduardo Rosa. & Ana I. R. N. A. Barros. (2017). "Critical Review on the Significance of Olive Phytochemicals in Plant Physiology and Human Health." *Molecules*, *22* (11), doi:10.3390/molecules22111986.

Gu, Liwei, Mark a Kelm, John F Hammerstone, Gary Beecher, Joanne Holden, David Haytowitz, Susan Gebhardt. & Ronald L Prior. (2004). "Concentrations of Proanthocyanidins in Common Foods and Estimations of Normal Consumption." *The Journal of Nutrition*, *134*, (December 2003), 613–617.

Gu, Qiang, Xiaojuan Yi, Jinling Shen, Jing Shi, Yijun Liu. & Jingdong Shao. (2016). "A Highly Automated Method for Simultaneous Determination of Sterols, Erythrodiol and Uvaol in Olive Oils." *Analytical Methods*, *8* (20), Royal Society of Chemistry: 4022–4027. doi:10.1039/c6ay00357e.

Guo, Zebin, Xiangze Jia, Zhichang Zheng, Xu Lu, Yafeng Zheng, Baodong Zheng. & Jianbo Xiao. (2018). "Chemical Composition and Nutritional Function of Olive (Olea Europaea L.): A Review." *Phytochemistry Reviews*, *17* (5), 1091–1110. doi:10.1007/ s11101-017-9526-0.

Heleno, Sandrina A., Patrícia Diz, Prieto, M. A., Lillian Barros, Alírio Rodrigues, Maria Filomena Barreiro. & Isabel C. F. R. Ferreira. (2016). "Optimization of Ultrasound-Assisted Extraction to Obtain Mycosterols from Agaricus Bisporus L. by Response Surface Methodology and Comparison with Conventional Soxhlet Extraction." *Food Chemistry*, *197*. Elsevier Ltd: 1054–1063. doi:10.1016/j.foodchem.2015.11.108.

Hernáez, Álvaro, Olga Castañer, Roberto Elosua, Xavier Pintó, Ramón Estruch, Jordi Salas-Salvadó, Dolores Corella., et al. (2017). "Mediterranean Diet Improves High-Density Lipoprotein Function in High-Cardiovascular-Risk Individuals." *Circulation*, *135* (7), 633–643. doi:10.1161/CIRCULATIONAHA.116.023712.

Imran, Muhammad, Muhammad Nadeem, Syed Amir Gilani, Shaista Khan, Muhammad Wasim Sajid. & Rai Muhammad Amir. (2018). "Antitumor Perspectives of Oleuropein and Its Metabolite Hydroxytyrosol: Recent Updates." *Journal of Food Science*, *83* (7), 1781–1791. doi:10.1111/1750-3841.14198.

Incalza, Maria Angela, Rossella D'Oria, Annalisa Natalicchio, Sebastio Perrini, Luigi Laviola. & Francesco Giorgino. (2018). "Oxidative Stress and Reactive Oxygen Species in Endothelial Dysfunction Associated with Cardiovascular and Metabolic Diseases." *Vascular Pharmacology*, *100*, 1–19. doi:10.1016/j.vph.2017.05.005.

Kotsiou, Kali. & Maria Tasioula-Margari. (2016). "Monitoring the Phenolic Compounds of Greek Extra-Virgin Olive Oils during Storage." *Food Chemistry*, *200*, 255–262. doi:10.1016/j.foodchem.2015.12.090.

Kyçyk, Onejda, Maria Paz Aguilera, José Juan Gaforio, Antonio Jiménez. & Gabriel Beltrán. (2016). "Sterol Composition of Virgin Olive Oil of Forty-Three Olive Cultivars from the World Collection Olive Germplasm Bank of Cordoba." *Journal of the Science of Food and Agriculture*, *96* (12), 4143–4150. doi:10.1002/jsfa.7616.

Laincer, Firdousse, Nunzia Iaccarino, Jussara Amato, Bruno Pagano, Alessia Pagano, Giancarlo Tenore, Abderezak Tamendjari., et al. (2016). "Characterization of Monovarietal Extra Virgin Olive Oils from the Province of Béjaïa (Algeria)." *Food Research International*, *89*, 1123–1133, doi:10.1016/ j.foodres.2016.04.024.

Li, Xianhua, Yanhong Liu, Qian Jia, Virginia LaMacchia, Kathryn O'Donoghue. & Zuyi Huang. (2016). "A Systems Biology Approach to Investigate the Antimicrobial Activity of Oleuropein." *Journal of Industrial Microbiology and Biotechnology*, *43* (12), 1705–1717. doi:10.1007/s10295-016-1841-8.

López-Villodres, Juan Antonio, Miriam Abdel-Karim, José Pedro De La Cruz, María Dolores Rodríguez-Pérez, José Julio Reyes, Rocío Guzmán-Moscoso, Guillermo Rodriguez-

Gutierrez, Juan Fernández-Bolaños. & José Antonio González-Correa. (2016). "Effects of Hydroxytyrosol on Cardiovascular Biomarkers in Experimental Diabetes Mellitus." *Journal of Nutritional Biochemistry*, *37*, 94–100. doi:10.1016/j.jnutbio.2016.07.015.

Luisi, Maria Luisa Eliana, Laura Lucarini, Barbara Biffi, Elena Rafanelli, Giacomo Pietramellara, Mariaconcetta Durante, Sofia Vidali., et al. (2019). "Effect of Mediterranean Diet Enriched in High Quality Extra Virgin Olive Oil on Oxidative Stress, Inflammation and Gut Microbiota in Obese and Normal Weight Adult Subjects." *Frontiers in Pharmacology*, *10*, (November), 1–9. doi:10.3389/ fphar.2019.01366.

Mansouri, Neda, Mohammad Reza Alivand, Sahar Bayat, Mahmoud Shekari Khaniani. & Sima Mansoori Derakhshan. (2019). "The Hopeful Anticancer Role of Oleuropein in Breast Cancer through Histone Deacetylase Modulation." *Journal of Cellular Biochemistry*, *120* (10), 17042–17049. doi:10.1002/jcb.28965.

Marcelino, Gabriela, Priscila Aiko Hiane, Karine de Cássia Freitas, Lidiani Figueiredo Santana, Arnildo Pott, Juliana Rodrigues Donadon. & Rita de Cássia Avellaneda Guimarães. (2019). "Effects of Olive Oil and Its Minor Components on Cardiovascular Diseases, Inflammation, and Gut Microbiota." *Nutrients*, *11* (8), 4–6. doi:10.3390/nu11081826.

Marchio, Patricia, Sol Guerra-Ojeda, José M. Vila, Martín Aldasoro, Victor M. Victor. & Maria D. Mauricio. (2019). "Targeting Early Atherosclerosis: A Focus on Oxidative Stress and Inflammation." *Oxidative Medicine and Cellular Longevity*, 2019:8563845. doi:10.1155/2019/8563845.

Martín-Peláez, Sandra, Juana Ines Mosele, Neus Pizarro, Marta Farràs, Rafael de la Torre, Isaac Subirana, Francisco José Pérez-Cano., et al. (2017). "Effect of Virgin Olive Oil and Thyme Phenolic Compounds on Blood Lipid Profile: Implications of Human Gut Microbiota." *European Journal of Nutrition*, *56* (1), 119–131, doi:10.1007/s00394-015-1063-2.

Mateos, Raquel, Sara Martínez-López, Gema Baeza Arévalo, Miryam Amigo-Benavent, Beatriz Sarriá. & Laura Bravo-Clemente. (2016). "Hydroxytyrosol in Functional Hydroxytyrosol-Enriched Biscuits Is Highly Bioavailable and Decreases Oxidised Low Density Lipoprotein Levels in Humans." *Food Chemistry*, *205*, 248–256. doi:10.1016/j.foodchem.2016.03.011.

Mischley, Laurie K., Richard C. Lau. & Rachel D. Bennett. (2017). "Role of Diet and Nutritional Supplements in Parkinson's Disease Progression." *Oxidative Medicine and Cellular Longevity*, 2017: 6405278. doi:10.1155/2017/6405278.

Navajas-porras, B., Pérez-burillo, S., Morales-pérez, J., Rufián-henares, J. A. & Pastoriza, S. (2020). "Relationship of Quality Parameters, Antioxidant Capacity and Total Phenolic Content of EVOO with Ripening State and Olive Variety." *Food Chemistry*, *325*:126926. doi:10.1016/j.foodchem.2020.126926.

Nediani, Chiara, Jessica Ruzzolini, Annalisa Romani. & Lido Calorini. (2019). "Oleuropein, a Bioactive Compound from Olea Europaea L., as a Potential Preventive and Therapeutic Agent in Non-Communicable Diseases." *Antioxidants*, *8* (12). doi:10.3390/ antiox8120578.

Niedzielska, Ewa, Irena Smaga, Maciej Gawlik, Andrzej Moniczewski, Piotr Stankowicz, Joanna Pera. & Małgorzata Filip. (2016). "Oxidative Stress in Neurodegenerative Diseases." *Molecular Neurobiology*, *53* (6), 4094–4125. doi:10.1007/s12035-015-9337-5.

Nocella, Cristina, Vittoria Cammisotto, Luca Fianchini, Alessandra D'Amico, Marta Novo, Valentina Castellani, Lucia Stefanini, Francesco Violi. & Roberto Carnevale. (2017). "Extra Virgin Olive Oil and Cardiovascular Diseases: Benefits for Human Health." *Endocrine, Metabolic & Immune Disorders - Drug Targets*, *18* (1), 4–13. doi:10.2174/1871530317666171114121533.

Ogun, Metin, Ayla Ozcan, Musa Karaman, Oguz Merhan, Hasan Ozen, Abdulsamed Kukurt. & Mahmut Karapehlivan. (2016). "Oleuropein Ameliorates Arsenic Induced Oxidative Stress in Mice." *Journal of Trace Elements in Medicine and Biology*, *36*, 1–6. doi:10.1016/j.jtemb.2016.03.006.

Pandolfi, Assunta. (2013). "Biology of Carotenoids and Their Potential Cardiovascular Health Benefits." ISBN: 978-1-62808-622-5, Nova Science Publishers, Inc.

Parkinson, Lisa. & Sara Cicerale. (2016). "The Health Benefiting Mechanisms of Virgin Olive Oil Phenolic Compounds." *Molecules*, *21* (12), 1–12. doi:10.3390/molecules21121734.

Peng, Yunhua, Chen Hou, Ziqi Yang, Caixia Li, Liyan Jia, Jing Liu, Ying Tang., et al. (2016). "Hydroxytyrosol Mildly Improve Cognitive Function Independent of APP Processing in APP/PS1 Mice." *Molecular Nutrition and Food Research*, *60* (11), 2331–2342. doi:10.1002/mnfr.201600332.

Perdomo, Liliana, Nuria Beneit, Yolanda F. Otero, Óscar Escribano, Sabela Díaz-Castroverde, Almudena Gómez-Hernández. & Manuel Benito. (2015). "Protective Role of Oleic Acid against Cardiovascular Insulin Resistance and in the Early and Late Cellular Atherosclerotic Process." *Cardiovascular Diabetology*, *14* (1), Cardiovascular Diabetology, 1–12. doi:10.1186/s12933-015-0237-9.

Pérez-Jiménez, Francisco, Juan Ruano, Pablo Perez-Martinez, Fernando Lopez-Segura. & Jose Lopez-Miranda. (2007). "The Influence of Olive Oil on Human Health: Not a Question of Fat Alone." *Molecular Nutrition and Food Research*, *51* (10), 1199–1208. doi:10.1002/mnfr.200600273.

Perona, Javier S., Rosana Cabello-Moruno. & Valentina Ruiz-Gutierrez. (2006). "The Role of Virgin Olive Oil Components in the Modulation of Endothelial Function." *Journal of Nutritional Biochemistry*, *17* (7), 429–445. doi:10.1016/ j.jnutbio.2005.11.007.

Perrone, A. Marco., Gualtieri, Gratteri, P. S., Ali, W., Sergi, D., Muscoli, S., Cammarano, A., Bernardini, S., Di Renzo, L. & Romeo, F. (2019). "Effects of Postprandial Hydroxytyrosol and Derivates on Oxidation of LDL, Cardiometabolic State and Gene Expression a Nutrigenomic Approach for Cardiovascular Prevention." *Journal of Cardiovascular Medicine*, *20* (7), 419–426. doi: 10.2459/JCM.0000000000000816.

Peyrol, Julien, Catherine Riva. & Marie Josèphe Amiot. (2017). "Hydroxytyrosol in the Prevention of the Metabolic Syndrome and Related Disorders." *Nutrients*, *9* (3), 1–18. doi:10.3390/nu9030306.

Piccinin, Elena, Marica Cariello, Stefania De Santis, Simon Ducheix, Carlo Sabbà, James M. Ntambi. & Antonio Moschetta. (2019). "Role of Oleic Acid in the Gut-Liver Axis: From Diet to the Regulation of Its Synthesis via Stearoyl-CoA Desaturase 1 (SCD1)." *Nutrients*, *11* (10), 1–22. doi:10.3390/nu11102283.

Piroddi, Marta, Adriana Albini, Roberto Fabiani, Lisa Giovannelli, Cristina Luceri, Fausta Natella, Patrizia Rosignoli., et al. (2017). "Nutrigenomics of Extra-Virgin Olive Oil: A Review." *BioFactors*, *43* (1), 17–41. doi:10.1002/biof.1318.

Piscopo, Amalia, Alessandra De Bruno, Angela Zappia, Carmine Ventre. & Marco Poiana. (2016). "Characterization of Monovarietal Olive Oils Obtained from Mills of Calabria Region (Southern Italy)." *Food Chemistry*, *213*, 313–318. doi:10.1016/j.foodchem.2016.06.080.

Poudyal, Hemant, Nikolaos Lemonakis, Panagiotis Efentakis, Evangelos Gikas, Maria Halabalaki, Ioanna Andreadou, Leandros Skaltsounis. & Lindsay Brown. (2017). "Hydroxytyrosol Ameliorates Metabolic, Cardiovascular and Liver Changes in a Rat Model of Diet-Induced Metabolic Syndrome: Pharmacological and Metabolism-Based Investigation." *Pharmacological Research*, *117*, 32–45. doi:10.1016/j.phrs.2016.12.002.

Pourkhodadad, Soheila, Masoud Alirezaei, Mehrnoush Moghaddasi, Hassan Ahmadvand, Manizheh Karami, Bahram Delfan. & Zahra Khanipour. (2016). "Neuroprotective Effects of Oleuropein against Cognitive Dysfunction Induced by Colchicine in Hippocampal CA1 Area in Rats." *Journal of Physiological Sciences*, *66* (5), 397–405. doi:10.1007/s12576-016-0437-4.

Quintero-Flórez, Angélica, Laura Sinausia Nieva, Araceli Sánchez-Ortíz, Gabriel Beltrán. & Javier S. Perona. (2015). "The Fatty Acid Composition of Virgin Olive Oil from Different Cultivars Is Determinant for Foam Cell Formation by Macrophages." *Journal of Agricultural and Food Chemistry*, *63* (30), 6731–6738. doi:10.1021/acs.jafc.5b01626.

Rajagopala, Seesandra V., Sanjay Vashee, Lauren M. Oldfield, Yo Suzuki, J. Craig Venter, Amalio Telenti. & Karen E. Nelson. (2017). "The Human Microbiome and Cancer." *Cancer Prevention Research*, *10* (4), 226–234. doi:10.1158/1940-6207.CAPR-16-0249.

Rohman, A., Riyanta, A. B., Lukitaningsih, E. & Riyanto, S. (2020). "Olive (Olea Europea) Oil: Physico-Chemical Characterization and Antioxidant Activities *in Vitro* and *in Vivo*." *Food Research*, *4* (3), 563–570. doi:10.26656/fr.2017.4(3).220.

Romani, Annalisa, Francesca Ieri, Silvia Urciuoli, Annalisa Noce, Giulia Marrone, Chiara Nediani. & Roberta Bernini. (2019). "Nutrients Health E Ff Ects of Phenolic Compounds Found in Extra-Virgin Olive Oil, By-Products, and Leaf of *Olea Europaea* L." *Nutrients*, *11* (1176), 1–33.

Ryu, Su Jung, Hyeon Son Choi, Kye Yoon Yoon, Ok Hwan Lee, Kui Jin Kim. & Boo Yong Lee. (2015). "Oleuropein Suppresses LPS-Induced Inflammatory Responses in RAW 264.7 Cell and Zebrafish." *Journal of Agricultural and Food Chemistry*, *63* (7), 2098–2105. doi:10.1021/jf505894b.

Saha, Subbroto Kumar, Soo Bin Lee, Jihye Won, Hye Yeon Choi, Kyeongseok Kim, Gwang Mo Yang, Ahmed Abdal Dayem. & Ssang Goo Cho. (2017). "Correlation between Oxidative Stress, Nutrition, and Cancer Initiation." *International Journal of Molecular Sciences*, *18* (7), doi:10.3390/ijms18071544.

Saibandith, Bandhita, Jeremy P. E. Spencer, Ian R. Rowland. & Daniel M. Commane. (2017). "Olive Polyphenols and the Metabolic Syndrome." *Molecules (Basel, Switzerland)*, *22* (7), doi:10.3390/molecules22071082.

Salazar, D. M., López-Cortés, I. & Salazar-García, D. C. (2017). *Olive Oil: Composition and Health Benefits. Olive Oil: Sensory Characteristics, Composition and Importance in Human Health.*

Sanchez-Rodriguez, Estefania, Sara Biel-Glesson, Jose R Fernandez-Navarro, Miguel A Calleja, Juan A Espejo-Calvo, Blas Gil-extremera, Rafael De Torre., et al. (2019). "Effects of Virgin Olive Oils Differing in Their Bioactive Compound Contents on

Biomarkers of Oxidative Stress and Inflammation in Healthy Adults: A Randomized Double-Blind Controlled Trial." *Nutrients*, *11* (561), doi:10.3390/nu11030561.

Santangelo, Carmela, Vari Rosariax, Scazzocchio Beatrice, De Sanctis Patrizia, Giovannini Claudio, D'Archivio Massimo, Masella Robertaet. "Anti-Inflammatory Activity of Extra Virgin Olive Oil Polyphenols: Which Role in the Prevention and Treatment of Immune-Mediated Inflammatory Diseases?" *Endocr Metab Immune Disord Drug Targets* 18 (1): 36–50. doi:doi:10.2174/1871530317666171114114321.

Schwingshackl, Lukas, Marina Christoph. & Georg Hoffmann. (2015). "Effects of Olive Oil on Markers of Inflammation and Endothelial function—A Systematic Review and Meta-Analysis." *Nutrients*, *7* (9), 7651–7675. doi:10.3390/nu7095356.

Silva, S., Combet, E., Figueira, M. E., Koeck, T., Mullen, W. & Bronze, M. R. (2015). "New Perspectives on Bioactivity of Olive Oil: Evidence from Animal Models, Human Interventions and the Use of Urinary Proteomic Biomarkers." *Proceedings of the Nutrition Society*, *74* (3), 268–281. doi:10.1017/S0029665115002323.

Soni, Manisha, Chandra Prakash, Sfurti Sehwag. & Vijay Kumar. (2017). "Protective Effect of Hydroxytyrosol in Arsenic-Induced Mitochondrial Dysfunction in Rat Brain." *Journal of Biochemical and Molecular Toxicology*, *31* (7). doi:10.1002/jbt.21906.

Storniolo, Carolina Emilia, Joan Roselló-Catafau, Xavier Pintó, María Teresa Mitjavila. & Juan José Moreno. (2014). "Polyphenol Fraction of Extra Virgin Olive Oil Protects against Endothelial Dysfunction Induced by High Glucose and Free Fatty Acids through Modulation of Nitric Oxide and Endothelin-1." *Redox Biology*, *2* (1), 971–977. doi:10.1016/j.redox.2014.07.001.

Sun, Wenyan, Xin Wang, Chen Hou, Liang Yang, Hongbao Li, Jing Guo, Chanjuan Huo., et al. (2017). "Oleuropein Improves Mitochondrial Function to Attenuate Oxidative Stress by Activating the Nrf2 Pathway in the Hypothalamic Paraventricular Nucleus of Spontaneously Hypertensive Rats." *Neuropharmacology*, *113*, 556–566. doi:10.1016/j.neuropharm.2016.11.010.

Termentzi, Aikaterini, Maria Halabalaki. & Alexios Leandros Skaltsounis. (2015). *From Drupes to Olive Oi: An Exploration of Olive. Olive and Olive Oil Bioactive Constituents*. AOCS Press. doi:10.1016/B978-1-63067-041-2.50012-4.

Terzuoli, Erika, Antonio Giachetti, Marina Ziche. & Sandra Donnini. (2016). "Hydroxytyrosol, a Product from Olive Oil, Reduces Colon Cancer Growth by Enhancing Epidermal Growth Factor Receptor Degradation." *Molecular Nutrition and Food Research*, *60* (3), 519–529. doi:10.1002/mnfr.201500498.

Tomasello, Giovanni, Margherita Mazzola, Angelo Leone, Emanuele Sinagra, Giovanni Zummo, Felicia Farina, Provvidenza Damiani., et al. (2016). "Nutrition, Oxidative Stress and Intestinal Dysbiosis: Influence of Diet on Gut Microbiota in Inflammatory Bowel Diseases." *Biomedical Papers*, *160* (4), 461–466. doi:10.5507/bp.2016.052.

Toric, Jelena, Ana Karkovic Markovic, Cvijeta Jakobušic Brala. & Monika Barbaric. (2019). "Anticancer Effects of Olive Oil Polyphenols and Their Combinations with Anticancer Drugs." *Acta Pharmaceutica Sciencia*, *69* (4), 461–482. doi:10.2478/acph-2019-0052.

Torres-Peña, Jose D., Antonio Garcia-Rios, Nieves Delgado-Casado, Purificacion Gomez-Luna, Juan F. Alcala-Diaz, Elena M. Yubero-Serrano, Francisco Gomez-Delgado., et al. (2018). "Mediterranean Diet Improves Endothelial Function in Patients with Diabetes and Prediabetes: A Report from the CORDIOPREV Study." *Atherosclerosis*, *269*, 50–56. doi:10.1016/j.atherosclerosis.2017.12.012.

Tuberoso, Carlo I. G., Igor Jerković, Marialuce Maldini. & Gabriele Serreli. (2016). "Phenolic Compounds, Antioxidant Activity, and Other Characteristics of Extra Virgin Olive Oils from Italian Autochthonous Varieties Tonda Di Villacidro, Tonda Di Cagliari, Semidana, and Bosana." *Journal of Chemistry*, ID 8462741, 7 pages.

Varadharaj, Saradhadevi, Owen J. Kelly, Rami N. Khayat, Purnima S. Kumar, Naseer Ahmed. & Jay L. Zweier. (2017). "Role of Dietary Antioxidants in the Preservation of Vascular Function and the Modulation of Health and Disease." *Frontiers in Cardiovascular Medicine*, *4*, (November), 1–11. doi:10.3389/fcvm.2017.00064.

Wani, Touseef Ahmed, Masoodi, F. A., Adil Gani, Waqas Nabi Baba, Neda Rahmanian, Rehana Akhter, Idrees Ahmed Wani. & Mudasir Ahmad. (2018). "Olive Oil and Its Principal Bioactive Compound: Hydroxytyrosol – A Review of the Recent Literature." *Trends in Food Science and Technology*, *77*, 77–90. doi:10.1016/j.tifs.2018.05.001.

Yang, Ping, Chunxiao Su, Xuan Luo, Han Zeng, Lei Zhao, Li Wei, Xiaoyu Zhang., et al. (2018). "Dietary Oleic Acid-Induced CD36 Promotes Cervical Cancer Cell Growth and Metastasis via up-Regulation Src/ERK Pathway." *Cancer Letters*, *438,* 76–85. doi:10.1016/j.canlet.2018.09.006.

Yorulmaz, Hasan Ozkan. & Dilsat Bozdogan Konuskan. (2017). "Antioxidant Activity, Sterol and Fatty Acid Compositions of Turkish Olive Oils as an Indicator of Variety and Ripening Degree." *Journal of Food Science and Technology*, *54* (12), 4067–4077. doi:10.1007/s13197-017-2879-y.

Yu, Guohua, Ajun Deng, Wanbin Tang, Junzhi Ma, Chonggang Yuan. & Jiyan Ma. (2016). "Hydroxytyrosol Induces Phase II Detoxifying Enzyme Expression and Effectively Protects Dopaminergic Cells against Dopamine- and 6-Hydroxydopamine Induced Cytotoxicity." *Neurochemistry International*, *96*, 113–120. doi:10.1016/j.neuint.2016.03.005.

Yu, Hailong, Peipei Liu, Hui Tang, Jian Jing, Xiang Lv, Lanlan Chen, Li Jiang, Jun Xu. & Jun Li. (2016). "Oleuropein, a Natural Extract from Plants, Offers Neuroprotection in Focal Cerebral Ischemia/reperfusion Injury in Mice." *European Journal of Pharmacology*, *775*, 113–119. doi:10.1016/j.ejphar.2016.02.027.

In: Super and Nutraceutical Foods
Editors: Sandra Teresita Martín-del-Campo et al.
ISBN: 978-1-53619-082-3

Chapter 16

NOPAL, COMPOSITION AND PROCESSING TECHNOLOGIES

M. Perucini-Avendaño, C. Jiménez-Martínez, A. Acosta -Caudillo, M. Nicolás-García and Gloria Dávila-Ortiz*

Instituto Politécnico Nacional, Escuela Nacional de Ciencias Biológicas, Av. Wilfrido Massieu s/n, Unidad Profesional Adolfo López Mateos, Zacatenco, Delegación Gustavo A. Madero, Mexico City, Mexico

ABSTRACT

Nopal (*Opuntia ficus-indica*) is part of the Mexican diet due to its nutritional composition and it is an important source of fiber and vitamins. The traditional consumption of this plant is raw or cooked. In the food industry, heat treatments are used to reduce the microbial load, non-nutritional compounds, and inactivate enzymes, thus prolonging storage life. Due to these treatments, some compounds also change. In the case of nopal, the effect is reflected in the content of phenolic compounds, whose concentration decreases, thereby, affecting their biological activity. This chapter focuses on the chemical composition and processing technologies used for nopal.

Keywords: *Opuntia ficus-indica*, morphology, chemical composition, processing technologies

INTRODUCTION: HISTORY, DISTRIBUTION, AND HEALTH BENEFITS

Nopal (*Opuntia ficus-indica*) belongs to the Cactaceae family, which includes about 1500 species (El-Mostafa et al., 2014). It is a tropical or subtropical plant native to Mexico that is used as a food source because of its high nutritional value (Ciriminna, Chavarría-Hernández, Rodríguez-Hernández, & Pagliaro, 2019). Its presence has been described in Mesoamerica since ancient times (20,000 years ago), specifically in the desert and semi-desert regions,

* Corresponding Author's E-mail: gdavilao@yahoo.com.

where they were essential sources of food, beverages, and medicines for indigenous peoples (Angulo-Bejarano, Martínez-Cruz, & Paredes-López, 2014). It is also grown in Africa, Australia, southern Europe, and Asia (Betatache, Aouabed, Drouiche, & Lounici, 2014; Khouloud, Abedelmalek, Chtourou, & Souissi, 2018), that is, these plants exist all over the the world (El Finti, El Boullani, Fallah, Msanda, & El Mousadik, 2013). The economic and social importance of nopal in Mexico lies in the area occupied by both wild and cultivated nopals, in the type and number of producers involved, in the regions where it is grown, and in the diversity of the generated products (Araya, Contreras, Alvina, Vera, & Pak, 2002; Blanco-Macias et al., 2011; Flores-Hernández, Orona-Castillo, Murillo-Amador, Cepeda, & García-Hernández, 2004). Young nopales are consumed as vegetables at different stages of maturation, ranging from 30 to 90 days or more (Marin-Bustamante et al., 2018).

The taxonomy of cladodes is complex, because there are environmental, climatic, and plant-specific factors that can modify it (Matsuhiro, Lillo, Sáenz, Urzúa, & Zárate, 2006; Nobel, 2002; Rodríguez-Felix & Cantwell, 1988). For example, some phenotypes present in the cladodes show differences due to environmental conditions, such as temperature, humidity, soil pH, etc.

In recent years, the effects of nopal (*Opuntia ficus-indica*) components and their potential to have a beneficial influence on human health have been demonstrated. The latter has been generating progressively a scientific knowledge based on numerous experimental models dedicated to the evaluation and extraction of different nopal compounds to treat various chronic-degenerative diseases (El-Mostafa et al., 2014). Its cultivation is currently considered an economic income option for communities in semi-arid regions (Bayar, Kriaa, & Kammoun, 2016).

General Aspects: |Agriculture, Botany, and Morphological Properties

The nopal trade has become a substantial economic industry over the years. One of the reasons for its introduction was its use as a drought-resistant crop due to its easy adaptation (Walters, Figueiredo, Zimmermann, & Mashope, 2011). In an attempt to minimize the risks of global climate change, land degradation, and decreased food security, the Food and Agriculture Organization has revived interest in nopal cultivation for agricultural purposes in developing countries (Nefzaoui, 2007). Mexico is the leading producer and exporter of nopal, accounting for 90% of the world production and exports to the United States, Canada, Japan, and European countries. The FAO (2017) reported the extensive output of nopal and the natural existence of the plant in the wild; however, in some countries, nopal plants are considered an environmental issue. Nopal is widely distributed in Mexico and throughout the American hemisphere, as well as in Africa and the Mediterranean basin (Zorgui, Ayed-Boussema, Ayed, Bacha, & Hassen, 2009). Mexico's Ministry of Agriculture, Livestock, Rural Development, Fisheries and Food (SAGARPA-Mexico), through the Food, Agriculture and Fisheries Information Service (SIAP), reported, in 2016, a national production of 810,228 ton of nopal, 180,250 ton of fodder nopal, and 2'454,557 ton of prickly pears, with a nopal cultivation area of 62,862.60, 84,094.50, and 262,304.64 ha for human consumption, cladodes for livestock feed, and prickly pears, respectively.

Figure 1. Overview of cladode (*Opuntia ficus-indica*) a), SEM image of a nopal spine b), optic image of a transversal cut of cladode, c) SEM image of calcium oxalate crystal d), flower e) and fruit f).

Most of the nopal species are concentrated in Mexico, on the other hand, different species have spread to some African and Asian regions, showing significant biodiversity within and among species. Some of these species are *Opuntia: Basilaris, chlorotica, engelmannii, Opuntia fragilis, humifusa, leucotricha, macrocentra, macrorhiza, dillenii, santa-rita, stricta*. One of the most consumed species of nopal is *Opuntia ficus-indica*, its original location is the Center-South of Mexico; it is also grown in warm parts of the world such as Algeria, Tunisia, Morocco, Kenya, and Egypt (Aruwa, Amoo, & Kudanga, 2018). The main factor affecting the composition of nopal is its biodiversity, followed by growing conditions such as climate and soil (Ammar, Ennouri, & Attia, 2015; Butera et al., 2002).

The nopal is a succulent plant that measures between 1.5 and 3 m in height, and it is divided into three main parts: cladodes, flowers, and fruits (Andreu, Nuncio-Jáuregui, Carbonell-Barrachina, Legua, & Hernández, 2018). The main constituent of the nopal is the cladodes (Figure 1a), which are characterized by their ovoid racquet shape, reaching up to 60 cm in length, depending on the water and nutrients available (Salehi, Emam-Djomeh, Askari, & Fathi, 2019). From the outside to the inside the cladodes present different types of tissues, the first is the surface where the cladodes are covered with spines (Figure 1b), which provide mechanical protection against herbivores and reduce water loss (Marin-Bustamante et al., 2018), this surface contains stomata which participate in photosynthesis and also in respiration (Salem-Fnayou, Zemni, Nefzaoui, & Ghorbel, 2014). The internal part (Figure 1c) of the cladodes is made up of a tissue of living cells, called parenchyma, and their function is to store water for the plant, which allows it to withstand long periods of drought (Saag, Sanderson, Moyna, & Ramos, 1975). The tissues along a cross-section were described by Ginestra et al., (2009) and are called surface or chlorenchyma (CH), vascular tissue (V), and nucleus (N). Several studies, utilizing scanning electron microscopy (SEM) (Monje & Baran, 2002), have found calcium oxalate crystals (Figure 1d) in the different tissues of nopal, which decrease in content according to the maturation of the cladode, making calcium more available in young cladodes (Marin-Bustamante et al., 2018).

Flowers (Figure 1e) often emerge in one-year cladodes, while buds are mainly released in two- or more year cladodes. Cladodes can bloom a second time during the same year if environmental conditions are favorable or through the use of advanced agricultural techniques (Arba, Falisse, Choukr-Allah, & Sindic, 2017). The fruits (Figure 1f) are known as prickly pears. They are ovoid and spherical, whose color depends on the species of nopal, they can be found in shades of red, purple, yellow, or green. Its pulp is juicy, sweet, and has many seeds, the exocarp is soft and protected with small thorns. The weight is variable and can go from 48 to 251 g (Carrillo, Gómez-Cuaspud, & Suarez, 2017; Ciriminna et al., 2019; Reale, Fichera, Ferri, Cerri, & Ferranti).

Chemical Composition of Nopal

The main component of nopal is water (95%), and they are considered a good source of fiber, carbohydrates, proteins, and minerals (Table 1) (Astello-García et al., 2015; Guevara-Figueroa et al., 2010). Its composition includes mucilaginous components containing carbohydrate polymers, such as chains of β-D-galacturonic acid bound in positions (1-4) and R-linked L-rhamnose traces (1-2) (McGarvie & Parolis, 1979). Mucilage is a complex polysaccharide, which is part of dietary fiber and is produced by specialized cells located in the cell wall. It constitutes approximately 14% of the dry weight (DW) of cladodes. It may contain more than 30% of the total water of the reserve parenchyma (Ventura-Aguilar, Bosquez-Molina, Bautista-Baños, & Rivera-Cabrera, 2017). The main monosaccharides that have been identified in nopal are glucose, arabinose, xylose, galacturonic acid, glucuronic acid, galactose, rhamnose, uronic acid, mannose, and fructose (Akanni, du Preez, Steyn, & Kilian, 2015; Rodríguez-González et al., 2014).

Table 1. Chemical composition of nopal cladodes (Dry weight basis)

Composition (%)				References
Carbohydrates	Protein	Lipid	Ash	
32.26	11.2	0.69	14.4	Astello-García et al. (2015)
67.00	14.4	2.5	14.9	Nuñez-López, Paredes-López, & Reynoso-Camacho et al. (2013)
25.38	12.90	1.10	18.38	Ramírez-Moreno, Córdoba-Díaz, de Cortes Sánchez-Mata, Díez-Marqués, & Goñi, et al. (2013)
91.72	11.73	1.89	23.05	Cornejo-Villegas et al., (2010)
39.00	6.7	0.1	17.3	Guevara-Figueroa et al. (2010)
33.95	7.07	2.16	17.65	Hernández-Urbiola et al. (2010)

The physiological role of mucilage in nopal cladodes is to regulate both the cell water content during prolonged drought and the plant's calcium flows (Hernández-Urbiola et al., 2010; Rodríguez-García et al., 2007). The components obtained from cladodes contain many active ingredients, particularly antioxidant constituents that include vitamin C, vitamin E, carotenoids, glutathione, flavonoids, and phenolic acids, among others (Ginestra et al., 2009; Panico et al., 2005).

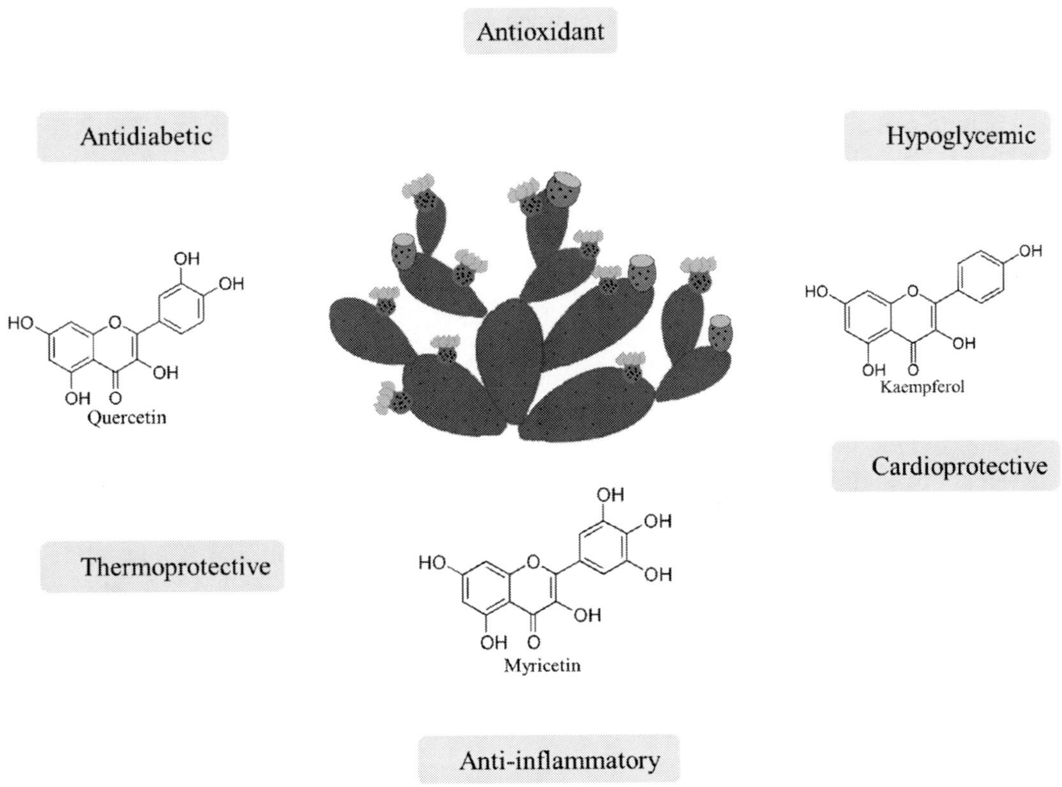

Figure 2. Biological activities reported in nopal.

In Mexico, nopales have been traditionally used to treat different diseases such as diabetes, hyperlipidemia, and obesity (Cejudo-Bastante, Durán-Guerrero, Natera-Marín, Castro-Mejías, & García-Barroso, 2013; Nuñez-López et al., 2013). Figure 2 shows some of the different biological activities that have been reported for the nopal: Antioxidant, hypoglycemic, cardioprotective, antidiabetic, thermoprotective, and anti-inflammatory (Andreu et al.,2018; Avila-Nava et al.,2017; Lee et al., 2002; López-Romero et al., 2014; Msaddak et al., 2017; Salehi et al., 2019; Valente et al., 2010; Abdel-Hameed et al., 2014; Amaya-Cruz et al., 2019; Cayupán et al., 2011; Onakpoya et al., 2015; Nuñez-López et al., 2013; Hassan, et al., 2011; Antunes-Ricardo et al., 2015; Slimen et al., 2019). Phenolic compounds are the main bioactive components in plants. They have a wide range of health benefits, mainly due to their antioxidant properties, such as the elimination and inhibition of reactive oxygen species, removal of electrophiles, and chelation of metals (Balasundram, Sundram, & Samman, 2006; Huyut, Beydemir, & Gülçin, 2017; Kim, Kim, Yang, & Sung, 2016; López-Romero et al., 2014; Petruk et al., 2017).

Several studies have demonstrated beneficial activities associated with the phenolic compounds of the nopal in humans (Kaur, Kaur, & Sharma, 2012; Smida et al., 2017; Valente et al., 2010). In the nopal, phenolic compounds are generally present in soluble form conjugated with sugars or organic acids and are typically components of complex structures, including lignins and hydrolysable tannins (Ouchemoukh et al., 2017; Ramawat & Mérillon, 2013; Singh, Singh, Kaur, & Singh, 2017). One of the main phenolic acids reported in nopal cladodes is gallic acid, which has high antioxidant activity, as well as cytotoxic activity against tumor cells in leukemia, lung and prostate cancer (Medina-Torres et al., 2011). Petruk et al. (2017) carried out studies with nopal extracts finding three phenolic acid derivatives: Piscidic, eucomic, and 2-hydroxy-4- (4-hydroxyphenyl)-butanoic acid. Hydroxybenzoic, piscidic, and caffeic acids, derivatives of ferulic and piscidic acids, were reported by Mata et al. (2016). Phenolic compounds play a crucial role in plants, such as UV protection, pigmentation, and stimulation of nitrogen-fixing nodules, and resistance against some diseases (Rivera et al., 2018).

Currently, researchers are using various high-resolution analytical techniques for the separation, identification, and quantification of phenolic compounds present in nopal cladodes (Antunes-Ricardo et al., 2014; Astello-García et al., 2015; De Santiago, Domínguez-Fernández, Cid, & De Peña, 2018; Guevara-Figueroa et al., 2010; Msaddak et al., 2017; Rocchetti, Pellizzoni, Montesano, & Lucini, 2018; Santos-Zea, Gutiérrez-Uribe, & Serna-Saldivar, 2011).

Nopal flavonoids have been identified more in cladodes than in nopal fruits (Mata et al., 2016; Matias et al., 2014; Moussa-Ayoub et al., 2014). Some of the flavonoids identified are quercetin 3-O-rutinoside (Ginestra et al., 2009; Guevara-Figueroa et al., 2010), quercetin 3-O-glucoside, quercetin 3-O-arabinofuranoside, quercetin, kaempferol, isorhamnetin 3-O-glucoside, isorhamnetin 3-O-neohesperidoside, coumaric acid and zataroside-A, quercetin and kaempferol derivatives, 3,3', 4', 5,7-pentahydroxy-flavanone isorhamnetin, indicaxanthin, and sterols (Astello-García et al., 2015). Flavonols such as quercetin, kaempferol, and myricetin are the most common and have antimutagenic, anticancer, and antihypertensive activities.

PROCESSING TECHNOLOGIES

In the food industry, heat treatments are used to reduce the microbial load and inactivate enzymes, thus extending storage life (Barba et al., 2017). However, conventional heat treatment in food matrices such as fruits and vegetables causes significant losses of nutritional and health-promoting phytochemical components due to the susceptibility of these compounds to high temperatures. The high temperatures can accelerate chemical reactions, producing new compounds related to color and flavor (Jiménez-Aguilar, López-Martínez, Hernández-Brenes, Gutiérrez-Uribe, & Welti-Chanes, 2015; Yamamoto, 2017). Nopal has antioxidant activity that is mostly attributed to its polyphenolic composition, as described above. However, for human consumption, nopal is generally consumed fresh or cooked, subjected to a thermal process (boiled), grilled, fried, or consumed as juices or in sauces (Figure 3) (De Santiago et al., 2018).

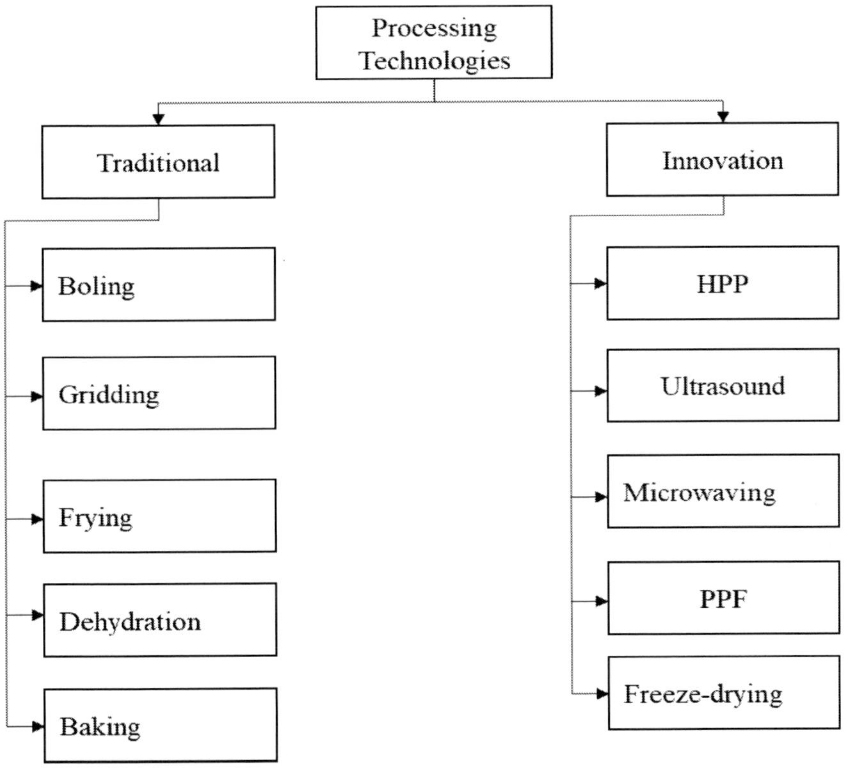

Figure 3. Principal processing technologies for nopal.

De Santiago et al. (2018) evaluated the influence of different thermal treatments: boiling, microwave, grill, and frying, on the chemical composition, antioxidant capacity, and phenolic compounds of nopal cladodes. They reported that the microwave treatment increased up to 3.2 times the presence of flavonoids, phenolic acids, and antioxidant capacity.

Dehydration is one of the most effective methods of preserving highly perishable foods. Taking into account that nopal has a water content of 95%, studies traditionally carried out for nopal drying consist mainly of exposing the cladodes directly to the sun for several days, after which they are milled and sifted to obtain flour (Patel et al. 2014). Due to cross-contamination by insects, animals, and birds, fluidized bed drying was implemented (65–70°C for 72 h). However, it was observed that some spore-forming bacteria are activated above 65°C (Ciriminna et al., 2019; Patel, 2014). Therefore, new works were developed to minimize the microbial load present in flours.

Ciriminna et al. (2019) reported a total carotenoid content in nopal cladodes of 232 mg/kg on a dry basis; of this, 46% is lutein, 36% β-carotene, and 18% α-cryptoxanthin. These results were obtained from nopal dried in a solar air-drying system, which controlled solar irradiation and airflow temperature. On the other hand, Rodriguez et al. (2019) optimized osmotic dehydration (OD) as a treatment prior to hot air drying (HAD) of nopal cladodes. Combined drying (OD + ADH) reduced moisture content in the range of 23 to 30% compared to ADH drying alone. It was found that optimizing OD allowed preserving quality properties, being an excellent alternative to give added value to nopal cladodes. Gallegos-Infante et al. (2009) studied the effect of air-drying flow rates on the amount and antioxidant capacity of cladode extracts. Nopal was dried at 458 °C with airflow rates of 3 and 5 m/sec. Drying of cladodes at an airflow rate of 3 m/sec showed higher values of phenols, flavonoids, and flavonols. Martínez-Soto et al. (2015) obtained nopal flour by hot air, and lyophilization as an excellent source of dietary fiber; wild material showed the highest values (54.2% DW). Phenolic compounds and flavonoids were more abundant in nature than in the powder of commercial cladodes, as well as the production of short-chain fatty acids. Contreras-Padilla et al. (2012) studied the effect of three different drying processes: Forced air oven drying, freeze-drying, and tunnel drying on the physicochemical and nutritional attributes of nopal cladodes at three different stages of maturity. The samples obtained by the freeze-drying process retained the highest content of the nutrients studied and were greener than the other samples.

Díaz-Ayala et al. (2015) measured the drying kinetics of thin strips of 4 mm thick, 140 mm long, and 17 mm wide of nopal cladodes. The results showed that the effect of air temperature on the drying time of nopal strips was more significant than the airspeed. The drying time of the thin strips was notably lower (99%) compared to that of the whole cladode and 50% lower than that previously reported for 25 mm cylindrical nopal pieces. López et al. (2009) studied the drying kinetics of *Opuntia* cladodes by observing two conditions: a complete cladode with the protective cuticle of the product intact, and with the cuticle reduced, using a forced-flow drying tunnel. The air temperature was set at 35, 45, and 60°C with speeds of 1.5 and 3.0 m/s. Environmental conditions were controlled and maintained at 22°C and 30% relative humidity. The results showed that the drying time was considerably reduced when 30% of the cuticle was removed.

Currently, the application of innovative methods allows operation at low or moderate temperatures, which provides a promising alternative of non-thermal preservation for different foods (Figure 3). Among these non-thermal preservation methods is the use of high hydrostatic pressure (HHP) and pulsed electric fields (PEF) (Barba et al., 2017; Režek Jambrak, Donsì, Paniwnyk, & Djekic, 2019), these, in turn, are technologies applicable for the extraction of bioactive compounds from the nopal (Antunes-Ricardo, Gutiérrez-Uribe, & Guajardo-Flores, 2017).

The three main parameters in the use of HHP are temperature, pressure, and exposure time of the food matrix. The pressures used in this process range from 100-800 MPa and are uniformly distributed throughout the food product, regardless of size or geometry, which differentiates it from thermal processing where gradual heat dissipation occurs (Barba, Galanakis, Esteve, Frigola, & Vorobiev, 2015). HHP predominantly affects non-covalent bonds, such as hydrophobic, hydrogen, and ionic bonds, so the covalently bonded amino acid sequence of a protein is not affected by this treatment (Ekezie, Cheng, & Sun, 2018).

A study carried out by Jiménez-Aguilar et al. (2015) determined the impact of the use of HHP and UHT (ultrahigh-temperature) in prickly pear drinks, formulated with and without acids and preservatives, on their vitamin C, phenolic compounds, flavonoids, and betalains content, as well as on their antioxidant activity. The beverages subjected to UHT showed significant losses of vitamin C (46-76%), phenolic compounds (27-52%), flavonoids (0-52%), betalains (7-45%), and a decrease in antioxidant capacity (16-45%), whereas the application of HHP at pressures of 550 MPa for 2 to 4 min for the processing of drinks, including the pulp and rind of prickly pears, improves their nutraceutical properties due to a significant increase (16-35%) of total phenolic compounds.

The effect of HHP on bioactive compounds in *Opuntia* fruits at different pressures (100, 350, and 600 MPa) as well as its impact on anti-inflammatory and antioxidant activities were studied by Gómez-Maqueo et al. (2019). The antioxidant activity in prickly pear peels and pulps increased at 350 MPa/5 min, whereas the anti-inflammatory activity of pulps increased at 600 MPa/ Recovery time and in peels at 350 MPa /5 min. These authors concluded that the use of this methodology increases the extraction capacity of bioactive compounds, leading to an increase in antioxidant and anti-inflammatory activities in vitro.

On the other hand, García-García et al. (2015) evaluated the effect of pH reduction, the use of preservatives and PEF on the inhibition of E. coli and S. cerevisiae in beverages stored at 25°C, formulated with two varieties of prickly pears (*Opuntia albicarpa* and *Opuntia ficus-indica*). They concluded that PEF treatments, in combination with pH reduction and addition of preservatives, produce synergistic effects allowing for the inactivation of S. cerevisiae in prickly pears to obtain pasteurized drinks.

Conclusion

The nopal (*Opuntia ficus-indica*) belongs to the Cactaceae family. Its importance in Mexico lies mainly in its traditional consumption as a vegetable at different stages of maturation (30 to 90 days) due to its beneficial effects on human health, it is used to treat various diseases such as diabetes, hyperlipidemia, and obesity. Nopal is generally consumed fresh (in juices) or cooked. Although the general composition, the presence of phenolic compounds, and the antioxidant capacity of raw nopal cladodes have been studied, research on the effect of heat treatment on them is minimal. Therefore, innovative alternative technologies have been developed to preserve the chemical, biological, and physical composition of nopal. The leading nopal processing technologies are classified into traditional and innovative technologies, and have been discussed in this chapter. Freeze-drying is a drying system; this technology preserved the most significant amount of bioactive compounds as well as the color of the cladode.

References

Abdel-Hameed, E.-S. S., Nagaty, M. A., Salman, M. S., & Bazaid, S. A. (2014). Phytochemicals, Nutritionals, and antioxidant properties of two prickly pear cactus cultivars (*Opuntia ficus indica* Mill.) growing in Taif, KSA. *Food Chemistry*, 160, 31-38. DOI:10.1016/j.foodchem.2014.03.060.

Akanni, G. B., du Preez, J. C., Steyn, L., & Kilian, S. G. (2015). Protein enrichment of an *Opuntia ficus-indica* cladode hydrolysate by cultivation of Candida utilis and Kluyveromyces marxianus. *Journal of the Science of Food and Agriculture*, 95(5), 1094-1102. DOI:10.1002/jsfa.6985.

Amaya-Cruz, D. M., Pérez-Ramírez, I. F., Delgado-García, J., Mondragón-Jacobo, C., Dector-Espinoza, A., & Reynoso-Camacho, R. (2019). An integral profile of bioactive compounds and functional properties of prickly pear (*Opuntia ficus indica* L.) peel with different tonalities. *Food Chemistry*, 278, 568-578. doi:10.1016/j.foodchem.2018.11.031.

Ammar, I., Ennouri, M., & Attia, H. (2015). Phenolic content and antioxidant activity of cactus (*Opuntia ficus-indica* L.) flowers are modified according to the extraction method. *Industrial Crops and Products*, 64, 97-104. doi:10.1016/j.indcrop.2014.11.030.

Andreu, L., Nuncio-Jáuregui, N., Carbonell-Barrachina, Á. A., Legua, P., & Hernández, F. (2018). Antioxidant properties and chemical characterization of Spanish *Opuntia ficus-indica* Mill. cladodes and fruits. *Journal of the Science of Food and Agriculture*, 98(4), 1566-1573. DOI: 10.1002/jsfa.8628.

Angulo-Bejarano, P., Martínez-Cruz, O., & Paredes-López, O. (2014). Phytochemical content, nutraceutical potential, and biotechnological applications of an ancient Mexican plant: nopal (*Opuntia ficus-indica*). *Current Nutrition & Food Science*, 10(3), 196-217. DOI:10.2174/157340131003140828121015.

Antunes-Ricardo, M., Gutiérrez-Uribe, J. A., & Guajardo-Flores, D. (2017). Extraction of isorhamnetin conjugates from *Opuntia ficus-indica* (L.) Mill using supercritical fluids. *The Journal of Supercritical Fluids*, 119, 58-63. DOI:10.1016/j.supflu.2016.09.003.

Antunes-Ricardo, M., Gutiérrez-Uribe, J. A., Martínez-Vitela, C., & Serna-Saldívar, S. O. (2015). Topical anti-inflammatory effects of isorhamnetin glycosides isolated from *Opuntia ficus-indica. BioMed research international*, 2015. DOI:10.1155/2015/847320.

Antunes-Ricardo, M., Moreno-García, B. E., Gutiérrez-Uribe, J. A., Aráiz-Hernández, D., Alvarez, M. M., & Serna-Saldivar, S. O. (2014). Induction of apoptosis in colon cancer cells treated with isorhamnetin glycosides from *Opuntia ficus-indica* pads. *Plant foods for Human Nutrition*, 69(4), 331-336. DOI:10.1007/s11130-014-0438-5.

Araya, H., Contreras, P., Alvina, M., Vera, G., & Pak, N. (2002). A comparison between an in vitro method to determine carbohydrate digestion rate and the glycemic response in young men. *European journal of clinical nutrition*, 56(8), 735-739. DOI:10.1038/sj.ejcn.1601386.

Arba, M., Falisse, A., Choukr-Allah, R., & Sindic, M. (2017). Biology, flowering and fruiting of the cactus *Opuntia* spp.: a review and some observations on three varieties in Morocco. *Brazilian Archives of Biology and Technology*, 60. DOI:10.1590/1678-4324-2017160568.

Aruwa, C. E., Amoo, S. O., & Kudanga, T. (2018). *Opuntia* (Cactaceae) plant compounds, biological activities, and prospects–A comprehensive review. *Food research international*, 112, 328-344. DOI:10.1016/j.foodres.2018.06.047.

Astello-García, M. G., Cervantes, I., Nair, V., del Socorro Santos-Díaz, M., Reyes-Agüero, A., Guéraud, F., … de la Rosa, A. P. B. (2015). Chemical composition and phenolic compounds profile of cladodes from *Opuntia* spp. cultivars with different domestication gradient. *Journal of Food Composition and Analysis*, 43, 119-130. DOI:10.1016/j.jfca. 2015.04.016.

Avila-Nava, A., Noriega, L. G., Tovar, A. R., Granados, O., Perez-Cruz, C., Pedraza-Chaverri, J., & Torres, N. (2017). Food combination based on a pre-hispanic Mexican diet decreases metabolic and cognitive abnormalities and gut microbiota dysbiosis caused by a sucrose-enriched high-fat diet in rats. *Molecular nutrition & food research*, 61(1), 1501023. DOI:10.1002/mnfr.201501023.

Balasundram, N., Sundram, K., & Samman, S. (2006). Phenolic compounds in plants and agri-industrial by-products: Antioxidant activity, occurrence, and potential uses. *Food Chemistry*, 99(1), 191-203. DOI:10.1016/j.foodchem.2005.07.042.

Barba, F. J., Galanakis, C. M., Esteve, M. J., Frigola, A., & Vorobiev, E. (2015). Potential use of pulsed electric technologies and ultrasounds to improve the recovery of high-added value compounds from blackberries. *Journal of Food Engineering*, 167, 38-44. DOI:10.1016/j.jfoodeng.2015.02.001.

Barba, F. J., Putnik, P., Kovačević, D. B., Poojary, M. M., Roohinejad, S., Lorenzo, J. M., & Koubaa, M. (2017). Impact of conventional and non-conventional processing on prickly pear (*Opuntia* spp.) and their derived products: From preservation of beverages to valorization of by-products. *Trends in food science & technology*, 67, 260-270. DOI:10.1016/j.tifs.2017.07.012.

Bayar, N., Kriaa, M., & Kammoun, R. (2016). Extraction and characterization of three polysaccharides extracted from *Opuntia ficus indica* cladodes. *International journal of biological macromolecules*, 92, 441-450. DOI:10.1016/j.ijbiomac.2016.07.042.

Betatache, H., Aouabed, A., Drouiche, N., & Lounici, H. (2014). Conditioning of sewage sludge by prickly pear cactus (*Opuntia ficus Indica*) juice. *Ecological Engineering*, 70, 465-469. DOI:10.1016/j.ecoleng.2014.06.031.

Blanco-Macias, F., Magallanes-Quintanar, R., Vazquez-Alvarado, R. E., Méndez–Gallegos, S. d. J., Troyo–Diéguez, E., Gallegos-Vázquez, C., & Valdez–Cepeda, R. D. (2011). *Opuntia ficus-indica* (L.) Mill. yield depends on nutrients and nutrient ratios. JPACD, 13, 1-9.

Butera, D., Tesoriere, L., Di Gaudio, F., Bongiorno, A., Allegra, M., Pintaudi, A. M., ... Livrea, M. A. (2002). Antioxidant activities of Sicilian prickly pear (*Opuntia ficus indica*) fruit extracts and reducing properties of its betalains: betanin and indica xanthin. *Journal of agricultural and food chemistry*, 50(23), 6895-6901. DOI:10.1021/jf025696p.

Carrillo, C. H., Gómez-Cuaspud, J., & Suarez, C. M. (2017). Compositional, thermal and microstructural characterization of the Nopal (*Opuntia ficus indica*), for addition in commercial cement mixtures. *Paper presented at the Journal of Physics*: Conference Series.

Cayupán, Y. S. C., Ochoa, M. J., & Nazareno, M. A. (2011). Health-promoting substances and antioxidant properties of Opuntia sp. fruits. Changes in bioactive-compound contents during ripening process. *Food chemistry*, 126(2), 514-519. doi:10.1016/j.foodchem. 2010.11.033.

Cejudo-Bastante, M. J., Durán-Guerrero, E., Natera-Marín, R., Castro-Mejías, R., & García-Barroso, C. (2013). Characterisation of commercial aromatised vinegars: Phenolic compounds, volatile composition and antioxidant activity. *Journal of the Science of Food and Agriculture*, 93(6), 1284-1302. DOI:10.1002/jsfa.5885.

Ciriminna, R., Chavarría-Hernández, N., Rodríguez-Hernández, A. I., & Pagliaro, M. (2019). Toward unfolding the bioeconomy of nopal (*Opuntia* spp.). *Biofuels, Bioproducts and Biorefining*, 13(6), 1417-1427. DOI:10.1002/bbb.2018.

Contreras-Padilla, M., Gutiérrez-Cortez, E., del Carmen Valderrama-Bravo, M., Rojas-Molina, I., Espinosa-Arbeláez, D. G., Suárez-Vargas, R., & Rodríguez-García, M. E. (2012). Effects of drying process on the physicochemical properties of nopal cladodes at different maturity stages. *Plant foods for Human nutrition*, 67(1), 44-49. DOI: 10.1007/s11130-011-0265-x.

Cornejo-Villegas, M., Acosta-Osorio, A., Rojas-Molina, I., Gutiérrez-Cortéz, E., Quiroga, M., Gaytán, M.,. Rodríguez-García, M. (2010). Study of the physicochemical and pasting properties of instant corn flour added with calcium and fibers from nopal powder. *Journal of Food Engineering*, 96(3), 401-409. DOI:10.1016/j.jfoodeng.2009.08.014.

De Santiago, E., Domínguez-Fernández, M., Cid, C., & De Peña, M.-P. (2018). Impact of cooking process on nutritional composition and antioxidants of cactus cladodes (*Opuntia ficus-indica*). *Food chemistry*, 240, 1055-1062. DOI:10.1016/j.foodchem.2017.08.039.

Díaz-Ayala, F., del Socorro Alvarez-Garcia, G., & Simá-Moo, E. (2015). Drying kinetics of slices of nopal (*Opuntia ficus indica*) cladodes in a convective transversal flow dryer. *Agrociencia*, 49(8), 845-857.

Ekezie, F.-G. C., Cheng, J.-H., & Sun, D.-W. (2018). Effects of nonthermal food processing technologies on food allergens: A review of recent research advances. *Trends in food science & technology*, 74, 12-25. DOI:10.1016/j.tifs.2018.01.007.

El-Mostafa, K., El Kharrassi, Y., Badreddine, A., Andreoletti, P., Vamecq, J., El Kebbaj, M., … Cherkaoui-Malki, M. (2014). Nopal cactus (*Opuntia ficus-indica*) as a source of bioactive compounds for nutrition, health and disease. *Molecules*, 19(9), 14879-14901. DOI:10.3390/molecules190914879.

El Finti, A., El Boullani, R., Fallah, M., Msanda, F., & El Mousadik, A. (2013). Assessment of some agro-technological parameters of cactus pear fruit (*Opuntia ficus-indica* Mill.) in Morocco cultivars. *Journal of Medicinal Plants Research*, 7(35), 2574-2583. DOI:10.5897/JMPR12.1236.

FAO, S. (2017). *Food and Agricultural Organization of the United Nations*; Statistics Division. In.

Flores-Hernández, A., Orona-Castillo, I., Murillo-Amador, B., Cepeda, R. D. V., & García-Hernández, J. L. (2004). Producción y calidad de nopalito en la región de la Comarca Lagunera de México y su relación con el precio en el mercado nacional. *Journal of the Professional Association for Cactus Development*, 6, 23-34. [Production and quality of nopalito in the region of the Lagunera region of Mexico and its relationship with the price in the national market.]

Gallegos-Infante, J.-A., Rocha-Guzman, N.-E., González-Laredo, R.-F., Reynoso-Camacho, R., Medina-Torres, L., & Cervantes-Cardozo, V. (2009). Effect of air flow rate on the polyphenols content and antioxidant capacity of convective dried cactus pear cladodes (*Opuntia ficus indica*). *International journal of food sciences and nutrition*, 60(sup2), 80-87. DOI:10.1080/09637480802477691.

García-García, R., Escobedo-Avellaneda, Z., Tejada-Ortigoza, V., Martín-Belloso, O., Valdez-Fragoso, A., & Welti-Chanes, J. (2015). Hurdle technology applied to prickly pear beverages for inhibiting *Saccharomyces cerevisiae* and *Escherichia coli*. *Letters in applied microbiology*, 60(6), 558-564. DOI:10.1111/lam.12406.

Ginestra, G., Parker, M. L., Bennett, R. N., Robertson, J., Mandalari, G., Narbad, A., Waldron, K. W. (2009). Anatomical, chemical, and biochemical characterization of cladodes from prickly pear [*Opuntia ficus-indica* (L.) Mill.]. *Journal of agricultural and food chemistry*, 57(21), 10323-10330. DOI:10.1021/jf9022096.

Gómez-Maqueo, A., García-Cayuela, T., Welti-Chanes, J., & Cano, M. P. (2019). Enhancement of anti-inflammatory and antioxidant activities of prickly pear fruits by high hydrostatic pressure: A chemical and microstructural approach. *Innovative Food Science & Emerging Technologies*, 54, 132-142. DOI:10.1016/j.ifset.2019.04.002.

Guevara-Figueroa, T., Jiménez-Islas, H., Reyes-Escogido, M. L., Mortensen, A. G., Laursen, B. B., Lin, L.-W., de la Rosa, A. P. B. (2010). Proximate composition, phenolic acids, and flavonoids characterization of commercial and wild nopal (*Opuntia* spp.). *Journal of Food Composition and Analysis*, 23(6), 525-532. DOI:doi.org/10.1016/j.jfca.2009.12.003.

Hassan, F., El-Razek, A., & Hassan, A. (2011). Nutritional value and hypoglycemic effect of prickly cactus pear (*Opuntia ficus-indica*) fruit juice in alloxan-induced diabetic rats. *Australian Journal of basic and applied sciences*, 5(10), 356-377.

Hernández-Urbiola, M., Contreras-Padilla, M., Pérez-Torrero, E., Hernández-Quevedo, G., Rojas-Molina, J., Cortes, M., & Rodríguez-García, M. (2010). Study of nutritional composition of nopal (*Opuntia ficus indica* cv. Redonda) at different maturity stages. *The Open Nutrition Journal*, 4(1). DOI:10.2174/1874288201004010011.

Huyut, Z., Beydemir, Ş., & Gülçin, İ. (2017). Antioxidant and antiradical properties of selected flavonoids and phenolic compounds. *Biochemistry research international*, 2017. DOI:10.1155/2017/7616791.

Jiménez-Aguilar, D. M., López-Martínez, J. M., Hernández-Brenes, C., Gutiérrez-Uribe, J. A., & Welti-Chanes, J. (2015). Dietary fiber, phytochemical composition and antioxidant activity of Mexican commercial varieties of cactus pear. *Journal of Food Composition and Analysis*, 41, 66-73. DOI:10.1016/j.jfca.2015.01.017.

Kaur, M., Kaur, A., & Sharma, R. (2012). Pharmacological actions of *Opuntia ficus indica*: A Review. *Journal of applied pharmaceutical science*, 2(7), 15-18. DOI:10.7324/JAPS.2012.2703.

Khouloud, A., Abedelmalek, S., Chtourou, H., & Souissi, N. (2018). The effect of *Opuntia ficus-indica* juice supplementation on oxidative stress, cardiovascular parameters, and biochemical markers following yo-yo Intermittent recovery test. *Food science & nutrition*, 6(2), 259-268. DOI:10.1002/fsn3.529.

Kim, J. W., Kim, T. B., Yang, H., & Sung, S. H. (2016). Phenolic compounds isolated from *Opuntia ficus-indica* fruits. *Natural Product Sciences*, 22(2), 117-121. DOI:10.20307/nps.2016.22.2.117.

Lee, J.-C., Kim, H.-R., Kim, J., & Jang, Y.-S. (2002). Antioxidant property of an ethanol extract of the stem of *Opuntia ficus-indica* var. saboten. *Journal of agricultural and food chemistry*, 50(22), 6490-6496. DOI:10.1021/jf020388c.

López-Romero, P., Pichardo-Ontiveros, E., Avila-Nava, A., Vázquez-Manjarrez, N., Tovar, A. R., Pedraza-Chaverri, J., & Torres, N. (2014). The effect of nopal (*Opuntia ficus indica*) on postprandial blood glucose, incretins, and antioxidant activity in Mexican patients with type 2 diabetes after consumption of two different composition breakfasts. *Journal of the Academy of Nutrition and Dietetics*, 114(11), 1811-1818. DOI:10.1016/j.jand.2014.06.352.

López, R., De Ita, A., & Vaca, M. (2009). Drying of prickly pear cactus cladodes (Opuntia ficus indica) in a forced convection tunnel. *Energy Conversion and Management*, 50(9), 2119-2126. DOI:10.1016/j.enconman.2009.04.014.

Marin-Bustamante, M., Chanona-Pérez, J., Guemes-Vera, N., Arzate-Vázquez, I., Perea-Flores, M., Mendoza-Pérez, J., Casarez-Santiago, R. (2018). Evaluation of physical, chemical, microstructural and micromechanical properties of nopal spines (*Opuntia ficus-indica*). *Industrial Crops and Products*, 123, 707-718. DOI:10.1016/j.indcrop. 2018.07.030.

Martínez-Soto, G., Celis-Fabián, F., Hernández-Pérez, T., & Paredes-López, O. (2015). Effect of drying methods on the nutraceutical potential of cactus cladodes (*Opuntia* spp.). *International Journal of Food and Nutrition Science* 2(6), 1. DOI:10.15436/2377-0619.15.023.

Mata, A., Ferreira, J., Semedo, C., Serra, T., Duarte, C., & Bronze, M. (2016). Contribution to the characterization of *Opuntia* spp. juices by LC–DAD–ESI-MS/MS. *Food chemistry*, 210, 558-565. DOI:10.1016/j.foodchem.2016.04.033.

Matias, A., Nunes, S., Poejo, J., Mecha, E., Serra, A., Madeira, P. J. A., … Duarte, C. (2014). Antioxidant and anti-inflammatory activity of a flavonoid-rich concentrate recovered from *Opuntia ficus-indica* juice. *Food & function*, 5(12), 3269-3280. DOI:10.1039/ C4FO00071D.

Matsuhiro, B., Lillo, L. E., Sáenz, C., Urzúa, C. C., & Zárate, O. (2006). Chemical characterization of the mucilage from fruits of *Opuntia ficus indica*. *Carbohydrate polymers*, 63(2), 263-267. DOI:10.1016/j.carbpol.2005.08.062.

McGarvie, D., & Parolis, H. (1979). The mucilage of *Opuntia ficus-indica*. *Carbohydrate Research*, 69(1), 171-179. DOI:10.1016/S0008-6215(00)85762-6.

Medina-Torres, L., Vernon-Carter, E. J., Gallegos-Infante, J. A., Rocha-Guzman, N. E., Herrera-Valencia, E., Calderas, F., & Jiménez-Alvarado, R. (2011). Study of the antioxidant properties of extracts obtained from nopal cactus (*Opuntia ficus-indica*) cladodes after convective drying. *Journal of the Science of Food and Agriculture*, 91(6), 1001-1005. DOI:10.1002/jsfa.4271.

Ministry of Agriculture, *Livestock, Rural Development, Fisheries and Food (SAGARPA) and Agri-food and Fishing Information Service (SIAP)* (2018) Available online: A http:// infosiap.siap.gob.mx/aagricola_siap_gb/ientidad/index.jsp. Accesed 30 March 2020

Monje, P. V., & Baran, E. J. (2002). Characterization of calcium oxalates generated as biominerals in cacti. *Plant Physiology*, 128(2), 707-713. DOI:10.1104/pp.010630.

Moussa-Ayoub, T. E., El-Hady, E.-S. A. A., Omran, H. T., El-Samahy, S. K., Kroh, L. W., & Rohn, S. (2014). Influence of cultivar and origin on the flavonol profile of fruits and cladodes from cactus *Opuntia ficus-indica*. *Food research international*, 64, 864-872. DOI:10.1016/j.foodres.2014.08.021.

Msaddak, L., Abdelhedi, O., Kridene, A., Rateb, M., Belbahri, L., Ammar, E., Zouari, N. (2017). Opuntia ficus-indica cladodes as a functional ingredient: bioactive compounds profile and their effect on antioxidant quality of bread. *Lipids in health and disease,* 16(1), 32. DOI:10.1186/s12944-016-0397-y.

Nefzaoui, A. (2007). Cactus to improve livestock feeding and income sources of the rural poor. Role of the FAO-Cactusnet. *Advanced nutrition and feeding strategies to improve sheep and goat,* 301-302. Retrieved from http://om.ciheam.org/om/pdf/a74/00800 395.pdf.

Nobel, P. S. (2002). *Cacti: biology and uses:* Univ of California Press.

Nuñez-López, M. A., Paredes-López, O., & Reynoso-Camacho, R. (2013). Functional and hypoglycemic properties of nopal cladodes (*O. ficus-indica*) at different maturity stages using in vitro and in vivo tests. *Journal of agricultural and food chemistry*, 61(46), 10981-10986. DOI: 10.1021/jf403834x.

Onakpoya, I. J., O'Sullivan, J., & Heneghan, C. J. (2015). The effect of cactus pear (*Opuntia ficus-indica*) on body weight and cardiovascular risk factors: a systematic review and meta-analysis of randomized clinical trials. *Nutrition*, 31(5), 640-646. DOI:10. 1016/j.nut.2014.11.015.

Ouchemoukh, S., Amessis-Ouchemoukh, N., Gómez-Romero, M., Aboud, F., Giuseppe, A., Fernández-Gutiérrez, A., & Segura-Carretero, A. (2017). Characterisation of phenolic compounds in Algerian honeys by RP-HPLC coupled to electrospray time-of-flight mass spectrometry. *LWT-Food Science and Technology*, 85, 460-469. DOI:10.1016/j. lwt.2016.11.084.

Panico, A., Cardile, V., Garufi, F., Puglia, C., Bonina, F., & Ronsisvalle, G. (2005). Protective effect of Capparis spinosa on chondrocytes. *Life sciences*, 77(20), 2479-2488. DOI:10.1016/j.lfs.2004.12.051.

Patel, S. (2014). *Opuntia* cladodes (nopal): Emerging functional food and dietary supplement. *Mediterranean Journal of Nutrition and Metabolism*, 7(1), 11-19. DOI:10.3233/MNM-140003.

Petruk, G., Di Lorenzo, F., Imbimbo, P., Silipo, A., Bonina, A., Rizza, L., … Lanzetta, R. (2017). Protective effect of Opuntia ficus-indica L. cladodes against UVA-induced oxidative stress in normal human keratinocytes. *Bioorganic & medicinal chemistry letters*, 27(24), 5485-5489. DOI:10.1016/j.bmcl.2017.10.043.

Ramawat, K. G., & Mérillon, J.-M. (2013). *Natural products: phytochemistry, botany and metabolism of alkaloids, phenolics and terpenes*: Springer.

Ramírez-Moreno, E., Córdoba-Díaz, D., de Cortes Sánchez-Mata, M., Díez-Marqués, C., & Goñi, I. (2013). Effect of boiling on nutritional, antioxidant and physicochemical characteristics in cladodes (*Opuntia ficus indica*). *LWT-Food Science and Technology*, 51(1), 296-302. DOI:10.1016/j.lwt.2012.10.005.

Reale, L., Fichera, C., Ferri, V., Cerri, M., & Ferranti, F. Morpho-histological characterization and nutritional properties of prickly pear (*Opuntia ficus indica* L. Mill). DOI:10.14787/ijprb.2016 Spl. 4-11.

Režek Jambrak, A., Donsì, F., Paniwnyk, L., & Djekic, I. (2019). Impact of novel nonthermal processing on food quality: Sustainability, modelling, and negative aspects. *Journal of Food Quality*, 2019. DOI:10.1155/2019/2171375.

Rivera, P. E. B., Rebaza, G. A. F., Peñaloza, L. E. B., Martel, L. E. C., Rodriguez, M. X. C., Lorenzo, M. L. H., … Gallegos, B. M. Y. (2018). Determinacion estructural de flavonoides en el extracto etanolico de cladodios de *Opuntia ficus-indica* (L.) Mill. Tuna Verde. *Revista Peruana de Medicina Integrativa*, 2(4), 835-840. [Structural determination of flavonoids in the ethanol extract of cladodes of Opuntia ficus-indica (L.) Mill. Green Tuna. *Peruvian Journal of Integrative Medicine*]

Rocchetti, G., Pellizzoni, M., Montesano, D., & Lucini, L. (2018). Italian *Opuntia ficus-indica* cladodes as rich source of bioactive compounds with health-promoting properties. *Foods*, 7(2), 24. DOI:10.3390/foods7020024.

Rodríguez-Felix, A., & Cantwell, M. (1988). Developmental changes in composition and quality of prickly pear cactus cladodes (nopalitos). *Plant foods for Human nutrition*, 38(1), 83-93. DOI:10.1007/bf01092314.

Rodríguez-García, M., De Lira, C., Hernández-Becerra, E., Cornejo-Villegas, M., Palacios-Fonseca, A., Rojas-Molina, I., … Zepeda, T. (2007). Physicochemical characterization of nopal pads (*Opuntia ficus indica*) and dry vacuum nopal powders as a function of the maturation. *Plant foods for Human nutrition*, 62(3), 107-112. DOI:10.1007/s11130-007-0049-5.

Rodríguez-González, S., Martínez-Flores, H. E., Chávez-Moreno, C. K., Macías-Rodríguez, L. I., Zavala-Mendoza, E., Garnica-Romo, M., & Chacón-García, L. (2014). Extraction and Characterization of Mucilage from Wild Species of *Opuntia*. *Journal of food process engineering*, 37(3), 285-292. DOI:10.1111/jfpe.12084.

Rodriguez, A., Sancho, A. M., Barrio, Y., Rosito, P., & Gozzi, M. S. (2019). Combined drying of Nopal pads (*Opuntia ficus-indica*): Optimization of osmotic dehydration as a pretreatment before hot air drying. *Journal of Food Processing and Preservation*, 43(11), e14183. DOI:10.1111/jfpp.14183.

Saag, L. M. K., Sanderson, G. R., Moyna, P., & Ramos, G. (1975). Cactaceae mucilage composition. *Journal of the Science of Food and Agriculture*, 26(7), 993-1000. DOI:10.1002/jsfa.2740260716.

Salehi, E., Emam-Djomeh, Z., Askari, G., & Fathi, M. (2019). Opuntia ficus indica fruit gum: Extraction, characterization, antioxidant activity and functional properties. Carbohydrate polymers, 206, 565-572. DOI:10.1016/j.carbpol.2018.11.035.

Salem-Fnayou, A. B., Zemni, H., Nefzaoui, A., & Ghorbel, A. (2014). Micromorphology of cactus-pear (*Opuntia ficus-indica* (L.) Mill) cladodes based on scanning microscopies. *Micron*, 56, 68-72. DOI:10.1016/j.micron.2013.10.010.

Santos-Zea, L., Gutiérrez-Uribe, J. A., & Serna-Saldivar, S. O. (2011). Comparative analyses of total phenols, antioxidant activity, and flavonol glycoside profile of cladode flours from different varieties of *Opuntia* spp. *Journal of agricultural and food chemistry*, 59(13), 7054-7061. DOI:10.1021/jf200944y.

Singh, B., Singh, J. P., Kaur, A., & Singh, N. (2017). Phenolic composition and antioxidant potential of grain legume seeds: A review. *Food research international*, 101, 1-16. DOI:10.1016/j.foodres.2017.09.026.

Slimen, I. B., Chabaane, H., Chniter, M., Mabrouk, M., Ghram, A., Miled, K., ... Najar, T. (2019). Thermoprotective properties of *Opuntia ficus-indica* f. inermis cladodes and mesocarps on sheep lymphocytes. *Journal of thermal biology*, 81, 73-81. DOI: 10.1016/j.jtherbio.2019.02.018.

Smida, A., Ncibi, S., Taleb, J., Saad, A. B., Ncib, S., & Zourgui, L. (2017). Immunoprotective activity and antioxidant properties of cactus (*Opuntia ficus indica*) extract against chlorpyrifos toxicity in rats. *Biomedicine & Pharmacotherapy*, 88, 844-851. DOI:10.1016/j.biopha.2017.01.105.

Valente, L. M., da Paixão, D., Do Nascimento, A. C., dos Santos, P. F., Scheinvar, L. A., Moura, M. R., ... da Silva, J. F. (2010). Antiradical activity, nutritional potential and flavonoids of the cladodes of *Opuntia monacantha* (Cactaceae). *Food chemistry*, 123(4), 1127-1131. DOI:10.1016/j.foodchem.2010.05.074.

Ventura-Aguilar, R. I., Bosquez-Molina, E., Bautista-Baños, S., & Rivera-Cabrera, F. (2017). Cactus stem (*Opuntia ficus-indica* Mill): anatomy, physiology and chemical composition with emphasis on its biofunctional properties. *Journal of the Science of Food and Agriculture*, 97(15), 5065-5073. DOI:10.1002/jsfa.8493 Accepted.

Walters, M., Figueiredo, E., Zimmermann, H., & Mashope, B. (2011). *Naturalised and invasive succulents of southern Africa.*

Yamamoto, K. (2017). Food processing by high hydrostatic pressure. *Bioscience, biotechnology, and biochemistry,* 81(4), 672-679. DOI:10.1080/09168451.2017.1281723.

Zorgui, L., Ayed-Boussema, I., Ayed, Y., Bacha, H., & Hassen, W. (2009). The antigenotoxic activities of cactus (*Opuntia ficus-indica*) cladodes against the mycotoxin zearalenone in Balb/c mice: prevention of micronuclei, chromosome aberrations and DNA fragmentation. *Food and chemical toxicology,* 47(3), 662-667. DOI:10.1016/j.fct. 2008.12.031.

About the Editors

Sandra Teresita Martín del Campo obtained the PhD degree in Food Science from the Institut National Agronomique Paris-Grignon, France, and a PhD degree in Biotechnological process from the University of Guadalajara, Mexico, with the evaluation of authenticity parameters in milk and dairy products. Dr. Martín del Campo is a professor-Researcher at Tecnológico de Monterrey Campus Querétaro. She teaches in the undergraduate programs of Food Industries Engineering and Biotechnology Engineering; and in the Master and Doctorate programs in Biotechnology. As a core researcher of the Nutriomics and Emerging Technologies group, her research focuses on the dairy products metabolomic studies, preservation of food properties during transformation, recuperation of biocompounds from food waste. She is a member of the National System of Researchers Level 1 in the Biotechnology area.

Jessica del Pilar Ramírez Anaya, biologist from the University of Guadalajara, Mexico. She has Master degrees in Food Biotechnology and in Food Technology and Quality. She gets PhD degree in Nutrition and Food Technology from Granada University, Spain, with the study of the effect of domestic processing in polyphenolic compounds of extra virgin olive oil and Mediterranean vegetables. She is a professor in the Nutrition undergraduate program at Guadalajara University, teaching Food Technology Appropriated to Towns. She also teaches in the Public Health Master program at the University of Guadalajara in the course Antioxidant Activity of the Diet. Her research activities centers on the Food Science and Technology area, ranging from formulation, improvement, processing, and sustainability of the food chain. Her work impacts from small local production to intense industrialization of regional crops. She has experience in the study of the changes in functional compounds profiles during the processing of pomegranate and avocado fruits. Recently, she is working in nanoparticles from food industry wastes.

Anaberta Cardador-Martínez is a researcher and professor at Tecnológico de Monterrey Campus Querétaro. Her research is related to the extraction, identification, and evaluation of biomolecule activities from food and/or agro-industrial wastes. She is a member of the National System of Researchers Level 1 and belongs to the Mexican Academy of Sciences.

INDEX

A

B

C

D

E

F

G

H

Q

R

S

T

U

V

W

X

Z